McCance and Widdowson's

The Composition of Foods

Sixth summary edition

© Crown copyright.

Crown copyright material reproduced under licence from the Controller of Her Majesty's Stationery Office and the Food Standards Agency.

All rights reserved

No part of this book may be reproduced or transmitted in any form or by any means – graphic, electronic, including photocopying, recording, taping, or information storage and retrieval systems – without written permission from Her Majesty's Stationery Office Copyright Unit, St Clements House, 2–16 Colegate, Norwich NR3 1BQ.

Published by The Royal Society of Chemistry, Cambridge and the Food Standards Agency, London.

The Royal Society of Chemistry is a registered charity (No. 207890).
For further information see the RSC website at www.rsc.org

This publication should be cited as follows:

Food Standards Agency (2002) McCance and Widdowson's The Composition of Foods, Sixth summary edition. Cambridge: Royal Society of Chemistry.

Reprinted 2004, 2006, 2008

Compiled by:
Mark A Roe, Institute of Food Research,
Paul M Finglas, Institute of Food Research
and
Susan M Church, Food Standards Agency

McCance and Widdowson's

The Composition of Foods

Sixth summary edition

Compiled by
Food Standards Agency
and
Institute of Food Research

For further information contact:

Sales and Customer Care Department
Royal Society of Chemistry
Thomas Graham House
Science Park
Milton Road
Cambridge CB4 0WF
UK

Tel: (01223) 432360; Fax: (01223) 426017; E-mail: sales@rsc.org

ISBN-10: 85404-428-0
ISBN-13: 978-0-85404-428-3

Photocomposed by Land & Unwin (Data Sciences) Ltd., Bugbrooke
Printed in the United Kingdom by Henry Ling Limited, at the Dorset Press, Dorchester, DT1 1HD

CONTENTS

3. Additional tables

4. Appendices

Dedicated to

Professor R A McCance (1898–1993)

And

Dr E M Widdowson (1906–2000)

FOREWORD TO THE SIXTH EDITION

The Composition of Foods is widely acknowledged as the key reference tool for all those who need to know the nutritional value of foods consumed in the UK, and for over 60 years it has proved invaluable to its many academic, professional and student users.

I therefore welcome this 6th summary edition, and am pleased that the Food Standards Agency has been able to play a part in its publication, working in close collaboration with the Institute of Food Research.

The series began with the vision of Professor McCance and Dr Widdowson in the 1930s. This edition is dedicated to their memory and I commend to you their foreword to the 5th edition of *The Composition of Foods*, reproduced overleaf.

John Krebs

Sir John Krebs
Chairman
Food Standards Agency

FOREWORD TO THE 5TH EDITION OF *The Composition of Foods*

By R.A. McCance and E.M. Widdowson

In 1926 I (R. A. McC) was a medical student at King's College Hospital, London. Dr R. D. Lawrence, himself a diabetic, was in charge of the diabetic patients, and he was writing a book 'The Diabetic Life'. He wanted to include some values for the carbohydrate content of fruits and vegetables, which were then an important part of diabetic diets, but there were problems with this. First the values that were being used were derived from Atwater and Bryant's tables published in America in 1906, and these were nearly all obtained 'by difference', that is, water, fat, nitrogen and ash were determined, nitrogen was multiplied by 6.25 to obtain protein, the percentages of these were added together and the sum subtracted from 100 to give the percentage of carbohydrate. Carbohydrate calculated in this way contained not only sugar and starch which were important to the diabetic, but also the 'unavailable carbohydrate' or dietary fibre. Another problem in using the American tables was that most of the analyses had been made on raw materials, whereas people eat most of their vegetables cooked and their composition is altered by cooking. So a grant of £30 a year was obtained from the Medical Research Council for me to analyse raw and cooked fruits and vegetables for total 'available carbohydrate', that is sugars plus starch, which was the value needed for calculating diabetic diets. I analysed 109 different plant materials, each on six separate occasions, in the time I had to spare from my medical studies and the results were published in 1929 as a Medical Research Council Special Report No. 35 'The Carbohydrate Content of Foods' by R. A. McCance and R. D. Lawrence.

When Professor Cathcart, Professor of Physiology at Glasgow University, read the report he suggested that the work should be extended, and that protein and fat should be determined in meat and fish. The Medical Research Council agreed to provide a grant to cover the salaries of a chemist, H. L. Shipp, and a technician, Alec Haynes, and a study of meat and fish began. Sixty-two varieties of fish were analysed, all except oysters cooked, 26 different cuts of meats, 9 varieties of poultry and game and 9 different kinds of 'offal', all cooked in standard ways. Besides total nitrogen, purine N, amino-N and extractive-N were determined and the analyses included fat, carbohydrate when present and minerals Na, K, Ca, Mg, Fe, P and Cl. We also investigated the losses of various constituents when meat and fish were cooked in various ways. Shrinkage caused most of the losses from meat, but not from fish. All the results were published in 1933 as a second Medical Research Council Special Report No. 187 'The Chemistry of Flesh Foods and their Losses on Cooking' by R. A. McCance and H. L. Shipp.

At the end of this study H. L. Shipp left and was replaced by L. R. B. Shackleton, and it was at this point that I (E. M. W.) joined the team. We four started again on fruits, vegetables and nuts. The analyses included 56 varieties of fruit, 9 of nuts, 28 of raw vegetables and 44 of vegetables after cooking. We analysed them for water, total nitrogen, glucose, fructose, sucrose and starch and for 'unavailable carbohydrate'. The same minerals were determined as in the meat and fish. Losses of sugars, nitrogen and minerals from vegetables while being boiled were also investigated. These results made a third Medical Research Council Special Report, No. 213, published in 1936 'The Nutritive Value of Fruits, Vegetables and Nuts' by R. A. McCance, E. M. Widdowson and L. R. B. Shackleton. The stock of all these reports was destroyed in a fire resulting from an air raid on London during World War II and they have been out of print ever since.

In 1938 we moved to Cambridge. L. R. B. Shackleton left but Alec Haynes came with us. We finished the analyses we had begun in London on cereals, dairy products, beverages and preserves and we put the results of all our analytical work together to make the first edition of 'The Chemical Composition of Foods' by R. A. McCance and E. M. Widdowson. This was published in 1940 as the fourth Medical Research Council Special Report No. 235. The working notebooks containing the details of all the analyses have been deposited with the Wellcome Institute for the History of Medicine.

Since one of the uses of the tables was likely to be the calculation of the composition of diets, and diets generally include cooked dishes we gave some information about their composition. Most of the recipes were taken from standard cookery books, and 90 are to be found in that first edition.

A second edition appeared in 1946, which included values for wartime foods, Household milk, dried eggs and National wheatmeal flour and bread made from it. Values for the composition of about 20 'economical' dishes were included.

In the 1950s we began to work on a third edition. By then many new foods had become available, and those introduced in wartime had disappeared from the market. Alec Haynes had left, and Dr. D. A. T. Southgate joined us. He, with the help of a technician, Janet Adams, was responsible for analysing more than 100 new foods for the same constituents as we had previously done.

By the 1950s methods for the determinations of vitamins had improved, and many foods had been analysed for one or more of them. We decided to depart from our original principle of including only the results of our own analyses in the tables, and to use values taken from the literature. Dr. W. I. M. Holman, who knew a great deal about the determination of vitamins in foods, undertook the task of reading every paper he could find on the vitamin content of foods published in the past 15 or 20 years. This involved abstracting well over 1000 papers. He selected those reporting results which he believed to be reliable, and then he left us and his abstracts to take up a post in South Africa. Miss I. M. Barrett joined us, and she constructed the tables of the vitamin content of foods from the information Dr. Holman had collected together.

Values for the amino-acid content of the main protein-containing foods, cereals, meat, fish, eggs, milk and its products, and of some nuts and vegetables were also included in the third edition. These were partly taken from the literature and partly from analyses made by Dr. B. P. Hughes who was working with us at the time. The third edition was published in 1960, with a change of title to 'The Composition of Foods'. As time had gone on some cookery experts had been rather critical of our original recipes, so the whole of the section on the composition of cooked dishes was revised with the help of members of the cookery department of King's College of Household and Social Science.

Up to the third edition we had the ultimate responsibility for the tables. I (R. A. McC) retired in 1966 and it became clear that a decision had to be made about the future of 'The Composition of Foods'. Tables such as these must be revised from time to time or they become obsolete and therefore useless. In the late sixties I (E. M. W) raised the matter at a meeting of the Interdepartmental Committee on Food Composition. It was unanimously agreed that the tables must not be allowed to die. The Interdepartmental Committee on Food Composition accepted responsibility for the revision of the tables, and appointed a Steering Panel under the chairmanship of Dr. D. A. T. Southgate to advise those responsible for the revision, leading to the fourth edition. In the event meats were completely reanalysed. The conformation of farm animals had altered and methods of butchering had changed since the 1930s when the original samples were collected. Cereals, milk and milk products were also extensively revised, but most other foods were not reanalsyed, and about a third of the values in the fourth edition, published in 1978, were our original figures, obtained by what are regarded nowadays as very primitive methods 40 years before. Those methods were no less accurate than the modern automated ones, but they took a much longer time.

Since 1978 several supplements to the tables have been published covering the composition of different groups of foodstuffs as these have been revised, and tables showing the composition of foods used by immigrants in the United Kingdom were made available. Now a fifth edition summary edition of 'The Composition of Foods' has been prepared. This represents the work of many people including those who were responsible for making the analyses as we had done half a century ago. We are happy that we are still part of it.

July 1991

ACKNOWLEDGEMENTS

The Food Standards Agency and the compilers are grateful to the numerous people who have helped during the preparation of this book.

Most of the values in this book are based on the detailed supplements to *The Composition of Foods*. The compilers of this summary edition are therefore indebted to all the people who have contributed towards the series of books. In particular, the major role of the primary authors (Bridie Holland and Dr Wynnie Chan) and the co-authors (Jane Brown, Ian Unwin, and Ailsa Welch) of the supplements is gratefully acknowledged. In addition, we are indebted to the late Dr David Buss for his role in leading the work on the series.

Most of the new analyses in this edition were undertaken by the Laboratory of the Government Chemist. Additional analyses were carried out by Agricultural Development and Advisory Service (ADAS), Aspland & James, Campden and Chorleywood Food Research Association (CCFRA), and RHM Technology.

We wish to thank numerous manufacturers, retailers and other organisations for information on the range and composition of their products. In particular, we would like to thank British Egg Information Service, British Egg Industry Council, Kelloggs, McVities, Nestlé UK, Quaker Oats Ltd, Procter & Gamble, Snack Nut and Crisp Manufacturers Association Ltd (SNACMA), Tesco Stores Ltd, and Weetabix Ltd for providing additional information. Thanks are also due to Mabel Blades at the Meat and Livestock Commission for her invaluable advice, and to Dr Caroline Bolton-Smith, of the Medical Research Council's Resource Centre for Human Nutrition Research (MRC HNR) for her input into the additional table on phylloquinone content of foods.

Many current and former professional and administrative staff at the Ministry of Agriculture, Fisheries and Food (MAFF) and the Food Standards Agency have been involved in the work leading to the production of this book, from design of the analytical projects on which most of the data are based, through data collation and checking, to the final compilation. In addition, Rosemary Bobbin, Yvonne Clements and Richard Faulks, all of the Institute of Food Research, warrant special mention for their contribution. The database used to compile this edition was developed by Vaughan McLintock, under contract to MAFF.

The preparation of this sixth summary edition was overseen by the Sub Group on Publication of Data, under the auspices of MAFF's, and latterly the Food Standards Agency's, Working Party on Nutrients in Food. In addition to the compilers, this group comprised Alison Paul (MRC HNR), Professor David Southgate and Rachel Abraham and Moya de Wet (both representing members of the British Dietetic Association). Secretariat support was provided by Judith Holden. *The Composition of Foods User Group*, set up in 1999, also provided advice on format and content of this edition, including the foods to be included.

INTRODUCTION

1.1 Background

"A knowledge of the chemical composition of foods is the first essential in the dietary treatment of disease or in any quantitative study of human nutrition" (McCance & Widdowson, 1940).

1.1.1 This sixth summary edition of the UK food composition tables extends and updates a series which began with the vision of R A McCance and E M Widdowson in the 1930s, under the auspices of the Medical Research Council. Following publication of the fourth edition of McCance and Widdowson's *The Composition of Foods* in 1978, the Ministry of Agriculture, Fisheries and Food (MAFF) took on the responsibility for maintaining and updating the official tables of food composition in the United Kingdom. In 1987, the Ministry joined with the Royal Society of Chemistry to begin production of a computerised UK National Nutrient Databank from which a number of detailed supplements (Table 1) and the fifth edition of *The Composition of Foods* (Holland *et al.*, 1991) were produced. Responsibility for data compilation returned to MAFF in 1997 and a detailed supplement on the fatty acid composition of foods was published in 1998 (MAFF, 1998). Responsibility for the maintenance of the UK National Nutrient Databank transferred to the Food Standards Agency on its establishment in April 2000. The data for this sixth summary edition were compiled, under contract, by the Institute of Food Research.

1.1.2 This sixth summary edition is intended to be a convenient book which includes in one volume the most recent values for a range of commonly-consumed foods. As such, it comprises a sub-set of published and new data with the range of both foods and nutrients being limited. It replaces the fifth edition, but not the detailed supplements (Table 1), which make up the UK National Nutrient Databank.

1.1.3 Computer-readable files of the data for most of the supplements and the fifth and sixth editions are available. Details can be obtained from the Food Standards Agency.

1.1.4 Now that the series of supplements is complete, a comprehensive integrated dataset will be produced. However, prior to this, it was decided to publish this summary edition in response to the widely expressed need for a convenient book which includes in one volume the most recent nutrient values for the whole range of common foods.

Table 1 *Supplements to* ***'The Composition of Foods'***

Amino Acids and Fatty Acids	Paul *et al.*, 1980
Immigrant Foods	Tan *et al.*, 1985
Cereals and Cereal Products	Holland *et al.*, 1988
Milk Products and Eggs	Holland *et al.*, 1989
Vegetables, Herbs and Spices	Holland *et al.*, 1991
Fruit and Nuts	Holland *et al.*, 1992
Vegetable Dishes	Holland *et al.*, 1992
Fish and Fish Products	Holland *et al.*, 1993
Miscellaneous Foods	Chan *et al.*, 1994
Meat, Poultry and Game	Chan *et al.*, 1995
Meat Products and Dishes	Chan *et al.*, 1996
Fatty Acids	MAFF, 1998

1.2 Sources of data and methods of evaluation

1.2.1 It is essential that food composition tables are regularly updated for a number of reasons. Since the fifth summary edition was published, many new fresh and manufactured foods have become familiar items in our shops, and values for these have been included wherever possible. In addition, the nutritional value of many of the more traditional foods has changed. This can happen when there are new varieties or new sources of supply for the raw materials; with new farming practices which can affect the nutritional value of both plant and animal products; with new manufacturing practices including changes in the type and amounts of ingredients (including reductions in the amount of fat, sugar and salt added or new fortification practices); and with new methods of preparation and cooking in the home.

1.2.2 To ensure that the UK food composition tables could continue to have as wide a coverage and be as up to date as possible, the Ministry of Agriculture, Fisheries and Food (MAFF) decided in the early 1980s to set up a rolling programme of food analysis. Responsibility for this programme transferred to the Food Standards Agency on its establishment in April 2000. The analytical reports from recent studies (1990 onwards) are available from the Food Standards Agency library. (A small charge will be made to cover copying and postage.) A few reports are available on the Food Standards Agency website (www.food.gov.uk). These reports comprise raw laboratory data and have not been evaluated to the same extent as data incorporated into *The Composition of Foods*.

1.2.3 Most of the values included in these Tables have been taken from the detailed supplements, themselves mainly derived from MAFF's series of analytical studies. This edition also includes new, and previously unpublished, analytical data for a number of key foods, particularly cereals and cereal products, and milk and milk products. Further details are given in the introduction to each food group. In addition, foods for which new data are included can easily be identified by the inclusion of a new food code in the food index. Reports from which new data for this summary edition were taken are included in the *References* section.

1.2.4 Where new analytical data were not available the values have been taken from a number of sources including the scientific literature, manufacturers' data and by calculation. All recipes have been recalculated, using the most recent available data for ingredients.

1.2.5 Where the values in the Tables were derived by direct analysis of the foods, great care was taken when designing sampling protocols to ensure that the foods analysed were representative of those used by the UK population. For most foods a number of samples were purchased at different shops, supermarkets or other retail outlets. The samples were not analysed separately but were pooled before analysis. When the composite sample was made up from a number of different brands of food, the numbers of the individual brands purchased were related to their relative shares of the retail market. If the food required preparation prior to analysis, techniques such as washing, soaking, cooking, etc. were as similar as possible to normal domestic practices.

1.2.6 A summary of the analytical techniques used for this edition is given in Section 4.1.

1.2.7 Where data from literature sources were included in the Tables preference was given to reports where the food was similar to that in the UK, where the publication gave full details of the sample and its method of preparation and analysis, and where the results were presented in a detailed and acceptable form. The criteria for assessing literature values are summarised in Table 2.

1.2.8 Where manufactured foods with proprietary names are included in the database they are restricted to leading brands with an established composition. It should be noted that manufacturers can change their products from time to time and this will influence nutrient content. This is particularly relevant for foods where nutrients are added for fortification purposes, or for technological purposes, such as antioxidants or as colouring agents. The inclusion of a particular brand does not imply that it has a special nutritional value.

1.2.9 The final selection of values published here is dependent on the judgement of the compilers and their interpretation of the available data. There can be no guarantee that a particular item will have precisely the same composition as that in these Tables because of the natural variability of foods.

1.2.10 Users are advised to consult other sources of data (e.g. product labels, manufacturers' data), where appropriate. For example, users who require data on the nutrient content of foods consumed by South Asians in the UK are advised to refer to Judd *et al.* (2000).

Table 2 *Criteria applied before acceptance of literature values*[a]

Name of food	Common name, with local and foreign synonyms Systematic name with variety where known.
Origin	*Plants:* Country of origin Locality, with details of growth conditions if available
	Animals: Country of origin Locality and method of husbandry and slaughter (if available)
Sampling	Place and time of collection Number of samples and how these were obtained Nature of sample (e.g. raw, prepared, deep frozen, prepacked etc.) Ingredient details
Treatment of samples before analysis	Conditions and length of storage Preparative treatment e.g. material discarded as waste and whether washed or drained Cooking details (where applicable) e.g. length of cooking, temperature and the cooking medium.
Analysis	Details of material analysed Methods used, with appropriate reference and details of any modifications
Methods of expression of results	Statistical treatment of analytical values Whether expressed on an 'as purchased', 'edible matter' or 'dry matter', etc. basis

[a] Modified from Southgate (1974), Greenfield and Southgate (1994)

1.3 Arrangement of the Tables

1.3.1 This book is composed of three parts, the Introduction, the Tables and a number of Additional Tables and Appendices.

1.3.2 The **Tables** contain four pages of information for each food.

The **first page** gives the food number, name and description along with data for edible proportion and the major constituents (water, nitrogen, protein, fat, carbohydrate and energy).

Food number

For ease of reference, each food has been assigned a consecutive publication number for the purposes of this edition only. In addition, each food has a unique food code number which is given in the index and will allow read-across to the supplements or the fifth edition, where appropriate. For foods that have already been included in supplements or in the fifth edition and for which there are no new data, their food code number (including the unique two digit prefix) has been repeated. These prefixes are 11 – *Cereals and Cereal Products*, 12 – *Milk Products and Eggs*, 13 – *Vegetables, Herbs and Spices*, 14 – *Fruit and Nuts*, 15 – *Vegetable Dishes*, 16 – *Fish and Fish Products*, 17 – *Miscellaneous Foods*, 18 – *Meat, Poultry and Game*, 19 – *Meat Products and Dishes*, and 50 – *Fifth Edition*. Foods that have not previously been included have been given a new food code number in the supplement using that prefix (e.g. plain bagel (11-534)). Where new data have been incorporated for an existing food, a new food code has also been allocated but with the same supplement prefix (e.g. beef bourguignonne was 19-161, now 19-330). For ease of use, the original food code number is given alongside the new one in the index for the foods concerned. These are the numbers that will be used in nutrient databank applications.

Food name

The food name has been chosen as that most recognisable and descriptive of the food referenced.

Description

Information given under the description and number of samples describes the number and nature of the samples taken for analysis. Sources of values derived either from the literature or by calculation are also indicated under this heading. Further summary information on the sources of data used for each food are given in the computer-readable files for this edition.

The **second page** gives starch, total and individual sugars (glucose, fructose, sucrose, maltose, lactose), dietary fibre (expressed as non-starch polysaccharide), fatty acid totals, and cholesterol.

The **third page** gives data for inorganic elements and the **fourth page** data for the vitamin composition of the foods.

All nutrients are quoted per 100g edible portion of food with the exception of the alcoholic beverages group where they are per 100ml.

Foods have been arranged in groups with common characteristics. The arrangement of the food groups in the Tables is as follows:- cereals and cereal products, milk and milk products, eggs, fats and oils, meat and meat products, fish and fish products, vegetables, herbs and spices, fruits, nuts, sugars, preserves and snacks, beverages, alcoholic beverages and sauces, soups, and miscellaneous foods. Generally the order within the groups is similar to that in the corresponding supplement. A few foods have been placed in different groups from those in which they previously appeared where this is more appropriate for a general work covering all food groups. Each food group is preceded by text covering points of specific relevance to the foods in that group.

1.3.3 **Additional tables** cover alternative methods for determining dietary fibre, phytosterols, carotenoid and vitamin E fractions, and vitamin K_1 (phylloquinone).

1.3.4 Information contained in the **Appendices** includes a summary of analytical techniques, weight changes on the preparation of foods, cooked foods and dishes, the recipes, calculation of nutrient content for foods 'as purchased' or 'as served', a table of alternative and taxonomic names for foods and references to the Tables and Introduction. These sections provide useful supporting information for the data in the Tables.

1.3.5 A combined food index and coding list is provided at the end of the appendices. This also includes cross-references from alternative food names and taxonomic names to the food names used in the Tables.

1.4 The definition and expression of nutrients

1.4.1 *The expression of nutrient values*

For this summary edition, all foods are expressed per 100g edible portion. The primary reason for this was to maximise the number of different foods that could be included in the book, while ensuring that it did not become unduly large. For foods that are generally purchased or served with waste, guidance for calculating nutrient content 'as purchased' or 'as served' is given in Section 4.2.

Generally the values have been expressed to a constant number of decimal places for each nutrient. However, exceptions have been made where appropriate, either within groups of foods or for individual values. For example, the iron content of liquid milks has been expressed to two decimal places, because the amounts that can be drunk render this value significant. The values of the more variable vitamins such as biotin have been expressed to less than their usual number of places where large values render the extra places non-significant.

Many foods are purchased or served with inedible material and an edible conversion factor is given which shows the proportion of the edible matter in the food. For raw food this refers to the edible material remaining after the inedible waste has been trimmed away, e.g. the outer leaves of a cabbage. For canned foods such as vegetables the factor refers to the edible contents after the liquid has been drained off.

1.4.2 *Protein*

For most foods, protein has been calculated by multiplying the total nitrogen value by the factors shown in Table 3.

Table 3 *Factors for converting total grams of nitrogen in foods to protein*[a]

Food	Factor	Food	Factor
Cereals		Nuts	
Wheat		Peanuts, Brazil nuts	5.41
Wholemeal flour	5.83	Almonds	5.18
Flours, except wholemeal	5.70	All other nuts	5.30
Pasta	5.70		
Bran	6.31	Milk and milk products	6.38
Maize	6.25	Gelatin	5.55
Rice	5.95	All other foods	6.25
Barley, oats, rye	5.83		
Soya	5.70		

[a] FAO/WHO (1973)

The proportion of non-protein nitrogen is high in many foods, notably fish, fruits and vegetables. In most of these, however, this is amino acid in nature and therefore little error is involved in the use of a factor applied to the total nitrogen, although protein in the strictest sense is overestimated. For those foods which contain a measurable amount of non-protein nitrogen in the form of urea, purines and pyrimidines (e.g. mushrooms) the non-protein nitrogen has been subtracted before multiplication by the appropriate factor.

1.4.3 *Fat*

The fat in most foods is a mixture of triglycerides, phospholipids, sterols and related compounds. The values in the Tables refer to this total fat and not just to the triglycerides.

1.4.4 *Carbohydrates*

Total carbohydrate and its components, starch and total and individual sugars (glucose, fructose, sucrose, maltose, lactose), but not fibre, are wherever possible expressed as their monosaccharide equivalent. The values for total carbohydrate in the Tables have generally been obtained from the sum of analysed values for these components of 'available carbohydrate', contrasting with figures for carbohydrate 'by difference' which are sometimes used in other food tables or on the labels of manufactured foods. Such figures are obtained by subtracting the measured weights of the other proximates from the total weight and many include the contribution from any dietary fibre present as well as errors from the other analyses. A few values have been included from other tables, or from manufacturers, and are printed in italics to distinguish them from direct analyses.

Available carbohydrate is the sum of the free sugars (glucose, fructose, galactose, sucrose, maltose, lactose and oligosaccharides) and complex carbohydrates (dextrins, starch and glycogen). These are the carbohydrates which are digested and absorbed, and are glucogenic in man. This corresponds to the term 'glycaemic carbohydrates' proposed in the FAO/WHO report on Carbohydrates in Human Nutrition (FAO, 1998).

Carbohydrate values expressed as monosaccharide equivalents can exceed 100g per 100g of food because on hydrolysis 100g of a disaccharide such as sucrose gives 105g monosaccharide (glucose and fructose). 100g of a polysaccharide such as starch gives 110g of the corresponding monosaccharide (glucose). Thus white sugar appears to contain 105g carbohydrate (expressed as monosaccharide) per 100g sugar. For conversion between carbohydrate weights and monosaccharide equivalents, the values shown in Table 4 should be used.

Table 4 *Conversion of carbohydrate weights to monosaccharide equivalents*

Carbohydrate	Equivalents after hydrolysis g/100g	Conversion to monosaccharide equivalents
Monosaccharides e.g. glucose, fructose and galactose	100	no conversion necessary
Disaccharides e.g. sucrose, lactose and maltose	105	x 1.05
Oligosaccharides e.g.		
raffinose (trisaccharide)	107	x 1.07
stachyose (tetrasaccharide)	108	x 1.08
verbascose (pentasaccharide)	109	x 1.09
Polysaccharides e.g. starch	110	x 1.10

Any known or measured contribution from oligosaccharides and/or maltodextrins has been included in the total carbohydrate value but not in the columns for starch or total sugars. In most foods oligosaccharides are present in relatively low quantities. In vegetables, however, and some processed foods where glucose syrups and maltodextrins are added, oligosaccharides will make a significant contribution to carbohydrate content. Because of this the sum of starch and total sugars will be less than the total carbohydrate for these foods and where this occurs the values have been marked in the Tables with footnotes.

1.4.5 *Dietary fibre*

Different methods give different estimates of the total fibre content of food. The values shown in the main Tables are total non-starch polysaccharides (NSP) (Englyst and Cummings, 1988). An additional table comparing values obtained by the NSP (Englyst) method and the AOAC method (AOAC, 2000), for the very few foods for which analytical data on the same samples are available, is also included. For nutritional labelling purposes, it is recommended that fibre values obtained by AOAC methodology are used.

1.4.6 *Alcohol*

The values for alcohol are given as g/100ml of alcoholic beverages. Pure ethyl alcohol has a specific gravity of 0.79 and dividing the values by 0.79 converts them to alcohol by volume (i.e. ml/100ml). The specific gravities of the alcoholic beverages are given in the introduction to that section of the Tables so that calculations can be made if the beverages are measured by weight. The alcohol contents of a range of strengths 'by volume' are also given in the introduction to the section on Alcoholic Beverages in the Tables.

1.4.7 *Energy value – kcal and kJ*

The metabolisable energy values of all foods are given in both kilocalories (kcal) and kilojoules (kJ). These energy values have been calculated from the amounts of protein, fat, carbohydrate and alcohol in the foods using the energy conversion factors shown in Table 5.

Table 5 *Metabolisable energy conversion factors used in these Tables*[a,b]

	kcal/g	kJ/g
Protein	4	17
Fat	9	37
Available carbohydrate expressed as monosaccharide	3.75	16
Alcohol	7	29

a Royal Society (1972)

b See Section 1.9 for the conversion factors that should be used in food labelling

These factors permit the calculation of the metabolisable energy of a typical United Kingdom mixed diet with a level of accuracy which compares well with values obtained in human subjects using calorimetry (Southgate and Durnin, 1970). No contribution from NSP or sugar alcohols is included in these calculations. There is currently some debate about the use of these factors (Livesey *et al.*, 2000).

The energy value of foods in kilojoules can also be calculated from the kilocalorie value using the conversion factor 4.184 kJ/kcal. Whilst it is more accurate to apply the kilojoule factors in Table 5 to protein, fat, carbohydrate and alcohol, a direct kcal/kJ conversion produces differences of little dietetic significance (1–2 per cent).

1.4.8 *Fatty acids*

For this edition, only total saturated, monosaturated, and polyunsaturated and total *trans* unsaturated fatty acids are given. More detailed information on individual fatty acids is available in the *Fatty Acids* supplement (MAFF, 1998).

The fat in most foods contains non fatty acid material such as phospholipids and sterols. To allow the calculation of the total fatty acids in a given weight of food, the conversion factors shown in Table 6 were applied.

A worked example is shown below (TFA = total fatty acids; taken from MAFF, 1998):

	Total fat in Beef, lean only	= 5.1g/100g
	Conversion factor	= 0.916
	Total fatty acids in beef = 5.1 x 0.916	= **4.7**g/100g
Saturates	at 43.7g/100g TFA x 4.7 ÷ 100	= 2.0g/100g food
Monounsaturates	at 47.9g/100g TFA x 4.7 ÷ 100	= 2.2g/100g food
Polyunsaturates	at 3.8g/100g TFA x 4.7 ÷ 100	= 0.2g/100g food

N.B. The values do not add up to the total fatty acids because branched-chain and *trans* fatty acids have been excluded from the saturated and unsaturated fatty acids respectively.

Table 6 *Conversion factors to give total fatty acids in fat*[a]

Wheat, barley and rye[b]		Beef lean[d]	0.916
whole grain	0.720	Beef fat[d]	0.953
flour	0.670	Lamb, take as beef	
bran	0.820	Pork lean[e]	0.910
		Pork fat[e]	0.953
Oats, whole[b]	0.940	Poultry	0.945
Rice, milled[b]	0.820	Heart[e]	0.789
Milk and milk products	0.945	Kidney[e]	0.747
Eggs	0.830	Liver[e]	0.741
		Fish, fatty[f]	0.900
Fats and oils		white[f]	0.700
all except coconut oil	0.956	Vegetables and fruit	0.800
coconut oil	0.942	Avocado pears	0.956
		Nuts	0.956

[a] Paul & Southgate (1978)
[b] Weihrauch *et al.* (1976)
[c] Posati *et al.* (1975)
[d] Anderson *et al.* (1975)
[e] Anderson (1976)
[f] Exler *et al.* (1975)

1.4.9 *Cholesterol*

Cholesterol values are included for all foods in this publication and are expressed as mg/100g food. To convert to mmol cholesterol, divide the values by 386.6.

1.4.10 *Inorganic constituents*

Details of the inorganic constituents covered in the Tables are given in Table 7. Further information on variability can be found in Section 1.5 and on bioavailability in Section 1.6.

Table 7 *Inorganic constituents*

Atomic symbol	Name	Units	Atomic weight[a]
Na	Sodium	mg/100g	23
K	Potassium	mg/100g	39
Ca	Calcium	mg/100g	40
Mg	Magnesium	mg/100g	24
P	Phosphorus[b]	mg/100g	31
Fe	Iron	mg/100g	56
Cu	Copper	mg/100g	64
Zn	Zinc	mg/100g	65
Cl	Chloride	mg/100g	35
Mn	Manganese	mg/100g	55
Se	Selenium	µg/100g	79
I	Iodine	µg/100g	127

[a] To convert the weight of a mineral to mmol or µmol divide by the atomic weight
[b] To convert mg P to mg PO_4 multiply by 3.06

Selenium

Many new values for selenium have been incorporated into this edition, taken from the analytical programme and from a specially commissioned analytical study (Barclay *et al.*, 1995). The selenium content of soil has a large effect on the foods harvested from it. The levels of selenium in UK soils are low and analysed values reflect this. Data from literature sources have been taken from those countries with similar soil profiles to the UK. Where the values selected are of non-UK origin (or a food is from an overseas source) the values appear in brackets.

1.4.11 *Vitamins*

Details of vitamins covered in the Tables are given in Table 8 (*see over*).

Vitamin A: retinol and carotene

The two main components of the vitamin are given separately in the Tables.

Retinol is found in many animal products, the main forms being all-*trans* retinol and 13-*cis* retinol. The latter has about 75% of the activity of the former (Sivell *et al.*, 1984). Eggs and fish roe also contain retinaldehyde which has 90% of the activity of all-*trans* retinol. Retinol is expressed in the Tables as the weight of all-*trans* retinol equivalent, i.e. the sum of all-trans retinol plus contributions from the other two forms after correction to account for their relative activities.

Approximately 600 carotenoids are found in plant products and milks but few have vitamin A activity (Olson, 1989). Of these, the most important is β-carotene.

Table 8 *Vitamins*

Vitamin	Units	International Units (IU)[a]
Vitamin A		
Retinol	μg/100g	0.3μg
Carotene (β-carotene equivalents)	μg/100g	0.6μg
Vitamin D	μg/100g	0.025μg
Cholecalciferol, ergocalciferol		
Vitamin E	mg/100g	0.67mg
α-Tocopherol equivalents		
Vitamin K1 (phylloquinone)	μg/100g	
(additional table only)		
Thiamin	mg/100g	
Riboflavin	mg/100g	
Niacin		
Total preformed niacin	mg/100g	
Tryptophan (mg) divided by 60	mg/100g	
Vitamin B_6	mg/100g	
All forms (pyridoxine, pyridoxal, pyridoxamine and phosphates of these)		
Vitamin B_{12}	μg/100g	
Folate	μg/100g	
Total folate		
Pantothenate	mg/100g	
Biotin	μg/100g	
Vitamin C	mg/100g	
Total ascorbic and dehydroascorbic acid		

[a] Amount equivalent to one International Unit

The other main forms with vitamin A activity are α-carotene and α- and β-cryptoxanthins, which have approximately half the activity of β-carotene. Carotene is expressed in the Tables in the form of β-carotene equivalents, that is the sum of the β-carotene and half the amounts of α-carotene and α- and β-cryptoxanthins present. Where the carotenoid profile was incomplete, it has been assumed that all is β-carotene. This may result in an overestimate but as α-carotene and cryptoxanthin are usually present in low levels in foods without complete carotenoid profiles, it is likely that any error is small.

Retinol equivalents

In the UK the requirement for vitamin A is expressed as retinol equivalents (Department of Health, 1991). This measure of the overall potency of vitamin A relates to the lower biological efficiency of carotenoids compared with retinol. The absorption and utilisation of carotenes vary, for example with the amount of fat in the diet and β-carotene concentration (Brubacher and Weiser, 1985), and there is currently much debate about use of retinol equivalents (Scott & Rodriquez-Amaya, 2000). However, the generally accepted relationship is still that 6μg β-carotene or 12μg of all other active carotenoids are equivalent to 1μg retinol (Department of Health, 1991), so that:-

$$\text{Vitamin A potency as } \mu\text{g retinol equivalents} = \mu\text{g retinol} + \frac{\mu\text{g } \beta\text{-carotene equivalents}}{6}$$

Recent work suggests that this convention may need revision in the future.

The relationship between the different units used to express vitamin A is shown in Table 9.

Table 9 *Relationship and conversion between the units used to express retinol and carotene*

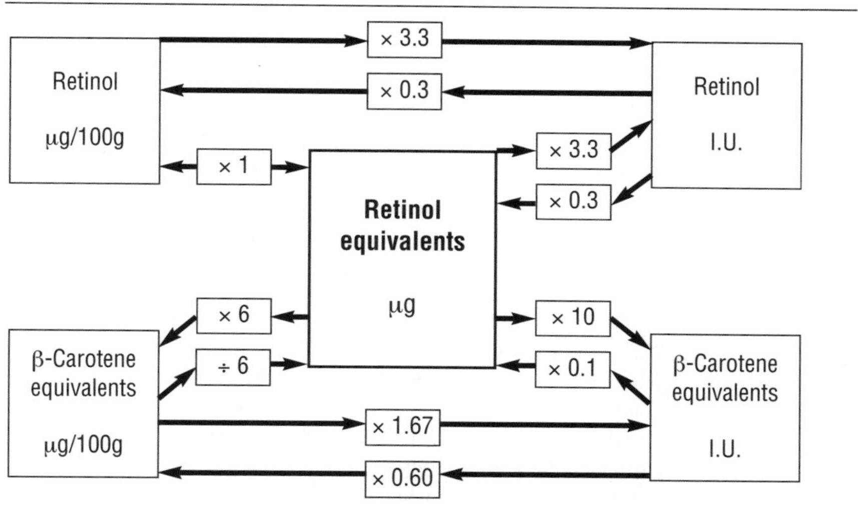

Vitamin D

Few foods contain vitamin D. All those which do so naturally are animal products and contain D_3 (cholecalciferol) derived, as in humans, from the action of sunlight on the animal's skin or from its own food. Vitamin D_2 (ergocalciferol) made commercially has the same potency in man. Both vitamin D_2 and vitamin D_3 are used to fortify a number of foods.

Meat can contain vitamin D_3 (cholecalciferol) derived from the action of sunlight or, for pigs and poultry, from the feed. This may be present in the form of the more active 25-hydroxy vitamin D_3. For meat, meat products, and poultry, therefore, the total vitamin D activity has been taken as the sum of vitamin D_3 (cholecalciferol) and five times 25-hydroxy vitamin D_3 (25-hydroxy cholecalciferol), where data are available. There is, however, some debate about the factor that should be used for 25-hydroxy vitamin D_3 when estimating total vitamin D activity.

Vitamin E

The vitamin E in food is present as various tocopherols and tocotrienols, each having a different level of vitamin E activity. In most animal products the α-form is the only significant form present but in plant products, especially seeds and their oils, γ-tocopherol and other forms are present in significant amounts. The values for vitamin E are expressed as α-tocopherol equivalents, using the factors shown in Table 10.

Table 10 *Conversion factors for vitamin E activity*[a]

α-tocopherol	x	1.00
β-tocopherol	x	0.40
γ-tocopherol	x	0.10
δ-tocopherol	x	0.01
α-tocotrienol	x	0.30
β-tocotrienol	x	0.05
γ-tocotrienol	x	0.01

[a] McLaughlin and Weihrauch (1979)

Vitamin K_1

The predominant, naturally occurring, vitamin K that occurs in foods is phylloquinone (vitamin K_1) and it is this that is reported in the Additional Table (Section 3.5). Phylloquinone is lipid soluble and is found in the photosynthetic tissue of plants. As such, the darker green the plant leaves, the more phylloquinone is present (Shearer *et al.*, 1996; Bolton-Smith *et al.*, 2000). Certain vegetable oils, namely rapeseed, soybean and olive oils are also relatively high in phylloquinone compared to corn (maize) and sunflower seed oil. The phylloquinone content of plants (and therefore presumably plant oils) also varies by climate and soil conditions (Ferland & Sadowski, 1992).

Hydrogenation of oils results in the conversion of phylloquinone to 1,3-dihydro-phylloquinone (Davidson *et al.*, 1996) and this may be a significant proportion of total vitamin K present in some foods, such as biscuits and margarines. In the USA, estimates of 2,3-dihydro-phylloquinone intake suggest it may be the major dietary form of vitamin K in some population groups (Booth *et al.*, 1996b). The biological activity of the dihydro form may be less than that of native phylloquinone; however, the precise relationship is unclear, and food content data for the UK are currently unavailable.

A second family of naturally occurring, functional vitamin K compounds, the menaquinones (MK*n*, where *n* represents the number of isoprene units in the side chain) are formed by bacteria. They are likely to occur in variable quantities in fermented foods, and to a minor extent in some cheese, as a result of the bacterial inoculation during their production. Menaquinones may also be found in some meats, such as chicken, as a result of feeding with the synthetic form of vitamin K, menadione, which is activated *in vivo* by conversion to MK4. Inadequate information on the MK content of foods is available for inclusion in the current table.

Thiamin

The majority of values for thiamin are expressed as thiamin chloride hydrochloride using either the direct thiochrome method, HPLC with fluorimetric detection or microbiological assay (see Section 4.1).

Niacin

The values are the sum of nicotinic acid and nicotinamide which are collectively known as niacin.

Tryptophan is converted in the body to nicotinic acid with varying efficiency. On average, 60mg tryptophan is equivalent to 1mg niacin, so the tryptophan content of the protein in each food has been shown after division by 60. This may be added to the amount of niacin to give the niacin equivalent for the food.

Vitamin B_6

Vitamin B_6 occurs in foods as pyridoxine, pyridoxal, pyridoxamine and their phosphates. However, the active form in the tissues is pyridoxal phosphate. In the main, pyridoxine is expressed in the Tables as pyridoxine hydrochloride by microbiological assay, or the sum of the individual forms by HPLC, and expressed as the sum of pyridoxine hydrochloride, pyridoxal hydrochloride and pyridoxamine dihydrochloride (see Section 4.1). The newer HPLC values for vitamin B_6 do not always agree closely with total B_6 values obtained by microbiological assay. This can be due to the different extraction procedures employed for the methods, and the varying response of the organism to the vitamers in the microbiological assay (Ollilainen *et al.*, 2001).

Folate

For folates, the value refers to total folates measured by microbiological assay after deconjugation of the polyglutamyl forms. Folic acid (PteGlu) is the predominant form used for fortification purposes. Other major folates present in food are 5-methyltetrahydrofolates (5-$CH_3H_4PteGlu_n$; mainly plant- and dairy-based foods), 5- and 10-formyltetrahydrofolates (5- and 10-$CHOH_4PteGlu_n$; mainly animal-based foods) and tetrahydrofolates ($H_4PteGlu_n$). Some HPLC-derived values are available for 5-methyltetrahydrofolate (Laboratory of the Government Chemist, 1996), but the values for other folates are much less reliable.

Pantothenate

The majority of values for pantothenate are expressed as calcium D-pantothenate.

Vitamin C

Values include both ascorbic and dehydroascorbic acids, as both forms are biologically active. In fresh foods the reduced form is the major one present but the amount of the dehydro-form increases during cooking and processing. The older values for vitamin C (prior to the 4th edition of *The Composition of Foods*) are based on the titrimetric procedure which only determines ascorbic acid. For the newer data, total ascorbate (ascorbic acid + dehydroascorbic acids) has been determined using either the fluorimetric procedure or HPLC with UV or fluorescence detection (see Section 4.1).

1.5 The variability of nutrients in foods

1.5.1 Although values in these Tables have been derived from careful analyses of representative samples of each food, it is important to appreciate that the composition of any individual sample may differ considerably from this. There are two main reasons for the variability, apart from the apparent differences caused by analytical variations.

1.5.2 *Natural variation*

All natural products vary in composition. Two samples from the same animal or plant may well be different, but the compositions of meat, milk and eggs are also affected by season and by the feeding regime and age of the animal. Different varieties of the same plant may differ in composition, and their nutritional value will also vary with the country of origin, growing conditions and subsequent storage. In general, those nutrients that are closely associated with structure and metabolic function show rather less variation than those which accumulate in particular locations of the plant or animal or those which are unstable. For instance, nitrogen and phosphorus tend to show less variation than vitamin A, iron or vitamin C.

A major influence on the nutrient concentration in foods is the water content and this is particularly important in plant foods where water is the main constituent. As the length and conditions of food storage affect the water content of foods, these will have an effect on their nutrient content per 100g. Many individual nutrients will also be affected by storage conditions with the greatest effect being on the more labile vitamins such as vitamin C, vitamin E and folate. Thus if the storage conditions of a food item differ from those for the samples analysed for the Tables, the nutrient values may differ from those given.

The level of fat in food can vary greatly and result in large variations in the nutrient content of each 100g of the food. It will also influence energy and the level of fat-soluble vitamins. An example of how fat and moisture content vary (in minced meat) is given in Table 11.

Table 11 *Fat and moisture content of minced meats*

	No. samples	Moisture % mean (range)	Fat % mean (range)
Beef mince, raw[a]	10	64.0 (57.3–70.0)	16.6 (7.8–26.5)
Beef mince, extra lean, raw[a]	10	69.6 (63.8–72.6)	8.3 (3.9–16.9)
Lamb mince, raw[b]	10	66.8 (58.1–71.6)	13.5 (8.1–22.8)
Pork, mince, raw[c]	10	70.6 (64.3–73.2)	9.3 (5.4–19.5)

[a] Laboratory of the Government Chemist (1993a)
[b] Laboratory of the Government Chemist (1993b)
[c] Laboratory of the Government Chemistry (1994b)

1.5.3 *Extrinsic differences*

Further differences in composition can be introduced by food manufacturers, caterers, and in the home. For example, manufacturers may change both their recipes and their fortification practices, and dishes prepared in the home or by caterers may vary widely in the amounts and types of ingredient used and thus differ in nutritional value from those included here.

Examples of some external influences on nutrient contents are shown below:

Sodium — The level found in many foods will depend upon the amount of salt and other sodium-containing compounds used in cooking or added by manufacturers, and can therefore be very variable. The majority of vegetables analysed for the food tables were cooked in distilled water without salt, although there are a few to which salt was added and these are indicated in the Tables. For planning low-sodium diets the Table values are adequate.

Potassium — The potassium content of boiled vegetables is dependent on the amount of water, length of cooking time and the state of preparation of the vegetable. The user should refer to the description and main data sources for the foods in the Tables to ensure sample foods are comparable.

Calcium — Most vegetables in the Tables were cooked in distilled water. Foods cooked in and prepared with tap water, which contains variable amounts of calcium, may not have the same levels as in these Tables.

The concentration of calcium in baking powder is high and variations in the quantity used will affect the calcium content of some cereal products.

Iron	Food can become contaminated with iron from knives, pans, soil particles and processing machinery. This has most effect on the iron content of ground foods such as spices.
Chloride	Chloride variation will be similar to that of sodium.
Iodine	Iodine levels in milk are affected by the levels in animal feedstuffs, and to a lesser degree by the iodine levels in the solutions used for teat dips, sanitisers and the lactation promoter iodinated casein (Phillips *et al.,* 1988).
β-Carotene	This is sometimes used as a food colouring additive (E160a). In certain manufactured foods such as orange squash, samples may contained added β-carotene.
Vitamin C	This is added to a number of foods for fortification or antioxidant purposes (E300, L-ascorbic acid) and so may be present in unexpectedly high levels in some foods, including some meat products and soft drinks.

1.6 Bioavailability of nutrients

1.6.1 The term bioavailability (biological availability) is a term used to describe the proportion of a nutrient in food that is utilised for normal body function (Fairweather-Tait, 1998). There are many factors, both dietary and physiological, that influence nutrient bioavailability and because these interactions are so variable, it is not possible to provide an accurate measure of bioavailability in these Tables.

Dietary-related factors include

- the physical form of the nutrient within the food structure, and the ease with which the nutrient can be released from that structure,
- the chemical form of the nutrient in a food and its solubility in the lumen,
- the presence of enhancers of absorption (e.g. ascorbic acid for iron, some organic acids, sugars, amino acids, bulk lipid for fat-soluble vitamins and specific fatty acids), and
- the presence of inhibitors (primarily of inorganic absorption, e.g. phosphates (especially phytate), polyphenols including tannins, oxalate and carbohydrate (especially dietary fibre)).

Physiological factors include the composition and volume of gastric and intestinal secretions, and a number of host-related variables, many of which are essential parts of the body's homeostatic regulatory mechanism (e.g. nutritional status, development state, mucosal cell regulation and gut microflora (Fairweather-Tait, 1998)).

Allowance has been made for reduced biological activities of different forms of three of the vitamins given in the Tables: 13-*cis*-retinol and retinaldehyde (vitamin A), carotenes other than β-carotene, and tocopherols and tocotrienols other than α-tocopherol (vitamin E), as described in Section 1.4. Other nutrients in the

Tables which are absorbed and utilised with varying degrees of efficiency include iron, calcium, magnesium, zinc, copper, manganese, selenium, folate, niacin and vitamin B_6. For all these, no allowance is made in these Tables for the potential lower availability, and the values quoted represent the actual content in foods.

Some additional information on the bioavailability of selected micronutrients follows.

1.6.2 *Iron*

Dietary iron occurs in two major forms, haem (found in haemoglobin and myoglobin in foods derived from animal tissues), and non-haem iron. These forms exhibit different levels of absorption via separate pathways. Haem iron is always relatively well absorbed (20–30%) and only marginally affected by dietary factors or iron status of an individual. Non-haem iron easily forms complexes which are less readily solubilised and absorbed than non-haem iron. The absorption of non-haem iron is highly variable depending on the nature of the meal. Dietary factors which enhance absorption of non-haem iron include; meat, ascorbic acid and certain other organic acids. Polyphenols (including tannins from tea), phytate and calcium decrease bioavailability. Non-haem iron bioavailability is also profoundly influenced by physiological variables, notably body iron status. Previous dietary iron will also affect the bioavailability of subsequent iron (Fairweather-Tait, 1999).

1.6.3 *Zinc*

Zinc is absorbed more efficiently than non-haem iron and it is affected by fewer dietary factors. Phytate is probably the most important zinc antagonist, especially in the presence of calcium, as it forms a chelate with zinc which is unavailable for absorption. Copper, cadmium and iron can also reduce zinc bioavailability. Some proteins have been shown to improve bioavailability but the mechanisms for the effect are not yet clear. Body zinc status plays an important role in determining dietary zinc bioavailability (Fairweather-Tait, 1999).

1.6.4 *Calcium*

The amount of calcium absorbed is dependent on individual vitamin D status, the customary level of calcium intake and needs of the individual and the presence of binding substances in the food (Allen, 1982). Dietary inhibitors include phytate and oxalate. High levels of dietary protein and sodium increase urinary calcium excretion which is accompanied by an increase in intestinal absorption. However, this results in a reduction of calcium utilised by the body and thus lower bioavailability. Lactose promotes absorption and calcium from milk and milk products has a relatively high bioavailability (Fairweather-Tait & Hurrell, 1996).

1.6.5 *Selenium*

Selenium bioavailability depends to a great extent on the chemical form present. Selenium is present as organic and inorganic forms. The two main organic forms are selenomethionine (Se-Met), principally in plant foods, and selenocysteine in foods of animal origin. Se-Met is readily absorbed and results in higher blood selenium concentration than inorganic selenium. The inorganic forms (selenite and selenate) do not occur naturally in foods but are often used as supplements. As with other elements, solubility is the key factor in determining absorption. The main dietary factors which influence selenium bioavailability are methionine, thiols, heavy metals and vitamin C (Fairweather-Tait & Hurrell, 1996).

1.6.6 *Vitamins A & E and Carotenoids*

These compounds need to be dissolved and carried in lipid and lipid+bile salt systems (micelles) in order to be absorbed at the brush border. Protein or protein-calorie malnutrition is often associated with malabsorption of vitamin A. Zinc deficiency, alcohol and some food constituents (e.g. nitrites) are associated with malabsorption of vitamin A (Biesalski, 1997). A number of other dietary factors influence the carotenoid bioavailability especially food structure and the physical form of the carotenoid within the food matrix. The absorption of carotenoids from raw foods can be very low, but cooking, chopping and other types of food preparation enhance absorption by increasing the ease with which carotenoids are extracted from the food matrix (Faulks *et al.*, 1997). *In vitro* results indicate that gastric pH is an important physiological determinant of carotenoid availability (Rich *et al.*, 1998).

α-Tocopherol accounts for almost all of the vitamin E activity in foods of animal origin. Under normal dietary conditions about 20–80% of ingested vitamin E is absorbed, depending on dose and lipid content of the meal. High intakes of pectin, wheat bran, alcohol and polyunsaturated fatty acids also reduce vitamin E absorption. Dietary constituents such as vitamin A, iron, selenium and zinc may affect vitamin E utilisation (Cohn, 1997).

1.6.7 *Folate*

Folate exists in nature primarily as reduced one-carbon substituted forms of pteroylpolyglutames. About 80% of dietary folate occurs as the polyglutamyl form of folate and must be cleaved to the monoglutamate form for absorption. There appears to be little or no difference in the extent of absorption of the various monoglutamyl forms, although stability in the gastro-intestinal tract and *in vivo* retention may differ. Numerous dietary and physiological factors influence the deconjugation and absorption of folate, and its subsequent utilisation in the body. Dietary factors may reduce folate bioavailability and include conjugase inhibitors (e.g. in pulses), milk folate binding proteins and dietary fibre (e.g. wheat bran). Physiological factors include intraluminal pH, decreased conjugase enzyme activity associated with ageing, and certain deficiencies (e.g. Zn, B_{12} and folate). The action of endogenous conjugase enzymes during food preparation may increase the bioavailability of naturally occurring polyglutamyl forms (Gregory, 1997a).

1.6.8 *Vitamin B_6*

Vitamin B_6 exists in foods as either the free or phosphorylated forms of pyridoxine, pyridoxamine and pyridoxal. Plant foods contain glucoside bound forms of pyrodoxine which may be unavailable for absorption. Orange juice and wheat bran reduce vitamin B_6 absorption (Gregory, 1997b).

1.6.9 *Niacin*

Much of the niacin occurring naturally in cereals (especially maize) is in a bound form and may be unavailable for absorption. However, alkali treatment, as used in some traditional processing methods, renders the niacin more bioavailable. Sorghum, wheat, barley and rice contain niacin in chemically bound forms (van den Berg, 1997).

1.7 Calculation of nutrient intakes using the Tables

1.7.1 *Calculation*

There are several steps involved in the calculation of nutrient intake from the Tables. The first is to choose the item in the Tables which corresponds most closely with the food consumed. The index includes many alternative names and it should be noted that a food may be found in a different food group from the one in which it is expected.

If the food consumed is not in the Tables then it is necessary to choose a suitable alternative by consideration of the food type, general characteristics and likely nutrient profile. The results, however, are likely to be less accurate. Alternatively, users might wish to seek other sources of data (e.g. manufacturers).

Once the food has been chosen, calculation of nutrient intake is achieved by multiplying the nutrient figure quoted in the Tables by the weight of the food consumed (nutrients are expressed either per 100g of the edible portion of the food, or per 100ml for alcoholic beverages), e.g. if 80g food has been consumed, the nutrient should be multiplied by 0.8, and if 120g consumed multiplied by 1.2. The results from these calculations are then summed to give the total intake.

1.7.2 *Computerised calculation*

Calculation of nutrient intake 'by hand' is a time consuming process which has largely been superseded by the use of computers. Information concerning the datafiles and packages available for personal computers and mainframes can be obtained from the Food Standards Agency.

Recipes

If the sample of food consumed is a cooked dish prepared with a different recipe from any of those in this book, the nutrients for the new recipe can be calculated using the methods given in Section 4.3.

Portion sizes

If the weight of food consumed has not been recorded or if an estimate is required, publications such as Bingham and Day (1987), MAFF (1993), and Davies and Dickerson (1991) may be used to provide information on portion sizes. In fieldwork, representations such as pictures (Nelson *et al.*, 1997), models or household measures may also be used to obtain estimates of portion size.

1.8 Potential pitfalls when using the Tables

"There are two schools of thought about food tables. One tends to regard the figures in them as having the accuracy of atomic weight determinations; the other dismisses them as valueless on the ground that a foodstuff may be so modified by the soil, the season, or its rate of growth that no figure can be a reliable guide to its composition. The truth, of course, lies somewhere between these two points of view."

(Widdowson & McCance, 1943).

Those who are unfamiliar with the uses of these Tables should note the following points which can reflect on the accuracy of the information obtained from them. Further details are available in Greenfield & Southgate (1994).

- When comparing the nutrient values in these Tables with those of other countries or literature reports, the expression of units and conversion factors used in calculation may vary.
- As nutrients are increasingly added to foods for fortification, antioxidant and colorant purposes, users should check the labels of manufactured products.
- Missing nutrient values in food composition tables should not be treated as zero values during calculation otherwise an underestimation of nutrient intake will result. However, the major sources of any nutrient are likely to have been analysed and included in these Tables.
- Errors will arise if food is classified incorrectly: for instance it may be assumed that milk has been consumed in the full fat form when it was in fact skimmed.
- Misclassification of foods may arise as a result of a food having several names. It is therefore important to be familiar with local and alternative names when using food tables, e.g. roast potatoes are known as baked potatoes in some parts of the country.
- In manual coding systems incorrect food code numbers may be used. Computerised systems which avoid the use of numbers and input information only by the food name tend to reduce this problem. However, it is still possible for names to be identified incorrectly during the use of the Tables and calculation software.
- It is possible that errors can be made both in the measurement and recording of food weights which will affect the calculation of nutrient intakes.
- Sources of estimated weight are more prone to error than the recorded weight of food because the portion size chosen by an investigator may not give a true indication of the actual amount eaten or an individual may misinterpret the amount shown in a representation of a portion size.
- There are several methods for collecting food intake data which range from weighed intakes to food frequency questionnaires giving information which is either quantitative or qualitative. It is worthwhile consulting appropriate publications (e.g. Bingham (1987), Cameron and van Staveren (1988), Nelson and Bingham (1997)) to find which method is the most suitable for the level of information required.

- As any one person exhibits a great deal of variation in diet, varied lengths of recording time are needed to assess representative intakes of nutrients. For example a 7-day weighed record collection (not necessarily consecutive days) may be necessary to assess energy and protein intakes assuming that an accuracy of ±10% standard error is acceptable. It may be possible to observe people with very stable eating habits for a shorter time but those with greater variation may require longer. For most other nutrients the recording period would need to be longer than for energy and protein, particularly for those concentrated in only a few foods. For example, vitamin C may require 36 days of recording to be within ±10% of the true intake. This topic is covered in greater detail in Bingham (1987), Cameron and van Staveren (1988) and Nelson *et al.* (1989).

1.9 Food labelling

Nutrition information is increasingly being given on food labels. Values from these food composition tables may be used for this purpose, but only if certain conditions are met. Values that meet the criteria below are included in the computer-readable files, where possible.

The rules which govern nutrition labelling are contained in Directive 90/496/EEC on Nutrition Labelling for Foodstuffs. In Great Britain these rules are implemented by the Food Labelling Regulations 1996 (as amended). Northern Ireland has similar but separate legislation. These rules are there to ensure consistency and accuracy, and to prevent misleading claims. Nutrition labelling is not compulsory unless a nutrition claim is made, but when such information is given the details in one of the following groups must be shown per 100g or per 100ml of the food as sold:

Either

energy value in kJ and kcal, **and**
protein, carbohydrate and fat, in grams, **and**
the amount of any other nutrients for which a claim is made

Or

energy value in kJ and kcal, **and**
protein, carbohydrate, sugars, fat, saturates, fibre and sodium, all in grams, **and**
the amount of any other nutrient for which a claim is made

Preference should be given to values derived from analyses of representative samples of the food. However, if the product or its ingredients are similar to those described in this book or the supplements, these values may be used instead. Nevertheless, it is important to note the following differences:

1. Protein should be given as total nitrogen × 6.25 for every food, whereas more specific factors have been used in this book.
2. Carbohydrate is to be declared as the weight of the carbohydrates themselves and not their monosaccharide equivalents.

The following factors may be used to convert monosaccharide equivalents from these Tables to actual weights:

Total carbohydrate	Divide by 1.05 unless it is known to be mainly starch
Starch	Divide by 1.10
Sucrose and lactose	Divide by 1.05
Glucose, etc.	As given

3. Different factors are to be used to calculate energy values. These are shown in Table 12.

Table 12 Energy conversion factors to be used in food labelling

	kcal/g	**kJ/g**
Protein	4	17
Carbohydrate expressed as weight	4	17
Fat	9	37
Alcohol	7	29
All organic acids	3	13
Sorbitol and other polyols	2.4	10

The

Tables

Symbols and abbreviations used in the Tables

Symbols

0	None of the nutrient is present
Tr	Trace
N	The nutrient is present in significant quantities but there is no reliable information on the amount
( )	Estimated value
Italic text	Carbohydrate estimated 'by difference' and energy values based upon these quantities

Abbreviations

IFR	Institute of Food Research, Norwich
LGC	Laboratory of The Government Chemist, Teddington
calcd.	calculated
Gluc	Glucose
Fruct	Fructose
Sucr	Sucrose
Malt	Maltose
Lact	Lactose
NSP	Non-starch polysaccharides
Satd	Saturated
Monounsatd	Monounsaturated (total)
Polyunsatd	Polyunsaturated (total)
Trypt	Tryptophan

Section 2.1

Cereals and cereal products

Much of the data and foods in this section of the Tables have been taken from the *Cereals and Cereal Products* (1988) supplement. However, new analytical values have been incorporated for bread, fresh pasta, some biscuits, cakes, buns, pastries and puddings, and pizzas. In addition, some new values for breakfast cereals have been provided by manufacturers. Foods for which new data are incorporated have been allocated a new food code and can thus easily be identified in the food index.

Values from the literature for wheat flours and their products were restricted to those from the UK because flours are required to be fortified by law (The Bread and Flour Regulations, 1998). UK flour should contain at least 1.65mg iron, 0.24mg thiamin and 1.60mg niacin per 100g and so these nutrients are added to all white flours and most brown flours in this country. Calcium carbonate must also be added to all flours except wholemeal and certain self-raising flours at a rate equivalent to 94–156mg calcium per 100g flour.

Sources of variation pertinent to cereals and cereal products include soil type and fertiliser use (which particularly affects inorganics) and the practice of allowing for losses of nutrients added during handling and storage ('overages'). In addition, the range and levels of added nutrients do change with time (e.g. calcium is now added to some breakfast cereals). Users requiring details of possible recent changes in fortification practices may wish to contact manufacturers directly.

Losses of labile vitamins assigned on recipe calculation were estimated using figures in Section 4.3. Changes in weight on toasting bread and boiling rice and pastas are shown in Section 4.3. Taxonomic names for foods included in this part of the Tables can be found in Section 4.5

Composition of food per 100g edible portion

No.	Food	Description and main data sources	Edible conversion factor	Water	Total nitrogen	Protein	Fat	Carbo-hydrate	Energy value	
				g	g	g	g	g	kcal	kJ
	Flours, grains and starches									
1	**Bran**, wheat	Analytical and literature sources	1.00	8.3	2.24	14.1	5.5	26.8	206	872
2	**Chapati flour**, brown	1 sample, single supplier	1.00	12.2	2.02	11.5	1.2	73.7	333	1419
3	white	2 samples, different suppliers, same weights	1.00	12.0	1.72	9.8	0.5	77.6	335	1426
4	**Cornflour**	3 samples from different shops	1.00	12.5	0.09	0.6	0.7	92.0	354	1508
5	**Custard powder**	Taken as cornflour except Na, Cl and Cu	1.00	12.5	0.09	0.6	0.7	92.0	354	1508
6	**Oatmeal**, quick cook, *raw*	10 samples, 8 brands	1.00	8.2	1.92	11.2	9.2	66.0	375	1587
7	**Rye flour**, whole	Analytical and literature sources	1.00	15.0	1.40	8.2	2.0	75.9	335	1428
8	**Soya flour**, full fat	Analytical and literature sources	1.00	7.0	6.45	36.8	23.5	23.5	447	1871
9	low fat	Analytical and literature sources	1.00	7.0	7.94	45.3	7.2	28.2	352	1488
10	**Wheat flour**, brown	VFSS, 1977–81, and literature sources	1.00	14.0	2.20	12.6	2.0	68.5	324	1384
11	white, breadmaking	Data from Voluntary Flour Sampling Scheme	1.00	14.0	2.02	11.5	1.4	75.3	341	1451
12	white, plain	(VFSS), 1977–81 plus literature sources.	1.00	14.0	1.64	9.4	1.3	77.7	341	1450
13	white, self-raising	Biscuit and cake flours are similar in composition	1.00	14.0	1.56	8.9	1.2	75.6	330	1407
14	wholemeal	to plain flour	1.00	14.0	2.18	12.7	2.2	63.9	310	1318
15	**Wheatgerm**	Literature sources	1.00	11.7	4.54	26.7	9.2	(44.7)	357	1509
	Rice									
16	**Brown rice**, *raw*	5 assorted samples	1.00	13.9	1.10	6.7	2.8	81.3	357	1518
17	*boiled*	Water content weighed, other nutrients calculated from *raw*	1.00	66.0	0.43	2.6	1.1	32.1	141	597
18	**Egg fried rice**, *takeaway*	10 samples from different outlets	1.00	57.5	0.68	4.3	4.9	33.3	186	787

Composition of food per 100g edible portion

No.	Food	Starch	Total sugars	Individual sugars					Dietary fibre	Fatty acids				Cholest-erol
				Gluc	Fruct	Sucr	Malt	Lact	NSP	Satd	Mono-unsatd	Poly-unsatd	Trans	
		g	g	g	g	g	g	g	g	g	g	g	g	mg
	Flours, grains and starches													
1	**Bran**, wheat	23.0	3.8	0.2	0.1	3.5	0.1	0	36.4	0.9	0.7	2.9	0	0
2	**Chapati flour**, brown	70.5	3.2[a]	N	N	N	N	0	(10.3)	0.2	0.1	0.5	0	0
3	white	75.5	2.1[a]	N	N	N	N	0	(4.1)	0.1	Tr	0.2	0	0
4	**Cornflour**	92.0	Tr	Tr	(0)	Tr	(0)	0	0.1	0.1	0.1	0.3	0	0
5	**Custard powder**	92.0	Tr	Tr	(0)	Tr	(0)	0	(0.1)	0.1	0.1	0.3	0	0
6	**Oatmeal**, quick cook, *raw*	64.9	1.1	Tr	Tr	0.8	0.3	0	7.1	1.6	3.3	3.7	0	0
7	**Rye flour**, whole	75.9	Tr	Tr	Tr	Tr	Tr	0	11.7	0.3	0.2	0.9	0	0
8	**Soya flour**, full fat	12.3	11.2	N	N	N	N	0	11.2	2.9	4.5	11.4	0	0
9	low fat	14.8	13.4	N	N	N	N	N	(13.5)	0.9	1.4	3.5	0	0
10	**Wheat flour**, brown	66.8	1.7[a]	Tr	Tr	0.7	0	0	6.4	0.3	0.2	0.9	0	0
11	white, breadmaking	73.9	1.4[a]	Tr	Tr	0.3	Tr	0	(3.1)	0.2	0.1	0.6	0	0
12	white, plain	76.2	1.5[a]	Tr	Tr	0.3	0.2	0	3.1	0.2	0.1	0.6	0	0
13	white, self-raising	74.3	1.3[a]	0.1	Tr	0.2	0	0	(3.1)	0.2	0.1	0.5	0	0
14	wholemeal	61.8	2.1[a]	0.1	Tr	1.0	0	0	9.0	0.3	0.3	1.0	0	0
15	**Wheatgerm**	(28.7)	(16.0)	(0.7)	(0.5)	(14.8)	0	0	15.6	1.3	1.1	4.2	0	0
	Rice													
16	**Brown rice**, *raw*	80.0	1.3	0.5	Tr	0.8	0	0	1.9	0.7	0.7	1.0	0	0
17	*boiled*	31.6	0.5	0.2	Tr	0.3	0	0	0.8	0.3	0.3	0.4	0	0
18	**Egg fried rice**, *takeaway*	33.1	Tr	Tr	Tr	Tr	Tr	Tr	0.8	0.6	2.3	1.3	0	19

[a] Includes the glucofructan levosin

Inorganic constituents per 100g edible portion

No.	Food	mg										µg	
		Na	K	Ca	Mg	P	Fe	Cu	Zn	Cl	Mn	Se	I
	Flours, grains and starches												
1	Bran, wheat	28	1160	110	520	1200	12.9	1.34	16.2	150	9.0	(2)	N
2	**Chapati flour**, brown	39	280	86	69	250	3.4	0.33	2.1	67	2.0	4	N
3	white	15	200	84	29	140	2.5	0.25	1.3	68	1.0	3	N
4	**Cornflour**	52	61	15	7	39	1.4	0.13	0.3	71	N	Tr	N
5	**Custard powder**	320	61	15	7	39	1.4	0.05	0.3	480	N	N	N
6	**Oatmeal**, quick cook, *raw*	9	350	52	110	380	3.8	0.49	3.3	25	3.9	3	N
7	**Rye flour**, whole	(1)	410	32	92	360	2.7	0.42	3.0	N	0.7	N	N
8	**Soya flour**, full fat	9	1660	210	240	600	6.9	2.92	3.9	110	2.3	9	N
9	low fat	14	2030	240	290	640	9.1	3.12	3.2	N	2.9	(11)	N
10	**Wheat flour**, brown	4	250	130[a]	80	230	3.2[a]	0.32	1.9	45	1.9	4	N
11	white, breadmaking	3	130	140[b]	31	120	2.1[b]	0.18	0.9	62	0.7	3	N
12	white, plain	3	150	140[b]	20	110	2.0[b]	0.15	0.6	81	0.6	2	10
13	white, self-raising	360[c]	150	350[c]	20	450[c]	2.0[b]	0.17	0.6	88	0.6	2	10
14	wholemeal	3	340	38	120	320	3.9	0.45	2.9	38	3.1	6	N
15	**Wheatgerm**	5	950	55	270	1050	8.5	0.90	17.0	80	12.3	(3)	N
	Rice												
16	**Brown rice**, *raw*	3	250	10	110	310	1.4	0.85	1.8	230	2.3	10	N
17	*boiled*	1	99	4	43	120	0.5	0.33	0.7	91	0.9	4	N
18	**Egg fried rice**, *takeaway*	417	37	12	6	47	0.5	0.07	0.8	N	0.5	N	N

[a] These are levels for fortified flour. Unfortified brown flour would contain about 20mg Ca and 2.5mg Fe per 100g

[b] These are levels for fortified flour. Unfortified white flours would contain about 15mg Ca and 1.5mg Fe per 100g

[c] The amount present will depend on the nature and level of the raising agent used

Vitamins per 100g edible portion

No.	Food	Retinol µg	Carotene µg	Vitamin D µg	Vitamin E mg	Thiamin mg	Riboflavin mg	Niacin mg	Trypt/60 mg	Vitamin B_6 mg	Vitamin B_{12} µg	Folate µg	Pantothenate mg	Biotin µg	Vitamin C mg
	Flours, grains and starches														
1	**Bran**, wheat	0	0	0	2.60	0.89	0.36	29.6	3.0	1.38	0	260	2.4	45	0
2	**Chapati flour**, brown	0	0	0	(0.60)	0.26	0.05	3.8	2.4	0.29	0	29	(0.4)	(3)	0
3	white	0	0	0	(0.30)	0.36	0.06	1.9	2.0	0.17	0	20	(0.3)	(1)	0
4	**Cornflour**	0	0	0	Tr	Tr	Tr	Tr	0.1	Tr	0	Tr	Tr	Tr	0
5	**Custard powder**	0	0	0	Tr	Tr	Tr	Tr	0.1	Tr	0	Tr	Tr	Tr	0
6	**Oatmeal**, quick cook, *raw*	0	0	0	1.50	0.90	0.09	0.8	2.6	0.33	0	60	1.2	21	0
7	**Rye flour**, whole	0	0	0	1.60	0.40	0.22	1.0	1.6	0.35	0	78	1.0	6	0
8	**Soya flour**, full fat	0	N	0	1.50	0.75	0.28	2.0	8.6	0.46	0	345	1.6	N	0
9	low fat	0	N	0	N	0.90	0.29	2.4	10.6	0.52	0	410	1.8	N	0
10	**Wheat flour**, brown	0	0	0	0.60	0.39[a]	0.07	4.0[a]	2.6	(0.30)	0	51	(0.4)	(3)	0
11	white, breadmaking	0	0	0	(0.30)	0.32[b]	0.03	2.0[b]	2.3	0.15	0	31	0.3	1	0
12	white, plain	0	0	0	0.30	0.31[b]	0.03	1.7[b]	1.9	0.15	0	22	0.3	1	0
13	white, self-raising	0	0	0	(0.30)	0.30[b]	0.03	1.5[b]	1.8	0.15	0	19	0.3	1	0
14	wholemeal	0	0	0	1.40	0.47[b]	0.09	5.7[b]	2.5	0.50	0	57	0.8	7	0
15	**Wheatgerm**	0	0	0	22.00	2.01	0.72	4.5	5.3	3.30	0	N	1.9	25	0
	Rice														
16	**Brown rice**, *raw*	0	0	0	0.80	0.59	0.07	5.3	1.5	N	0	49	N	N	0
17	*boiled*	0	0	0	0.30	0.14	0.02	1.3	0.6	N	0	10	N	N	0
18	**Egg fried rice**, *takeaway*	6	Tr	0.3	0.88	0.03	0.08	0.3	1.1	0.06	0.4	8	0.4	5	Tr

[a] These are levels for fortified flour. Unfortified brown flour would contain 0.30mg thiamin and 1.7mg niacin per 100g

[b] These are levels for fortified flour. Unfortified white flours would contain 0.10mg thiamin and 0.7mg niacin per 100g

Composition of food per 100g edible portion

No.	Food	Description and main data sources	Edible conversion factor	Water	Total nitrogen	Protein	Fat	Carbo-hydrate	Energy value	
				g	g	g	g	g	kcal	kJ
***Rice** continued*										
19	**Pilau**, plain	Recipe from a personal collection	1.00	69.4	0.38	2.3	4.6	24.8	142	599
20	**Savoury rice**, *raw*	10 samples, 5 varieties, meat and vegetable	1.00	7.0	1.41	8.4	10.3	77.4	415	1755
21	*cooked*	Calculation from raw, boiled in water	1.00	68.7	0.48	2.9	3.5[a]	26.3	142	599
22	**White rice**, basmati, *raw*	Ref. Chughtai and Khan (1960)	1.00	10.5	1.30	7.4	0.5	79.8	359	1502
23	easy cook, *raw*	10 samples, 9 different brands, parboiled	1.00	11.4	1.23	7.3	3.6	85.8	383	1630
24	easy cook, *boiled*	Calculation from raw	1.00	68.0	0.44	2.6	1.3	30.9	138	587
25	*fried*	Recipe, fried with onions in vegetable oil	1.00	68.5	0.42	2.5	4.1	25.9	144	609
Pasta										
26	**Macaroni**, *raw*	10 samples, 7 brands; literature sources	1.00	9.7	2.11	12.0	1.8	75.8	348	1483
27	*boiled*	10 samples, 7 brands boiled in water	1.00	78.1	0.52	3.0	0.5	18.5	86	365
28	**Macaroni cheese**	Recipe	1.00	70.2	1.07	6.7	9.9	12.2	162	677
29	**Noodles**, egg, *raw*	10 samples, 8 brands	1.00	9.1	2.12	12.1	8.2	71.7	391	1656
30	egg, *boiled*	10 samples, 8 brands boiled in water	1.00	84.3	0.40	2.2	0.5	13.0	62	264
31	**Pasta**, plain, fresh, *raw*	12 samples, assorted types e.g. Spaghetti, Tagliatelle, Lasagne, Linguine and Fusilli	1.00	33.0	1.75	11.3	2.4	55.5	274	1164
32	plain, fresh, *cooked*	12 samples, 8 brands including Spaghetti, Tagliatelle, Lasagne, Linguine and Fusilli, boiled in water	1.00	61.5	1.06	6.6	1.5	31.8	159	677
33	**Pasta**, fresh, cheese and vegetable stuffed, *cooked*	10 samples, Tortellini, Agnolotti, Ravioli	1.00	61.3	1.23	7.7	4.6	25.8	169	714
34	**Ravioli**, canned in tomato sauce	10 samples, 4 brands	1.00	79.9	0.53	3.0	2.2	10.3	70	297

[a] Calculated assuming only water added; savoury rice cooked with fat contains approximately 2.7g protein, 8.8g fat, 24.2g carbohydrate, 181 kcal energy and 758kJ energy per 100g

Composition of food per 100g edible portion

No.	Food	Starch	Total sugars	Individual sugars Gluc	Fruct	Sucr	Malt	Lact	Dietary fibre NSP	Fatty acids Satd	Mono-unsatd	Poly-unsatd	Trans	Cholest-erol
		g	g	g	g	g	g	g	g	g	g	g	g	mg
	***Rice** continued*													
19	**Pilau**, plain	23.8	0.7	0.3	0.2	0.2	0	Tr	0.3	2.6	1.1	0.5	0	10
20	**Savoury rice**, *raw*	73.8	3.6	0.2	0.5	2.5	0.2	0.1	N	3.2	3.7	1.8	0	1
21	*cooked*	25.1	1.2	0.1	0.2	0.9	0.1	Tr	1.4	1.1	1.3	0.6	0	Tr
22	**White rice**, basmati, *raw*	*79.8*	Tr	Tr	Tr	Tr	(0)	(0)	n	N	N	N	N	0
23	easy cook, *raw*	85.8	Tr	Tr	Tr	Tr	0	0	0.4	0.9	0.9	1.3	0	0
24	easy cook, *boiled*	30.9	Tr	Tr	Tr	Tr	0	0	0.1	0.3	0.3	0.5	0	0
25	*fried*	24.0	1.4	0.5	0.4	0.5	0	0	0.5	0.6	1.9	1.3	N	0
	Pasta													
26	**Macaroni**, *raw*	73.6	2.2	0.2	0.1	0.6	1.2	0	3.1[a]	0.3	0.1	0.8	Tr	0
27	*boiled*	18.2	0.3	Tr	Tr	0.1	0.2	0	0.9[a]	0.1	Tr	0.2	Tr	0
28	**Macaroni cheese**	9.7	2.5	Tr	Tr	Tr	0.1	2.3	0.5	4.9	2.8	1.6	0.3	21
29	**Noodles**, egg, *raw*	69.8	1.9	0.1	Tr	0.6	1.1	0	(2.9)	2.3	3.5	0.9	Tr	30
30	egg, *boiled*	12.8	0.2	Tr	Tr	0.1	0.1	0	(0.6)	0.1	0.2	0.1	Tr	6
31	**Pasta**, plain, fresh, *raw*	53.5	1.5	0.2	0.1	0.4	0.8	Tr	N	N	N	N	Tr	N
32	plain, fresh, *cooked*	30.7	0.6	0.1	0	0.1	0.4	Tr	1.9	0.3	0.3	0.4	Tr	N
33	**Pasta**, fresh, cheese and vegetable stuffed, *cooked*	24.9	0.9	0.1	0.1	0.1	0.4	0.3	1.3	N	N	N	Tr	N
34	**Ravioli**, canned in tomato sauce	8.1	2.2	0.5	0.7	0.7	0.3	0	0.9	0.8	0.8	0.3	Tr	6

[a] Wholemeal macaroni contains 8.3g (raw) and 2.8g (boiled) NSP per 100g

Inorganic constituents per 100g edible portion

No.	Food	mg										µg	
		Na	K	Ca	Mg	P	Fe	Cu	Zn	Cl	Mn	Se	I
	***Rice** continued*												
19	**Pilau**, plain	275	74	22	12	48	0.6	0.12	0.5	425	0.4	4	5
20	**Savoury rice**, *raw*	1440	340	73	45	200	1.5	0.14	1.3	2520	1.2	N	N
21	*cooked*	490	110	25	15	67	0.5	0.05	0.4	860	0.4	N	N
22	**White rice**, basmati, *raw*	N	N	19	N	73	1.3	N	N	N	N	N	N
23	easy cook, *raw*	4	150	51	32	150	0.5	0.37	1.8	10	1.2	13	(14)
24	easy cook, *boiled*	1	54	18	11	54	0.2	0.13	0.7	4	0.2	5	5
25	*fried*	111	99	21	11	53	0.3	0.11	0.6	176	0.2	4	5
	Pasta												
26	**Macaroni**, *raw*	11	230	25	53	180	1.6	0.30	1.5	20	0.9	20	Tr
27	*boiled*	1	25	6	14	42	0.5	0.09	0.5	5	0.3	5	Tr
28	**Macaroni cheese**	283	103	170	16	138	0.3	0.05	N	438	0.2	3	20
29	**Noodles**, egg, *raw*	180	260	28	43	200	1.5	0.24	1.3	180	0.8	N	N
30	egg, *boiled*	15	23	5	8	31	0.3	0.06	0.3	10	0.2	N	N
31	**Pasta**, plain, fresh, *raw*	(28)	(85)	(64)	(33)	(150)	(1.4)	(0.80)	(1.4)	(51)	(0.7)	(23)	(63)
32	plain, fresh, *cooked*	16	49	37	19	86	0.8	0.46	0.8	29	0.4	13	36
33	**Pasta**, fresh, cheese and vegetable stuffed, *cooked*	204	86	115	18	130	0.9	0.45	1.1	N	0.3	N	N
34	**Ravioli**, canned in tomato sauce	490	150	16	12	43	0.8	0.08	0.5	760	0.2	N	N

No.	Food	Retinol µg	Carotene µg	Vitamin D µg	Vitamin E mg	Thiamin mg	Riboflavin mg	Niacin mg	Trypt/60 mg	Vitamin B_6 mg	Vitamin B_{12} µg	Folate µg	Pantothenate mg	Biotin µg	Vitamin C mg
Rice *continued*															
19	**Pilau**, plain	23	22	0.1	0.18	0.08	0.01	0.8	0.5	0.07	Tr	4	0.2	1	Tr
20	**Savoury rice**, *raw*	0	N	Tr	N	0.46	0.06	5.2	1.9	0.37	Tr	25	N	N	0
21	*cooked*	0	N	Tr	N	0.10	0.01	1.1	0.6	0.07	Tr	4	N	N	0
22	**White rice**, basmati, *raw*	0	0	0	N	N	N	N	N	N	0	N	N	N	0
23	easy cook, *raw*	0	0	0	(0.10)	0.41	0.02	4.2	1.6	0.31	0	20	(0.6)	(3)	0
24	easy cook, *boiled*	0	0	0	Tr	0.01	Tr	0.9	0.6	0.07	0	7	(0.1)	(1)	0
25	*fried*	0	3	0	0.07	0.04	0	0.9	0.6	0.11	0	9	0.1	1	2
Pasta															
26	**Macaroni**, *raw*	0	0	0	Tr	0.18	0.05	2.9	2.5	0.10	(0)	23	(0.3)	(1)	0
27	*boiled*	0	0	0	Tr	0.03	Tr	0.5	0.6	0.01	(0)	3	Tr	Tr	0
28	**Macaroni cheese**	91	42	0.3	1.26	0.03	0.17	0.3	1.6	0.05	0.8	5	0.3	2	0
29	**Noodles**, egg, *raw*	37	0	0.3	N	0.26	0.10	2.2	2.5	0.10	Tr	29	N	N	0
30	egg, *boiled*	2	0	Tr	N	0.01	0.01	0.2	0.5	0.01	Tr	1	N	N	0
31	**Pasta**, plain, fresh, *raw*	0	0	0	Tr	(0.06)	(0.03)	(0.7)	(1.8)	(0.02)	0	(3)	Tr	Tr	0
32	plain, fresh, *cooked*	0	0	0	Tr	0.06	0.03	0.7	1.0	0.02	0	4	Tr	Tr	0
33	**Pasta**, fresh, cheese and vegetable stuffed, *cooked*	25	143	N	0.82	0.05	0.06	0.7	1.4	0.02	N	3	0.4	4	Tr
34	**Ravioli**, canned in tomato sauce	N	N	0	N	0.05	0.04	0.9	0.6	0.10	Tr	3	N	N	Tr

Composition of food per 100g edible portion

No.	Food	Description and main data sources	Edible conversion factor	Water g	Total nitrogen g	Protein g	Fat g	Carbo-hydrate g	Energy value kcal	Energy value kJ
	***Pasta** continued*									
35	**Spaghetti**, white, *raw*	10 samples, 7 brands	1.00	9.8	2.11	12.0	1.8	74.1	342	1456
36	white, *boiled*	10 samples, 7 brands boiled in water	1.00	73.8	0.63	3.6	0.7	22.2	104	442
37	wholemeal, *raw*	10 samples, 5 brands	1.00	10.5	2.30	13.4	2.5	66.2	324	1379
38	wholemeal, *boiled*	Water content weighed, other nutrients calculated from raw	1.00	69.1	0.81	4.7	0.9	23.2	113	485
39	canned in tomato sauce	10 samples, 3 brands	1.00	81.9	0.33	1.9	0.4	14.1	64	273
	Breads									
40	**Brown bread**, average	11 samples, 8 brands	1.00	41.2	1.40	7.9	2.0	42.1	207	882
41	**Chapatis**, *made with fat*	6 samples[a]	1.00	28.5	1.42	8.1	12.8	48.3	328	1383
42	*made without fat*	Analysed and calculated values	1.00	45.8	1.28	7.3	1.0	43.7	202	860
43	**Ciabatta**	8 samples	1.00	29.2	1.80	10.2	3.9	52.0	271	1150
44	**Currant bread**	10 samples, 10 different shops	1.00	29.4	1.32	7.5	7.6	50.7	289	1220
45	**Garlic bread**, pre-packed, *frozen,*	10 samples, 8 brands. Part baked	1.00	25.7	1.40	7.8	18.3	45.0	365	1530
46	**Granary bread**	24 samples, sliced and unsliced	1.00	34.9	1.65	9.6	2.3	47.4	237	1005
47	**Malt bread**, fruited	6 samples, 3 brands	1.00	24.2	1.40	7.8	2.3	64.9	295	1256
48	**Naan bread**	12 samples including garlic and coriander	1.00	30.8	1.40	7.8	7.3	50.2	285	1206
49	**Pappadums**, *takeaway*	10 samples from different outlets	1.00	3.9	1.84	11.5	38.8	28.3	501	2084
50	**Pitta bread**, white	10 samples, 8 brands	1.00	31.4	1.60	9.1	1.3	55.1	255	1084
51	**Rye bread**	15 samples, different shops; literature sources	1.00	37.4	1.46	8.3	1.7	45.8	219	932
52	**Wheatgerm bread**	7 samples, 4 brands, pre-packed, sliced	1.00	39.7	1.90	11.1	3.1	39.5	220	935

[a] Puris (deep fried chapatis) contain 19.1g water, 7.0g protein, 25.0g fat and 43.3g carbohydrate per 100g

No.	Food	Starch	Total sugars	Individual sugars					Dietary fibre	Fatty acids				Cholest-erol
				Gluc	Fruct	Sucr	Malt	Lact	NSP	Satd	Mono-unsatd	Poly-unsatd	Trans	
		g	g	g	g	g	g	g	g	g	g	g	g	mg
	Pasta *continued*													
35	**Spaghetti**, white, *raw*	70.8	3.3	0.3	0.1	0.8	1.8	0	2.9	0.2	0.2	0.8	Tr	0
36	white, *boiled*	21.7	0.5	Tr	Tr	0.1	0.3	0	1.2	0.1	0.1	0.3	Tr	0
37	wholemeal, *raw*	62.5	3.7	0.8	0.4	1.1	1.2	0	8.4	0.4	0.3	1.1	Tr	0
38	wholemeal, *boiled*	21.9	1.3	0.3	0.1	0.4	0.4	0	3.5	0.1	0.1	0.4	Tr	0
39	canned in tomato sauce	8.6	5.5	1.0	1.1	2.9	0.4	0	0.7[a]	0.1	0.1	0.2	Tr	0
	Breads													
40	**Brown bread**, average	38.7	3.4	Tr	0.3	Tr	3.0	Tr	3.5	0.4	0.4	0.7	Tr	0
41	**Chapatis**, *made with fat*	46.5	1.8	N	N	N	N	0	N	N	N	N	N	N
42	*made without fat*	42.1	1.6	N	N	N	N	0	N	0.1	0.1	0.4	Tr	0
43	**Ciabatta**	48.9	3.1	Tr	0.1	0.3	2.7	Tr	2.3	0.6	2.1	0.9	Tr	Tr
44	**Currant bread**	36.3	14.4	6.3	6.7	0	1.2	0.2	N	(1.6)	(1.5)	(2.0)	N	0
45	**Garlic bread**, pre-packed, *frozen*	42.3	2.7	Tr	0.3	Tr	2.4	0	N	9.7	5.5	1.5	0.9	37
46	**Granary bread**	44.5	2.9	0.1	0.3	Tr	2.5	Tr	3.3	0.6	0.6	0.8	Tr	0
47	**Malt bread**, fruited	42.3	22.6	7.4	6.3	1.0	7.2	0.7	2.6	0.5	1.0	1.0	0.2	0
48	**Naan bread**	47.0	3.1	0.7	0.8	Tr	1.3	0.4	2.0	1.0	3.1	2.4	0.1	5
49	**Pappadums**, *takeaway*	28.3	Tr	Tr	Tr	Tr	Tr	Tr	5.8	8.0	16.5	12.5	0	2
50	**Pitta bread**, white	52.2	3.0	0.5	0.5	Tr	2.0	0	2.4[b]	0.2	0.1	0.5	Tr	0
51	**Rye bread**	44.0	1.8	N	N	N	N	0	4.4[c]	0.3	0.3	0.6	Tr	0
52	**Wheatgerm bread**	35.8	3.8	0.2	0.6	Tr	2.9	0	4.0	0.7	0.7	1.1	Tr	0

[a] Wholemeal types contain 2.0g NSP per 100g
[b] Wholemeal pitta bread contains 5.2g NSP per 100g
[c] Pumpernickel contains approximately 7.5g NSP per 100g

Inorganic constituents per 100g edible portion

No.	Food	mg										µg	
		Na	K	Ca	Mg	P	Fe	Cu	Zn	Cl	Mn	Se	I
***Pasta** continued*													
35	**Spaghetti**, white, *raw*	3	250	25	56	190	2.1	0.32	1.5	25	0.9	(17)	Tr
36	white, *boiled*	Tr	24	7	15	44	0.5	0.10	0.5	Tr	0.3	(5)	Tr
37	wholemeal, *raw*	130	390	31	120	330	3.9	0.51	3.0	210	2.6	(16)	N
38	wholemeal, *boiled*	45	140	11	42	110	1.4	0.18	1.1	73	0.9	(6)	N
39	canned in tomato sauce	420	110	12	10	29	0.3	0.06	0.3	500	0.1	N	N
Breads													
40	**Brown bread**, average	443	216	186	45	157	2.2	0.17	1.3	787	1.1	4	6
41	**Chapatis**, *made with fat*	130	160	66	41	130	2.3	0.20	1.1	250	(1.4)	4	N
42	*made without fat*	120	150	60	37	120	2.1	0.20	1.0	230	(1.2)	4	N
43	**Ciabatta**	538	(152)	(121)	(22)	(100)	(1.4)	(0.09)	(0.7)	(830)	(0.6)	19	(10)
44	**Currant bread**	290	220	86	26	93	1.6	0.32	0.7	480	0.4	N	29
45	**Garlic bread**, pre-packed, *frozen*	644	N	N	N	N	N	N	N	N	N	N	N
46	**Granary bread**	545	191	209	39	138	1.9	0.18	1.1	796	0.8	6	8
47	**Malt bread**, fruited	246	234	104	37	125	1.7	0.14	0.7	408	0.7	10	27
48	**Naan bread**	604	172	187	21	299	1.6	0.09	0.7	N	0.5	Tr	N
49	**Pappadums**, *takeaway*	1460	609	N	121	196	4.4	0.36	1.7	2070	1.0	15	40
50	**Pitta bread**, white	439[a]	178	138[a]	22	99	1.9[a]	0.12	0.8[a]	(678)	0.5	2	N
51	**Rye bread**	580	190	80	48	160	2.5	0.18	1.3	1410	1.0	N	N
52	**Wheatgerm bread**	578	269	212	64	219	2.9	0.26	2.3	828	2.1	12	(22)

[a] Wholemeal pitta bread contains 460mg Na, 48mg Ca, 2.7mg Fe and 1.8mg Zn per 100g

Vitamins per 100g edible portion

No.	Food	Retinol µg	Carotene µg	Vitamin D µg	Vitamin E mg	Thiamin mg	Ribo-flavin mg	Niacin mg	Trypt/60 mg	Vitamin B_6 mg	Vitamin B_{12} µg	Folate µg	Panto-thenate mg	Biotin µg	Vitamin C mg
	***Pasta** continued*														
35	**Spaghetti**, white, *raw*	0	0	0	Tr	0.22	0.03	3.1	2.5	0.17	(0)	23	(0.3)	(1)	0
36	white, *boiled*	0	0	0	Tr	0.01	0.01	0.5	0.7	0.02	(0)	7	Tr	Tr	0
37	wholemeal, *raw*	0	0	0	Tr	0.99	0.11	6.2	2.7	0.39	(0)	33	(0.8)	(1)	0
38	wholemeal, *boiled*	0	0	0	Tr	0.21	0.02	1.3	1.0	0.08	(0)	7	(0.2)	Tr	0
39	canned in tomato sauce	N	N	0	N	0.07	0.01	0.6	0.4	0.07	Tr	5	Tr	Tr	Tr
	Breads														
40	**Brown bread**, average	0	0	0	0.01	0.22	0.07	2.8	2.1	0.17	0	45	0.5	3	0
41	**Chapatis**, *made with fat*	N	0	N	N	0.26	0.04	1.7	1.7	(0.20)	0	15	(0.2)	(2)	0
42	*made without fat*	0	0	0	Tr	0.23	0.04	1.5	1.5	(0.18)	0	14	(0.2)	(2)	0
43	**Ciabatta**	0	0	0	0.47	0.24	(0.06)	2.0	(2.2)	(0.07)	0	21	(0.4)	(1)	0
44	**Currant bread**	0	Tr	0	Tr	0.19	0.09	1.5	1.5	0.09	Tr	19	N	N	0
45	**Garlic bread**, pre-packed, *frozen*	142	54	Tr	0.16	0.24	0.07	1.5	(2.2)	0.07	Tr	20	(0.4)	(1)	Tr
46	**Granary bread**	0	0	0	0.23	0.24	0.09	2.7	2.4	0.19	0	88	0.5	1	0
47	**Malt bread**, fruited	Tr	Tr	Tr	0.18	0.24	0.32	2.4	1.9	0.11	0	34	N	N	Tr
48	**Naan bread**	5	11	0.1	0.64	0.27	0.05	3.0	1.8	N	Tr	15	N	N	Tr
49	**Pappadums**, *takeaway*	N	N	0	3.70	0.35	(0.09)	0.6	1.8	0.05	0	23	0.7	5	0
50	**Pitta bread**, white	0	0	0	N	0.34	0.08	2.2	2.2	N	0	20	N	N	0
51	**Rye bread**	0	0	0	1.20	0.29	0.05	2.3	1.7	0.09	0	24	0.5	N	0
52	**Wheatgerm bread**	0	0	0	0.48	0.34	0.11	3.6	2.5	0.09	0	38	0.5	2	0

Composition of food per 100g edible portion

No.	Food	Description and main data sources	Edible conversion factor	Water	Total nitrogen	Protein	Fat	Carbo-hydrate	Energy value	
				g	g	g	g	g	kcal	kJ
	Breads *continued*									
53	**White bread**, sliced	15 samples	1.00	38.6	1.40	7.9	1.6	46.1	219	931
54	*fried in lard*	Calculated from white sliced bread using analysed fat and water changes	1.00	7.4	1.42	8.1	32.2[a]	46.8	498	2078
55	*toasted*	Calculated using weight loss of 18%	1.00	25.1	1.71	9.7	2.0	56.2	267	1137
56	farmhouse or split tin	20 samples	1.00	36.8	1.60	9.0	2.0	48.4	236	1001
57	French stick	18 samples, baguette and flute, thick and thin	1.00	29.0	1.60	9.0	1.9	56.1	263	1121
58	premium	15 samples, 13 brands	1.00	38.4	1.50	8.3	2.3	47.0	230	978
59	Danish style	8 samples, 4 brands	1.00	37.5	1.60	9.1	2.7	44.5	228	967
60	'with added fibre'	Manufacturer's data for Mighty White (Allied Bakeries) and Champion (British Bakeries)	1.00	40.0	1.33	7.6	1.5	49.6	230	978
61	'with added fibre', *toasted*	Calculated using weight loss of 16%	1.00	26.2	1.58	9.0	1.8	59.0	273	1164
62	**Wholemeal bread**, average	21 samples, sliced and unsliced	1.00	41.2	1.65	9.4	2.5	42.0	217	922
63	*toasted*	Calculated using weight loss of 14.6%	1.00	31.1	1.93	11.2	2.9	49.2	255	1084
	Rolls									
64	**Brown rolls**, crusty	12 samples of 6 rolls, different shops	1.00	30.5	1.81	10.3	2.8	50.4	255	1085
65	soft	14 samples of 6 rolls, different shops	1.00	36.4	1.70	9.9	3.2	44.8	236	1004
66	**Croissants**	10 samples, 7 brands	1.00	24.8	1.50	8.3	19.7	43.3	373	1563
67	**Granary rolls**	10 samples, pre-packed, freshly backed	1.00	34.5	1.70	10.0	4.2	42.7	238	1009
68	**Hamburger buns**	5 packets of 6 buns including frozen	1.00	32.9	1.60	9.1	5.0	48.8	264	1121
69	**White rolls**, crusty	10 samples	1.00	29.7	1.60	9.2	2.2	54.9	262	1116
70	soft	10 samples	1.00	35.6	1.60	9.3	2.6	51.5	254	1078
71	**Wholemeal rolls**	10 samples	1.00	37.2	1.80	10.4	3.3	46.1	244	1037

[a] The fat content depends on the conditions of frying; thin slices pick up proportionately more fat than thick ones

Composition of food per 100g edible portion

No.	Food	Starch	Total sugars	Individual sugars					Dietary fibre	Fatty acids				Cholest-erol
				Gluc	Fruct	Sucr	Malt	Lact	NSP	Satd	Mono-unsatd	Poly-unsatd	Trans	
		g	g	g	g	g	g	g	g	g	g	g	g	mg
Breads *continued*														
53	**White bread**, sliced	42.7	3.4	Tr	0.2	Tr	3.2	Tr	1.9	0.3	0.3	0.5	Tr	0
54	*fried in lard*	43.3	3.5	Tr	0.2	Tr	3.2	Tr	1.9	12.5	13.4	2.9	N	Tr
55	*toasted*	52.1	4.1	Tr	0.2	Tr	3.9	0	2.3	0.4	0.4	0.6	Tr	Tr
56	farmhouse or split tin	47.0	2.9	Tr	0.2	Tr	2.7	Tr	2.1	0.5	0.4	0.6	0.1	0
57	French stick	53.3	2.8	Tr	0.2	Tr	2.7	Tr	2.4	0.3	0.3	0.7	Tr	0
58	premium	44.4	2.7	Tr	0.2	Tr	2.5	Tr	1.9	N	N	N	Tr	N
59	Danish style	41.6	3.0	0.1	0.2	Tr	2.7	0	2.4	(0.5)	(0.5)	(0.9)	(Tr)	0
60	'with added fibre'	46.3	3.3	N	N	N	N	0	3.1	0.4	0.6	0.3	Tr	0
61	'with added fibre', *toasted*	55.1	3.9	N	N	N	N	0	3.7	0.5	0.7	0.3	Tr	0
62	**Wholemeal bread**, average	39.3	2.8	0.2	0.4	Tr	2.2	0	5.0	0.5	0.6	0.8	Tr	0
63	*toasted*	46.0	3.2	0.2	0.5	Tr	2.5	0	5.9	0.5	0.7	1.0	0.1	0
Rolls														
64	**Brown rolls**, crusty	48.5	1.9	N	N	N	N	0	(3.5)	0.6	0.6	0.7	N	0
65	soft	42.0	2.8	0.5	0.4	Tr	2.0	0	3.8	1.1	1.0	1.0	N	0
66	**Croissants**	38.0	5.3	1.2	2.0	Tr	1.9	0.2	1.6	9.8	4.6	1.6	1.5	52
67	**Granary rolls**	39.7	3.0	0.3	0.6	Tr	2.0	0	3.6	(1.1)	(1.1)	(1.4)	N	0
68	**Hamburger buns**	46.6	2.2	N	N	N	N	0	(1.5)	1.1	1.3	1.1	N	0
69	**White rolls**, crusty	52.1	2.7	Tr	0.2	Tr	2.6	0	2.4	0.5	0.5	0.7	0.1	0
70	soft	48.8	2.6	0.2	0.2	Tr	2.2	0	2.0	0.6	0.6	0.8	0.1	0
71	**Wholemeal rolls**	43.5	2.6	0.5	0.5	Tr	1.7	0	4.4	0.8	0.9	0.7	N	0

Inorganic constituents per 100g edible portion

No.	Food	mg										µg	
		Na	K	Ca	Mg	P	Fe	Cu	Zn	Cl	Mn	Se	I
	Breads *continued*												
53	**White bread**, sliced	461	137	177	23	95	1.6	0.14	0.8	829	0.5	6	4
54	*fried in lard*	468	139	180	23	96	1.6	0.14	0.8	842	0.5	6	4
55	*toasted*	562	167	216	28	116	2.0	0.17	1.0	1011	0.6	5	5
56	farmhouse or split tin	590	135	172	19	89	1.3	0.12	0.8	812	0.5	5	4
57	French stick	616	152	121	22	100	1.4	0.09	0.7	974	0.6	5	10
58	premium	530	138	177	25	101	1.5	0.19	0.8	831	0.5	10	N
59	Danish style	537	110	130	21	93	1.7	0.14	0.6	950	0.4	N	N
60	'with added fibre'	450	160	150	30	100	2.3	0.10	0.9	790	0.5	4	(6)
61	'with added fibre', *toasted*	540	190	180	36	120	2.7	0.12	1.1	940	0.6	5	(7)
62	**Wholemeal bread**, average	487	253	106	66	202	2.4	0.23	1.6	800	1.8	7	Tr
63	*toasted*	570	296	124	77	237	2.8	0.27	1.9	937	2.1	11	N
	Rolls												
64	**Brown rolls**, crusty	570	200	100	65	190	3.2	0.34	1.5	1040	(1.4)	4	(7)
65	soft	494	(234)	(201)	(49)	(170)	(2.4)	(0.18)	(1.4)	(762)	(1.2)	6	(6)
66	**Croissants**	419	126	75	19	93	1.1	0.05	0.7	646	0.4	8	N
67	**Granary rolls**	566	(191)	(209)	(39)	(138)	(1.9)	(0.18)	(1.1)	(796)	(0.8)	4	(8)
68	**Hamburger buns**	550	110	130	31	150	2.3	0.13	0.7	890	0.5	9	(19)
69	**White rolls**, crusty	656	164	177	22	104	1.7	0.13	0.9	888	0.5	4	(19)
70	soft	535	145	184	23	99	1.5	0.13	0.9	(813)	0.5	6	19
71	**Wholemeal rolls**	526	248	87	61	197	2.4	0.26	1.7	850	1.5	7	Tr

No.	Food	Retinol µg	Carotene µg	Vitamin D µg	Vitamin E mg	Thiamin mg	Riboflavin mg	Niacin mg	Trypt/60 mg	Vitamin B_6 mg	Vitamin B_{12} µg	Folate µg	Pantothenate mg	Biotin µg	Vitamin C mg
	Breads *continued*														
53	**White bread**, sliced	0	0	0	Tr	0.24	0.08	1.6	2.0	0.08	0	25	0.4	1	0
54	*fried in lard*	Tr	0	N	Tr	0.18	0.07	1.5	2.0	0.06	Tr	13	0.3	1	0
55	*toasted*	0	0	0	Tr	0.25	0.10	2.0	2.4	0.10	0	30	0.5	1	0
56	farmhouse or split tin	0	0	0	0.22	0.19	0.06	1.4	2.0	0.06	0	23	0.4	3	0
57	French stick	0	0	0	Tr	0.21	0.06	1.7	2.2	0.07	0	29	0.4	1	0
58	premium	0	0	0	0.17	0.23	0.07	1.5	2.1	0.07	0	23	0.5	1	0
59	Danish style	0	0	0	Tr	0.25	0.07	2.0	1.7	0.07	0	44	N	N	0
60	'with added fibre'	0	0	0	Tr	0.20[a]	(0.05)	1.6[a]	1.5	(0.07)	0[a]	(17)[a]	(0.3)	(1)	0
61	'with added fibre', *toasted*	0	0	0	Tr	0.20[a]	(0.06)	1.9[a]	1.8	(0.08)	0[a]	(20)[a]	(0.4)	(1)	0
62	**Wholemeal bread**, average	0	0	0	0.28	0.25	0.05	3.8	2.3	0.11	Tr	40	0.6	6	0
63	*toasted*	0	0	0	0.33	0.25	0.06	4.5	2.7	0.13	0.7	46	0.7	7	0
	Rolls														
64	**Brown rolls**, crusty	0	0	0	Tr	(0.32)	(0.15)	(3.8)	(2.8)	(0.20)	0	(63)	(0.5)	(4)	0
65	soft	0	0	0	Tr	0.29	0.14	3.5	2.6	(0.18)	0	57	(0.5)	(3)	0
66	**Croissants**	163	19	0.1	0.99	0.19	(0.16)	1.5	1.8	(0.11)	Tr	47	(0.5)	(9)	0
67	**Granary rolls**	0	0	0	(0.23)	(0.24)	(0.09)	(2.7)	(2.4)	(0.19)	0	78	(0.5)	(1)	0
68	**Hamburger buns**	N	0	N	Tr	0.23	0.10	1.5	1.9	0.06	Tr	48	(0.3)	(1)	0
69	**White rolls**, crusty	0	0	0	(0.23)	0.22	0.07	2.0	2.2	0.03	0	31	0.4	1	0
70	soft	0	0	0	(0.20)	(0.20)	(0.06)	(1.8)	(2.0)	(0.03)	0	27	(0.4)	(1)	0
71	**Wholemeal rolls**	0	0	0	(0.30)	0.30	0.09	4.1	(2.5)	0.10	0	57	(0.6)	(6)	0

[a] May be present at higher levels as a result of fortification

Composition of food per 100g edible portion

No.	Food	Description and main data sources	Edible conversion factor	Water	Total nitrogen	Protein	Fat	Carbo-hydrate	Energy value	
				g	g	g	g	g	kcal	kJ
Sandwiches										
72	**Sandwich**, Bacon, lettuce and tomato, white bread	Recipe	1.00	(50.6)	1.39	8.2	12.4	24.1	235	983
73	Cheddar cheese and pickle, whitebread	Recipe	1.00	(39.9)	1.97	12.0	14.9	28.7	290	1216
74	Chicken salad, white bread	Recipe	1.00	(58.0)	1.78	10.7	5.3	22.6	175	739
75	Egg mayonnaise, white bread	Recipe	1.00	(46.7)	1.43	8.4	12.0	28.5	248	1042
76	Ham salad, white bread	Recipe	1.00	(57.7)	1.38	8.2	4.5	25.0	167	705
77	Tuna mayonnaise, white bread	Recipe	1.00	(48.1)	2.01	12.1	10.5	25.3	237	998
Breakfast cereals										
78	**All-Bran**	Analysis and manufacturer's data (Kelloggs, All Bran Plus)	1.00	3.0	2.06	13.0	4.0	*48.5*	*270*	*1144*
79	**Bran Flakes**	Manufacturer's data (Kelloggs)	1.00	3.0	1.79	10.2	2.5	*71.2*	*330*	*1406*
80	**Cheerios**	Manufacturer's data (Cereal Partners UK)	1.00	N	1.26	7.9	3.8	*80.7*	*368*	*1566*
81	**Clusters**	Manufacturer's data (Cereal Partners UK)	1.00	N	1.65	10.3	8.5	*71.9*	*387*	*1639*
82	**Coco Pops**	Manufacturer's data (Kelloggs)	1.00	3.0	0.76	4.5	2.5	*91.5*	*383*	*1632*
83	**Corn Flakes**	Analysis and manufacturer's data (Kelloggs)	1.00	3.0	1.26	7.9	0.9	*89.6*	*376*	*1601*
84	**Crunchy Nut Corn Flakes**	Manufacturer's data (Kelloggs)	1.00	3.0	1.18	7.4	3.5	*91.6*	*405*	*1721*
85	**Frosties**	Manufacturer's data (Kelloggs)	1.00	3.0	0.85	5.3	0.6	*94.6*	*381*	*1626*
86	**Fruit 'n Fibre**	Manufacturer's data (Kelloggs)	1.00	5.7	1.58	9.0	5.0	*72.5*	*353*	*1498*
87	**Muesli**, Swiss style	Analysis and manufacturers' data (Kelloggs, Weetabix)[a]	1.00	7.2	1.57	9.8	5.9	*72.2*	*363*	*1540*
88	with no added sugar	Analysis and manufacturers' data (Kelloggs, Weetabix)	1.00	7.6	1.68	10.5	7.8	*67.1*	*366*	*1552*

[a] Muesli composition is very variable

Composition of food per 100g edible portion

No.	Food	Starch	Total sugars	Individual sugars					Dietary fibre	Fatty acids				Cholest-erol
				Gluc	Fruct	Sucr	Malt	Lact	NSP	Satd	Mono-unsatd	Poly-unsatd	Trans	
		g	g	g	g	g	g	g	g	g	g	g	g	mg
	Sandwiches													
72	**Sandwich**, Bacon, lettuce and tomato, white bread	21.8	2.4	0.2	0.4	0.1	1.6	0	1.2	2.8	3.8	4.9	0.1	18
73	Cheddar cheese and pickle, white bread	23.0	5.7	1.3	1.4	1.2	1.7	0.1	1.1	7.4	4.1	2.3	0.5	30
74	Chicken salad, white bread	20.4	2.2	0.2	0.4	0	1.5	0	1.1	1.2	1.8	1.9	Tr	26
75	Egg mayonnaise, white bread	26.3	2.2	0	0.1	0.1	2.0	0	1.2	2.4	3.7	4.9	0.1	111
76	Ham salad, white bread	22.5	2.5	N	N	N	N	0	1.2	0.9	1.4	1.8	Tr	12
77	Tuna mayonnaise, white bread	23.3	2.0	0	0.1	0.1	1.7	0	1.0	1.7	2.7	5.4	0.1	23
	Breakfast cereals													
78	**All-Bran**	*28.6*	19.9	1.0	1.0	15.8	2.1	0	24.5	0.7	0.5	2.0	Tr	0
79	**Bran Flakes**	*48.4*	22.8	2.5	3.0	17.3	Tr	0	13.0	0.4	0.3	1.5	Tr	0
80	**Cheerios**	*58.3*	22.4	N	N	N	N	N	(6.2)	1.5	N	N	Tr	0
81	**Clusters**	*50.2*	21.7	N	N	N	N	N	(8.9)	2.7	N	N	Tr	0
82	**Coco Pops**	*49.5*	42.0	0.5	0.5	41.0	Tr	0	0.6	1.0	0.6	0.5	Tr	0
83	**Corn Flakes**	*81.4*	8.2	1.5	1.5	4.2	1.0	0	0.9	0.2	0.2	0.4	Tr	0
84	**Crunchy Nut Corn Flakes**	*53.9*	37.7	1.5	1.5	34.7	Tr	0	0.8	0.7	1.5	1.0	Tr	0
85	**Frosties**	*50.6*	44.0	1.0	1.0	42.0	Tr	0	0.6	0.1	0.1	0.4	Tr	0
86	**Fruit 'n Fibre**	*49.5*	23.0	6.0	6.0	11.0	Tr	0	7.0	2.5	1.0	0.7	Tr	0
87	**Muesli**, Swiss style	*46.0*	26.2	N	N	N	N	N	6.4	0.8	2.8	1.6	Tr	Tr
88	with no added sugar	*51.4*	15.7	N	N	N	N	N	7.6	1.5	3.5	2.4	Tr	Tr

Inorganic constituents per 100g edible portion

No.	Food	mg										µg	
		Na	K	Ca	Mg	P	Fe	Cu	Zn	Cl	Mn	Se	I
	Sandwiches												
72	**Sandwich**, Bacon, lettuce and tomato, white bread	683	180	95	17	87	1.1	0.08	0.7	1008	0.3	5	6
73	Cheddar cheese and pickle, white bread	753	106	321	22	206	1.0	0.08	1.7	1016	0.3	5	11
74	Chicken salad, white bread	330	197	91	19	107	1.1	0.09	0.8	470	0.3	7	4
75	Egg mayonnaise, white bread	463	121	125	18	115	1.5	0.11	0.9	657	0.3	7	19
76	Ham salad, white bread	594	191	98	19	128	1.1	0.10	0.8	805	0.3	6	4
77	Tuna mayonnaise, white bread	494	151	100	21	110	1.2	0.09	0.7	746	0.3	29	10
	Breakfast cereals												
78	**All-Bran**	850	950	340	240	950	8.8	0.44	6.0	1310	N	4	N
79	**Bran Flakes**	800	600	40	120	450	24.3	0.35	2.5	1240	N	(4)	N
80	**Cheerios**	800	N	450	N	N	11.9	N	N	N	N	N	N
81	**Clusters**	500	N	N	N	N	11.9	N	N	N	N	N	N
82	**Coco Pops**	450	250	453	40	120	7.9	0.20	1.0	700	N	2	N
83	**Corn Flakes**	1000	90	5	10	50	7.9	0.03	0.2	1540	0.1	5	10
84	**Crunchy Nut Corn Flakes**	600	140	15	20	60	7.9	0.08	0.6	930	N	(5)	N
85	**Frosties**	600	60	453	5	30	7.9	Tr	0.2	930	N	(2)	N
86	**Fruit 'n Fibre**	600	400	40	60	240	8.8	0.24	1.5	930	N	N	N
87	**Muesli**, Swiss style	380	440	110	85	280	5.8	0.10	2.3	790	N	(4)	N
88	with no added sugar	47	530	47	90	330	3.5	0.36	2.1	10	2.6	(4)	N

Vitamins per 100g edible portion

No.	Food	Retinol µg	Carotene µg	Vitamin D µg	Vitamin E mg	Thiamin mg	Riboflavin mg	Niacin mg	Trypt/60 mg	Vitamin B_6 mg	Vitamin B_{12} µg	Folate µg	Pantothenate mg	Biotin µg	Vitamin C mg
	Sandwiches														
72	**Sandwich**, Bacon, lettuce and tomato, white bread	N	N	N	2.88	0.34	0.07	2.2	1.7	0.15	0.2	22	0.5	1	2
73	Cheddar cheese and pickle, white bread	N	N	N	N	0.14	0.16	0.9	3.1	0.09	0.7	23	(0.4)	2	Tr
74	Chicken salad, white bread	N	N	N	1.66	0.15	0.08	3.2	2.3	0.15	Tr	21	0.6	1	2
75	Egg mayonnaise, white bread	N	N	N	3.15	0.17	0.15	1.0	2.3	0.08	0.4	26	0.6	N	Tr
76	Ham salad, white bread	N	N	N	1.77	0.31	0.08	2.3	1.7	0.19	0.2	24	0.5	1	2
77	Tuna mayonnaise, white bread	N	N	N	3.22	0.14	0.09	5.6	2.6	0.20	1.4	15	0.3	1	Tr
	Breakfast cereals														
78	**All-Bran**	0	0	0	4.30	0.90	1.00	11.3	3.0	1.30	0.7	125	1.7	25	25
79	**Bran Flakes**	0	0	4.2	8.30	0.80	1.10	14.2	2.4	2.30	1.7	333	0.9	11	66
80	**Cheerios**	0	0	N	N	1.20	1.40	15.3	N	1.70	0.9	170	5.1	N	51
81	**Clusters**	0	0	N	N	1.20	1.40	15.3	N	1.70	0.9	170	5.1	N	51
82	**Coco Pops**	0	0	0	N	1.20	1.30	15.0	1.2	1.70	0.9	167	N	N	0
83	**Corn Flakes**	0	0	0	0.40	1.20	1.30	15.0	0.9	1.70	0.9	333	0.3	2	0
84	**Crunchy Nut Corn Flakes**	0	0	0	0.44	1.20	1.30	15.0	0.8	1.70	0.9	167	N	N	N
85	**Frosties**	0	0	0	N	2.30	1.30	30.0	0.6	3.30	0.9	167	(0.3)	(1)	0
86	**Fruit 'n Fibre**	0	Tr	0	1.40	0.90	1.00	11.3	1.7	1.30	0.6	125	3.8	Tr	0
87	**Muesli**, Swiss style	Tr	Tr	0	3.20	0.50	0.70	6.5	2.3	1.60	0	(140)	1.2	15	Tr
88	with no added sugar	Tr	Tr	0	(2.90)	0.25	0.40	4.6	2.2	0.30	0	N	N	N	Tr

Composition of food per 100g edible portion

No.	Food	Description and main data sources	Edible conversion factor	Water	Total nitrogen	Protein	Fat	Carbo-hydrate	Energy value	
				g	g	g	g	g	kcal	kJ
Breakfast cereals *continued*										
89	**Nutri-Grain**	Average of strawberry, apple, blueberry, cherry and chocolate cereal bars. Manufacturer's data (Kelloggs)	1.00	14.2	0.66	4.1	8.4	*71.5*	*360*	*1525*
90	**Oat Bran Flakes,** with raisins	Manufacturer's data (Kelloggs)	1.00	6.0	1.58	10.0	5.0	*69.7*	*346*	*1469*
91	**Porridge**, *made with water*	Recipe. Ref. Wiles et al. (1980)	1.00	87.3	0.24	1.4	1.1	8.1	46	195
92	*made with whole milk*	Recipe	1.00	74.8	0.76	4.8	5.1	12.6	113	472
93	**Puffed Wheat**	Analytical and literature sources	1.00	2.5	2.44	14.2	1.3	67.3	321	1366
94	**Ready Brek**	6 packets of the same brand and manufacturer's data (Weetabix)	1.00	8.3	1.86	11.6	8.3	*65.4*	*366*	*1550*
95	**Rice Krispies**	Analysis and manufacturer's data (Kelloggs)	1.00	3.0	1.03	6.1	1.0	*92.9*	*382*	*1628*
96	**Ricicles**	Manufacturer's data (Kelloggs)	1.00	3.0	0.64	4.0	0.7	*94.9*	*378*	*1612*
97	**Shredded Wheat**	Analysis and manufacturer's data (Nestlé)	1.00	7.6	1.92	11.2	2.1	*71.7*	*332*	*1415*
98	**Shreddies**	Analysis and manufacturer's data (Nestlé)	1.00	4.0	1.72	9.8	1.9	*77.3*	*346*	*1474*
99	**Special K**	Analysis and manufacturer's data (Kelloggs)	1.00	3.0	2.46	15.3	1.0	*81.6*	*376*	*1603*
100	**Sugar Puffs**	Manufacturer's data (Quaker) and analysis (6 packets of the same brand)	1.00	1.8	1.04	6.1	1.0	92.7	381	1623
101	**Sultana Bran**	Manufacturer's data (Kelloggs)	1.00	7.0	1.58	9.0	2.0	*69.8*	*316*	*1344*
102	**Weetabix**	Manufacturer's data (Weetabix)	1.00	5.6	1.80	11.2	2.7	*75.5*	*352*	*1498*
Biscuits										
103	**Chocolate biscuits**, full coated	7 samples, 5 brands including Breakaway, United and chocolate fingers	1.00	1.2	1.28	7.3	24.3	67.6	501	2105
104	cream filled, full coated	9 samples of different brands including Club, Penguin, Trio and Hob Nob bars	1.00	0.9	1.12	6.4	28.4	57.3	496	2076

No.	Food	Starch	Total sugars	Individual sugars					Dietary fibre	Fatty acids				Cholest-erol
				Gluc	Fruct	Sucr	Malt	Lact	NSP	Satd	Mono-unsatd	Poly-unsatd	Trans	
		g	g	g	g	g	g	g	g	g	g	g	g	mg
***Breakfast cereals** continued*														
89	**Nutri-Grain**	*35.0*	30.7	10.8	10.0	10.0	0	0	(3.0)	1.6	3.3	0.9	Tr	0
90	**Oat Bran Flakes**, with raisins	*47.3*	22.4	6.0	8.0	7.4	1.0	0	10.0	0.8	1.8	2.0	Tr	0
91	**Porridge**, *made with water*	8.0	0.1	Tr	Tr	0.1	0	0	0.9	0.2	0.4	0.5	Tr	0
92	*made with whole milk*	7.8	4.8	Tr	Tr	0.1	0	4.6	0.9	2.8	1.5	0.6	0.14	14
93	**Puffed Wheat**	67.0	0.3	Tr	0.1	0.2	0	0	5.6	0.2	0.2	0.6	Tr	0
94	**Ready Brek**	*63.5*	1.9	0.2	Tr	1.2	0.5	0	8.0	2.0	3.0	3.3	Tr	0
95	**Rice Krispies**	*82.5*	10.4	1.0	0.5	8.9	Tr	0	0.7	0.3	0.2	0.4	Tr	0
96	**Ricicles**	*53.9*	41.9	N	N	N	N	0	0.4	0.2	0.1	0.2	Tr	0
97	**Shredded Wheat**	*71.1*	0.6	Tr	Tr	0.6	0	0	9.8	0.3	0.3	1.0	Tr	0
98	**Shreddies**	*61.9*	15.4	N	N	N	N	0	9.5	0.4	0.3	0.9	Tr	0
99	**Special K**	*63.8*	17.8	0.5	0.5	15.2	0.5	1.1	2.0	0.3	0.2	0.4	Tr	0
100	**Sugar Puffs**	41.3	51.5	(4.1)	(1.8)	(41.5)	(4.0)	0	3.2	0.2	0.1	0.4	Tr	0
101	**Sultana Bran**	*36.3*	33.5	11.0	12.0	10.5	Tr	0	10.0	0.4	0.2	1.0	Tr	0
102	**Weetabix**	*70.6*	4.9	0.7	0.7	2.6	0.8	0	9.7	0.6	0.3	1.8	Tr	0
Biscuits														
103	**Chocolate biscuits**, full coated	21.8	45.8	0.5	0.4	39.3	Tr	5.6	1.5	13.2	8.4	1.5	3.4	(22)
104	cream filled, full coated	19.5	37.8	0.4	0.2	32.3	Tr	4.9	1.5	16.3	9.2	1.6	N	11

Inorganic constituents per 100g edible portion

No.	Food	mg										µg	
		Na	K	Ca	Mg	P	Fe	Cu	Zn	Cl	Mn	Se	I
	Breakfast cereals *continued*												
89	**Nutri-Grain**	300	230	540	30	120	6.4	(0.10)	0.6	460	N	N	N
90	**Oat Bran Flakes,** with raisins	600	500	50	100	350	6.0	N	2.0	930	N	N	N
91	**Porridge**, *made with water*	565	44	7	15	47	0.5	0.06	0.4	863	0.5	Tr	N
92	*made with whole milk*	595	202	128	25	142	0.5	0.06	0.8	932	0.5	1	N
93	**Puffed Wheat**	4	390	26	140	350	4.6	0.56	2.8	50	N	N	N
94	**Ready Brek**	12	390	1200	120	420	11.9	0.41	2.7	18	N	(3)	N
95	**Rice Krispies**	650	160	453	40	140	7.9	0.10	1.0	1000	1.0	(2)	N
96	**Ricicles**	450	100	5	30	100	7.9	0.13	0.8	700	N	(2)	N
97	**Shredded Wheat**	8	330	38	130	340	4.2	0.40	2.3	53	N	3	N
98	**Shreddies**	550	210	40	88	320	7.8	0.44	2.5	220	2.3	N	N
99	**Special K**	800	250	70	50	220	23.3	0.13	2.0	1240	N	N	N
100	**Sugar Puffs**	9	160	14	55	140	8.0	0.23	1.5	41	N	N	N
101	**Sultana Bran**	600	700	50	100	350	18.2	0.13	2.5	930	N	N	N
102	**Weetabix**	270	370	35	120	290	11.9	0.54	2.0	420	N	2	N
	Biscuits												
103	**Chocolate biscuits**, full coated	235	240	130	38	150	1.3	0.22	0.8	285	0.4	N	N
104	cream filled, full coated	175	240	125	34	150	1.4	0.18	0.7	220	0.4	N	110

Vitamins per 100g edible portion

No.	Food	Retinol	Carotene	Vitamin D	Vitamin E	Thiamin	Ribo-flavin	Niacin	Trypt/60	Vitamin B_6	Vitamin B_{12}	Folate	Panto-thenate	Biotin	Vitamin C
		µg	µg	µg	mg	mg	mg	mg	mg	mg	µg	µg	mg	µg	mg
	***Breakfast cereals** continued*														
89	**Nutri-Grain**	0	Tr	0	N	1.00	1.10	12.0	N	1.40	0.7	135	N	N	0
90	**Oat Bran Flakes,** with raisins	0	0	0	N	0.90	1.00	11.3	2.3	1.30	0.7	125	N	N	0
91	**Porridge**, *made with water*	0	0	0	0.18	0.07	0.01	0.1	0.3	0.02	0	4	0.1	3	0
92	*made with whole milk*	34	20	Tr	0.25	0.09	0.22	0.2	0.9	0.08	0.9	10	0.6	4	1
93	**Puffed Wheat**	0	0	0	2.00	Tr	0.06	5.2	2.9	0.14	0	19	0.5	7	0
94	**Ready Brek**	0	0	0	1.20	1.20	1.40	15.3	2.3	1.70	0.9	170	5.1	23	27
95	**Rice Krispies**	0	0	0	0.60	1.20	1.30	15.0	1.4	1.70	0.9	333	0.7	2	0
96	**Ricicles**	0	0	4.2	N	1.20	1.30	15.0	1.0	1.70	0.9	167	(0.4)	(1)	0
97	**Shredded Wheat**	0	0	0	1.20	0.27	0.05	4.5	2.1	0.24	0	42	0.8	9	0
98	**Shreddies**	0	0	2.8	N	0.80	0.90	10.0	2.0	1.10	0.6	111	3.3	7	0
99	**Special K**	0	0	8.3	0.55	2.30	2.70	30.0	2.8	3.30	1.7	333	0.5	3	100
100	**Sugar Puffs**	*0*	*0*	*0*	*0.34*	*1.00*	*1.00*	*10.0*	*1.2*	*0.05*	*0*	*12*	*N*	*N*	*0*
101	**Sultana Bran**	0	Tr	3.1	6.30	0.60	0.80	10.6	2.0	1.80	1.3	250	N	N	50
102	**Weetabix**	0	0	0	1.31	1.20	1.40	15.3	2.2	0.22	0	170	0.7	8	0
	Biscuits														
103	**Chocolate biscuits**, full coated	24	17	Tr	2.60	0.14	0.21	0.9	2.0	0.09	Tr	9	0.6	3	0
104	cream filled, full coated	Tr	Tr	Tr	1.16	0.28	0.24	0.7	1.7	0.04	0	10	N	N	0

Composition of food per 100g edible portion

No.	Food	Description and main data sources	Edible conversion factor	Water	Total nitrogen	Protein	Fat	Carbo-hydrate	Energy value	
				g	g	g	g	g	kcal	kJ
***Biscuits** continued*										
105	**Chocolate chip cookies**	16 samples, 8 brands	1.00	3.3	1.02	5.8	22.9	65.2	474	1989
106	**Cream crackers**	10 samples of the same brand	1.00	4.3	1.66	9.5	13.3	68.3	414	1746
107	**Crispbread**, rye	Analytical and literature sources	1.00	6.4	1.61	9.4	0.6	70.6	308	1312
108	**Crunch biscuits**, cream filled	5 samples, 2 brands of crunch creams	1.00	2.4	0.91	5.2	24.6	67.9	497	2086
109	**Digestive biscuits**, chocolate	22 samples, 4 brands; plain and milk	1.00	2.5	1.17	6.8	24.1	66.5	493	2071
110	plain	10 samples, 4 brands	1.00	2.5	1.10	6.3	20.3	68.6	465	1956
111	**Flapjacks**	Recipe	1.00	6.6	0.83	4.8	27.0	62.4	493	2064
112	**Gingernut biscuits**	10 packets, 5 brands	1.00	3.4	0.98	5.6	13.0	79.1	436	1842
113	**Oat based biscuits**	10 samples, 3 brands including Hob Nobs, Snapjacks, Oatbakes and Barnstormers	1.00	1.9	1.30	7.6	21.4	65.2	468	1964
114	**Oatcakes**, *retail*	6 packets, 4 brands, fats analysed on 3 brands	1.00	5.5	1.71	10.0	15.1	63.0	412	1737
115	**Sandwich biscuits**, cream filled	20 samples, 5 brands including custard creams and bourbon	1.00	1.7	1.04	5.9	20.7	72.5	482	2026
116	jam filled	6 samples, 3 brands including Jammy Dodgers and jam rings	1.00	4.4	0.98	5.6	17.3	69.5	439	1847
117	**Semi-sweet biscuits**	10 samples, 3 brands including Osborne, Rich Tea and Marie	1.00	2.5	1.18	6.7	13.3	74.8	427	1803
118	**Short sweet biscuits**	10 samples, 2 brands including Lincoln and Shortcake	1.00	2.6	1.08	6.2	21.8	62.2	454	1907
119	**Shortbread**	4 samples	1.00	3.3	1.05	6.0	27.5	63.3	509	2133
120	**Wafer biscuits**, filled	9 packets, assorted	1.00	2.3	0.82	4.7	30.1	66.0	537	2250

Composition of food per 100g edible portion

No.	Food	Starch	Total sugars	Individual sugars					Dietary fibre	Fatty acids				Cholest-erol
				Gluc	Fruct	Sucr	Malt	Lact	NSP	Satd	Mono-unsatd	Poly-unsatd	Trans	
		g	g	g	g	g	g	g	g	g	g	g	g	mg
	Biscuits *continued*													
105	**Chocolate chip cookies**	30.6	31.5	0.3	0.3	30.9	Tr	Tr	2.0	10.6	8.6	2.6	0.3	1
106	**Cream crackers**	68.3	Tr	Tr	Tr	Tr	Tr	0	2.2	5.4	5.8	1.5	1.2	N
107	**Crispbread**, rye	67.4	3.2	0.5	0.9	1.3	0.5	0	11.7[a]	Tr	0.1	0.2	Tr	0
108	**Crunch biscuits**, cream filled	26.8	41.1	0.8	0.5	37.1	1.5	1.2	1.0	15.0	6.8	1.7	0.9	3
109	**Digestive biscuits**, chocolate	38.0	28.5	Tr	Tr	26.0	0	2.5	2.2	12.2	8.9	1.6	1.6	51
110	plain	55.0	13.6	0.3	0.3	13.0	0	0	2.2	9.0	8.3	2.0	1.0	41
111	**Flapjacks**	28.0	34.5	4.5	4.4	25.5	Tr	0.1	2.6	4.9	7.6	10.3	N	1
112	**Gingernut biscuits**	43.3	35.8	2.1	0.9	32.8	0	0	1.4	6.0	5.1	1.3	0.5	N
113	**Oat based biscuits**	39.7	25.5	0.3	0.3	24.9	0	0	3.5	9.2	8.3	2.5	0.9	N
114	**Oatcakes**, *retail*	59.9	3.1	Tr	Tr	2.0	1.1	0	N	5.1	6.3	2.9	0.4	(51)
115	**Sandwich biscuits**, cream filled	37.3	35.2	0.9	0.2	32.9	Tr	1.2	1.6	11.0	7.3	1.9	2.0	(51)
116	jam filled	40.4	29.1	5.2	4.2	18.1	1.2	0.4	1.5	7.2	7.4	1.9	1.3	N
117	**Semi-sweet biscuits**	52.5	22.3	0	0	19.1	3.2	0	1.7	6.3	5.1	1.3	0.8	(31)
118	**Short sweet biscuits**	38.1	24.1	1.4	0	22.7	0	0	1.5	11.1	8.1	1.5	0.9	(37)
119	**Shortbread**	47.6	15.7	0.2	0.1	15.3	Tr	0.1	1.9	18.2	6.7	1.3	N	(74)
120	**Wafer biscuits**, filled	21.3	44.7	1.4	0	42.9	0	0.4	N	20.7	6.8	1.0	0.4	N

[a] High fibre varieties contain approximately 17.9g NSP per 100g

No.	Food	mg										µg	
		Na	K	Ca	Mg	P	Fe	Cu	Zn	Cl	Mn	Se	I
	***Biscuits** continued*												
105	**Chocolate chip cookies**	350	175	83	29	125	1.3	0.10	0.6	310	0.5	N	N
106	**Cream crackers**	610	120	110	25	110	1.7	(0.20)	(0.7)	830	(0)	(4)	(13)
107	**Crispbread**, rye	220[a]	500	45[a]	100	310	3.5[a]	0.38	3.0[a]	370	3.5	(3)	15
108	**Crunch biscuits**, cream filled	260	120	83	16	100	2.1	Tr	0.5	295	0.4	N	65
109	**Digestive biscuits**, chocolate	450	210	84	41	130	2.1	0.24	1.0	410	N	N	N
110	plain	600	170	92	23	88	3.2	0.28	0.5	540	0.5	N	N
111	**Flapjacks**	264	186	36	47	150	2.3	0.12	1.3	330	1.6	1	N
112	**Gingernut biscuits**	330	220	130	25	87	4.0	0.16	0.5	320	(0.9)	N	N
113	**Oat based biscuits**	540	235	37	56	215	2.0	0.14	1.3	540	1.7	N	31
114	**Oatcakes**, *retail*	1230	340	54	100	420	4.5	0.37	2.3	1290	(3.2)	N	N
115	**Sandwich biscuits**, cream filled	145	180	83	23	81	1.4	0.14	0.5	190	0.4	N	N
116	jam filled	230	120	91	14	72	1.3	0.10	0.4	260	0.4	N	57
117	**Semi-sweet biscuits**	410	140	120	17	84	2.1	0.08	0.6	520	N	N	N
118	**Short sweet biscuits**	360	110	87	15	85	1.8	0.11	0.6	490	N	N	N
119	**Shortbread**	270	110	89	13	74	1.5	Tr	0.3	445	0.5	N	N
120	**Wafer biscuits**, filled	70	160	73	22	83	1.6	0.16	0.5	150	N	N	N

[a] Cracotte type crispbread contains 640mg Na, 80mg Ca, 2.1mg Fe and 0.6mg Zn per 100g

Vitamins per 100g edible portion

No.	Food	Retinol µg	Carotene µg	Vitamin D µg	Vitamin E mg	Thiamin mg	Riboflavin mg	Niacin mg	Trypt/60 mg	Vitamin B_6 mg	Vitamin B_{12} µg	Folate µg	Pantothenate mg	Biotin µg	Vitamin C mg
	Biscuits *continued*														
105	**Chocolate chip cookies**	Tr	Tr	0	3.16	0.13	0.08	0.9	1.1	0.03	0	36	N	N	0
106	**Cream crackers**	0	0	0	(1.30)	(0.23)	(0.05)	(1.7)	1.9	(0.12)	0	(22)	(0.3)	(2)	0
107	**Crispbread**, rye	0	0	0	0.86	0.28	0.14	1.1	1.8	0.29	0	35	(1.1)	(7)	0
108	**Crunch biscuits**, cream filled	0	0	0	1.63	0.23	0.15	1.6	1.0	0.07	0	11	0.4	1	0
109	**Digestive biscuits**, chocolate	Tr	Tr	Tr	2.90	0.08	0.11	1.3	1.4	0.08	Tr	N	N	N	0
110	plain	0	0	0	5.86	0.14	0.11	1.1	1.3	0.09	0	13	N	N	0
111	**Flapjacks**	193	101	2.3	9.95	0.14	0.03	0.4	1.1	0.04	Tr	11	0.3	8	0
112	**Gingernut biscuits**	0	N	0	4.24	0.10	0.03	0.9	1.1	0.07	0	(4)	(0.1)	(1)	0
113	**Oat based biscuits**	0	0	0	3.36	0.18	0.13	1.6	1.6	0.14	0	13	0.6	9	0
114	**Oatcakes**, *retail*	0	0	0	2.52	0.32	0.09	0.7	2.3	0.10	0	(26)	(1.0)	(17)	0
115	**Sandwich biscuits**, cream filled	0	0	0	2.83	0.15	0.08	0.9	1.2	0.05	0	18	N	N	0
116	jam filled	0	Tr	0	1.69	0.20	0.05	0.9	1.1	0.04	0	14	Tr	Tr	0
117	**Semi-sweet biscuits**	0	0	0	2.35	0.13	0.08	1.5	1.4	0.06	0	(13)	N	N	0
118	**Short sweet biscuits**	0	0	0	3.60	0.16	0.04	0.9	1.3	0.05	0	(13)	N	N	0
119	**Shortbread**	327	149	0.4	0.88	0.18	0.03	1.1	1.2	0.08	0	14	0.3	1	0
120	**Wafer biscuits**, filled	0	0	0	0.85	0.09	0.08	0.5	1.0	0.03	0	N	N	N	0

Composition of food per 100g edible portion

No.	Food	Description and main data sources	Edible conversion factor	Water	Total nitrogen	Protein	Fat	Carbo-hydrate	Energy value	
				g	g	g	g	g	kcal	kJ
	***Biscuits** continued*									
121	**Wafers**, filled, chocolate, full coated	11 samples, different brands including Taxi and Blue Riband	1.00	1.0	1.21	6.9	29.7	58.1	513	2146
122	**Water biscuits**	3 brands	1.00	4.5	1.90	10.8	12.5	75.8	440	1859
123	**Wholemeal crackers**	Recipe	1.00	4.4	1.76	10.1	11.5	72.0	414	1748
	Cakes									
124	**Banana bread**	Recipe	1.00	26.9	0.75	4.4	13.6	52.7	338	1421
125	**Battenburg cake**	Recipe. Ref. Wiles et al. (1980)	1.00	24.0	0.95	5.6	16.8	53.1	373	1567
126	**Cake mix**, *made up*	Made as packet directions. Fats analysed	1.00	31.5	0.89	5.3	11.6	52.4	322	1358
127	**Carrot cake**	With topping, recipe	1.00	35.1	0.73	4.3	22.7	37.0	359	1499
128	**Chocolate fudge cake**	12 samples, 6 brands. Mixture of fresh and frozen	1.00	24.7	0.83	5.2	14.3	55.7	358	1509
129	**Crispie cakes**	Recipe	1.00	1.5	0.93	5.7	18.0	73.8	461	1943
130	**Fancy iced cakes**, *individual*	10 samples including French and fondant fancies	1.00	12.7	0.66	3.8	9.1	68.8	355	1502
131	**Fruit cake**, plain, *retail*	10 cakes, 4 brands; sultana	1.00	19.5	0.89	5.1	14.8	57.9	371	1561
132	rich	Recipe	1.00	17.5	0.65	3.9	11.4	59.9	343	1447
133	rich, iced	Coated with marzipan and Royal icing; recipe	1.00	15.3	0.63	3.6	9.8	65.9	350	1478
134	wholemeal	Recipe	1.00	21.6	1.01	6.0	16.2	52.4	366	1546
135	**Gateau**, chocolate based, *frozen*	11 samples, including Black Forest gateau	1.00	43.1	0.55	3.5	15.7	37.2	295	1236
136	fruit, *frozen*	10 samples; fruit and cream sponge including strawberry, orange and lemon and tropical fruit	1.00	51.9	0.50	3.2	12.3	33.3	248	1042
137	**Jaffa cakes**	Manufacturer's data (McVities). Inorganics and vitamins calculated from recipe	1.00	13.2	0.70	4.0	8.1	76.9	377	1598

Composition of food per 100g edible portion

No.	Food	Starch	Total sugars	Individual sugars Gluc	Fruct	Sucr	Malt	Lact	Dietary fibre NSP	Fatty acids Satd	Mono-unsatd	Poly-unsatd	Trans	Cholest-erol
		g	g	g	g	g	g	g	g	g	g	g	g	mg
	***Biscuits** continued*													
121	**Wafers**, filled, chocolate, full coated	17.0	41.3	1.5	0.4	33.7	Tr	5.7	1.1	18.2	9.0	1.0	N	14
122	**Water biscuits**	73.5	2.3	0	0	0	(2.3)	0	3.1	N	N	N	N	N
123	**Wholemeal crackers**	70.3	1.6	Tr	Tr	0.2	0.1	0	4.4	2.3	3.4	5.0	N	N
	Cakes													
124	**Banana bread**	16.4	36.3	7.6	6.4	19.9	2.2	0	(1.5)	2.4	3.9	6.5	Tr	33
125	**Battenburg cake**	16.3	36.8	0.8	0.4	34.1	0.4	0.9	0.9	3.4	6.0	6.4	Tr	74
126	**Cake mix**, *made up*	24.2	28.3	Tr	Tr	26.6	1.6	0	N	6.0	4.3	0.8	1.1	67
127	**Carrot cake**	13.2	23.9	0.9	0.9	21.9	0	0	1.0	5.5	5.3	10.7	0.2	42
128	**Chocolate fudge cake**	11.3	44.4	1.7	0.4	39.1	1.3	1.0	0.9	4.6	6.4	2.7	0.4	17
129	**Crispie cakes**	31.0	42.8	0.5	0.4	41.7	0.2	0.1	1.9	10.7	5.7	0.8	Tr	4
130	**Fancy iced cakes**, *individual*	14.8	54.0	4.2	2.0	47.8	0	0	N	9.3	4.7	0.7	2.0	N
131	**Fruit cake**, plain, *retail*	14.8	43.1	11.3	11.3	20.5	0	0	N	6.9	5.9	1.2	1.3	N
132	rich	11.2	48.7	16.9	15.2	13.5	2.9	0	1.5	2.4	3.7	4.5	Tr	46
133	rich, iced	7.7	58.3	12.6	10.8	32.3	2.4	0	1.3	1.8	3.8	3.6	Tr	31
134	wholemeal	23.2	29.2	5.6	5.3	17.2	0.7	0.4	2.4	3.5	5.1	6.5	Tr	50
135	**Gateau**, chocolate based, *frozen*	17.6	17.4	2.4	1.8	11.5	0.7	0.9	1.0	9.0	3.8	1.2	0.4	56
136	fruit, *frozen*	17.8	14.9	2.0	1.5	10.1	0.5	0.7	0.9	7.0	2.9	0.9	0.3	53
137	**Jaffa cakes**	22.4	53.9	9.3	0	39.2	5.5	0	(1.3)	4.2	2.8	0.9	0	47

Inorganic constituents per 100g edible portion

No.	Food	mg										µg	
		Na	K	Ca	Mg	P	Fe	Cu	Zn	Cl	Mn	Se	I
	***Biscuits** continued*												
121	**Wafers**, filled, chocolate, full coated	94	300	140	42	155	1.4	0.24	0.8	135	0.4	N	N
122	**Water biscuits**	470	140	120	19	87	1.6	0.08	(0.7)	680	N	N	N
123	**Wholemeal crackers**	691	204	112	46	166	2.5	0.25	1.2	1014	1.2	3	N
	Cakes												
124	**Banana bread**	167	294	100	27	151	1.2	0.21	0.5	165	0.4	N	N
125	**Battenburg cake**	433	105	84	18	182	1.0	0.11	0.6	477	0.2	3	N
126	**Cake mix**, *made up*	370	82	59	9	260	0.9	0.14	0.5	110	0.2	N	17
127	**Carrot cake**	217	154	94	19	175	0.9	0.11	N	N	0.3	2	N
128	**Chocolate fudge cake**	265	357	67	45	156	3.1	0.38	0.8	274	0.4	4	N
129	**Crispie cakes**	310	235	106	65	123	4.4	0.47	1.0	477	0.6	4	N
130	**Fancy iced cakes**, *individual*	250	170	44	30	120	1.4	0.25	0.7	230	N	N	N
131	**Fruit cake**, plain, *retail*	250	390	60	25	110	1.7	0.25	0.5	320	N	N	N[a]
132	rich	155	466	79	22	79	2.1	0.27	0.4	199	0.3	2	N[a]
133	rich, iced	113	345	65	26	75	1.6	0.23	0.4	140	0.3	2	N[a]
134	wholemeal	227	285	79	34	176	1.9	0.21	0.9	220	0.7	3	N
135	**Gateau**, chocolate based, *frozen*	173	189	49	20	153	1.5	0.52	0.6	N	0.2	N	N
136	fruit, *frozen*	128	107	40	10	110	0.7	0.38	0.7	96	0.2	N	35
137	**Jaffa cakes**	130	170	55	34	130	1.5	0.30	0.3	170	N	N	48

[a] Iodine from erythrosine is present but largely unavailable

No.	Food	Retinol µg	Carotene µg	Vitamin D µg	Vitamin E mg	Thiamin mg	Ribo-flavin mg	Niacin mg	Trypt/60 mg	Vitamin B_6 mg	Vitamin B_{12} µg	Folate µg	Panto-thenate mg	Biotin µg	Vitamin C mg
	***Biscuits** continued*														
121	**Wafers**, filled, chocolate, full coated	Tr	Tr	0	1.03	0.07	0.26	0.9	1.6	0.05	0	15	N	N	0
122	**Water biscuits**	0	0	0	N	(0.11)	(0.03)	(0.9)	2.2	(0.06)	0	N	N	N	0
123	**Wholemeal crackers**	0	Tr	0	4.50	0.27	0.07	2.7	2.0	0.18	Tr	37	0.4	4	Tr
	Cakes														
124	**Banana bread**	88	45	1.0	3.96	0.09	0.06	0.6	0.9	0.14	0.2	9	0.3	4	3
125	**Battenburg cake**	147[a]	55[a]	1.6[a]	6.11	0.07	0.12	0.5	1.3	0.05	0.5	8	0.4	6	0
126	**Cake mix**, *made up*	31	0	N	N	0.14	0.07	0.5	1.3	0.03	Tr	8	N	N	0
127	**Carrot cake**	79	976	0.5	7.24	0.07	0.08	0.4	1.1	0.09	0.2	9	0.3	4	2
128	**Chocolate fudge cake**	30	30	0.4	1.45	0.05	0.06	0.5	1.5	Tr	Tr	7	0.4	3	0
129	**Crispie cakes**	9	9	0	1.09	0.47	0.52	5.8	0.9	0.65	0.3	131	0.4	3	0
130	**Fancy iced cakes**, *individual*	0	N	0	2.45	0.01	0.04	0.2	0.8	N	0	N	N	N	0
131	**Fruit cake**, plain, *retail*	N	N	N	1.69	0.08	0.07	0.6	1.0	(0.11)	0	(8)	(0.2)	(5)	0
132	rich	101	46	1.1	4.00	0.08	0.07	0.5	0.8	0.10	0.3	8	0.2	4	Tr
133	rich, iced	68	31	0.8	3.71	0.06	0.08	0.5	0.7	0.07	0.2	8	0.2	6	Tr
134	wholemeal	139	63	1.5	5.90	0.12	0.09	1.3	1.3	0.12	0.4	11	0.4	5	0
135	**Gateau**, chocolate based, *frozen*	102	80	N	1.01	0.05	0.09	0.3	0.5	0.01	0.2	2	0.2	3	0
136	fruit, *frozen*	109	55	Tr	1.06	0.03	0.06	0.4	0.5	0.02	0.2	6	0.1	2	4
137	**Jaffa cakes**	14	0	0.1	0.81	0.05	0.05	0.3	0.7	0.03	0	5	0.2	3	2

[a] Recipes calculated using soft polyunsaturated margarine, rather than (unfortified) catering margarine

Composition of food per 100g edible portion

No.	Food	Description and main data sources	Edible conversion factor	Water	Total nitrogen	Protein	Fat	Carbo-hydrate	Energy value	
				g	g	g	g	g	kcal	kJ
	Cakes *continued*									
138	**Madeira cake**	10 samples including lemon	1.00	20.2	0.94	5.4	15.1	58.4	377	1585
139	**Muffins**, American style, chocolate chip	Recipe	1.00	22.3	1.04	6.3	18.2	52.3	385	1617
140	**Reduced fat cake**	Manufacturers' data (Sainsburys and Entemann's) Average of 10 assorted cakes	1.00	(30.0)	0.71	4.0	4.2	60.5[a]	281	1191
141	**Rice cakes**	Manufacturers' data (Kallo and Sainsburys).	1.00	N	1.58	9.4	3.6	81.1	374	1591
142	**Sponge cake**	Basic recipe, creaming method	1.00	15.2	1.05	6.3	27.2	52.4	467	1951
143	*made without fat*	Basic recipe, whisking method	1.00	31.6	1.64	10.0	6.9	53.0	301	1271
144	*jam filled*	10 cakes, 3 brands; sandwich and Swiss roll	1.00	24.5	0.74	4.2	4.9	64.2	302	1280
145	*with dairy cream and jam*	10 samples, 8 brands. Frozen	1.00	38.4	0.69	4.3	10.9	43.9	280	1179
146	**Swiss roll**, chocolate, *individual*	10 samples, 5 brands, 4 bakeries	1.00	17.5	0.75	4.3	16.8	58.1	386	1624
	Pastry									
147	**Flaky pastry**, *raw*	Recipe	1.00	30.1	0.73	4.2	31.1	34.8	427	1777
148	*cooked*	Recipe	1.00	7.7	0.97	5.6	41.0	46.0	564	2347
149	**Shortcrust pastry**, *raw*	Recipe[b]	1.00	20.0	0.99	5.7	28.1	46.8	451	1884
150	*cooked*	Recipe	1.00	7.2	1.14	6.6	32.6	54.3	524	2186
151	**Wholemeal pastry**, *raw*	Recipe. Ref. Wiles *et al.* (1980)	1.00	20.0	1.31	7.7	28.7	38.5	433	1808
152	*cooked*	Recipe. Ref. Wiles *et al.* (1980)	1.00	7.4	1.52	8.9	33.2	44.6	501	2092

[a] Including polyols

[b] Filo pastry raw contains 9.2g protein, 4.1g fat, 68.2g carbohydrate (66.2g as starch), 329kcal and 1399kJ per 100g

Composition of food per 100g edible portion

No.	Food	Starch	Total sugars	Individual sugars					Dietary fibre	Fatty acids				Cholesterol
				Gluc	Fruct	Sucr	Malt	Lact	NSP	Satd	Mono-unsatd	Poly-unsatd	Trans	
		g	g	g	g	g	g	g	g	g	g	g	g	mg
	***Cakes** continued*													
138	**Madeira cake**	21.9	36.5	0.5	0.5	35.5	0	0	0.9	8.4	3.8	1.6	0.7	N
139	**Muffins**, American style, chocolate chip	23.8	28.4	0	Tr	26.9	Tr	1.2	1.6	10.7	5.2	0.9	0.4	68
140	**Reduced fat cake**	16.3	42.6	N	N	N	N	N	N	1.5	1.3	0.6	Tr	8
141	**Rice cakes**	79.2	1.9	N	N	N	N	N	N	N	N	N	N	N
142	**Sponge cake**	22.0	30.4	Tr	Tr	30.0	0.1	0.1	0.9	5.8	8.9	10.9	Tr	112
143	*made without fat*	22.1	30.9	Tr	Tr	30.5	0.1	0	0.9	1.9	2.6	1.2	0.1	227
144	*jam filled*	16.5	47.7	8.1	3.9	35.7	0	0	1.8	1.6	1.7	0.7	N	N
145	*with dairy cream and jam*	18.6	25.3	3.0	0.7	18.1	1.6	0.9	Tr	N	N	N	N	59
146	**Swiss roll**, chocolate, *individual*	16.3	41.8	4.8	0.2	32.5	2.4	1.9	N	7.0	7.2	1.5	2.2	86
	Pastry													
147	**Flaky pastry**, *raw*	34.1	0.7	0	0	0.1	0.1	0	1.4	9.7	11.8	8.0	N	16
148	*cooked*	45.0	1.0	0	0	0.2	0.1	0	1.8	12.8	15.6	10.5	N	21
149	**Shortcrust pastry**, *raw*	45.8	0.9	Tr	Tr	0.2	0.1	0	1.9	8.7	10.6	7.3	N	14
150	*cooked*	53.2	1.1	Tr	Tr	0.2	0.1	0	2.2	10.1	12.3	8.4	N	17
151	**Wholemeal pastry**, *raw*	37.2	1.3	0.1	Tr	0.6	0	0	5.4	8.8	10.7	7.5	N	14
152	*cooked*	43.1	1.5	0.1	Tr	0.7	0	0	6.3	10.2	12.4	8.7	N	17

Inorganic constituents per 100g edible portion

No.	Food	mg										µg	
		Na	K	Ca	Mg	P	Fe	Cu	Zn	Cl	Mn	Se	I
	***Cakes** continued*												
138	**Madeira cake**	380	120	42	12	120	1.1	0.10	0.5	500	N	N	N
139	**Muffins**, American style, chocolate chip	254	177	161	33	259	1.45	0.26	0.8	173	0.35	3	21
140	**Reduced fat cake**	327	N	N	N	N	N	N	N	N	N	N	N
141	**Rice cakes**	N[a]	N	N	N	N	N	N	N	N	N	N	N
142	**Sponge cake**	326	83	69	10	156	1.3	0.11	0.6	360	0.2	4	N
143	*made without fat*	84	120	77	13	148	1.7	0.12	1.0	116	0.2	7	34
144	*jam filled*	420	140	44	14	220	1.6	0.20	0.5	260	N	(10)	14
145	*with dairy cream and jam*	218	83	33	6	171	0.4	0.04	0.3	179	0.1	4	12
146	**Swiss roll**, chocolate, *individual*	350	210	77	19	200	1.1	0.25	0.5	510	0.2	N	13
	Pastry												
147	**Flaky pastry**, *raw*	335	71	64	10	52	1.0	0.08	0.3	543	0.3	1	N
148	*cooked*	443	94	84	13	69	1.3	0.10	0.3	717	0.4	1	N
149	**Shortcrust pastry**, *raw*	400	92	85	13	68	1.3	0.10	0.4	653	0.4	1	N
150	*cooked*	464	106	99	15	79	1.5	0.12	0.4	758	0.4	1	N
151	**Wholemeal pastry**, *raw*	341	206	24	73	195	2.4	0.28	1.8	538	1.9	4	N
152	*cooked*	395	239	27	85	226	2.8	0.32	2.0	623	2.2	4	N

[a]Unsalted rice cakes contain 10mg Na/100g, low salt contain 100mg/100g

No.	Food	Retinol µg	Carotene µg	Vitamin D µg	Vitamin E mg	Thiamin mg	Riboflavin mg	Niacin mg	Trypt/60 mg	Vitamin B_6 mg	Vitamin B_{12} µg	Folate µg	Pantothenate mg	Biotin µg	Vitamin C mg
	***Cakes** continued*														
138	**Madeira cake**	N	N	N	0.86	0.06	0.11	0.5	1.1	N	Tr	N	N	N	0
139	**Muffins**, American style, chocolate chip	132	72	0.3	0.79	0.09	0.12	0.6	1.3	0.06	0.5	10	0.4	4	0
140	**Reduced fat cake**	N	N	N	N	Tr	0.02	0.9	0.8	N	Tr	N	N	N	N
141	**Rice cakes**	N	N	N	N	N	N	N	N	N	N	N	N	N	N
142	**Sponge cake**	247	100	2.8	9.70	0.09	0.12	0.5	1.6	0.06	0.7	10	0.4	6	0
143	*made without fat*	110	Tr	1.0	0.73	0.11	0.24	0.5	2.7	0.08	1.5	18	0.8	12	0
144	*jam filled*	N	N	N	Tr	0.04	0.07	0.4	0.9	N	(1.0)	N	N	N	0
145	*with dairy cream and jam*	77	40	0.4	N	0.10	0.10	0.4	0.7	0.01	0.3	7	0.3	2	Tr
146	**Swiss roll**, chocolate, *individual*	N	N	N	2.40	0.12	0.19	0.3	0.9	0.03	Tr	10	N	N	0
	Pastry														
147	**Flaky pastry**, *raw*	113	59	1.3	5.77	0.14	0.01	0.8	0.8	0.07	Tr	10	0.1	1	1
148	*cooked*	149	78	1.8	7.62	0.14	0.02	1.0	1.1	0.07	Tr	7	0.1	1	1
149	**Shortcrust pastry**, *raw*	102	53	1.2	5.23	0.19	0.02	1.0	1.1	0.09	Tr	13	0.2	1	0
150	*cooked*	118	61	1.4	6.07	0.16	0.02	1.1	1.3	0.08	Tr	8	0.2	1	0
151	**Wholemeal pastry**, *raw*	102	53	1.2	5.90	0.28	0.05	3.4	1.5	0.30	Tr	34	0.5	4	0
152	*cooked*	118	61	1.4	6.83	0.25	0.05	3.8	1.7	0.26	Tr	20	0.4	5	0

Composition of food per 100g edible portion

No.	Food	Description and main data sources	Edible conversion factor	Water	Total nitrogen	Protein	Fat	Carbo-hydrate	Energy value	
				g	g	g	g	g	kcal	kJ
	Buns and pastries									
153	**Bagels**, plain	2 brands	1.00	32.1	1.60	10.0	1.8	57.8	273	1161
154	**Chelsea buns**	Recipe. Ref. Wiles *et al.* (1980)	1.00	20.5	1.33	7.8	14.2	55.8	368	1549
155	**Crumpets**, *toasted*	9 samples	1.00	40.3	1.20	6.9	1.0	45.4	207	881
156	**Currant buns**	9 samples, 5 supermarkets, 3 bakeries	1.00	27.9	1.40	8.0	5.6	52.6	280	1185
157	**Custard tarts**, *individual*	10 samples, 2 brands, 8 bakeries	1.00	44.7	1.00	6.3	14.5	32.4	277	1161
158	**Danish pastries**	10 samples, different shops	1.00	21.6	1.01	5.8	14.1	51.3	342	1441
159	**Doughnuts**, jam	10 samples, different shops	1.00	26.9	1.00	5.7	14.5	48.8	336	1414
160	ring	10 samples, different shops	1.00	23.8	1.07	6.1	22.4	47.2	403	1688
161	**Eccles cake**	Recipe.	1.00	15.3	0.68	4.0	17.8	56.3	387	1627
162	**Eclairs**, *frozen*	10 samples of the same brand (Birds Eye)	1.00	38.7	0.98	5.6	30.6	26.1	396	1647
163	**Greek pastries**	4 assorted samples, baclava, tangos, tsamika, shredded type	1.00	17.5	0.82	4.7	17.0	40.0	322	1349
164	**Hot cross buns**	Recipe	1.00	25.2	1.27	7.4	7.0	58.4	312	1319
165	**Jam tarts**, *retail*	10 samples, 6 brands, 4 bakeries	1.00	14.4	0.53	3.3	14.7	63.4	383	1614
166	**Mince pies**, *individual*	Recipe	1.00	11.9	0.72	4.2	21.3	60.5	435	1826
167	**Muffins**, English style, white	10 samples	1.00	40.5	1.80	10.0	1.9	44.2	223	948
168	English style, white, *toasted*	10 samples	1.00	31.9	2.00	11.3	2.7	51.0	261	1108
169	**Scones**, fruit, *retail*	10 samples, 5 brands, 3 bakeries	1.00	24.5	1.10	6.5	8.7	56.2	315	1332
170	plain	Recipe	1.00	22.6	1.24	7.2	14.8	53.7	364	1530
171	wholemeal	Recipe. Ref. Wiles *et al.* (1980)	1.00	26.9	1.48	8.8	14.6	43.0	328	1378
172	**Scotch pancakes**, *retail*	6 samples, 4 brands	1.00	39.7	1.00	5.6	9.6	43.0	270	1138
173	**Teacakes**, *toasted*	Calculated using weight loss of 10%	1.00	18.6	1.56	8.9	8.3	58.3	329	1392

No.	Food	Starch	Total sugars	Individual sugars					Dietary fibre	Fatty acids				Cholest-erol
				Gluc	Fruct	Sucr	Malt	Lact	NSP	Satd	Mono-unsatd	Poly-unsatd	Trans	
		g	g	g	g	g	g	g	g	g	g	g	g	mg
	Buns and pastries													
153	**Bagels**, plain	51.3	6.5	0.8	1.0	Tr	3.0	Tr	2.4	N	N	N	N	0
154	**Chelsea buns**	34.5	21.3	4.4	4.3	11.1	Tr	0.9	1.7	3.0	4.5	5.9	N	33
155	**Crumpets**, *toasted*	42.3	3.1	1.1	0.3	Tr	1.7	0	(0)	0.1	0.1	0.5	N	0
156	**Currant buns**	36.6	16.0	8.2	7.0	Tr	0.9	Tr	2.2	1.9	2.0	1.2	0.1	2
157	**Custard tarts**, *individual*	19.6	12.8	0.1	0.1	10.5	0.5	1.6	1.2	6.1	6.0	1.4	1.1	95
158	**Danish pastries**	22.8	28.5	6.3	5.5	13.9	2.2	0.6	1.6	8.6	2.0	1.9	0.8	41
159	**Doughnuts**, jam	30.0	18.8	5.3	4.2	6.4	2.6	0.2	N	4.3	5.4	3.6	N	15
160	ring	31.9	15.3	2.3	2.7	9.1	1.1	0.1	N	5.8	9.4	6.1	1.1	24
161	**Eccles cake**	18.1	38.2	10.0	8.8	17.7	2.1	0	1.5	N	N	N	N	13
162	**Eclairs**, *frozen*	19.5	6.6	Tr	Tr	5.3	0.3	0.9	0.8	16.1	10.2	1.9	N	150
163	**Greek pastries**	21.6	18.4	N	N	N	N	N	N	N	N	N	N	N
164	**Hot cross buns**	34.9	23.4	5.0	4.1	11.7	1.2	0.9	1.9	1.8	2.1	2.5	N	23
165	**Jam tarts**, *retail*	27.4	36.0	12.4	5.9	11.0	6.6	0.1	N	6.6	5.1	1.8	1.0	42
166	**Mince pies**, *individual*	31.5	29.1	14.1	14.1	0.4	0.1	0	1.9	N	N	N	N	12
167	**Muffins**, English style, white	40.8	3.4	0.5	0.8	Tr	2.1	Tr	1.9	0.4	0.5	0.6	Tr	0
168	English style, white, *toasted*	47.3	3.8	0.5	0.8	Tr	2.4	Tr	2.2	0.4	0.5	0.7	Tr	0
169	**Scones**, fruit, *retail*	37.3	18.9	3.0	2.7	9.9	0.4	2.8	2.0	2.4	4.1	1.7	1.1	6
170	plain	47.9	5.9	Tr	Tr	3.4	0.1	1.8	1.9	3.7	4.5	5.9	N	6
171	wholemeal	37.1	5.9	0.1	Tr	3.6	0	1.7	5.2	3.6	4.4	5.8	N	6
172	**Scotch pancakes**, *retail*	21.5	21.5	2.0	1.6	14.1	Tr	3.8	1.5	0.7	3.5	2.1	0.1	21
173	**Teacakes**, *toasted*	41.9	16.4	6.6	6.9	1.1	1.4	0.4	N	2.9	2.7	1.5	0.2	20

Inorganic constituents per 100g edible portion

No.	Food	mg										µg	
		Na	K	Ca	Mg	P	Fe	Cu	Zn	Cl	Mn	Se	I
	Buns and pastries												
153	**Bagels**, plain	(550)	N	N	N	N	N	N	N	N	N	N	N
154	**Chelsea buns**	311	219	109	24	116	1.6	0.27	0.8	496	0.4	3	N
155	**Crumpets**, *toasted*	1029	168	123	17	220	1.4	0.10	0.6	1212	0.4	6	(1)
156	**Currant buns**	317	210	110	27	100	1.9	0.18	0.6	210	0.4	N	N
157	**Custard tarts**, *individual*	130	110	95	14	98	0.8	0.07	0.5	390	0.2	N	N
158	**Danish pastries**	190	170	92	24	98	1.3	0.06	0.5	340	0.3	N	N
159	**Doughnuts**, jam	180	110	72	19	71	1.2	0.09	0.5	290	0.3	N	15
160	ring	230	87	76	21	81	1.2	0.14	0.6	360	0.3	N	(17)
161	**Eccles cake**	230	235	70	17	50	1.0	0.31	0.3	374	0.3	1	8
162	**Eclairs**, *frozen*	73	160	87	20	120	1.1	0.22	0.8	75	0.1	N	N
163	**Greek pastries**	310	90	44	22	70	0.9	0.14	0.4	390	0.3	N	N
164	**Hot cross buns**	94	206	114	24	107	1.6	0.26	0.7	146	0.4	2	N
165	**Jam tarts**, *retail*	130	120	72	14	50	1.7	0.18	0.6	160	0.3	N	2
166	**Mince pies**, *individual*	283	83	74	11	53	1.1	0.12	0.4	452	0.3	1	N
167	**Muffins**, English style, white	431	124	123	20	89	1.3	0.11	0.8	(665)	0.4	6	N
168	English style, white, *toasted*	493	142	141	23	102	1.5	0.13	0.9	(760)	0.5	7	N
169	**Scones**, fruit, *retail*	617	220	150	24	360	1.5	0.22	0.8	450	0.4	N	N
170	plain	840	154	186	17	512	1.3	0.10	0.7	469	0.4	2	N
171	wholemeal	704	256	112	74	565	2.3	0.27	1.9	375	1.8	4	N
172	**Scotch pancakes**, *retail*	418	233	84	20	240	1.0	0.10	0.4	351	0.3	3	10
173	**Teacakes**, *toasted*	300	240	98	32	110	2.9	0.27	0.8	490	0.5	N	N

Vitamins per 100g edible portion

No.	Food	Retinol µg	Carotene µg	Vitamin D µg	Vitamin E mg	Thiamin mg	Riboflavin mg	Niacin mg	Trypt/60 mg	Vitamin B_6 mg	Vitamin B_{12} µg	Folate µg	Pantothenate mg	Biotin µg	Vitamin C mg
	Buns and pastries														
153	**Bagels**, plain	Tr	0	N	N	N	N	N	N	N	0	N	N	N	0
154	**Chelsea buns**	118[a]	54[a]	1.3[a]	5.17	0.16	0.15	1.4	1.6	0.11	0.4	54	0.4	5	0
155	**Crumpets**, *toasted*	0	0	0	0.26	0.23	0.06	1.7	1.8	0.07	0	9	0.3	3	0
156	**Currant buns**	Tr	Tr	0	0.37	0.22	0.16	2.1	1.6	0.11	Tr	12	N	N	1
157	**Custard tarts**, *individual*	32	Tr	N	1.27	0.14	0.16	0.5	1.4	0.03	Tr	13	N	N	0
158	**Danish pastries**	N	N	N	1.29	0.13	0.07	0.9	1.2	0.07	Tr	20	(0.5)	(7)	0
159	**Doughnuts**, jam	N	N	N	Tr	0.22	0.07	1.3	1.2	0.03	Tr	21	N	N	N
160	ring	N	N	N	4.18	0.22	0.07	1.2	1.2	0.02	Tr	19	N	N	0
161	**Eccles cake**	60	41	0.8	0.90	0.09	0.03	0.6	0.6	0.07	0	3	0.1	2	Tr
162	**Eclairs**, *frozen*	240	Tr	(0)	(1.25)	0.10	0.19	0.3	1.1	0.03	1.0	11	(0.3)	(5)	Tr
163	**Greek pastries**	N	N	N	N	0.09	0.04	1.0	1.0	N	N	N	N	N	N
164	**Hot cross buns**	56	26	0.6	2.15	0.20	0.12	1.4	1.5	0.10	0.3	28	0.3	4	0
165	**Jam tarts**, *retail*	N	N	N	0.28	0.06	0.02	0.5	0.7	0.03	Tr	5	(0.1)	Tr	Tr
166	**Mince pies**, *individual*	70	40	0.8	N	0.14	0.02	0.9	0.8	0.10	Tr	11	0.1	0	Tr
167	**Muffins**, English style, white	65	Tr	Tr	0.09	0.24	0.16	2.0	2.1	0.12	Tr	41	0.5	7	Tr
168	English style, white, *toasted*	74	Tr	Tr	0.10	0.27	0.18	2.3	2.4	0.14	Tr	47	0.6	8	Tr
169	**Scones**, fruit, *retail*	Tr	Tr	Tr	0.11	0.22	0.10	1.2	1.5	0.05	Tr	6	N	N	Tr
170	plain	115	61	1.2	5.14	0.15	0.09	1.0	1.4	0.09	0.3	8	0.3	2	1
171	wholemeal	109	58	1.1	5.53	0.21	0.12	3.2	1.7	0.23	0.3	18	0.5	5	1
172	**Scotch pancakes**, *retail*	102	Tr	0.6	1.35	0.16	0.08	0.8	1.5	0.08	Tr	6	0.3	1	Tr
173	**Teacakes**, *toasted*	N	N	0	N	0.20	0.17	2.0	1.8	0.06	Tr	40	N	N	Tr

[a] Recipes calculated using soft polyunsaturated margarine, rather than (unfortified) catering margarine

Composition of food per 100g edible portion

No.	Food	Description and main data sources	Edible conversion factor	Water	Total nitrogen	Protein	Fat	Carbo-hydrate	Energy value	
				g	g	g	g	g	kcal	kJ
	Puddings									
174	**Bread pudding**	Recipe	1.00	30.6	0.98	5.9	9.5	48.0	289	1220
175	**Christmas pudding**, *retail*	10 samples, 4 brands	1.00	(23.6)	0.53	3.0	11.8	56.3	329	1388
176	**Crumble**, fruit	10 samples, including apple, apple and blackberry and rhubarb. Fresh and frozen	1.00	49.1	0.38	2.4	8.3	36.0	219	924
177	fruit, wholemeal	Recipe. Apple, gooseberry, plum, rhubarb	1.00	54.8	0.44	2.6	7.4	31.6	195	822
178	**Fruit pie**, *one crust*	Recipe. Apple, gooseberry, plum, rhubarb	1.00	58.6	0.36	2.1	8.2	28.8	190	798
179	*pastry top and bottom*	Recipe. Ref. Wiles et al. (1980)	1.00	47.8	0.53	3.1	13.6	33.9	262	1096
180	*individual*	10 pies, as purchased, 3 brands; apple, blackcurrant, blackberry, apricot	1.00	22.9	0.75	4.3	14.0	56.7	356	1498
181	blackcurrant, *pastry top and bottom*	Recipe. Ref. Wiles et al. (1980)	1.00	42.3	0.54	3.1	13.5	34.5	263	1104
182	**Fruit pie**, wholemeal, *one crust*	Recipe. Ref. Wiles et al. (1980). Apple, gooseberry, plum, rhubarb	1.00	58.6	0.45	2.7	8.3	26.5	185	777
183	wholemeal, *pastry top and bottom*	Recipe. Ref. Wiles et al. (1980). Apple, gooseberry, plum, rhubarb	1.00	47.9	0.68	4.0	13.8	30.0	253	1060
184	**Lemon meringue pie**	8 samples, 4 brands. Fresh and frozen	1.00	42.1	0.46	2.9	8.5	43.5	251	1060
185	**Pancakes**, sweet, *made with whole milk*	Recipe	1.00	43.4	0.98	6.0	16.3	34.9	302	1265
186	**Sponge pudding**, *canned*	10 assorted samples of the same brand	1.00	35.3	0.54	3.1	9.1	45.4	265	1116
187	**Treacle tart**	Recipe	1.00	19.1	0.68	3.9	14.2	62.8	379	1597

No.	Food	Starch	Total sugars	Individual sugars					Dietary fibre	Fatty acids				Cholest-erol
				Gluc	Fruct	Sucr	Malt	Lact	NSP	Satd	Mono-unsatd	Poly-unsatd	Trans	
		g	g	g	g	g	g	g	g	g	g	g	g	mg
	Puddings													
174	**Bread pudding**	14.7	33.3	8.9	8.5	12.2	1.7	2.0	1.2	N	N	N	N	52
175	**Christmas pudding**, *retail*	10.1	46.2	20.3	20.8	3.5	1.5	0.1	1.7	6.1	4.1	0.6	N	36
176	**Crumble**, fruit	14.0	22.0	4.1	4.2	11.3	2.2	0	1.3	4.0	3.1	0.7	0.9	12
177	fruit, wholemeal	10.3	21.4	1.3	1.6	18.2	0	0	2.7	1.5	2.3	3.2	N	0
178	**Fruit pie**, *one crust*	13.1	15.7	1.3	1.6	12.6	0	0	1.7	2.5	3.0	2.1	N	4
179	*pastry top and bottom*	22.0	12.0	Tr	Tr	9.0	0.1	0	1.7	4.2	5.1	3.5	N	7
180	*individual*	25.8	30.9	5.7	2.8	21.5	0.9	0	N	5.4	6.0	1.8	1.2	0
181	blackcurrant, *pastry top and bottom*	22.0	12.6	1.4	1.6	9.2	0.1	0	2.6	4.2	5.1	3.5	N	7
182	**Fruit pie**, wholemeal, *one crust*	10.6	15.8	1.3	1.6	12.8	0	0	2.7	2.5	3.1	2.2	N	4
183	wholemeal, *pastry top and* bottom	17.8	12.1	1.0	1.2	9.7	0	0	3.4	4.2	5.1	3.6	N	7
184	**Lemon meringue pie**	13.6	29.8	4.5	3.4	20.0	1.0	0.4	(0.7)	3.1	3.5	1.5	0.5	12
185	**Pancakes**, sweet, *made with* whole milk	18.8	16.2	Tr	Tr	13.0	Tr	2.9	0.8	7.0	6.6	1.7	0.1	68
186	**Sponge pudding**, *canned*	19.6	25.8	4.6	3.8	14.9	2.0	0.6	0.8	5.0	3.0	0.5	0.7	32
187	**Treacle tart**	29.2	33.6	9.6	9.6	13.8	0.1	0	1.1	4.4	5.3	3.6	N	7

Inorganic constituents per 100g edible portion

No.	Food	mg										µg	
		Na	K	Ca	Mg	P	Fe	Cu	Zn	Cl	Mn	Se	I
	Puddings												
174	**Bread pudding**	289	373	144	23	110	1.4	0.20	0.7	419	0.3	3	N
175	**Christmas pudding**, *retail*	170	340	35	18	92	1.2	0.14	0.7	180	0.5	N	N
176	**Crumble**, fruit	82	82	41	6	27	0.3	Tr	0.2	130	0.1	2	4
177	fruit, wholemeal	59	195	31	26	68	0.9	0.14	0.6	107	0.6	1	N
178	**Fruit pie**, *one crust*	116	160	48	9	33	0.6	0.08	0.2	202	0.2	Tr	N
179	*pastry top and bottom*	193	144	58	10	43	0.8	0.09	0.2	325	0.2	Tr	N
180	*individual*	210	120	51	12	64	1.2	0.10	0.5	260	2.0	N	N
181	blackcurrant, *pastry top and bottom*	194	222	70	15	53	1.3	0.13	0.3	320	0.3	Tr	N
182	**Fruit pie**, wholemeal, *one crust*	100	193	30	26	69	0.9	0.13	0.6	169	0.6	1	N
183	wholemeal, *pastry top and bottom*	165	199	29	39	103	1.3	0.17	0.9	(269)	0.9	2	N
184	**Lemon meringue pie**	113	70	38	6	40	Tr	0.04	0.2	237	0.1	2	8
185	**Pancakes**, sweet, *made with whole milk*	46	152	118	13	111	0.8	0.06	0.6	96	0.2	2	28
186	**Sponge pudding**, *canned*	340	160	50	13	170	1.2	0.31	0.4	220	0.2	N	4
187	**Treacle tart**	376	83	60	11	45	1.0	0.09	0.3	439	0.2	1	N

No.	Food	Retinol µg	Carotene µg	Vitamin D µg	Vitamin E mg	Thiamin mg	Riboflavin mg	Niacin mg	Trypt/60 mg	Vitamin B_6 mg	Vitamin B_{12} µg	Folate µg	Pantothenate mg	Biotin µg	Vitamin C mg
	Puddings														
174	**Bread pudding**	102	59	0.2	0.26	0.10	0.15	0.8	1.3	0.09	0.6	10	0.4	3	1
175	**Christmas pudding**, *retail*	N	N	0	N	Tr	0.03	0.4	0.6	0.07	Tr	9	N	N	Tr
176	**Crumble**, fruit	47	35	0.7	0.74	0.08	0.02	0.5	0.6	0.05	0	10	0.1	Tr	N
177	fruit, wholemeal	56	123	0.7	3.18	0.08	0.03	1.1	0.5	0.08	Tr	5	0.2	1	5
178	**Fruit pie**, *one crust*	29	106	0.3	1.73	0.07	0.02	0.5	0.4	0.05	Tr	5	0.1	Tr	5
179	*pastry top and bottom*	49	93	0.6	2.68	0.10	0.02	0.7	0.6	0.06	Tr	7	0.1	Tr	3
180	*individual*	0	Tr	0	4.10	0.05	0.02	0.4	0.9	N	0	N	N	N	Tr
181	blackcurrant, *pastry top and bottom*	49	73	0.6	2.99	0.10	0.03	0.6	0.6	0.07	Tr	N	0.2	1	72
182	**Fruit pie**, wholemeal, *one crust*	29	106	0.3	1.92	0.10	0.03	1.2	0.5	0.11	Tr	11	0.2	2	5
183	wholemeal, *pastry top and bottom*	49	93	0.6	3.00	0.15	0.04	1.8	0.8	0.16	Tr	17	0.3	2	4
184	**Lemon meringue pie**	46	10	Tr	0.96	0.08	0.04	0.4	0.7	0.02	Tr	6	0.2	1	5
185	**Pancakes**, sweet, *made with whole milk*	44	12	0.2	0.37	0.07	0.18	0.4	1.3	0.07	0.8	11	0.6	4	1
186	**Sponge pudding**, *canned*	N	N	N	0.05	0.05	0.16	0.4	0.6	0.09	Tr	3	0.2	1	0
187	**Treacle tart**	51	26	0.6	2.62	0.11	0.01	0.7	0.8	0.05	Tr	8	0.1	0	0

Composition of food per 100g edible portion

No.	Food	Description and main data sources	Edible conversion factor	Water	Total nitrogen	Protein	Fat	Carbo-hydrate	Energy value	
				g	g	g	g	g	kcal	kJ
Savouries										
188	**Cheese and onion rolls**, pastry	Oven-baked pastry, 3 brands	1.00	39.6	1.31	8.2	20.0	30.4	327	1366
189	**Dumplings**	Recipe	1.00	60.5	0.49	2.8	11.7	24.5	208	871
190	**Pancakes**, savoury, *made with whole milk*	Recipe	1.00	53.8	1.05	6.4	15.5	23.9	255	1065
191	**Prawn crackers**, *takeaway*	10 samples from different outlets	1.00	2.8	0.05	0.3	39.0	58.2	570	2379
192	**Risotto**, plain	Recipe	1.00	52.3	0.59	3.5	9.7	35.1	233	980
193	**Stuffing mix**, *dried*	10 samples, 4 brands; assorted flavours	1.00	5.9	1.58	9.9	5.2	67.2	338	1436
194	**Stuffing**, sage and onion	Recipe	1.00	48.1	1.04	6.1	15.1	29.0	269	1126
195	**Yorkshire pudding**	Recipe	1.00	57.3	1.10	6.7	10.1	24.6	210	881
Pizzas										
196	**Pizza base**, *raw*	Average of ambient and chilled, 12 brands	1.00	30.0	1.37	7.8	4.8	57.5	290	1229
197	**Pizza**, cheese and tomato, *deep pan*	10 samples, 7 brands, takeaway	1.00	43.5	1.99	12.4	7.5	35.1	249	1050
198	cheese and tomato, thin base	10 samples, 7 brands, takeaway	1.00	40.6	2.30	14.4	10.3	33.9	277	1168
199	cheese and tomato, french bread	Cooked, 10 samples, 3 brands	1.00	47.7	1.69	10.6	7.8	31.4	230	971
200	cheese and tomato, *frozen*	Cooked,10 samples, 6 brands	1.00	48.0	1.84	11.5	8.8	30.1	238	1003
201	chicken topped, *chilled*	9 samples, 4 brands, includes thin base and deep pan	1.00	50.8	2.15	13.4	8.3	31.3	246	1036
202	fish topped, *takeaway*	18 samples, 8 brands, prawn and tuna toppings, includes thin base and deep pan	1.00	50.1	2.13	13.3	7.5	28.0	226	952

Composition of food per 100g edible portion

No.	Food	Starch	Total sugars	Individual sugars					Dietary fibre	Fatty acids				Cholest-erol
				Gluc	Fruct	Sucr	Malt	Lact	NSP	Satd	Mono-unsatd	Poly-unsatd	Trans	
		g	g	g	g	g	g	g	g	g	g	g	g	mg
	Savouries													
188	**Cheese and onion rolls**, pastry	24.2	6.3	0.4	0.3	0.3	1.4	0.6	1.2	9.0	7.1	2.2	0.6	26
189	**Dumplings**	24.0	0.5	0	0	0.1	0.1	0	1.0	6.6	4.0	0.5	N	11
190	**Pancakes**, savoury, *made with whole milk*	20.2	3.7	Tr	Tr	0.1	0.1	3.3	0.8	4.5	4.8	5.2	0.1	57
191	**Prawn crackers**, *takeaway*	56.0	2.2	Tr	Tr	1.9	0.3	0	1.2	3.6	22.4	11.0	0.1	0
192	**Risotto**, plain	34.0	0.8	0.3	0.2	0.3	0	0	0.4	2.1	3.0	4.1	N	0
193	**Stuffing mix**, *dried*	62.8	4.4	0.8	1.6	2.0	0	0	4.7	2.4	1.6	0.1	N	5
194	**Stuffing**, sage and onion	23.1	4.3	N	N	N	N	0	1.5	3.1	4.7	5.9	N	54
195	**Yorkshire pudding**	21.0	3.6	Tr	Tr	0.1	0.1	3.2	0.9	5.2	3.5	0.6	N	69
	Pizzas													
196	**Pizza base**, *raw*	54.1	3.3	0.3	0.3	Tr	2.8	Tr	2.1	N	N	N	N	N
197	**Pizza**, cheese and tomato, *deep pan*	31.5	2.2	0.3	0.4	Tr	1.4	0.2	2.2	3.1	2.4	1.3	0.2	13
198	cheese and tomato, thin base	30.3	2.5	0.6	0.6	Tr	1.3	0.3	1.9	4.8	3.1	1.3	0.4	22
199	cheese and tomato, french bread	29.0	2.4	0.5	0.7	0.2	1.0	Tr	N	N	N	N	N	N
200	cheese and tomato, *frozen*	28.1	2.0	0.4	0.6	0.1	0.8	Tr	(1.5)	3.1	2.8	1.4	0.3	N
201	chicken topped, *chilled*	29.2	2.1	0.5	0.7	Tr	1.0	Tr	N	N	N	N	N	N
202	fish topped, *takeaway*	26.2	2.0	0.3	0.4	Tr	1.1	0.1	N	3.2	2.5	1.3	0.3	25

Inorganic constituents per 100g edible portion

No.	Food	mg										µg	
		Na	K	Ca	Mg	P	Fe	Cu	Zn	Cl	Mn	Se	I
	Savouries												
188	**Cheese and onion rolls**, pastry	567	169	121	19	119	1.3	0.08	0.8	833	0.3	4	19
189	**Dumplings**	422	45	54	6	129	0.6	0.04	0.2	457	0.2	1	4
190	**Pancakes**, savoury, *made with whole milk*	255	169	131	15	123	0.8	0.05	0.6	417	0.2	3	N
191	**Prawn crackers**, *takeaway*	767	17	21	5	6	1.6	0.12	0.1	1000	0.5	3	50
192	**Risotto**, plain	834	99	26	15	71	0.3	0.18	0.8	1239	0.5	5	N
193	**Stuffing mix**, *dried*	1460	240	960	41	130	5.1	0.17	0.8	2820	1.0	N	N
194	**Stuffing**, sage and onion	496	170	71	17	86	1.4	0.11	0.5	754	0.4	2	N
195	**Yorkshire pudding**	591	170	130	16	125	0.9	0.05	0.6	946	0.2	3	33
	Pizzas												
196	**Pizza base**, *raw*	272	124	86	18	85	1.5	0.09	0.6	N	0.5	N	N
197	**Pizza**, cheese and tomato, *deep pan*	247	165	213	24	170	1.2	0.11	1.3	726	0.3	7	35
198	cheese and tomato, thin base	282	180	279	25	230	1.1	0.12	1.8	937	0.3	6	34
199	cheese and tomato, french bread	270	185	240	23	150	1.1	0.08	1.1	898	0.3	N	30
200	cheese and tomato, *frozen*	248	202	230	23	180	1.0	0.09	1.4	(825)	0.3	(7)	(35)
201	chicken topped, *chilled*	272	270	217	27	190	1.3	0.12	1.6	(420)	0.3	N	N
202	fish topped, *takeaway*	248	179	173	24	170	1.2	0.10	1.3	847	0.3	10	59

Vitamins per 100g edible portion

No.	Food	Retinol µg	Carotene µg	Vitamin D µg	Vitamin E mg	Thiamin mg	Riboflavin mg	Niacin mg	Trypt/60 mg	Vitamin B_6 mg	Vitamin B_{12} µg	Folate µg	Pantothenate mg	Biotin µg	Vitamin C mg
Savouries															
188	**Cheese and onion rolls**, pastry	Tr	17	1.0	N	0.15	0.08	1.0	1.5	0.06	0.5	9	0.3	2	Tr
189	**Dumplings**	7	10	Tr	0.28	0.05	0.01	0.3	0.6	0.03	Tr	3	0.1	0	0
190	**Pancakes**, savoury, *made with whole milk*	136	61	1.3	4.58	0.08	0.20	0.4	1.4	0.08	0.9	12	0.6	4	1
191	**Prawn crackers**, *takeaway*	0	Tr	0	5.77	Tr	Tr	0.1	N	0.43	0	2	N	N	0
192	**Risotto**, plain	65	36	0.8	3.24	0.11	0.01	1.1	0.7	0.09	0	5	0.2	1	0
193	**Stuffing mix**, *dried*	Tr	Tr	Tr	N	1.42	0.90	1.8	1.8	N	Tr	N	N	N	0
194	**Stuffing**, sage and onion	130	75	1.5	5.38	0.11	0.06	0.7	1.4	0.10	0.3	15	0.4	4	2
195	**Yorkshire pudding**	51	14	0.3	0.31	0.09	0.20	0.6	1.5	0.08	1.0	9	0.6	5	1
Pizzas															
196	**Pizza base**, *raw*	N	N	N	N	0.41	0.03	1.7	N	0.03	0	8	N	N	0
197	**Pizza**, cheese and tomato, *deep pan*	60	84	0.3	1.68	0.19	0.11	1.3	2.1	0.04	0.7	7	0.4	4	1
198	cheese and tomato, thin base	59	123	1.8	2.52	0.16	0.13	1.2	2.9	0.04	1.0	7	0.2	4	1
199	cheese and tomato, french bread	N	N	0.9	0	0.16	0.10	1.3	N	0.03	0.3	7	0.4	4	N
200	cheese and tomato, *frozen*	21	133	Tr	3.39	0.18	0.09	1.5	1.5	0.05	0.2	8	0.5	3	4
201	chicken topped, *chilled*	N	N	N	N	0.28	0.14	2.9	N	0.07	0.4	9	N	N	3
202	fish topped, *takeaway*	N	N	N	N	0.15	0.10	2.3	2.5	0.05	1.1	5	0.3	3	2

No.	Food	Description and main data sources	Edible conversion factor	Water	Total nitrogen	Protein	Fat	Carbo-hydrate	Energy value	
				g	g	g	g	g	kcal	kJ
Pizzas *continued*										
203	ham and pineapple, *chilled*	10 samples, 5 brands, includes thin base and deep pan	1.00	43.3	2.16	13.5	8.6	34.4	260	1098
204	meat topped	39 samples including pepperoni, spicy beef, spicy pork. Includes frozen, chilled and takeaway, thin base and deep pan	1.00	44.9	2.11	13.2	10.3	29.3	255	1075
205	vegetarian	30 samples, 7 brands, includes chilled and takeaway, thin base and deep pan	1.00	51.1	1.72	10.8	6.9	29.6	216	913

No.	Food	Starch	Total sugars	Individual sugars					Dietary fibre	Fatty acids				Cholest-erol
				Gluc	Fruct	Sucr	Malt	Lact	NSP	Satd	Mono-unsatd	Poly-unsatd	Trans	
		g	g	g	g	g	g	g	g	g	g	g	g	mg
Pizzas *continued*														
203	ham and pineapple, *takeaway*	31.3	3.1	0.7	0.8	Tr	1.4	0.2	N	N	N	N	N	N
204	meat topped	27.3	1.8	0.4	0.5	Tr	0.9	Tr	N	4.0	3.7	1.5	0.3	19
205	vegetarian	26.9	2.3	0.6	0.7	Tr	1.0	0.2	1.9	N	N	N	N	N

Inorganic constituents per 100g edible portion

No.	Food	mg										µg	
		Na	K	Ca	Mg	P	Fe	Cu	Zn	Cl	Mn	Se	I
	***Pizzas** continued*												
203	ham and pineapple, *takeaway*	303	226	221	23	210	1.3	0.10	1.5	N	0.5	N	N
204	meat topped	306	196	203	23	190	1.2	0.10	1.6	N	0.3	7	N
205	vegetarian	241	192	196	22	160	1.2	0.11	1.3	867	0.3	N	34

Vitamins per 100g edible portion

No.	Food	Retinol µg	Carotene µg	Vitamin D µg	Vitamin E mg	Thiamin mg	Ribo-flavin mg	Niacin mg	$\frac{Trypt}{60}$ mg	Vitamin B_6 mg	Vitamin B_{12} µg	Folate µg	Panto-thenate mg	Biotin µg	Vitamin C mg
	***Pizzas** continued*														
203	ham and pineapple, *takeaway*	N	N	N	N	0.28	0.16	2.3	2.4	0.05	0.5	6	N	N	0
204	meat topped	N	N	N	0.93	0.23	0.13	1.9	2.3	0.05	0.8	7	3.2	4	0
205	vegetarian	42	114	N	1.52	0.23	0.11	1.4	2.1	0.05	Tr	7	N	N	4

Section 2.2

Milk and milk products

Some of the data in this section of the Tables have been taken from the *Milk Products and Eggs* (1989) supplement. However, new analytical data have been incorporated for pasteurised liquid milk, other milks and cream, yoghurt, fromage frais, ice cream, puddings and chilled desserts, and cheese. Foods for which new data are incorporated have been allocated a new food code and can thus easily be identified in the food index.

Variation of milk composition by season is pertinent to this section (e.g. carotenes, iodine). Where summer and winter values are given separately, summer are June/September and winter are January/March. Recipe calculations use the average of the values for summer and winter milks. Some loss of vitamins is inevitable when milk is stored. On the doorstep, milk exposed for several hours to bright sunlight can lose up to 70 per cent of its riboflavin. Vitamin C can also decline under these conditions from the 1–1.5mg per 100g in the original milk to almost zero. There will also be gradual losses of folate and vitamin B_{12} from UHT and sterilised milks even under ideal storage conditions because of the reactions with small amounts of oxygen in the pack.

As many products are sold or measured by volume, example specific gravities (densities) of some of these products are given below. More detailed information can be found in the *Milk Products and Eggs* supplement and in the appropriate analytical reports (see general introduction). For the majority of purposes, the values are given on a weight basis may be regarded as the same as those expressed by volume.

Specific gravities of selected dairy products

Skimmed milk	1.03	Double cream	0.94
Semi-skimmed milk	1.03	Yogurt, low fat, fruit	1.08
Whole milk	1.03	Ice creams:	
Evaporated milk	1.07	vanilla, dairy	0.61
Single cream	1.00	vanilla, non dairy	0.51
Whipping cream	0.96		

Losses of labile vitamins assigned on recipe calculation were estimated using the figures in Section 4.3.

Composition of food per 100g edible portion

No.	Food	Description and main data sources	Edible conversion factor	Water	Total nitrogen	Protein	Fat	Carbo-hydrate	Energy value	
				g	g	g	g	g	kcal	kJ
	Skimmed milk									
206	**Skimmed milk**, *average*	Weighted average of pasteurised, sterilised and UHT	1.00	90.9	0.54	3.4	0.2	4.4	32	136
207	pasteurised, *average*	Average of summer (June/Sep) and winter milk (Jan/Mar). Samples from 11 areas, in glass bottles (50%), plastic containers (30%) and cartons (20%)	1.00	90.8	0.55	3.5	0.3	4.8	34	144
208	pasteurised, *fortified plus SMP*	10 samples, own label and Vitapint,[a]	1.00	89.3	0.60	3.8	0.1	6.0	39	164
209	sterilised	21 samples, summer (June/Sep) and winter (Jan/Mar)	1.00	91.4	0.45	2.9	0.3	5.4	35	147
210	UHT	22 samples, summer (June/Sep) and winter (Jan/Mar)	1.00	91.3	0.44	2.8	0.1	3.9	27	114
	Semi-skimmed milk									
211	**Semi-skimmed milk**, *average*	Weighted average of pasteurised and UHT	1.00	89.6	0.54	3.4	1.7	4.7	46	195
212	pasteurised, *average*	Average of summer and winter milk. Samples from 11 areas, in glass bottles (50%), plastic containers (30%) and cartons (20%)	1.00	89.4	0.54	3.5	1.7	4.7	46	195
213	pasteurised, *summer*	Samples taken in June and September	1.00	89.5	0.55	3.5	1.7	4.5	46	194
214	pasteurised, *winter*	Samples taken in January and March	1.00	89.5	0.54	3.4	1.7	4.9	47	196
215	pasteurised, *fortified plus SMP*	10 samples, own label and Vitapint[a]	1.00	88.4	0.59	3.7	1.6	5.8	51	215
216	UHT	22 samples, average of winter and summer	1.00	90.9	0.50	3.3	1.6	4.9	46	194

[a] SMP = Skimmed milk proteins

Composition of food per 100g edible portion

No.	Food	Starch	Total sugars	Individual sugars: Gluc	Fruct	Sucr	Malt	Lact	Dietary fibre: NSP	Fatty acids: Satd	Mono-unsatd	Poly-unsatd	Trans	Cholest-erol
		g	g	g	g	g	g	g	g	g	g	g	g	mg
	Skimmed milk													
206	**Skimmed milk**, *average*	0	4.4	0	0	0	0	4.4	0	0.1	0.1	Tr	Tr	3
207	pasteurised, *average*	0	4.8	0	0	0	0	4.8	0	0.1	0.1	Tr	Tr	4
208	pasteurised, *fortified plus SMP*	0	6.0	0	0	0	0	6.0	0	0.1	Tr	Tr	Tr	2
209	sterilised	0	5.4	0	0	0	0	5.4	0	0.3	Tr	Tr	Tr	2
210	UHT	0	3.9	0	0	0	0	3.9	0	N	N	Tr	Tr	2
	Semi-skimmed milk													
211	**Semi-skimmed milk**, *average*	0	4.7	0	0	0	0	4.7	0	1.1	0.4	Tr	0.1	6
212	pasteurised, *average*	0	4.7	0	0	0	0	4.7	0	1.1	0.4	Tr	0.1	6
213	pasteurised, *summer*	0	4.5	0	0	0	0	4.5	0	1.0	0.4	0.1	0.1	6
214	pasteurised, *winter*	0	4.9	0	0	0	0	4.9	0	1.0	0.4	0.1	0.1	6
215	pasteurised, *fortified plus SMP*	0	5.8	0	0	0	0	5.8	0	1.0	0.5	Tr	(0.1)	7
216	UHT	0	4.9	0	0	0	0	4.9	0	1.1	0.4	Tr	(0.1)	7

Inorganic constituents per 100g edible portion

No.	Food	mg										µg	
		Na	K	Ca	Mg	P	Fe	Cu	Zn	Cl	Mn	Se	I
	Skimmed milk												
206	**Skimmed milk**, *average*	44	160	122	11	96	0.03	Tr	0.5	89	Tr	1	29
207	pasteurised, *average*	44	162	125	11	96	0.03	Tr	0.5	87	Tr	1	30[a]
208	pasteurised, *fortified plus SMP*	61	170	140	13	110	0.04	Tr	0.4	110	Tr	(1)	(15)
209	sterilised	38	140	105	10	91	Tr	0.01	0.3	94	Tr	2	20
210	UHT	40	148	102	10	92	Tr	0.01	0.3	102	Tr	2	25
	Semi-skimmed milk												
211	**Semi-skimmed milk**, *average*	43	156	120	11	94	0.02	Tr	0.4	87	Tr	1	30
212	pasteurised, *average*	43	156	120	11	94	0.02	Tr	0.4	87	Tr	1	30
213	pasteurised, *summer*	43	152	118	11	93	0.02	Tr	0.4	85	Tr	Tr	20
214	pasteurised, *winter*	43	161	123	11	96	0.02	Tr	0.4	89	Tr	1	41
215	pasteurised, *fortified plus SMP*	59	150	130	12	100	0.03	Tr	0.4	110	Tr	(1)	(15)
216	UHT	50	150	110	11	90	0.17	Tr	0.4	100	Tr	(1)	(31)

[a] Winter milk may contain slightly higher levels of Iodine than summer milk

Vitamins per 100g edible portion

No.	Food	Retinol µg	Carotene µg	Vitamin D µg	Vitamin E mg	Thiamin mg	Riboflavin mg	Niacin mg	Trypt/60 mg	Vitamin B_6 mg	Vitamin B_{12} µg	Folate µg	Pantothenate mg	Biotin µg	Vitamin C mg
	Skimmed milk														
206	**Skimmed milk**, *average*	1	Tr	Tr	Tr	0.03	0.22	0.1	0.7	0.06	0.8	9	0.48	2.4	1
207	pasteurised, *average*	1	Tr	Tr	Tr	0.03	0.22	0.1	0.7	0.06	0.8	9	0.50	2.5	1
208	pasteurised, *fortified plus SMP*	43	5	0.3	0.01	0.04	0.19	0.1	0.9	0.06	0.4	5	0.40	2.4	1
209	sterilised	Tr	7	Tr	0.03	0.03	0.16	0.1	0.5	0.02	0.3	Tr	0.38	1.6	Tr
210	UHT	Tr	Tr	0.1	0.02	0.04	0.17	0.1	0.8	0.02	0.6	1	0.41	1.8	Tr
	Semi-skimmed milk														
211	**Semi-skimmed milk**, *average*	19	9	Tr	0.04	0.04	0.24	0.1	0.8	0.06	0.4	6	0.32	2.0	1
212	pasteurised, *average*	19	9	Tr	0.04	0.03	0.24	0.1	0.6	0.06	0.9	9	0.68	3.0	2
213	pasteurised, *summer*	15	7	Tr	0.05	0.03	0.24	0.1	0.6	0.06	0.8	7	0.57	2.9	2
214	pasteurised, *winter*	22	11	Tr	0.03	0.03	0.24	0.1	0.7	0.06	0.9	12	0.80	3.0	2
215	pasteurised, *fortified plus SMP*	90	5	0.1	0.04	0.04	0.19	0.1	0.9	0.06	0.4	5	0.37	2.3	1
216	UHT	20	11	0	0.03	0.04	0.18	0.1	0.8	0.05	0.2	2	0.33	1.8	Tr

Composition of food per 100g edible portion

No.	Food	Description and main data sources	Edible conversion factor	Water	Total nitrogen	Protein	Fat	Carbo-hydrate	Energy value	
				g	g	g	g	g	kcal	kJ
	Whole milk									
217	**Whole milk**, *average*	Includes pasteurised, sterilised and UHT	1.00	87.6	0.52	3.3	3.9	4.5	66	274
218	pasteurised, *average*	Average of summer and winter milk. Samples from 11 areas, in glass bottles (50%), plastic containers (30%) and cartons (20%)[a]	1.00	87.3	0.52	3.3	3.9	4.6	66	274
219	pasteurised, *summer*	Samples taken in June and September	1.00	87.4	0.53	3.4	4.0	4.1	65	270
220	pasteurised, *winter*	Samples taken in January and March	1.00	87.3	0.52	3.3	3.9	5.0	67	289
221	sterilised	10 samples, 2 brands, polybottles	1.00	87.6	0.55	3.5	3.9	4.5	66	277
222	UHT	22 samples, summer (Jun/Sep) and winter (Jan/Mar)	1.00	90.2	0.50	3.2	3.9	4.8	66	276
	Channel Island milk									
223	**Channel Island milk**, whole, pasteurised	Samples from dairy and retail outlets. Fat from Milk Marketing Board	1.00	86.4	0.57	3.6	5.1	4.8	78	327
	Breakfast milk									
224	**Breakfast milk**, pasteurised,	Fresh pasteurised Channel Island (homogenised) milk.	1.00	86.6	0.56	3.5	4.7	4.3	72	302
	average	Average of 6 samples of summer and winter milk								
225	*summer*	3 samples	1.00	86.7	0.61	3.9	4.7	4.1	73	306
226	*winter*	3 samples	1.00	86.5	0.50	3.2	4.7	4.4	72	299

[a] All the values for pasteurised milk are equally applicable to unpasteurised milk

Composition of food per 100g edible portion

No.	Food	Starch	Total sugars	Individual sugars					Dietary fibre	Fatty acids				Cholest-erol
				Gluc	Fruct	Sucr	Malt	Lact	NSP	Satd	Mono-unsatd	Poly-unsatd	Trans	
		g	g	g	g	g	g	g	g	g	g	g	g	mg
	Whole milk													
217	**Whole milk**, *average*	0	4.5	0	0	0	0	4.5	0	2.5	1.0	0.1	0.1	14
218	pasteurised, *average*	0	4.6	0	0	0	0	4.6	0	2.5	1.0	0.1	0.1	14
219	pasteurised, *summer*	0	4.1	0	0	0	0	4.1	0	2.5	1.0	0.1	0.1	14
220	pasteurised, *winter*	0	5.0	0	0	0	0	5.0	0	2.5	1.0	0.1	0.1	14
221	sterilised	0	4.5	0	0	0	0	4.5	0	2.4	1.1	0.1	0.1	14
222	UHT	0	4.8	0	0	0	0	4.8	0	2.4	1.1	0.1	0.1	14
	Channel Island milk													
223	**Channel Island milk**, whole, pasteurised	0	4.8	0	0	0	0	4.8	0	3.3	1.3	0.1	(0.1)	16
	Breakfast milk													
224	**Breakfast milk**, pasteurised, *average*	0	4.3	0	0	0	0	4.3	0	3.0	1.1	0.2	0.2	16
225	*summer*	0	4.1	0	0	0	0	4.1	0	3.0	1.2	0.2	0.2	16
226	*winter*	0	4.4	0	0	0	0	4.4	0	3.3	1.0	0.1	0.1	16

Inorganic constituents per 100g edible portion

No.	Food	mg										µg	
		Na	K	Ca	Mg	P	Fe	Cu	Zn	Cl	Mn	Se	I
	Whole milk												
217	**Whole milk**, *average*	43	155	118	11	93	0.03	Tr	0.4	89	Tr	1	31
218	pasteurised, *average*	43	155	118	11	93	0.03	Tr	0.4	89	Tr	1	31
219	pasteurised, *summer*	43	150	116	11	91	0.03	Tr	0.4	90	Tr	1	20
220	pasteurised, *winter*	43	159	121	11	96	0.03	Tr	0.4	88	Tr	1	41
221	sterilised	57	140	120	13	91	0.18	Tr	0.3	100	Tr	(1)	(31)
222	UHT	55	140	110	11	87	0.23	0.01	0.4	93	Tr	(1)	(31)
	Channel Island milk												
223	**Channel Island milk**, whole, pasteurised	54	140	130	12	100	0.05	Tr	0.4	100	Tr	(1)	N
	Breakfast milk												
224	**Breakfast milk**, pasteurised, *average*	39	131	129	12	106	Tr	0.01	0.4	(100)	Tr	(1)	29
225	*summer*	39	131	129	12	106	Tr	0.01	0.4	(100)	Tr	(1)	24
226	*winter*	39	131	129	12	106	Tr	0.01	0.4	(100)	Tr	(1)	34

Vitamins per 100g edible portion

No.	Food	Retinol µg	Carotene µg	Vitamin D µg	Vitamin E mg	Thiamin mg	Ribo-flavin mg	Niacin mg	Trypt/60 mg	Vitamin B_6 mg	Vitamin B_{12} µg	Folate µg	Panto-thenate mg	Biotin µg	Vitamin C mg
	Whole milk														
217	**Whole milk**, *average*	33	20	Tr	0.08	0.03	0.23	0.2	0.6	0.06	0.9	8	0.58	2.5	2
218	pasteurised, *average*	30	19	Tr	0.08	0.03	0.23	0.2	0.6	0.06	0.9	8	0.58	2.5	2
219	pasteurised, *summer*	32	25	Tr	0.08	0.03	0.24	0.2	0.5	0.06	0.9	6	0.57	2.3	2
220	pasteurised, *winter*	28	13	Tr	0.07	0.03	0.22	0.2	0.7	0.07	0.8	9	0.60	2.8	2
221	sterilised	61	21	Tr	0.12	0.03	0.14	0.1	0.8	0.04	0.1	Tr	0.28	1.8	Tr
222	UHT	54	31	0	0.08	0.04	0.18	0.1	0.8	0.04	0.2	1	0.32	1.8	Tr
	Channel Island milk														
223	**Channel Island milk**, whole, pasteurised	46	71	Tr	0.11	0.04	0.19	0.1	0.9	0.06	0.4	6	0.36	1.9	1
	Breakfast milk														
224	**Breakfast milk**, pasteurised, *average*	35	41	0.1	0.17	0.04	0.22	0.1	0.6	0.03	0.8	6	0.38	1.9	(1)
225	*summer*	70	82	0.1	0.17	0.04	0.22	0.1	0.6	0.03	0.8	6	0.38	1.9	(1)
226	*winter*	Tr	Tr	0.1	0.17	0.04	0.22	0.1	0.6	0.03	0.8	6	0.38	1.9	Tr

Composition of food per 100g edible portion

No.	Food	Description and main data sources	Edible conversion factor	Water	Total nitrogen	Protein	Fat	Carbo-hydrate	Energy value	
				g	g	g	g	g	kcal	kJ
	Processed milks									
227	**Condensed milk**, skimmed, *sweetened*	10 cans (Fussells)	1.00	29.7	1.57	10.0	0.2	60.0	267	1137
228	whole, *sweetened*	10 cans, 2 brands	1.00	25.9	1.33	8.5	10.1	55.5	333	1406
229	**Dried skimmed milk**	20 samples, 7 brands, fortified	1.00	3.0	5.70	36.1	0.6	52.9	348	1482
230	*with vegetable fat*	12 samples, 5 brands, fortified	1.00	2.0	3.70	23.3	25.9	42.6	487	2038
231	**Evaporated milk**, whole	12 samples, Ideal, Carnation and own brands	1.00	69.1	1.32	8.4	9.4	8.5	151	629
232	light, *4% fat*	7 samples, 4 brands	1.00	75.9	1.22	7.8	4.1	10.3	107	449
233	**Flavoured milk**, pasteurised	10 samples, 6 brands; strawberry, banana	1.00	83.9	0.57	3.6	1.5	9.6[a]	64	270
234	pasteurised, chocolate	9 samples, 6 brands, including low fat, semi-skimmed	1.00	82.8	0.56	3.6	1.5	9.4[a]	63	267
235	**Milkshake**, *thick, takeaway*	10 samples, 3 brands including chocolate, vanilla, and banana	1.00	73.2	0.58	3.7	1.8	15.3	88	374
	Other milks									
236	**Goats milk**, pasteurised	20 samples from one herd and literature sources	1.00	88.9	0.49	3.1	3.7	4.4	62	260
237	**Human milk**, mature	Department of Health and literature sources	1.00	87.1	0.20	1.3[b]	4.1	7.2	69	289
238	**Sheeps milk**, *raw*	30 samples from 2 herds and literature sources	1.00	83.0	0.85	5.4	5.8	5.1	93	388
239	**Soya**, non-dairy alternative to milk, *sweetened, calcium enriched*	10 samples, 9 brands	1.00	90.1	0.55	3.1	2.4	2.5	43	182
240	**Soya**, non-dairy alternative to milk, *unsweetened*	10 samples, 8 brands	1.00	93.0	0.42	2.4	1.6	0.5	26	108

[a] Including oligosaccharides from the glucose syrup/maltodextrins in the product
[b] N x 6.38. True protein = 0.85g per 100g excluding the non-protein nitrogen

Composition of food per 100g edible portion

No.	Food	Starch	Total sugars	Individual sugars					Dietary fibre	Fatty acids				Cholest-erol
				Gluc	Fruct	Sucr	Malt	Lact	NSP	Satd	Mono-unsatd	Poly-unsatd	Trans	
		g	g	g	g	g	g	g	g	g	g	g	g	mg
	Processed milks													
227	**Condensed milk**, skimmed, *sweetened*	0	60.0	0	0	46.7	0	13.3	0	0.1	0.1	Tr	Tr	1
228	whole, *sweetened*	0	55.5	0	0	43.2	0	12.3	0	6.3	2.9	0.3	N	36
229	**Dried skimmed milk**	0	52.9	0	0	0	0	52.9	0	0.4	0.2	Tr	Tr	12
230	*with vegetable fat*	0	42.6	0	0	0	0	42.6	0	16.8	7.3	0.7	N	17
231	**Evaporated milk**, whole	0	8.5	0	0	0	0	8.5	0	5.9	2.7	0.3	N	34
232	light, *4% fat*	0	10.3	0	0	0	0	10.3	0	2.5	1.1	0.2	0.2	17
233	**Flavoured milk**, pasteurised	0.2	8.9[a]	Tr	Tr	3.9	0.1	4.9	0	1.0	0.3	0.1	Tr	7
234	pasteurised, chocolate	0.4	8.3[a]	Tr	1.3	2.1	0.8	4.1	0	1.0	0.4	0.1	0.1	7
235	**Milkshake**, *thick, takeaway*	0.3	11.1	0.3	4.2	1.4	0.7	4.5	Tr	1.2	0.4	0.1	0.1	11
	Other milks													
236	**Goats milk**, pasteurised	0	4.4	0	0	0	0	4.4	0	2.4	1.0	0.2	0.1	11
237	**Human milk**, mature	0	7.2	0	0	0	0	7.2	0	1.8	1.6	0.5	N	16
238	**Sheeps milk**, *raw*	0	5.1	0	0	0	0	5.1	0	3.6	1.5	0.3	0.4	12
239	**Soya**, non-dairy alternative to milk, *sweetened, calcium enriched*	0	2.2	0.3	1.2	0.7	0	0	Tr	0.4	0.5	1.4	N	0
240	**Soya**, non-dairy alternative to milk, *unsweetened*	0	0.2	0	0	0.2	0	0	0.2	0.2	0.3	1.1	Tr	0

[a] Not including oligosaccharides from the glucose syrup/maltodextrins in the product

Inorganic constituents per 100g edible portion

No.	Food	mg										µg	
		Na	K	Ca	Mg	P	Fe	Cu	Zn	Cl	Mn	Se	I
	Processed milks												
227	**Condensed milk**, skimmed, *sweetened*	150	450	330	33	270	0.33	Tr	1.2	300	Tr	(3)	(89)
228	whole, *sweetened*	140	360	290	29	240	0.23	Tr	1.0	230	Tr	(3)	74
229	**Dried skimmed milk**	550	1590	1280	130	970	0.27	Tr	4.0	1070	Tr	(11)	(150)
230	*with vegetable fat*	440	1030	840	74	680	0.19	Tr	0.6	760	Tr	(7)	N
231	**Evaporated milk**, whole	180	360	290	29	260	0.26	0.02	0.9	250	Tr	(3)	11
232	light, *4% fat*	115	336	260	25	233	Tr	Tr	1.0	222	Tr	3	47
233	**Flavoured milk**, pasteurised	52	168	120	12	102	0.13	Tr	0.4	110	Tr	N	N
234	pasteurised, chocolate	45	206	115	19	107	0.62	0.06	0.5	110	0.1	N	N
235	**Milkshake**, *thick, takeaway*	57	171	129	13	120	Tr	Tr	0.1	111	Tr	2	37
	Other milks												
236	**Goats milk**, pasteurised	42	170	100	13	90	0.12	0.03	0.5	150	Tr	N	N
237	**Human milk**, mature	15	58	34	3	15	0.07	0.04	0.3	42	Tr	1	7
238	**Sheeps milk**, *raw*	44	120	170	18	150	0.03	0.10	0.7	82	Tr	N	N
239	**Soya**, non-dairy alternative to milk, *sweetened, calcium enriched*	56	119	89	18	89	0.31	0.09	0.3	3	0.2	4	1
240	**Soya**, non-dairy alternative to milk, *unsweetened*	32	74	13	15	48	0.43	0.09	0.3	3	0.3	4	1

No.	Food	Retinol µg	Carotene µg	Vitamin D µg	Vitamin E mg	Thiamin mg	Riboflavin mg	Niacin mg	Trypt/60 mg	Vitamin B_6 mg	Vitamin B_{12} µg	Folate µg	Pantothenate mg	Biotin µg	Vitamin C mg
	Processed milks														
227	**Condensed milk**, skimmed, *sweetened*	28	20	0.9	0.04	0.11	0.51	0.3	2.3	0.09	0.9	16	1.03	5.2	5
228	whole, *sweetened*	110	70	5.4	0.19	0.09	0.46	0.3	2.0	0.07	0.7	15	0.85	3.9	4
229	**Dried skimmed milk**	350[a]	5[a]	2.1[a]	0.27[a]	0.38	1.63	1.0	8.5	0.60	2.6	51	3.28	20.1	13
230	*with vegetable fat*	395	15	10.5	1.32	0.23	1.20	0.6	5.5	0.35	2.3	36	2.15	15.0	11
231	**Evaporated milk**, whole	105	100	4.0[b]	0.19	0.07	0.42	0.2	2.0	0.07	0.1	11	0.75	4.0	1
232	light, *4% fat*	50	21	3.1	0.11	0.07	(0.42)	0.2	(2.0)	0.04	0.2	8	(0.75)	(4.0)	(1)
233	**Flavoured milk**, pasteurised	20	8	0	0.03	0.03	0.17	0.1	0.8	0.03	0.1	2	0.30	2.2	Tr
234	pasteurised, chocolate	20	8	0	0.03	0.03	0.17	0.1	0.8	0.03	0.1	2	0.30	2.2	Tr
235	**Milkshake**, *thick, takeaway*	35	11	Tr	0.10	0.03	0.23	0.1	0.7	0.03	0.5	4	0.31	2.0	1
	Other milks														
236	**Goats milk**, pasteurised	44	Tr	0.1	0.03	0.03	0.04	0.1	0.7	0.06	0.1	1	0.41	3.0	1
237	**Human milk**, mature	58	(24)	Tr	0.34	0.02	0.03	0.2	0.5	0.01	Tr	5	0.25	0.7	4
238	**Sheeps milk**, *raw*	83	Tr	0.2	0.11	0.08	0.32	0.4	1.3	0.08	0.6	5	0.45	2.5	5
239	**Soya**, non-dairy alternative to milk, *sweetened, calcium enriched*	0	Tr	0	0.32	0.06	0.05	0.1	0.7	0.03	0	9	Tr	N	1
240	**Soya**, non-dairy alternative to milk, *unsweetened*	Tr	Tr	0	0.32	0.06	0.05	0.1	0.7	0.03	0	14	Tr	1.0	0

[a] Unfortified skimmed milk powder contains approximately 8µg retinol, 3µg carotene, Tr vitamin D, and 0.01mg vitamin E per 100g. Some brands contain as much as 755µg retinol, 10µg carotene and 4.6µg vitamin D per 100g

[b] This is for fortified product. Unfortified evaporated milk contains approximately 0.09µg vitamin D per 100g

No.	Food	Description and main data sources	Edible conversion factor	Water	Total nitrogen	Protein	Fat	Carbo-hydrate	Energy value	
				g	g	g	g	g	kcal	kJ
Fresh creams (pasteurised)										
241	**Cream**, single	Average of 22 samples of summer and winter cream	1.00	77.0	0.52	3.3	19.1	2.2	193	798
242	soured	8 samples, 4 brands	1.00	72.5	0.45	2.9	19.9	3.8	205	845
243	whipping	Average of 22 samples of summer and winter cream	1.00	54.5	0.31	2.0	40.3	2.7	381	1568
244	double	Average of 22 samples of summer and winter cream. Includes Jersey cream	1.00	46.9	0.25	1.6	53.7[a]	1.7	496	2041
245	clotted	17 samples, 3 brands	1.00	32.2	0.25	1.6	63.5	2.3	586	2413
246	**Creme fraiche**	9 samples, 6 brands	1.00	55.8	0.34	2.2	40.0	2.4	378	1556
247	half fat	8 samples, 6 brands	1.00	76.5	0.42	2.7	15.0	4.4	162	671
248	**Dairy cream**, extra thick	16 samples, 4 brands, summer and winter	1.00	69.0	0.45	2.9	23.5	3.4	236	973
Sterilised creams										
249	**Cream**, sterilised, canned	13 cans, 6 brands	1.00	69.2	0.39	2.5	23.9	3.7	239	985
UHT creams										
250	**Dairy cream**, UHT, canned spray	10 samples, 6 brands	1.00	63.6	0.30	1.9	24.2	7.2	252	1043
251	UHT, canned spray, half fat	4 samples, 1 brand (Anchor Light)	1.00	71.9	0.44	2.8	17.3	7.6	196	811
Imitation creams										
252	**Dream Topping**, *made up with semi-skimmed milk*	Recipe	1.00	71.5	0.61	3.9	11.7	12.2	166	694

[a] Double cream with added alcohol contains 39.7g fat, 10.3g carbohydrate (8.0g sucrose, 2.0g lactose, 0.3g maltodextrins), 5.0g alcohol, 531kcal and 2186kJ energy per 100g

No.	Food	Starch	Total sugars	Individual sugars					Dietary fibre	Fatty acids				Cholest-erol
				Gluc	Fruct	Sucr	Malt	Lact	NSP	Satd	Mono-unsatd	Poly-unsatd	Trans	
		g	g	g	g	g	g	g	g	g	g	g	g	mg
Fresh creams (pasteurised)														
241	**Cream**, single	0	2.2	0	0	0	0	2.2	0	12.2	5.1	0.6	0.7	55
242	soured	0	3.8	0	0	0	0	3.8	0	12.5	5.8	0.6	N	60
243	whipping	0	2.7	0	0	0	0	2.7	0	25.2	11.7	1.1	N	105
244	double	0	1.7	0	0	0	0	1.7	0	33.4	13.8	1.9	1.8	137
245	clotted	0	2.3	0	0	0	0	2.3	0	39.7	18.4	1.8	N	170
246	**Creme fraiche**	0.3	2.1	0	0	0	0	2.1	0	27.1	8.6	1.1	0.8	113
247	half fat	1.4	3.0	0	0	0	0	3.0	0	10.2	3.2	0.4	0.3	N
248	**Dairy cream**, extra thick	0	3.4	0	0	0	0	3.4	0	15.3	6.0	0.8	0.8	74
Sterilised creams														
249	**Cream**, sterilised, canned	0	3.7	0	0	0	0	3.7	0	14.9	6.9	0.7	N	65
UHT creams														
250	**Dairy cream**, UHT, canned spray	0	7.2	0	0	0	0	3.3	0	15.2	6.1	0.8	0.8	(68)
251	UHT, canned spray, half fat	0	7.4	0	0	3.8	0	3.6	0	10.9	4.3	0.6	0.6	46
Imitation creams														
252	**Dream Topping**, *made up with semi-skimmed milk*	2.1	10.2	Tr	Tr	4.5	0.2	5.4	Tr	10.5	0.5	0.1	N	6

Inorganic constituents per 100g edible portion

No.	Food	mg										µg	
		Na	K	Ca	Mg	P	Fe	Cu	Zn	Cl	Mn	Se	I
Fresh creams (pasteurised)													
241	**Cream**, single	29	104	89	8	79	Tr	Tr	0.3	80	Tr	N	N
242	soured	41	110	93	10	81	0.4	Tr	0.5	81	Tr	Tr	N
243	whipping	25	86	58	6	59	Tr	Tr	0.2	59	Tr	N	N
244	double	22	65	49	5	52	0.1	Tr	0.2	36	Tr	3	35
245	clotted	18	55	37	5	40	0.1	0.09	0.2	40	Tr	Tr	Tr
246	**Creme fraiche**	22	81	58	6	58	0.1	Tr	0.2	55	Tr	0	8
247	half fat	36	122	95	9	81	0.1	Tr	0.3	N	Tr	(4)	(8)
248	**Dairy cream**, extra thick	29	100	95	8	81	0.1	0.01	0.3	N	Tr	N	N
Sterilised creams													
249	**Cream**, sterilised, canned	53	110	86	10	73	0.8	Tr	1.1	78	Tr	Tr	N
UHT creams													
250	**Dairy cream**, UHT, canned spray	31	107	54	7	57	Tr	Tr	0.2	66	Tr	1	11
251	UHT, canned spray, half fat	35	110	87	9	77	Tr	Tr	0.3	66	Tr	1	11
Imitation creams													
252	**Dream Topping**, *made up with semi-skimmed milk*	70	130	99	9	94	0.1	0.03	0.4	82	Tr	N	12

No.	Food	Retinol µg	Carotene µg	Vitamin D µg	Vitamin E mg	Thiamin mg	Riboflavin mg	Niacin mg	Trypt/60 mg	Vitamin B_6 mg	Vitamin B_{12} µg	Folate µg	Pantothenate mg	Biotin µg	Vitamin C mg
	Fresh creams (pasteurised)														
241	**Cream**, single	291	169	0.3	0.47	0.03	0.19	0.1	0.5	0.03	0.4	5	0.30	2.8	1
242	soured	330	105	0.2	0.44	0.03	0.17	0.1	0.7	0.04	0.2	12	0.24	1.5	Tr
243	whipping	399	247	0.3	1.32	0.02	0.17	Tr	0.5	0.04	0.2	7	0.22	1.4	1
244	double	779[a]	483[a]	0.3	1.64[a]	0.02	0.19	Tr	0.3	0.01	0.6	7	0.23	0.9	1
245	clotted	705	685	0.3	1.48	0.02	0.16	Tr	0.4	0.03	0.1	6	0.14	1.0	Tr
246	**Creme fraiche**	388	143	0.3	0.72	0.02	0.21	0.1	N	0.01	0.2	3	N	N	N
247	half fat	300	21	Tr	0.42	0.02	0.21	0.1	N	0.01	0.2	3	N	N	N
248	**Dairy cream**, extra thick	435	384	0.3	0.80	0.03	0.19	0.1	0.5	0.03	0.4	5	0.30	2.8	1
	Sterilised creams														
249	**Cream**, sterilised, canned	240	215	Tr	0.48	0.02	0.16	0.1	0.6	0.02	0.1	1	0.25	2.1	Tr
	UHT creams														
250	**Dairy cream**, UHT, canned spray	279	111	0.3	0.79	0.03	0.26	0.1	0.5	0.02	0.1	6	0.19	1.7	0
251	UHT, canned spray, half fat	147	39	Tr	0.46	0.03	0.26	0.1	0.5	0.02	0.1	6	0.19	1.7	0
	Imitation creams														
252	**Dream Topping**, made up with semi-skimmed milk	16	N[b]	0	N	0.04	0.19	0.1	0.9	0.05	0.5	4	N	N	1

[a] Double cream with added alcohol contains 390mg retinol, 187mg carotene and 1.08mg Vitamin E

[b] β-Carotene is added as a colouring agent

Composition of food per 100g edible portion

No.	Food	Description and main data sources	Edible conversion factor	Water	Total nitrogen	Protein	Fat	Carbo-hydrate	Energy value	
				g	g	g	g	g	kcal	kJ
	Imitation creams *continued*									
253	**Elmlea**, single	5 samples	1.00	76.8	0.49	3.1	14.5	4.0	158	654
254	whipping	4 samples	1.00	62.4	0.41	2.6	29.9	3.3	292	1204
255	double	4 samples	1.00	55.8	0.41	2.6	35.7	3.6	345	1423
256	**Tip Top dessert topping**	4 samples	1.00	78.1	0.77	4.9	6.5	9.0	112	468
	Cheeses									
257	**Brie**	20 samples, with outer rind removed	0.69	48.7	3.18	20.3	29.1	Tr	343	1422
258	**Camembert**	18 samples	1.00	54.4	3.37	21.5	22.7	Tr	290	1205
259	**Cheddar cheese**	20 samples of English cheddar including mild and mature, spring and autumn.	1.00	36.6	3.98	25.4	34.9	0.1	416	1725
260	**Cheddar type**, half fat	16 samples	1.00	47.4	5.13	32.7	15.8	Tr	273	1141
261	**Cheddar**, vegetarian	16 samples	1.00	37.2	4.00	25.5	32.0	Tr	390	1618
262	**Cheese spread**, plain	20 samples, 7 brands, portions and tubs	1.00	58.8	1.77	11.3	22.8	4.4	267	1106
263	reduced fat	13 samples, 9 brands, portions and tubs	1.00	61.4	2.35	15.0	9.5	7.9	175	733
264	**Cottage cheese**, plain	16 samples	1.00	78.6	1.97	12.6	4.3	3.1	101	423
265	plain, reduced fat	20 samples	1.00	81.9	2.09	13.3	1.5	3.3	79	334
266	plain, *with additions*	10 samples, mixed, e.g. with pineapple	1.00	76.9	2.00	12.8	3.8	2.6	95	400
267	**Cream cheese**	3 samples	1.00	45.5	0.49	3.1	47.4	Tr	439	1807
268	**Danish blue**	18 samples	1.00	46.3	3.22	20.5	28.9	Tr	342	1418
269	**Edam**	20 samples	1.00	43.8	4.18	26.7	26.0	Tr	341	1416
270	**Feta**	18 samples, made from sheeps and goats milk	1.00	56.5	2.45	15.6	20.2	1.5	250	1037

No.	Food	Starch	Total sugars	Individual sugars					Dietary fibre	Fatty acids				Cholest-erol
				Gluc	Fruct	Sucr	Malt	Lact	NSP	Satd	Mono-unsatd	Poly-unsatd	Trans	
		g	g	g	g	g	g	g	g	g	g	g	g	mg
	***Imitation creams** continued*													
253	**Elmlea**, single	0	4.0	0	0	0	0	4.0	0.3[a]	9.2	3.2	1.3	0.4	(4)
254	whipping	0	3.3	0	0	0	0	3.3	0.1[a]	26.4	2.8	0.9	N	8
255	double	0	3.6	0	0	0	0	2.3	0.1[a]	24.3	6.5	2.8	0.9	(11)
256	**Tip Top dessert topping**	1.9	7.1	0	0	0.3	0	6.8	Tr	5.9	0.2	0.1	0.1	4
	Cheeses													
257	**Brie**	0	Tr	0	0	0	0	Tr	0	18.2	6.7	0.6	1.3	93
258	**Camembert**	0	Tr	0	0	0	0	Tr	0	14.2	6.6	0.7	N	72
259	**Cheddar cheese**	0	0.1	0	0	0	0	0.1	0	21.7	9.4	1.1	1.4	97
260	**Cheddar type**, half fat	0	Tr	0	0	0	0	Tr	0	9.9	4.6	0.4	N	43
261	**Cheddar**, vegetarian	0	Tr	0	0	0	0	Tr	0	20.8	8.7	1.2	1.5	105
262	**Cheese spread**, plain	0	4.4	0	0	0	0	4.4	0	15.8	5.8	0.8	1.1	67
263	reduced fat	0.6	7.3	0	0	0	0	7.3	0	6.6	2.4	0.3	0.5	N
264	**Cottage cheese**, plain	0	3.1	0	0	0	0	3.1	0	2.3	1.2	0.2	0.2	16
265	plain, reduced fat	0	3.3	0	0	0	0	3.3	0	1.0	0.4	Tr	Tr	5
266	plain, *with additions*	0	2.6	0.6	0	0	0	2.0	Tr	2.4	1.1	0.1	N	13
267	**Cream cheese**	0	Tr	0	0	0	0	Tr	0	29.7	13.7	1.4	N	95
268	**Danish blue**	0	Tr	0	0	0	0	Tr	0	19.1	7.5	1.0	1.1	75
269	**Edam**	0	Tr	0	0	0	0	Tr	0	15.8	5.2	0.4	(0.7)	71
270	**Feta**	0	1.5	0	0	0	0	1.4	0	(13.7)	(4.1)	(0.6)	N	70

[a] Carob and guar gums are added as thickeners

Inorganic constituents per 100g edible portion

No.	Food	mg										µg	
		Na	K	Ca	Mg	P	Fe	Cu	Zn	Cl	Mn	Se	I
	***Imitation creams** continued*												
253	**Elmlea**, single	61	139	96	10	88	0.1	Tr	0.3	N	Tr	(2)	N
254	whipping	56	94	78	9	75	0.4	Tr	0.3	77	Tr	2	12
255	double	47	109	79	8	73	0.2	Tr	0.3	N	N	(2)	N
256	**Tip Top dessert topping**	110	205	173	18	171	0.2	Tr	0.6	147	Tr	(2)	N
	Cheeses												
257	**Brie**	556	91	256	15	232	Tr	Tr	2.0	900	Tr	5	16
258	**Camembert**	605	104	235	14	241	Tr	Tr	2.1	1120	Tr	7	N
259	**Cheddar cheese**	723	75	739	29	505	0.3	0.03	4.1	1040	Tr	6	30
260	**Cheddar type**, half fat	670	110	840	39	620	0.2	0.05	2.8	1110	Tr	11	N
261	**Cheddar**, vegetarian	670	67	690	31	490	0.2	Tr	1.9	990	0.1	5	26
262	**Cheese spread**, plain	1077	219	498	24	835	Tr	Tr	1.8	820	Tr	(4)	29
263	reduced fat	1035	235	485	24	850	0.3	0.05	1.7	775	Tr	4	29
264	**Cottage cheese**, plain	(300)	(161)	(127)	(13)	(171)	Tr	Tr	(0.6)	(490)	Tr	(4)	(24)
265	plain, reduced fat	300	161	127	13	171	Tr	Tr	0.6	490	Tr	4	24
266	plain, *with additions*	360	130	110	12	160	0.1	0.05	0.5	590	Tr	(4)	N
267	**Cream cheese**	300	160	98	10	100	0.1	(0.04)	0.5	480	Tr	4	N
268	**Danish blue**	1220	88	488	20	344	Tr	Tr	3.0	1950	Tr	7	12
269	**Edam**	996	89	795	34	508	0.3	Tr	3.8	1570	Tr	7	13
270	**Feta**	1440	95	360	20	280	0.2	0.07	0.9	2350	Tr	5	N

Vitamins per 100g edible portion

No.	Food	Retinol µg	Carotene µg	Vitamin D µg	Vitamin E mg	Thiamin mg	Riboflavin mg	Niacin mg	Trypt/60 mg	Vitamin B_6 mg	Vitamin B_{12} µg	Folate µg	Pantothenate mg	Biotin µg	Vitamin C mg
	***Imitation creams** continued*														
253	**Elmlea**, single	11	166	Tr	0.84	N	N	N	0.7	N	N	N	N	N	N
254	whipping	9	340	Tr	0.53	0.04	0.24	0.1	0.4	0.01	0.3	8	0.23	1.1	N
255	double	10	363	Tr	1.33	N	N	N	0.6	N	N	N	N	N	N
256	**Tip Top dessert topping**	Tr	Tr	Tr	0.14	0.04	0.33	0.1	1.2	0.02	0.1	7	N	N	Tr
	Cheeses														
257	**Brie**	297	192	0.2	0.81	0.03[a]	0.33	0.5	4.6	0.14	0.6	55	0.50	3.6	Tr
258	**Camembert**	230	315	0.1	0.65	0.05	0.52	0.9	4.9	0.23	1.1	83	0.80	7.5	Tr
259	**Cheddar cheese**	364	141	0.3	0.52	0.03	0.39	0.1	6.8	0.15	2.4	31	0.50	4.4	Tr
260	**Cheddar type**, half fat	190	121	0.1	0.47	0.03	0.53	0.1	7.4	0.13	1.3	56	0.51	3.8	Tr
261	**Cheddar**, vegetarian	356	203	0.3	0.80	0.03	0.41	0	6.2	0.11	1.2	30	0.30	2.6	Tr
262	**Cheese spread**, plain	262	119	0.2	0.30	0.05	0.36	0.1	3.2	0.08	0.6	19	0.51	3.6	Tr
263	reduced fat	119	90	N	0.40	0.06	0.53	0.1	3.1	0.07	2.0	7	0.42	3.0	Tr
264	**Cottage cheese**, plain	46	13	0	0.10	(0.05)	(0.24)	(0.2)	(3.4)	(0.05)	(0.6)	(22)	(0.30)	(5.1)	Tr
265	plain, reduced fat	16	4	0	0.03	0.05	0.24	0.2	3.4	0.05	0.6	22	0.30	5.1	Tr
266	plain, *with additions*	43	10	0	0.08	0.06	0.21	0.2	3.0	0.08	0.6	13	0.31	3.0	1
267	**Cream cheese**	385	220	0.3	1.00	0.03	0.13	0.1	0.7	0.04	0.3	11	0.27	1.6	Tr
268	**Danish blue**	244	283	(0.2)	0.71	0.03	0.41	0.6	5.5	0.10	1.3	55	0.53	2.7	Tr
269	**Edam**	188	182	(0.2)	(0.80)	0.03	0.35	0.1	6.1	0.09	2.1	39	0.38	1.8	Tr
270	**Feta**	220	33	0.5	0.37	0.04	0.21	0.2	3.5	0.07	1.1	23	0.36	2.4	Tr

[a] The rind alone contains 0.05mg thiamin per 100g

Composition of food per 100g edible portion

No.	Food	Description and main data sources	Edible conversion factor	Water	Total nitrogen	Protein	Fat	Carbo-hydrate	Energy value	
				g	g	g	g	g	kcal	kJ
***Cheeses** continued*										
271	**Goats milk soft cheese**, full fat, white rind	16 samples, English and French	1.00	50.8	3.30	21.1	25.8	1.0	320	1329
272	**Gouda**	18 samples	1.00	40.4	3.97	25.3	30.6	Tr	377	1562
273	**Hard cheese**, *average*	Average of English Cheddar, Red Leicester and Double Gloucester	1.00	37.3	3.91	24.9	34.5	0.1	411	1702
274	**Mozzarella**, *fresh*	18 samples[a]	1.00	57.4	2.91	18.6	20.3	Tr	257	1067
275	**Parmesan**, *fresh*	8 samples, wedges/freshly grated[b]	1.00	27.6	5.67	36.2	29.7	0.9	415	1729
276	**Processed cheese**, plain	20 samples, 7 brands	1.00	47.4	2.79	17.8	23.0	5.0	297	1234
277	slices, reduced fat	10 samples, 7 brands	1.00	52.5	3.51	22.4	13.3	5.0	228	953
278	**Spreadable cheese**, *soft white, low fat*	18 samples of extra light soft cheese spreads up to 10% fat	1.00	72.5	2.33	14.9	8.0	3.5	132	549
279	*soft white, medium fat*	20 samples, including Philadelphia light, 15% fat	1.00	69.1	1.54	9.8	16.3	3.5	199	826
280	*soft white, full fat*	20 samples, including Philadelphia, 30% fat	1.00	58.6	1.17	7.5	31.3	Tr	312	1286
281	**Stilton**, blue	20 samples	1.00	38.0	3.72	23.7	35.0	0.1	410	1698
282	**White cheese**, *average*	Average of Cheshire, Lancashire and Wensleydale	1.00	41.1	3.72	23.7	31.8	0.1	381	1580
Yogurts and Fromage frais										
283	**Whole milk yogurt**, plain	22 samples, 2 brands	1.00	81.9	0.89	5.7	3.0	7.8	79	333
284	fruit	9 samples, 5 brands, assorted flavours including bio varieties	1.00	76.0	0.63	4.0	3.0	17.7	109	463
285	infant, fruit flavour	8 samples, 4 brands, assorted flavours	1.00	78.4	0.59	3.8	3.7	11.1	90	378
286	twinpot, thick and creamy with fruit	11 samples, 8 brands, various flavours	1.00	74.7	0.71	4.1	3.2	16.2	106	446

[a] Grated in drums contains 48.8g water, 25g protein, 21.7g fat, 295kcal and 1228kJ energy per 100g

[b] Grated in drums contains 19g water, 43.3g protein, 34.6g fat, 0.9g carbohydrate, 488kcal and 2031kJ energy per 100g

Composition of food per 100g edible portion

No.	Food	Starch	Total sugars	Individual sugars: Gluc	Fruct	Sucr	Malt	Lact	Dietary fibre NSP	Fatty acids: Satd	Mono-unsatd	Poly-unsatd	Trans	Cholest-erol
		g	g	g	g	g	g	g	g	g	g	g	g	mg
	***Cheeses** continued*													
271	**Goats milk soft cheese**, full fat, white rind	0	1.0	0	0	0	0	0.9	0	17.9	6.1	1.0	1.0	93
272	**Gouda**	0	Tr	0	0	0	0	Tr	0	20.3	7.4	0.9	1.1	85
273	**Hard cheese**, *average*	0	0.1	0	0	0	0	0.1	0	21.6	10.1	1.0	N	100
274	**Mozzarella**, *fresh*	0	Tr	0	0	0	0	Tr	0	13.8	5.0	0.8	0.8	58
275	**Parmesan**, *fresh*	0	0.9	0	0	0	0	0.9	0	19.3	7.7	1.1	1.1	93
276	**Processed cheese**, plain	0	5.0	0	0	0	0	5.0	0	14.3	6.3	0.8	1.1	85
277	slices, reduced fat	0	5.0	0	0	0	0	5.0	0	8.1	3.6	0.5	0.4	(48)
278	**Spreadable cheese**, *soft white, low fat*	0	3.5	0	0	0	0	3.5	0	(5.2)	(2.0)	(0.3)	(0.3)	(24)
279	*soft white, medium fat*	0	3.5	0	0	0	0	3.5	0	(10.7)	(4.0)	(0.5)	(0.7)	(48)
280	*soft white, full fat*	0	Tr	0	0	0	0	Tr	0	20.5	7.8	1.0	1.2	92
281	**Stilton**, blue	0	0.1	0	0	0	0	0.1	0	23.0	9.2	1.2	1.5	95
282	**White cheese**, *average*	0	0.1	0	0	0	0	0.1	0	21.1	7.9	0.7	N	90
	Yogurts and fromage frais													
283	**Whole milk yogurt**, plain	0	7.8	0	0	0	0	4.7	N	1.7	0.9	0.2	N	11
284	fruit	1.1	16.6[a]	3.3	2.2	6.2	0.2	4.0	N	2.0	0.7	0.1	0.1	3
285	infant, fruit flavour	0.7	10.4	Tr	1.5	4.5	0.1	3.6	0.1	2.5	0.9	0.1	0.1	4
286	twinpot, thick and creamy with fruit	0.6[b]	15.6	2.3	2.2	6.9	0.2	3.5	N	N	N	N	N	N

[a] 'Real' fruit yogurts contain 12.1g total sugars per 100g
[b] includes maltodextrins

Inorganic constituents per 100g edible portion

No.	Food	mg										µg	
		Na	K	Ca	Mg	P	Fe	Cu	Zn	Cl	Mn	Se	I
	***Cheeses** continued*												
271	**Goats milk soft cheese**, full fat, white rind	601	132	133	14	229	Tr	0.06	1.0	(1060)	Tr	6	51
272	**Gouda**	925	82	773	32	498	0.3	Tr	3.9	1440	Tr	8	N
273	**Hard cheese**, *average*	687	76	731	29	500	0.3	0.05	4.1	1005	Tr	6	30
274	**Mozzarella**, *fresh*	395	51	362	15	267	Tr	Tr	2.7	650	Tr	6	18
275	**Parmesan**, *fresh*	756	152	1025	41	680	0.8	0.84	5.1	(1260)	Tr	12	72
276	**Processed cheese**, plain	1351	178	610	27	768	0.5	Tr	2.6	1080	Tr	5	27
277	slices, reduced fat	1390	185	800	31	640	0.3	0.07	3.0	(1080)	Tr	(7)	(27)
278	**Spreadable cheese**, *soft white, low fat*	(438)	(135)	(116)	(11)	(148)	Tr	Tr	(1.1)	(745)	Tr	(3)	(11)
279	*soft white, medium fat*	346	120	99	10	129	Tr	Tr	0.7	(590)	Tr	(4)	(11)
280	*soft white, full fat*	288	89	76	7	97	Tr	Tr	0.7	490	Tr	3	11
281	**Stilton**, blue	788	96	326	15	314	0.2	0.04	2.9	1230	Tr	7	40
282	**White cheese**, *average*	502	82	544	22	408	0.3	0.03	3.5	810	Tr	3	41
	Yogurts and fromage frais												
283	**Whole milk yogurt**, plain	80	280	200	19	170	0.1	Tr	0.7	170	Tr	(2)	(63)
284	fruit	58	170	122	13	96	0.1	Tr	0.4	179	Tr	2	27
285	infant, fruit flavour	46	176	120	12	114	0.2	0.02	0.5	(179)	Tr	(2)	(27)
286	twinpot, thick and creamy with fruit	53	175	130	13	106	0.2	Tr	0.4	N	Tr	N	N

Vitamins per 100g edible portion

No.	Food	Retinol µg	Carotene µg	Vitamin D µg	Vitamin E mg	Thiamin mg	Riboflavin mg	Niacin mg	Trypt/60 mg	Vitamin B_6 mg	Vitamin B_{12} µg	Folate µg	Pantothenate mg	Biotin µg	Vitamin C mg
	***Cheeses** continued*														
271	**Goats milk soft cheese**, full fat, white rind	333	Tr	(0.5)	0.63	0.03	0.39	0.7	6.0	0.10	0.5	22	0.40	5.1	Tr
272	**Gouda**	258	139	(0.2)	0.57	0.03	0.30	0.1	7.0	0.08	1.7	43	0.32	1.4	Tr
273	**Hard cheese**, average	330	215	0.3	0.52	0.03	0.41	0.1	6.8	0.15	2.4	31	0.50	3.0	Tr
274	**Mozzarella**, *fresh*	258	152	0.2	0.31	0.03	0.40	0.1	5.0	0.10	1.7	20	0.25	2.2	Tr
275	**Parmesan**, *fresh*	371	233	0.3	0.76	0.03	0.32	0.1	9.0	0.11	3.3	12	0.43	3.3	Tr
276	**Processed cheese**, plain	270	95	0.2	0.55	0.06	0.25	0.1	4.7	0.07	1.2	15	0.60	5.6	Tr
277	slices, reduced fat	157	197	N	0.54	0.06	0.25	0.1	4.7	0.07	1.2	15	0.60	5.6	Tr
278	**Spreadable cheese**, *soft white, low fat*	86	158	N	0.42	(0.06)	(0.52)	(0.1)	(5.3)	(0.02)	(0.6)	(47)	(0.40)	(10.1)	Tr
279	*soft white, medium fat*	(195)	(175)	0.1	(0.12)	(0.04)	(0.34)	(0.1)	(3.5)	(0.01)	(0.4)	(30)	(0.26)	(6.7)	Tr
280	*soft white, full fat*	260	199	0.1	0.24	0.03	0.26	0.1	2.7	0.01	0.3	23	0.20	5.1	Tr
281	**Stilton**, blue	360	182	0.2	0.60	0.03	0.47	0.7	5.9	0.13	1.2	78	0.90	3.3	Tr
282	**White cheese**, *average*	351	231	0.2	0.62	0.02	0.46	0.1	6.3	0.08	1.6	39	0.29	3.9	Tr
	Yogurts and fromage frais														
283	**Whole milk yogurt**, plain	28	21	0	0.05	0.06	0.27	0.2	1.3	0.10	0.2	18	0.50	2.6	1
284	fruit	36	Tr	0.1	0.18	0.12	0.16	0.1	0.7	0.01	0.3	10	0.40	1.1	1
285	infant, fruit flavour	(36)	Tr	(0.1)	(0.18)	0.12	0.15	(0.1)	(0.7)	0.01	0.3	10	(0.40)	(1.1)	Tr
286	twinpot, thick and creamy with fruit	(20)	(15)	0	(0.12)	(0.06)	(0.19)	(0.2)	(0.9)	(0.08)	0	(13)	(0.36)	(2.0)	(2)

No.	Food	Description and main data sources	Edible conversion factor	Water	Total nitrogen	Protein	Fat	Carbo-hydrate	Energy value	
				g	g	g	g	g	kcal	kJ
Yogurts and Fromage frais *continued*										
287	**Low fat yogurt**, plain	8 samples, 5 brands	1.00	87.2	0.75	4.8	1.0	7.4	56	237
288	fruit	21 samples, 9 brands, including French set	1.00	78.9	0.66	4.2	1.1	13.7	78	331
289	**Virtually fat free/diet yogurt**, plain	6 samples, 4 brands, including bio varieties	1.00	86.9	0.84	5.4	0.2	8.2	54	230
290	fruit	14 samples, 10 brands, including bio varieties, flavours include strawberry, raspberry, black cherry and rhubarb	1.00	85.4	0.75	4.8	0.2	7.0	47	201
291	**Greek style yogurt**, plain	7 samples, 6 brands, made with whole milk	1.00	78.2	0.90	5.7	10.2	4.8	133	551
292	fruit	6 samples, 4 brands, including peach, apricot, strawberry and blackcurrant, made with whole milk	1.00	73.5	0.76	4.8	8.4	11.2	137	572
293	**Drinking yogurt**	5 samples (Ambrosia), UHT	1.00	84.4	0.48	3.1	Tr	13.1	62	263
294	**Greek yogurt**, sheep	3 samples (Total), 'set' variety and manufacturer's data	1.00	80.9	0.75	4.8	6.0	5.0	92	384
295	**Lassi**, *sweetened*[a]	5 samples	1.00	83.3	0.41	2.6	0.9	11.6	62	263
296	**Soya**, alternative to yogurt, fruit	3 samples, Soja sun, strawberry	1.00	81.1	0.37	2.1	1.8	12.9	73	309
297	**Tzatziki**	Yogurt-based Greek starter. Recipe	1.00	85.8	0.60	3.8	4.9	1.9	66	275
298	**Fromage frais**, plain	5 samples, 3 brands	1.00	81.8	0.96	6.1	8.0	4.4	113	470
299	fruit	7 samples, 4 brands, including strawberry, peach, apricot and raspberry flavour	1.00	74.7	0.83	5.3	5.6	13.9	124	520
300	virtually fat free, natural	7 samples, 6 brands	1.00	87.2	1.20	7.7	0.1	4.6	49	208
301	virtually fat free, fruit	11 samples, 6 brands, including strawberry, raspberry, apricot and blackcherry flavour	1.00	86.7	1.07	6.8	0.2	5.6	50	213

[a] Yakult (fermented skimmed milk drink) contains 1.2g protein, 0.1g fat, 18.0g carbohydrate, 77kcal and 322kJ energy per 100ml

Composition of food per 100g edible portion

No.	Food	Starch	Total sugars	Individual sugars					Dietary fibre	Fatty acids				Cholest-erol
				Gluc	Fruct	Sucr	Malt	Lact	NSP	Satd	Mono-unsatd	Poly-unsatd	Trans	
		g	g	g	g	g	g	g	g	g	g	g	g	mg
	Yogurts and fromage frais *continued*													
287	**Low fat yogurt**, plain	0.3	7.1	1.5	Tr	4.0	Tr	Tr	N	0.7	0.2	Tr	Tr	1
288	fruit	1.0	12.7	Tr	1.0	6.1	0.3	4.4	0.2	(0.8)	(0.3)	Tr	Tr	(0)
289	**Virtually fat free/diet yogurt**, plain	0.3	7.9	1.6	Tr	0.1	Tr	4.6	0	(0.1)	(0.1)	Tr	Tr	N
290	fruit	0.7	6.3	0.2	0.7	0.1	0	4.0	Tr	(0.1)	(0.1)	Tr	Tr	N
291	**Greek style yogurt**, plain	0.3	4.5	0.1	Tr	Tr	Tr	3.5	0	6.8	2.5	0.3	0.2	17
292	fruit	0.7	10.5	Tr	1.0	3.8	0.4	4.0	Tr	5.6	2.2	0.2	0.2	14
293	**Drinking yogurt**	0	13.1	1.3	1.1	6.0	0	2.8	Tr	Tr	Tr	Tr	Tr	Tr
294	**Greek yogurt**, sheep	0	5.0	0	0	0	0	4.5	0	4.2	1.6	0.2	N	(14)
295	**Lassi**, *sweetened*	0.2	11.2	0.1	0.1	2.3	0.1	8.6	0	0.6	0.2	Tr	Tr	N
296	**Soya**, alternative to yogurt, fruit	0.7	12.1	2.7	2.0	7.4	0	0	0.3	0.3	0.4	1.1	0	0
297	**Tzatziki**	0.2	1.7	0.3	0.3	0.1	0	0.3	0.3	2.8	1.4	0.3	N	N
298	**Fromage frais**, plain	0.3	4.1	0	0	0	0	4.0	0	5.5	1.8	0.2	0.1	9
299	fruit	0.6	13.3	1.1	0.8	8.3	0.1	3.0	Tr	3.5	1.6	0.2	0.1	20
300	virtually fat free, natural	0.2	4.4	0.2	0	0	0	4.1	0	0.1	Tr	Tr	Tr	1
301	virtually fat free, fruit	0.7	4.9	0.7	0.9	0.2	0	2.9	0.4	0.1	0.1	Tr	Tr	1

Inorganic constituents per 100g edible portion

No.	Food	mg										µg	
		Na	K	Ca	Mg	P	Fe	Cu	Zn	Cl	Mn	Se	I
	***Yogurts and fromage frais** continued*												
287	**Low fat yogurt**, plain	63	228	162	16	143	0.1	0.03	0.6	235	Tr	2	34
288	fruit	62	204	140	15	120	0.1	Tr	0.5	(130)	Tr	(2)	(48)
289	**Virtually fat free/diet yogurt**, plain	71	247	160	16	151	0.1	0.03	0.6	252	Tr	2	53
290	fruit	73	180	130	13	110	0.1	Tr	0.4	120	Tr	(1)	N
291	**Greek style yogurt**, plain	66	184	126	13	138	0.1	Tr	0.5	159	Tr	3	39
292	fruit	64	218	141	14	136	0.2	Tr	0.6	(159)	Tr	(3)	(39)
293	**Drinking yogurt**	47	130	100	11	81	0.1	0.01	0.3	75	Tr	(1)	N
294	**Greek yogurt**, sheep	150	190	150	16	140	Tr	Tr	0.5	220	Tr	1	N
295	**Lassi**, *sweetened*	45	109	92	9	74	Tr	Tr	0.3	85	Tr	N	N
296	**Soya**, alternative to yogurt, fruit	24	94	14	15	72	0.5	Tr	0.2	22	0.2	2	10
297	**Tzatziki**	372	150	88	11	93	0.3	0.01	0.3	569	0.1	1	N
298	**Fromage frais**, plain	36	143	110	11	123	0.1	0.03	0.4	137	Tr	3	17
299	fruit	35	110	86	8	110	0.1	0.02	0.4	78	Tr	(3)	(17)
300	virtually fat free, natural	37	155	127	12	120	0.1	0.03	0.6	(137)	Tr	(3)	23
301	virtually fat free, fruit	(33)	(110)	(87)	(8)	(110)	(0.1)	(0.01)	(0.3)	(89)	Tr	(2)	N

No.	Food	Retinol µg	Carotene µg	Vitamin D µg	Vitamin E mg	Thiamin mg	Riboflavin mg	Niacin mg	Trypt/60 mg	Vitamin B_6 mg	Vitamin B_{12} µg	Folate µg	Pantothenate mg	Biotin µg	Vitamin C mg
	***Yogurts and fromage frais** continued*														
287	**Low fat yogurt**, plain	8	Tr	0.1	Tr	0.12	0.22	0.1	1.0	0.01	0.3	18	0.56	1.5	1
288	fruit	(10)	Tr	Tr	0.28	0.12	0.21	0.1	1.0	Tr	0.3	16	0.33	2.3	1
289	**Virtually fat free/diet yogurt**, plain	Tr	Tr	Tr	Tr	(0.04)	(0.29)	(0.1)	(1.0)	(0.07)	(0.2)	(8)	N	N	(1)
290	fruit	Tr	Tr	Tr	0.03	0.04	0.29	0.1	1.0	0.07	(0.2)	8	N	N	1
291	**Greek style yogurt**, plain	115	Tr	0.1	0.38	0.12	0.13	(0.1)	(1.5)	0.01	0.2	(6)	N	N	Tr
292	fruit	115	Tr	0.1	0.39	(0.12)	(0.13)	(0.1)	(1.5)	Tr	0	(6)	N	N	Tr
293	**Drinking yogurt**	Tr	Tr	Tr	Tr	0.03	0.16	0.1	0.7	0.05	0.2	12	0.19	0.9	0
294	**Greek yogurt**, sheep	86	(11)	0.2	0.73	0.05	0.33	0.2	1.0	0.08	0.2	3	N	N	Tr
295	**Lassi**, *sweetened*	9	Tr	Tr	N	N	0.21	N	N	N	N	N	N	N	N
296	**Soya**, alternative to yogurt, fruit	23	(3)	0	1.91	0.11	0.02	N	0.7	Tr	0	N	0.12	1.0	0
297	**Tzatziki**	61	46	0	0.23	0.03	0.20	0.1	0.9	0.05	0.1	7	N	N	1
298	**Fromage frais**, plain	82	Tr	0	0.15	0.13	0.20	0.1	1.2	0.01	0.5	15	0.47	Tr	Tr
299	fruit	82	Tr	0	(0.01)	0.12	0.13	0.1	1.2	0.01	0.5	15	0.38	0.6	Tr
300	virtually fat free, natural	(3)	Tr	Tr	Tr	(0.03)	(0.37)	(0.1)	1.8	(0.07)	(1.4)	(15)	N	N	Tr
301	virtually fat free, fruit	(3)	Tr	Tr	Tr	(0.03)	(0.37)	(0.1)	1.8	(0.07)	(1.4)	(15)	N	N	Tr

Composition of food per 100g edible portion

No.	Food	Description and main data sources	Edible conversion factor	Water	Total nitrogen	Protein	Fat	Carbo-hydrate	Energy value	
				g	g	g	g	g	kcal	kJ
	Ice creams									
302	**Choc ice**	10 samples, 5 brands, non-dairy	1.00	44.4	0.51	3.2	21.7	23.2[a]	295	1229
303	**Chocolate nut sundae**	Recipe	1.00	53.3	0.44	2.6	14.9	26.2	243	1016
304	**Cornetto-type ice-cream cone**	10 samples, 5 brands, chocolate and nut and mint choc chip flavours[b]	1.00	42.5	0.64	4.0	17.8	28.8[a]	284	1187
305	**Frozen ice-cream desserts**	10 samples, 7 brands e.g. Viennetta, Romantica, After Eight	1.00	51.4	0.56	3.5	17.6	21.0[a]	251	1046
306	**Ice-cream bar**, chocolate coated	10 samples, different brands including Mars, Bounty and Snickers	1.00	33.1	0.80	5.0	23.3	21.8	311	1295
307	**Ice-cream wafers**	6 samples, 2 brands	1.00	2.8	1.77	10.1	0.7	78.8	342	1458
308	**Ice-cream**, dairy, vanilla	11 samples	1.00	62.5	0.56	3.6	9.8	19.8[a]	177	741
309	dairy, premium	10 samples, 5 brands	1.00	60.9	0.61	3.9	15.1	16.8[a]	215	894
310	**Ice-cream**, non-dairy, vanilla	14 samples, hard and soft scoop	1.00	66.5	0.48	3.0	7.8	18.8[a]	153	640
311	**Lollies**, containing ice-cream	3 samples	1.00	75.2	0.19	1.4	3.8	20.9	118	499
312	with real fruit juice	10 samples, 6 brands, assorted flavours	1.00	77.8	0.02	0.1	0.3	18.6[a]	73	310
313	**Sorbet**, fruit	10 samples, assorted flavours	1.00	68.9	0.03	0.2	0.3	24.8[a]	97	411
	Puddings and chilled desserts									
314	**Banoffee pie**	10 samples, 6 brands including 2 Mississippi mud pie	1.00	41.4	0.61	3.8	20.0	32.9[a]	319	1331
315	**Cheesecake**, *frozen*	10 samples, assorted flavours, fruit topping	1.00	43.6	0.64	4.0	16.2	35.2[a]	294	1231
316	fruit, *individual*	8 samples, 3 brands including strawberry, apricot, blackcurrant and cherry	1.00	46.6	0.97	6.1	12.3	34.5	264	1111

[a] Including oligosaccharides from the glucose syrup/maltodextrins in the product

[b] Strawberry flavour contains 46.8g water, 3.3g protein, 11.8g fat, 34.3g carbohydrate (9.0g starch, 23.8g sugars), 248kcal and 1042kJ energy per 100g

Composition of food per 100g edible portion

No.	Food	Starch	Total sugars	Individual sugars: Gluc	Fruct	Sucr	Malt	Lact	Dietary fibre NSP	Fatty acids: Satd	Mono-unsatd	Poly-unsatd	Trans	Cholest-erol
		g	g	g	g	g	g	g	g	g	g	g	g	mg
	Ice creams													
302	**Choc ice**	0.7	20.5[a]	0.3	Tr	13.1	2.4	4.7	Tr	18.4	1.9	0.4	0.1	7
303	**Chocolate nut sundae**	1.3	24.5	7.0	4.8	10.1	0	2.7	0.2	8.6	4.3	0.9	N	28
304	**Cornetto-type ice-cream cone**	8.7	18.9[a]	0.8	Tr	14.4	0.6	3.1	0.3	13.2	3.2	0.6	0.7	15
305	**Frozen ice-cream desserts**	0.7	19.7[a]	0.5	Tr	13.8	1.3	4.1	Tr	14.2	2.2	0.5	0.2	(4)
306	**Ice-cream bar**, chocolate coated	0.6	21.2	2.2	0.1	18.9	0	0	Tr	12.5	7.8	1.3	0.7	N
307	**Ice-cream wafers**	77.7	1.1	0.1	0.1	0.7	0.2	0	N	N	N	N	N	0
308	**Ice-cream**, dairy, vanilla	Tr	18.7[a]	2.0	0	11.5	Tr	5.2	Tr[b]	6.1	2.8	0.3	0.8	24
309	dairy, premium	Tr	16.7[a]	Tr	Tr	12.0	Tr	4.7	Tr	9.1	4.4	0.6	0.7	N
310	**Ice-cream**, non-dairy, vanilla	0	18.0[a]	3.8	Tr	9.4	Tr	4.8	Tr[b]	4.8	2.2	0.4	0.3	7
311	**Lollies**, containing ice-cream	0	20.9	2.9	0.4	12.6	0	5.0	0	2.1	1.2	0.3	Tr	4
312	with real fruit juice	0	17.8[a]	1.0	0.9	15.9	0	0	0	N	N	N	N	N
313	**Sorbet**, fruit	0	23.3[a]	3.6	1.6	17.2	0.8	0	Tr	N	N	N	N	0
	Puddings and chilled desserts													
314	**Banoffee pie**	11.4	20.9[a]	1.4	1.3	15.0	0.8	2.3	2.5	N	N	N	N	N
315	**Cheesecake**, *frozen*	10.0	25.0[a]	1.8	1.6	19.4	0.9	1.3	0.8	9.4	5.0	0.8	0.7	92
316	fruit, *individual*	9.1	25.4	2.3	2.0	18.1	0.4	2.6	1.0	7.5	3.5	0.5	0.2	15

[a] Not including oligosaccharides from the glucose syrup/maltodextrins in the product
[b] Gums and cellulose derivatives are added as stablilisers

Inorganic constituents per 100g edible portion

No.	Food	mg										µg	
		Na	K	Ca	Mg	P	Fe	Cu	Zn	Cl	Mn	Se	I
	Ice creams												
302	**Choc ice**	70	189	84	22	87	0.3	0.10	0.3	N	0.1	N	N
303	**Chocolate nut sundae**	85	134	50	16	65	0.3	0.05	0.2	84	0.1	2	24
304	**Cornetto-type ice-cream cone**	69	181	84	20	94	0.7	0.13	0.4	209	0.3	2	26
305	**Frozen ice-cream desserts**	62	234	93	21	94	0.2	0.13	0.4	110	0.1	N	20
306	**Ice-cream bar**, chocolate coated	91	250	140	31	145	0.7	0.04	0.7	175	0.2	N	160
307	**Ice-cream wafers**	93	190	170	46	130	2.0	0.11	0.7	130	0.7	N	N
308	**Ice-cream**, dairy, vanilla	60	174	100	12	91	Tr	Tr	0.3	110	Tr	2	32
309	dairy, premium	(60)	(174)	(100)	(12)	(91)	Tr	Tr	(0.3)	N	Tr	(2)	57
310	**Ice-cream**, non-dairy, vanilla	62	164	72	11	74	0.1	Tr	0.2	107	Tr	2	36
311	**Lollies**, containing ice-cream	31	69	49	6	39	0.2	Tr	0.1	53	Tr	Tr	84
312	with real fruit juice	11	28	5	2	3	Tr	Tr	0	114	Tr	Tr	Tr
313	**Sorbet**, fruit	10	41	8	4	4	0.1	0.04	0	16	0.2	1	Tr
	Puddings and chilled desserts												
314	**Banoffee pie**	164	163	84	15	89	0.3	0.08	0.4	N	0.2	N	N
315	**Cheesecake**, *frozen*	146	96	56	9	64	0.5	0.06	0.4	220	0.2	(2)	9
316	fruit, *individual*	151	165	78	13	100	0.4	Tr	0.5	258	0.1	2	26

No.	Food	Retinol µg	Carotene µg	Vitamin D µg	Vitamin E mg	Thiamin mg	Riboflavin mg	Niacin mg	Trypt/60 mg	Vitamin B_6 mg	Vitamin B_{12} µg	Folate µg	Pantothenate mg	Biotin µg	Vitamin C mg
	Ice creams														
302	**Choc ice**	Tr	Tr	Tr	0.27	N	N	N	0.8	N	N	N	N	N	N
303	**Chocolate nut sundae**	140	89	0	0.75	0.08	0.17	0.3	0.5	0.02	0.4	7	0.29	3.8	1
304	**Cornetto-type ice-cream cone**	27	15	1.2	0.06	0.07	0.18	0.4	0.9	0.03	0.7	9	0.47	2.5	N
305	**Frozen ice-cream desserts**	2	5	0.3	Tr	0.04	0.20	0.2	0.8	0.06	0.4	3	N	N	0
306	**Ice-cream bar**, chocolate coated	78	47	0.2	(1.10)	0.05	0.29	0.6	1.4	0.04	0.4	12	0.50	6.5	0
307	**Ice-cream wafers**	0	0	0	N	0.20	0.04	2.3	2.1	0.15	0	15	N	N	0
308	**Ice-cream**, dairy, vanilla	91	45	0.5	0.49	0.10	0.28	0.2	0.9	0.04	0.5	6	1.05	2.2	1
309	dairy, premium	164	80	0.3	0.26	(0.10)	(0.28)	(0.2)	(0.9)	(0.04)	(0.5)	(6)	(1.05)	(2.2)	(1)
310	**Ice-cream**, non-dairy, vanilla	1	5	Tr	0.60	0.14	0.26	0.2	0.7	Tr	0.7	8	0.43	3.0	1
311	**Lollies**, containing ice-cream	14	9	0.5	0.51	0.02	0.09	0.1	0.4	0.04	0.2	8	0.30	3.0	Tr
312	with real fruit juice	Tr	Tr	Tr	Tr	0.04	N	N	N	Tr	Tr	N	N	N	7
313	**Sorbet**, fruit	Tr	95	0	Tr	0.04	Tr	0.2	0.2	Tr	Tr	5	0.08	0	12
	Puddings and chilled desserts														
314	**Banoffee pie**	105	70	N	1.11	0.09	0.12	0.5	0.8	Tr	0.3	5	N	N	N
315	**Cheesecake**, *frozen*	97	50	0.2	1.19	0.07	0.09	0.5	0.9	0.02	0.5	7	N	N	6
316	fruit, *individual*	N	Tr	N	1.29	0.12	0.14	0.5	0.8	0.04	Tr	(7)	0.35	1.4	Tr

Composition of food per 100g edible portion

No.	Food	Description and main data sources	Edible conversion factor	Water	Total nitrogen	Protein	Fat	Carbo-hydrate	Energy value	
				g	g	g	g	g	kcal	kJ
	Puddings and chilled desserts *continued*									
317	**Chocolate dairy desserts**	8 samples, 4 brands including milk chocolate and caramel and white chocolate dessert pots, chilled	1.00	58.8	0.68	4.3	10.7	26.7	214	896
318	**Creme caramel**	9 samples, 4 brands	1.00	72.0	0.47	3.0	1.6	20.6	104	440
319	**Custard**, *made up with whole milk*	Recipe	1.00	75.2	0.60	3.9	4.5	16.2	118	494
320	*made up with semi-skimmed milk*	Recipe	1.00	77.5	0.63	4.0	2.0	16.4	95	404
321	**Custard**, ready to eat	10 samples, 3 brands, canned and tetra-pak; ambient	1.00	77.5	0.43	2.7	2.9	16.3	98	414
322	**Instant dessert powder**	10 samples, 2 types, assorted flavours	1.00	1.0	0.39	2.4	17.3	60.1	391	1643
323	**Jelly**, *made with water*	Recipe	1.00	84.0	0.21	1.2	0	15.1	61	260
324	**Meringue**	Recipe	1.00	1.7	0.84	5.3	Tr	96.0	381	1625
325	with cream	Recipe. Ref. Wiles et al. (1980)	1.00	46.5	0.41	2.6	24.2	27.2	330	1375
326	**Milk pudding**, *made with whole milk*	e.g. rice, sago, semolina, tapioca; recipe	1.00	72.2	0.64	4.1	4.3	19.6	130	545
327	**Mousse**, chocolate	10 samples, 4 brands, fresh	1.00	67.3	0.63	4.0	6.5	19.9	149	627
328	chocolate, reduced fat	7 samples, 4 brands	1.00	69.0	0.86	5.5	3.7	18.0	123	518
329	fruit	8 samples, assorted flavours, fresh	1.00	71.7	0.71	4.5	6.4	18.0	143	601
330	**Pavlova**, with fruit and cream	12 samples, 7 brands, including raspberry, strawberry and tropical fruits, frozen	1.00	39.6	0.43	2.7	13.2	42.2[a]	288	1210
331	no fruit	10 samples, 6 brands, frozen	1.00	26.0	0.61	3.8	19.7	47.4[a]	370	1552
332	**Profiteroles with sauce**	10 samples, 7 brands, frozen	1.00	39.6	0.88	5.5	25.7	24.6[a]	345	1436

[a] Including oligosaccharides from the glucose syrup/maltodextrins in the product

Composition of food per 100g edible portion

No.	Food	Starch	Total sugars	Individual sugars					Dietary fibre	Fatty acids				Cholest-erol
				Gluc	Fruct	Sucr	Malt	Lact	NSP	Satd	Mono-unsatd	Poly-unsatd	Trans	
		g	g	g	g	g	g	g	g	g	g	g	g	mg
	Puddings and chilled desserts *continued*													
317	**Chocolate dairy desserts**	2.7	24.0	0.4	0.2	17.3	0.7	5.4	Tr	6.3	3.3	0.4	0.5	21
318	**Creme caramel**	2.6	18.0	2.3	1.3	10.3	0.5	3.5	N	0.9	0.5	0.1	0.2	N
319	**Custard**, *made up with whole milk*	5.1	11.1	Tr	0	5.9	0	5.2	0	2.9	1.2	0.2	0.2	16
320	*made up with semi-skimmed milk*	5.1	11.3	Tr	0	5.9	0	5.4	0	1.2	0.5	0.1	0.1	7
321	**Custard**, canned	3.5	12.8	Tr	Tr	8.2	Tr	4.6	(0.1)	0	0.8	0.1	0.1	2
322	**Instant dessert powder**	19.4	40.7	Tr	Tr	38.3	0	2.2	(1.0)	15.9	0.3	0.2	N	1
323	**Jelly**, *made with water*	0	15.1	3.5	1.7	8.6	1.3	0	0	0	0	0	0	0
324	**Meringue**	0	96.0	Tr	0	96.0	0	0	0	Tr	Tr	Tr	Tr	0
325	with cream	0	27.2	Tr	0	25.6	0	1.6	0	15.1	7.0	0.7	N	63
326	**Milk pudding**, *made with whole milk*	9.2	10.4	Tr	Tr	5.5	0	4.9	0.1	2.7	1.1	0.2	0.2	15
327	**Mousse**, chocolate	2.4	17.5	1.1	1.8	10.8	0	3.8	N	3.3	2.7	0.1	1.3	N
328	chocolate, reduced fat	2.2	15.8	Tr	6.3	3.8	Tr	5.7	N	2.5	0.9	0.1	0	1
329	fruit	Tr	18.0	3.1	2.9	7.6	0.3	3.7	N	4.1	1.8	0.1	0.7	N
330	**Pavlova**, with fruit and cream	1.1	41.0[a]	2.5	1.7	35.9	Tr	0.9	0.3	7.3	4.6	0.7	0.9	30
331	no fruit	0.3	45.3[a]	2.1	0.8	40.1	1.1	1.1	(0.3)	10.8	6.8	1.0	1.3	45
332	**Profiteroles with sauce**	6.4	17.0[a]	1.8	0.8	11.7	1.2	1.5	N	14.0	8.7	1.7	1.3	N

[a] Not including oligosaccharides from the glucose syrup/maltodextrins in the product

Inorganic constituents per 100g edible portion

No.	Food	mg										µg	
		Na	K	Ca	Mg	P	Fe	Cu	Zn	Cl	Mn	Se	I
	***Puddings and chilled desserts** continued*												
317	**Chocolate dairy desserts**	74	195	135	20	125	0.4	0.07	0.5	N	0.1	1	26
318	**Creme caramel**	70	150	94	9	77	Tr	Tr	0.3	100	Tr	N	33
319	**Custard**, *made up with whole milk*	67	182	138	13	110	0.1	0.01	0.5	129	N	N	N
320	*made up with semi-skimmed milk*	67	184	140	13	111	0.1	0.01	0.5	127	N	N	N
321	**Custard**, canned	41	129	91	9	83	0.1	Tr	0.3	137	Tr	1	26
322	**Instant dessert powder**	1100	64	20	11	650	0.5	0.20	0.4	45	0.1	N	Tr
323	**Jelly**, *made with water*	5	5	7	Tr	1	0.4	0.01	N	6	N	N	N
324	**Meringue**	116	92	12	8	20	0.2	0.12	0.1	99	Tr	4	2
325	with cream	46	76	38	6	41	0.1	0.03	0.2	62	Tr	N	N
326	**Milk pudding**, *made with whole milk*	47	176	130	13	109	0.1	0.02	0.5	98	N	N	N
327	**Mousse**, chocolate	67	220	97	28	100	0.2	0.12	0.6	86	0.2	N	N
328	chocolate, reduced fat	69	301	126	33	133	1.2	0.12	0.8	191	0.1	2	66
329	fruit	62	150	120	12	96	Tr	Tr	0.4	110	Tr	N	N
330	**Pavlova**, with fruit and cream	41	80	26	6	27	0.3	Tr	0.1	N	0.1	N	N
331	no fruit	67	133	44	15	48	0.5	0.11	0.3	N	0.3	N	N
332	**Profiteroles with sauce**	130	190	58	25	114	1.5	0.18	0.6	209	0.2	5	12

No.	Food	Retinol µg	Carotene µg	Vitamin D µg	Vitamin E mg	Thiamin mg	Riboflavin mg	Niacin mg	Trypt/60 mg	Vitamin B_6 mg	Vitamin B_{12} µg	Folate µg	Pantothenate mg	Biotin µg	Vitamin C mg
	Puddings and chilled desserts *continued*														
317	**Chocolate dairy desserts**	83	52	Tr	0.52	0.04	0.28	0.2	1.0	0.03	0.6	1	0.50	1.9	Tr
318	**Creme caramel**	37	8	0.1	0.03	0.03	0.20	0.1	0.7	0.03	0.3	8	N	N	0
319	**Custard**, *made up with whole milk*	38	23	Tr	0.07	0.03	0.24	0.2	0.7	0.06	1.0	7	0.60	2.9	1
320	*made up with semi-skimmed milk*	21	10	Tr	0.04	0.04	0.25	0.1	0.9	0.06	0.4	6	0.33	2.3	1
321	**Custard**, canned	36	376	Tr	0.29	0.12	0.19	0.1	0.3	0.01	0.2	2	0.43	1.3	0
322	**Instant dessert powder**	N	N	N	N	Tr	0.01	Tr	0.5	Tr	0.3	Tr	N	N	0
323	**Jelly**, *made with water*	0	0	0	0	0	0	0	0	0	0	0	0	0	0
324	**Meringue**	0	0	0	0	0	0.21	0.1	1.5	0.01	0.1	4	0.13	4.1	0
325	with cream	239	148	0.2	0.79	0.01	0.16	0	0.7	0.03	0.1	5	0.17	1.9	1
326	**Milk pudding**, *made with whole milk*	36	21	Tr	0.09	0.03	0.22	0.3	0.8	0.06	1.0	5	0.49	2.8	2
327	**Mousse**, chocolate	46	11	Tr	1.01	0.04	0.21	0.2	0.9	0.04	0.2	6	(0.74)	(2.1)	0
328	chocolate, reduced fat	N	Tr	Tr	0.79	0.12	0.26	(0.2)	0.8	0.01	0	(0)	0.74	2.1	0
329	fruit	36	16	0.1	0.39	0.04	0.23	0.2	1.1	0.05	0.2	6	N	N	Tr
330	**Pavlova**, with fruit and cream	119	70	N	0.33	0.03	0.18	0.2	1.0	Tr	Tr	10	0.38	1.3	5
331	no fruit	155	50	Tr	1.00	0.06	0.12	0.2	0.8	Tr	Tr	8	N	N	0
332	**Profiteroles with sauce**	114	90	0.3	1.18	0.07	0.14	0.3	1.2	0.02	0.3	9	0.63	3.7	Tr

Composition of food per 100g edible portion

No.	Food	Description and main data sources	Edible conversion factor	Water	Total nitrogen	Protein	Fat	Carbo-hydrate	Energy value	
				g	g	g	g	g	kcal	kJ
Puddings and chilled desserts *continued*										
333	**Rice pudding**, canned	10 cans, 7 brands	1.00	79.2	0.53	3.3	1.3	16.1	85	362
334	canned, low fat	10 samples, 6 brands	1.00	82.1	0.56	3.5	0.8	13.4	71	304
335	**Torte**, fruit	8 samples, 5 brands including lemon, raspberry and passion fruit	1.00	53.0	0.60	3.8	15.5	27.7	258	1080
336	**Trifle**	Recipe	1.00	66.5	0.43	2.6	8.1	21.0	166	696
337	fruit	12 samples, 7 brands	1.00	67.9	0.41	2.6	9.0	19.5	164	689

No.	Food	Starch	Total sugars	Individual sugars					Dietary fibre	Fatty acids				Cholest-erol
				Gluc	Fruct	Sucr	Malt	Lact	NSP	Satd	Mono-unsatd	Poly-unsatd	Trans	
		g	g	g	g	g	g	g	g	g	g	g	g	mg
	Puddings and chilled desserts *continued*													
333	**Rice pudding**, canned	7.3	8.7	Tr	Tr	4.9	Tr	3.9	0.1	0.8	0.3	0.1	Tr	(9)
334	canned, low fat	7.2	6.1	Tr	Tr	1.8	Tr	4.3	(0.1)	(0.5)	(0.2)	(0.1)	Tr	N
335	**Torte**, fruit	9.9	17.4	1.3	0.6	13.0	0.7	1.8	0.5	9.4	4.7	1.2	0.6	42
336	**Trifle**	4.3	16.7	2.7	2.0	8.9	0.8	2.2	0.4	2.4	2.6	1.7	N	21
337	fruit	4.2	15.3	1.8	4.6	6.2	Tr	2.7	2.1	5.6	2.5	0.4	0.2	13

Inorganic constituents per 100g edible portion

No.	Food	mg										µg	
		Na	K	Ca	Mg	P	Fe	Cu	Zn	Cl	Mn	Se	I
	Puddings and chilled desserts *continued*												
333	**Rice pudding**, canned	43	130	88	12	86	0.1	0.13	0.5	93	0.1	N	28
334	canned, low fat	(43)	(130)	(88)	(12)	(86)	(0.1)	(0.13)	(0.5)	(93)	(0.1)	N	(28)
335	**Torte**, fruit	88	116	66	10	77	0.3	0.34	0.5	N	0.1	N	N
336	**Trifle**	70	119	57	11	71	0.3	0.04	0.3	125	0.1	1	N
337	fruit	65	137	73	10	84	0.2	0.02	0.3	95	Tr	2	37

Vitamins per 100g edible portion

No.	Food	Retinol µg	Carotene µg	Vitamin D µg	Vitamin E mg	Thiamin mg	Ribo-flavin mg	Niacin mg	Trypt/60 mg	Vitamin B_6 mg	Vitamin B_{12} µg	Folate µg	Panto-thenate mg	Biotin µg	Vitamin C mg
	Puddings and chilled desserts *continued*														
333	**Rice pudding**, canned	16	10	Tr	0.16	0.01	0.13	0.2	0.6	0.01	Tr	0	0.30	2.0	0
334	canned, low fat	(16)	(10)	Tr	(0.10)	(0.01)	(0.13)	(0.2)	(0.7)	(0.01)	Tr	Tr	(0.30)	(2.0)	Tr
335	**Torte**, fruit	99	77	Tr	1.43	0.03	0.09	0.2	0.5	0.02	Tr	3	N	N	Tr
336	**Trifle**	82	216	0.3	1.68	0.08	0.12	0.3	0.5	0.03	0.2	6	0.31	3.1	5
337	fruit	N	Tr	N	0.66	0.12	0.08	0.1	0.5	0.01	0.2	(8)	0.29	0.7	Tr

Section 2.3

Eggs and egg dishes

The eggs and egg dishes in this section of the Tables are taken from the *Milk Products and Eggs* (1989) supplement.

Although most of the nutrients in eggs have been analysed, a few of the values for cooked eggs were derived by calculation from the amounts in raw eggs. Allowances have been made for any water loss or fat uptake in cases where eggs were cooked with fat.

The nutrient content of eggs may vary by rearing method (e.g. battery, deep litter, free range) and by the type of feed used (e.g. for vitamin D).

Losses of labile vitamins assigned on recipe calculation were estimated using the figures in Section 4.3.

Composition of food per 100g edible portion

No.	Food	Description and main data sources	Edible conversion factor	Water	Total nitrogen	Protein	Fat	Carbo-hydrate	Energy value	
				g	g	g	g	g	kcal	kJ
Eggs										
338	**Eggs**, chicken, *raw*	Analysis of battery, deep litter and free range[a]	1.00	75.1	2.01	12.5	11.2	Tr	151	627
339	white, *raw*	34 eggs and literature sources	1.00	88.3	1.44	9.0	Tr	Tr	36	153
340	yolk, *raw*	34 eggs and literature sources	1.00	51.0	2.58	16.1	30.5	Tr	339	1402
341	chicken, *boiled*	10 eggs	1.00	75.1	2.01	12.5	10.8	Tr	147	612
342	*fried in vegetable oil*	12 eggs, shallow fried	1.00	70.1	2.18	13.6	13.9	Tr	179	745
343	*poached*[b]	10 eggs, no fat added	1.00	75.1	2.01	12.5	10.8	Tr	147	612
344	*scrambled, with milk*	Recipe	1.00	63.4	1.75	10.9	23.4	0.7	257	1062
345	duck, whole, *raw*	Analytical and literature sources. Ref. Posati and Orr (1976)	1.00	70.6	2.29	14.3	11.8	Tr	163	680
Egg dishes										
346	**Omelette**, plain	Recipe	1.00	68.9	1.75	10.9	16.8	Tr	195	808
347	cheese	Recipe. Ref. Wiles *et al.* (1980)	1.00	57.8	2.51	15.9	23.0	Tr	271	1121
348	**Quiche**, cheese and egg	Recipe. Ref. Wiles *et al.* (1980)	1.00	47.1	2.00	12.4	22.3	17.1	315	1310
349	cheese and egg, wholemeal	Recipe	1.00	47.1	2.11	13.1	22.5	14.3	309	1284
350	**Quiche**, Lorraine	Recipe	1.00	39.8	2.20	13.7	25.5	19.6	358	1488

[a] An average egg is composed of 11% shell, 58% white and 31% yolk

[b] Eggs poached with fat added contain 74.4g water, 12.4g protein, 11.7g fat, Tr carbohydrate, 155 kcals and 644 kJ per 100g

Composition of food per 100g edible portion

No.	Food	Starch	Total sugars	Individual sugars					Dietary fibre	Fatty acids				Cholest-erol
				Gluc	Fruct	Sucr	Malt	Lact	NSP	Satd	Mono-unsatd	Poly-unsatd	Trans	
		g	g	g	g	g	g	g	g	g	g	g	g	mg
	Eggs													
338	**Eggs**, chicken, *raw*	0	Tr	Tr	0	0	0	0	0	3.2	4.4	1.7	0.1	391
339	white, *raw*	0	Tr	Tr	0	0	0	0	0	Tr	Tr	Tr	Tr	0
340	yolk, *raw*	0	Tr	Tr	0	0	0	0	0	8.7	13.2	3.4	N	1120
341	chicken, *boiled*	0	Tr	Tr	0	0	0	0	0	3.1	4.7	1.2	N	385
342	*fried in vegetable oil*	0	Tr	Tr	0	0	0	0	0	4.0	6.0	1.5	N	435
343	*poached*	0	Tr	Tr	0	0	0	0	0	3.1	4.7	1.2	N	385
344	*scrambled, with milk*	0	0.7	Tr	0	0	0	0.7	0	11.6	7.3	1.9	0.6	361
345	duck, whole, *raw*	0	Tr	Tr	0	0	0	0	0	2.9	4.9	2.0	0.1	680
	Egg dishes													
346	**Omelette**, plain	0	0	Tr	0	0	0	0	0	7.2	5.6	1.7	N	357
347	cheese	0	Tr	Tr	0	0	0	Tr	0	12.2	6.9	1.5	N	268
348	**Quiche**, cheese and egg	15.6	1.5	Tr	Tr	0.1	0	1.2	0.6	9.9	7.4	3.2	N	133
349	cheese and egg, wholemeal	12.6	1.7	0	Tr	0.2	0	1.2	1.8	10.0	7.4	3.3	N	133
350	**Quiche**, Lorraine	17.4	2.1	Tr	Tr	0.1	0	1.8	0.7	10.6	9.0	4.0	N	116

Inorganic constituents per 100g edible portion

No.	Food	mg										µg	
		Na	K	Ca	Mg	P	Fe	Cu	Zn	Cl	Mn	Se	I
	Eggs												
338	**Eggs**, chicken, *raw*	140	130	57	12	200	1.9	0.08	1.3	160	Tr	11	53
339	white, *raw*	190	150	5	11	33	0.1	0.02	0.1	170	Tr	6	(3)
340	yolk, *raw*	50	120	130	15	500	6.1	0.15	3.9	140	0.1	20	(140)
341	chicken, *boiled*	140	130	57	12	200	1.9	0.08	1.3	160	Tr	11	53
342	*fried in vegetable oil*	160	150	65	14	230	2.2	0.09	1.5	180	Tr	12	60
343	*poached*	140	130	57	12	200	1.9	0.08	1.3	160	Tr	11	53
344	*scrambled, with milk*	222	137	67	12	182	1.6	0.07	1.1	308	Tr	9	54
345	duck, whole, *raw*	120	190	63	16	200	2.9	N	1.4	N	(0.1)	N	N
	Egg dishes												
346	**Omelette**, plain	1024	117	51	12	175	1.7	0.07	1.1	1521	Tr	10	50
347	cheese	921	103	287	18	288	1.2	0.06	2.1	1356	Tr	9	43
348	**Quiche**, cheese and egg	366	124	262	18	227	1.0	0.06	1.6	550	0.1	5	N
349	cheese and egg, wholemeal	347	163	242	38	270	1.4	0.12	2.1	511	0.6	6	N
350	**Quiche**, Lorraine	572	182	231	20	223	1.0	0.07	1.6	796	0.2	5	N

Vitamins per 100g edible portion

No.	Food	Retinol µg	Carotene µg	Vitamin D µg	Vitamin E mg	Thiamin mg	Riboflavin mg	Niacin mg	Trypt/60 mg	Vitamin B_6 mg	Vitamin B_{12} µg	Folate µg	Pantothenate mg	Biotin µg	Vitamin C mg
Eggs															
338	**Eggs**, chicken, *raw*	190	Tr	1.8[a]	1.11	0.09	0.47	0.1	3.7	0.12	2.5	50	1.77	20.0	0
339	white, *raw*	0	0	0	0	0.01	0.43	0.1	2.6	0.02	0.1	13	0.30	7.0	0
340	yolk, *raw*	535	Tr	4.9[a]	3.11	0.30	0.54	0.1	4.7	0.30	6.9	130	4.60	50.0	0
341	chicken, *boiled*	190	Tr	1.8[a]	1.11	0.07	0.35	0.1	3.7	0.12	1.1	39	1.30	16.0	0
342	*fried in vegetable oil*	215	Tr	2.0[a]	N	0.07	0.31	0.1	4.0	0.14	1.6	40	1.30	18.0	0
343	*poached*	190	Tr	1.8[a]	1.11	0.07	0.36	0.1	3.7	0.12	1.0	45	1.30	15.0	0
344	*scrambled, with milk*	320	104	1.6[a]	1.23	0.07	0.35	0.1	3.1	0.09	2.2	30	1.32	16.9	Tr
345	duck, whole, *raw*	540	120	5.0	N	0.16	0.47	0.2	4.2	0.25	5.4	80	N	N	0
Egg dishes															
346	**Omelette**, plain	247	53	1.6	1.12	0.07	0.33	0.1	3.2	0.09	2.2	30	1.31	17.3	Tr
347	cheese	287	83	1.2	0.91	0.06	0.35	0.1	4.4	0.11	2.3	30	1.03	12.9	Tr
348	**Quiche**, cheese and egg	184	59	0.9	2.21	0.09	0.23	0.4	3.2	0.09	1.5	16	0.61	7.1	1
349	cheese and egg, wholemeal	184	59	0.9	2.44	0.13	0.23	1.3	3.3	0.17	1.5	23	0.71	8.3	1
350	**Quiche**, Lorraine	157	55	1.0	2.34	0.17	0.24	1.2	3.2	0.14	1.4	15	0.70	6.2	1

[a] If the hens have been fed a supplement, values may be considerably higher

Section 2.4

Fats and oils

Most data in this section are derived from the *Miscellaneous Foods* (1994) supplement, although there are a few new values obtained from analysis or from manufacturers. Foods for which new data are incorporated have been allocated a new food code and can thus easily be identified in the food index.

Most oils show a wide range of fatty acid composition depending on the variety, growing conditions and maturity of the seed. In addition, the blend of fats and oils used in many of the foods included in this section can frequently be adjusted by manufacturers and this will alter the fatty acid composition. If accurate fatty acid data are required for specific products, and analytical facilities are not available, it is advisable to contact the manufacturer directly.

The profile for fatty acids in 'vegetable oil' was calculated from the values for the component soya, rape, and corn oils, the proportions of which may vary, and this profile, which is of doubtful value, has been included only to aid recipe calculation and survey work where unidentified oil has been consumed.

Composition of food per 100g edible portion

No.	Food	Description and main data sources	Edible conversion factor	Water	Total nitrogen	Protein	Fat	Carbo-hydrate	Energy value	
				g	g	g	g	g	kcal	kJ
Spreading fats										
351	**Butter**	Average of UK/Irish, Danish, French, New Zealand, salted and unsalted	1.00	14.9	0.10	0.6	82.2	0.6	744	3059
352	spreadable	8 samples, different brands	1.00	15.5	0.08	0.5	82.5	Tr	745	3061
353	**Blended spread** (70–80% fat)	30 samples including Clover, Golden Crown and Willow	1.00	21.0	0.10	0.6	74.8	1.1	680	2795
354	(40% fat)	20 samples including Anchor half fat butter and Clover Extra Light	1.00	51.4	1.02	6.5	40.3	0.4	390	1608
355	**Dairy spread** (40% fat)	Manufacturers' data on own brands	1.00	(52.9)	1.10	7.0	40.0	0.1	388	1601
356	**Margarine**, hard, animal and vegetable fats	10 samples of Echo and Stork	1.00	16.0	0.03	0.2	79.3	1.0	718	2954
357	hard, vegetable fats only	4 samples of Tomor. Analysis and manufacturer's data (Rakusens Ltd)	1.00	16.0	0.03	0.2	82.3	0	742	3049
358	soft, not polyunsaturated	20 samples of a mixture of Stork SB and own brands soft margarine	1.00	16.0	0.03	0.2	81.7	1.0	740	3042
359	soft, polyunsaturated	20 samples of a mixture of Blue Band and own brands soya margarine	1.00	16.0	Tr	Tr	82.8	0.2	746	3067
360	**Fat spread** (70–80% fat), not polyunsaturated	10 samples including Krona Gold	1.00	22.0	0.06	0.4	71.2	Tr	642	2641
361	(70% fat), polyunsaturated	Data from TRANSFAIR; 5 samples including Vitalite	1.00	(26.6)	(0.08)	(0.5)	68.5	(0.8)	622	2556

No.	Food	Starch	Total sugars	Individual sugars					Dietary fibre	Fatty acids				Cholest-erol
				Gluc	Fruct	Sucr	Malt	Lact	NSP	Satd	Mono-unsatd	Poly-unsatd	Trans	
		g	g	g	g	g	g	g	g	g	g	g	g	mg
	Spreading fats													
351	**Butter**	0	0.6	0	0	0	0	0.6	0	52.1	20.9	2.8	2.9	213
352	spreadable	0	Tr	0	0	0	0	Tr	0	45.4	22.7	3.5	2.8	280
353	**Blended spread** (70–80% fat)	0	1.1	0	0	0	0	1.1	0	25.5	37.5	8.5	4.3	67
354	(40% fat)	0	0.4	0	0	0	0	0.4	0	18.1	13.4	7.3	4.9	46
355	**Dairy spread** (40% fat)	0	0.1	0	0	0	0	0.1	0	26.8	10.0	1.2	N	N
356	**Margarine**, hard, animal and vegetable fats	0	1.0	0	0	0	0	1.0	0	34.6	36.2	5.4	12.2	285
357	hard, vegetable fats only	0	0	0	0	0	0	0	0	40.0	21.0	21.3	1.0	15
358	soft, not polyunsaturated	0	1.0	0	0	0	0	1.0	0	27.2	38.9	12.4	8.9	275
359	soft, polyunsaturated	0	0.2	0	0	0	0	0.2	0	17.0	26.6	36.0	6.7	2
360	**Fat spread** (70–80% fat), not polyunsaturated	0	0	0	0	0	0	0	0	30.4	31.2	6.5	11.5	86
361	(70% fat), polyunsaturated	0	(0.8)	0	0	0	0	(0.8)	0	16.2	15.2	33.6	0.3	Tr

Inorganic constituents per 100g edible portion

No.	Food	mg										µg	
		Na	K	Ca	Mg	P	Fe	Cu	Zn	Cl	Mn	Se	I
	Spreading fats												
351	**Butter**	606[a]	27	18	2	23	Tr	0.01	0.1	994	Tr	Tr	38
352	spreadable	390[b]	(15)	(15)	(2)	(24)	(0.2)	(0.03)	(0.1)	640	Tr	Tr	(38)
353	**Blended spread** (70–80% fat)	670	43	14	2	18	Tr	Tr	Tr	1010	Tr	N	N
354	(40% fat)	510	N	N	N	N	N	N	N	780	Tr	N	N
355	**Dairy spread** (40% fat)	600	N	N	N	N	N	N	N	(930)	Tr	N	N
356	**Margarine**, hard, animal and vegetable fats	940	5	4	1	12	0.3	0.04	N	1200	Tr	Tr	N
357	hard, vegetable fats only	590	5	4	1	12	0.3	0.04	N	1200	Tr	Tr	N
358	soft, not polyunsaturated	880	5	4	1	12	0.3	0.04	N	(1320)	Tr	Tr	N
359	soft, polyunsaturated	680	5	4	1	12	0.3	0.04	N	(1020)	Tr	Tr	N
360	**Fat spread** (70–80% fat), not polyunsaturated	1060	43	14	2	18	Tr	Tr	Tr	1270	Tr	N	N
361	(70% fat), polyunsaturated	(800)	N	N	N	N	Tr	Tr	N	(1200)	Tr	N	N

[a] Unsalted butter contains 9mg Na and 19mg Cl per 100g

[b] Average of salted and unsalted. Salted versions contain between 800 to 1500mg Na per 100g

No.	Food	Retinol µg	Carotene[a] µg	Vitamin D µg	Vitamin[b] E mg	Thiamin mg	Ribo-flavin mg	Niacin mg	Trypt/60 mg	Vitamin B_6 mg	Vitamin B_{12} µg	Folate µg	Panto-thenate mg	Biotin µg	Vitamin C mg
	Spreading fats														
351	**Butter**	958	608	0.9	1.85	Tr	0.07	Tr	0.1	Tr	0.3	Tr	0.05	0.2	Tr
352	spreadable	(815)	670	Tr	2.90	Tr	(0.02)	Tr	(0.1)	Tr	Tr	Tr	(0.04)	Tr	0
353	**Blended spread** (70–80% fat)	565	445	4.1	11.28	Tr	Tr	Tr	0.1	Tr	Tr	Tr	Tr	Tr	0
354	(40% fat)	160	430	0.2	3.88	Tr	Tr	Tr	Tr	Tr	Tr	Tr	Tr	Tr	0
355	**Dairy spread** (40% fat)	N	N	N	N	Tr	Tr	Tr	Tr	Tr	Tr	Tr	Tr	Tr	0
356	**Margarine**, hard, animal and vegetable fats	(665)	(750)	(7.9)	4.44	Tr	Tr	Tr	Tr	Tr	Tr	Tr	Tr	Tr	0
357	hard, vegetable fats only	(665)	(750)	(7.9)	N	Tr	Tr	Tr	Tr	Tr	0	Tr	Tr	Tr	0
358	soft, not polyunsaturated	745	445	7.8	12.34	Tr	Tr	Tr	Tr	Tr	Tr	Tr	Tr	Tr	0
359	soft, polyunsaturated	675	350	(7.9)	32.60	Tr	Tr	Tr	Tr	Tr	Tr	Tr	Tr	Tr	0
360	**Fat spread** (70–80% fat), not polyunsaturated	40	330	5.8	2.53	Tr	Tr	Tr	0.1	Tr	Tr	Tr	Tr	Tr	0
361	(70% fat), polyunsaturated	N	N	N	(38.00)	Tr	Tr	Tr	Tr	Tr	Tr	Tr	Tr	Tr	0

[a] Some brands may not contain β-carotene
[b] The vitamin E content will vary according to the type of oil

No.	Food	Description and main data sources	Edible conversion factor	Water	Total nitrogen	Protein	Fat	Carbo-hydrate	Energy value	
				g	g	g	g	g	kcal	kJ
***Spreading fats** continued*										
362	**Fat spread** (60% fat), polyunsaturated	10 samples including Vitalite Light	1.00	(37.7)	0.03	0.2	60.8	1.3	553	2274
363	(60% fat), with olive oil	5 samples including Olivio and own brands	1.00	38.2	0.02	0.1	62.7	1.1	569	2339
364	(40% fat), not polyunsaturated	20 samples including Gold and Delight. Fat data from TRANSFAIR	1.00	52.3	1.02	6.5	37.5	1.3	368	1519
365	(35–40% fat), polyunsaturated	Manufacturers' data (Gold Sunflower and Flora Extra Light)	1.00	53.8	0.76	4.9	37.6	1.8	365	1503
366	(20–25% fat), not polyunsaturated	20 samples including Gold Lowest and Outline. Fat data from TRANSFAIR	1.00	59.0	0.92	5.9	25.5	2.5	262	1084
367	polyunsaturated	Manufacturers' data on own brands	1.00	(78.7)	0	0	20.0	0.8	183	753
368	(5% fat)	Manufacturer's data (Tesco)	1.00	N	0.63	4.0	5.0	14.5[a]	115	484
Cooking fats										
369	**Compound cooking fat**	10 samples of a mixture of Cookeen and White Cap	1.00	Tr	Tr	Tr	99.9	0	899	3696
370	**Dripping**, beef	Data from TRANSFAIR; 5 samples, different brands	1.00	1.0	Tr	Tr	99.0	Tr	891	3663
371	**Ghee**, butter	5 assorted samples	1.00	0.1	Tr	Tr	99.8	Tr	898	3693
372	vegetable	5 samples; different types	1.00	0.1	Tr	Tr	99.4	Tr	895	3678
373	**Lard**	6 samples; 3 brands	1.00	1.0	Tr	Tr	99.0	0	891	3663
374	**Suet**, shredded	6 samples of the same brand	1.00	1.5	Tr	Tr	86.7	12.1	826	3402
375	vegetable	10 samples; 5 brands	1.00	0.8	0.19	1.2	87.9	10.1	836	3444

[a]Including maltodextrin

Composition of food per 100g edible portion

No.	Food	Starch	Total sugars	Individual sugars					Dietary fibre	Fatty acids				Cholest-erol
				Gluc	Fruct	Sucr	Malt	Lact	NSP	Satd	Mono-unsatd	Poly-unsatd	Trans	
		g	g	g	g	g	g	g	g	g	g	g	g	mg
Spreading fats *continued*														
362	**Fat spread** (60% fat), polyunsaturated	0	1.3	0	0	0	0	1.3	0	11.3	18.1	28.6	3.3	3
363	(60% fat), with olive oil	0	1.1	0	0	0	0	1.1	0	11.3	36.4	12.5	6.0	0
364	(40% fat), not polyunsaturated	0	1.3	0	0	0	0	1.3	0	8.4	21.0	6.2	4.4	6
365	(35–40% fat), polyunsaturated	0.6	1.3	0	0	0	0	1.3	0	8.9	9.4	18.0	0.7	Tr
366	(20–25% fat), not polyunsaturated	Tr	1.1	0	0	0	0	1.1	0	6.8	14.0	3.4	3.9	8
367	polyunsaturated	0.8	0	0	0	0	0	0	0	3.7	7.2	9.1	N	Tr
368	(5% fat)	0	4.8	0	0	0	0	4.8	(5.0)	6.0	1.1	3.2	N	N
Cooking fats														
369	**Compound cooking fat**	0	0	0	0	0	0	0	0	49.5	41.2	5.3	16.4	425
370	**Dripping**, beef	0	Tr	0	0	0	0	0	0	50.6	38.0	2.4	4.4	94
371	**Ghee**, butter	0	Tr	0	0	0	0	0	0	66.0	24.1	3.4	N	280
372	vegetable	0	Tr	0	0	0	0	0	0	48.4	37.0	9.7	1.1	0
373	**Lard**	0	0	0	0	0	0	0	0	40.3	43.4	10.0	Tr	93
374	**Suet**, shredded	11.9	0.2	0.1	0.1	Tr	Tr	0	0.5	49.9	30.4	2.2	(4.0)	82
375	vegetable	10.1	0	0	0	0	0	0	0	45.0	26.3	12.8	21.8	0

Inorganic constituents per 100g edible portion

No.	Food	mg										µg	
		Na	K	Ca	Mg	P	Fe	Cu	Zn	Cl	Mn	Se	I
	***Spreading fats** continued*												
362	**Fat spread** (60% fat), polyunsaturated	710	N	N	N	N	Tr	Tr	N	1070	Tr	N	N
363	(60% fat), with olive oil	600	N	N	N	N	Tr	Tr	N	910	Tr	N	N
364	(40% fat), not polyunsaturated	650	110	39	4	82	Tr	Tr	0.2	800	Tr	N	N
365	(35–40% fat), polyunsaturated	650	N	N	N	N	Tr	Tr	N	990	Tr	N	N
366	(20–25% fat), not polyunsaturated	540	630	N	N	N	Tr	Tr	N	(830)	Tr	N	N
367	polyunsaturated	500	N	N	N	N	Tr	Tr	N	(770)	Tr	N	N
368	(5% fat)	500	N	N	N	N	Tr	Tr	N	(770)	Tr	N	N
	Cooking fats												
369	**Compound cooking fat**	Tr	Tr	Tr	Tr	Tr	Tr	Tr	Tr	Tr	Tr	Tr	Tr
370	**Dripping**, beef	5	4	1	Tr	13	0.2	N	N	2	Tr	Tr	(5)
371	**Ghee**, butter	2	3	Tr	Tr	Tr	0.2	Tr	Tr	28	Tr	Tr	N
372	vegetable	1	1	Tr	Tr	Tr	Tr	0.14	Tr	N	Tr	N	N
373	**Lard**	2	1	1	1	3	0.1	0.02	N	4	Tr	Tr	Tr
374	**Suet**, shredded	Tr	Tr	Tr	Tr	Tr	Tr	Tr	Tr	Tr	Tr	N	5
375	vegetable	10	Tr	Tr	Tr	Tr	Tr	Tr	Tr	N	Tr	Tr	Tr

No.	Food	Retinol µg	Carotene[a] µg	Vitamin D µg	Vitamin[b] E mg	Thiamin mg	Riboflavin mg	Niacin mg	Trypt/60 mg	Vitamin B_6 mg	Vitamin B_{12} µg	Folate µg	Pantothenate mg	Biotin µg	Vitamin C mg
	***Spreading fats** continued*														
362	**Fat spread** (60% fat), polyunsaturated	980	Tr	N	30.75	Tr	Tr	Tr	Tr	Tr	Tr	Tr	Tr	Tr	0
363	(60% fat), with olive oil	N	N	N	N	Tr	Tr	Tr	Tr	Tr	0	Tr	Tr	Tr	0
364	(40% fat), not polyunsaturated	650	805	8.4	8.01	Tr	Tr	Tr	1.4	Tr	Tr	Tr	Tr	Tr	0
365	(35–40% fat), polyunsaturated	N	N	N	N	Tr	Tr	Tr	Tr	Tr	Tr	Tr	Tr	Tr	0
366	(20–25% fat), not polyunsaturated	470	575	7.8	5.11	Tr	Tr	Tr	1.9	Tr	Tr	Tr	Tr	Tr	0
367	polyunsaturated	N	N	N	N	Tr	Tr	Tr	Tr	Tr	Tr	Tr	Tr	Tr	0
368	(5% fat)	N	800	5.0	N	Tr	Tr	Tr	Tr	Tr	Tr	Tr	Tr	Tr	0
	Cooking fats														
369	**Compound cooking fat**	0	0	0	Tr	0	0	0	0	0	0	0	0	0	0
370	**Dripping**, beef	N	N	Tr	0.40	Tr	Tr	Tr	Tr	Tr	Tr	Tr	Tr	Tr	0
371	**Ghee**, butter	675	500	1.9	3.31	0	Tr	Tr	Tr	Tr	Tr	0	Tr	Tr	0
372	vegetable	Tr	Tr	0	10.27	0	0	Tr	Tr	Tr	0	0	Tr	Tr	0
373	**Lard**	Tr	0	N	1.00	Tr	Tr	Tr	Tr	Tr	Tr	Tr	Tr	Tr	0
374	**Suet**, shredded	52	73	Tr	1.50	Tr	Tr	Tr	Tr	Tr	Tr	Tr	Tr	Tr	0
375	vegetable	0	0	0	17.97	Tr	Tr	Tr	Tr	Tr	0	Tr	Tr	Tr	0

[a] Some brands may not contain β-carotene

[b] The vitamin E content will vary according to the type of oil

Composition of food per 100g edible portion

No.	Food	Description and main data sources	Edible conversion factor	Water	Total nitrogen	Protein	Fat	Carbo-hydrate	Energy value	
				g	g	g	g	g	kcal	kJ
Oils										
376	**Coconut oil**	Mean of 35 samples	1.00	Tr	Tr	Tr	99.9	0	899	3696
377	**Cod liver oil**	Mean of 20 samples	1.00	Tr	Tr	Tr	99.9	0	899	3696
378	**Corn oil**	Mean of 42 samples	1.00	Tr	Tr	Tr	99.9	0	899	3696
379	**Evening primrose oil**	Mean of 35 samples	1.00	Tr	Tr	Tr	99.9	0	899	3696
380	**Olive oil**	Mean of 35 samples; including virgin and extra virgin olive oil	1.00	Tr	Tr	Tr	99.9	0	899	3696
381	**Palm oil**	Mean of 55 samples	1.00	Tr	Tr	Tr	99.9	0	899	3696
382	**Peanut (Groundnut) oil**	Mean of 71 samples	1.00	Tr	Tr	Tr	99.9	0	899	3696
383	**Rapeseed oil**	Mean of 100 samples	1.00	Tr	Tr	Tr	99.9	0	899	3696
384	**Safflower oil**	Mean of 28 samples	1.00	Tr	Tr	Tr	99.9	0	899	3696
385	**Sesame oil**	Mean of 22 samples and literature sources	1.00	0.1	0.03	0.2	99.7	0	898	3692
386	**Soya oil**	Mean of 39 samples	1.00	Tr	Tr	Tr	99.9	0	899	3696
387	**Sunflower oil**	Mean of 46 samples	1.00	Tr	Tr	Tr	99.9	0	899	3696
388	**Vegetable oil**, blended, average	Data from the Institute of Human Nutrition and Brain Chemistry	1.00	Tr	Tr	Tr	99.9	0	899	3696
389	**Walnut oil**	Mean of 13 samples	1.00	Tr	Tr	Tr	99.9	0	899	3696
390	**Wheatgerm oil**	Mean of 35 samples	1.00	Tr	Tr	Tr	99.9	0	899	3696

Composition of food per 100g edible portion

No.	Food	Starch	Total sugars	Individual sugars Gluc	Fruct	Sucr	Malt	Lact	Dietary fibre NSP	Fatty acids Satd	Mono-unsatd	Poly-unsatd	Trans	Cholest-erol
		g	g	g	g	g	g	g	g	g	g	g	g	mg
Oils														
376	**Coconut oil**	0	0	0	0	0	0	0	0	86.5	6.0	1.5	Tr	0
377	**Cod liver oil**	0	0	0	0	0	0	0	0	21.1	44.6	30.5	Tr	(570)
378	**Corn oil**	0	0	0	0	0	0	0	0	14.4	29.9	51.3	Tr	0
379	**Evening primrose oil**	0	0	0	0	0	0	0	0	7.8	10.6	76.6	Tr	0
380	**Olive oil**	0	0	0	0	0	0	0	0	14.3	73.0	8.2	0	0
381	**Palm oil**	0	0	0	0	0	0	0	0	47.8	37.1	10.4	Tr	0
382	**Peanut (Groundnut) oil**	0	0	0	0	0	0	0	0	20.0	44.4	31.0	Tr	0
383	**Rapeseed oil**	0	0	0	0	0	0	0	0	6.6	59.3	29.3	Tr	0
384	**Safflower oil**	0	0	0	0	0	0	0	0	9.7	12.0	74.0	Tr	0
385	**Sesame oil**	0	0	0	0	0	0	0	0	14.6	37.5	43.4	Tr	0
386	**Soya oil**	0	0	0	0	0	0	0	0	15.6	21.3	58.8	Tr	0
387	**Sunflower oil**	0	0	0	0	0	0	0	0	12.0	20.5	63.3	Tr	0
388	**Vegetable oil**, blended, average	0	0	0	0	0	0	0	0	11.7[a]	53.2[a]	29.8[a]	Tr	0
389	**Walnut oil**	0	0	0	0	0	0	0	0	9.1	16.5	69.9	Tr	0
390	**Wheatgerm oil**	0	0	0	0	0	0	0	0	18.6	16.6	60.4	Tr	0

[a] The fatty acid profile will depend on the blend of oil used

Inorganic constituents per 100g edible portion

No.	Food	mg										µg	
		Na	K	Ca	Mg	P	Fe	Cu	Zn	Cl	Mn	Se	I
Oils													
376	**Coconut oil**	Tr	Tr	Tr	Tr	Tr	Tr	Tr	Tr	Tr	Tr	Tr	Tr
377	**Cod liver oil**	Tr	Tr	Tr	Tr	Tr	Tr	Tr	Tr	Tr	Tr	Tr	Tr
378	**Corn oil**	Tr	Tr	Tr	Tr	Tr	0.1	0.01	Tr	Tr	Tr	Tr	Tr
379	**Evening primrose oil**	Tr	Tr	Tr	Tr	Tr	Tr	Tr	Tr	Tr	Tr	Tr	Tr
380	**Olive oil**	Tr	Tr	Tr	Tr	Tr	0.4	0.01	Tr	Tr	Tr	Tr	Tr
381	**Palm oil**	Tr	Tr	Tr	Tr	Tr	0.4	Tr	Tr	Tr	Tr	Tr	Tr
382	**Peanut (Groundnut) oil**	Tr	Tr	Tr	Tr	Tr	Tr	Tr	Tr	Tr	Tr	Tr	Tr
383	**Rapeseed oil**	Tr	Tr	Tr	Tr	Tr	0.1	0.01	Tr	Tr	Tr	Tr	Tr
384	**Safflower oil**	Tr	Tr	Tr	Tr	Tr	Tr	Tr	Tr	Tr	Tr	Tr	Tr
385	**Sesame oil**	2	20	10	Tr	N	0.1	Tr	Tr	Tr	Tr	Tr	Tr
386	**Soya oil**	Tr	Tr	Tr	Tr	Tr	0.1	0.01	Tr	Tr	Tr	Tr	Tr
387	**Sunflower oil**	Tr	Tr	Tr	Tr	Tr	0.1	0.01	Tr	Tr	Tr	Tr	Tr
388	**Vegetable oil**, blended, average	Tr	Tr	Tr	Tr	Tr	Tr	Tr	Tr	Tr	Tr	Tr	11
389	**Walnut oil**	Tr	Tr	Tr	Tr	Tr	Tr	Tr	Tr	Tr	Tr	Tr	Tr
390	**Wheatgerm oil**	Tr	Tr	Tr	Tr	Tr	Tr	Tr	Tr	Tr	Tr	Tr	Tr

Vitamins per 100g edible portion

No.	Food	Retinol µg	Carotene µg	Vitamin D µg	Vitamin E mg	Thiamin mg	Riboflavin mg	Niacin mg	Trypt/60 mg	Vitamin B_6 mg	Vitamin B_{12} µg	Folate µg	Pantothenate mg	Biotin µg	Vitamin C mg
	Oils														
376	**Coconut oil**	0	Tr	0	0.66	Tr	Tr	Tr	Tr	Tr	0	Tr	Tr	Tr	0
377	**Cod liver oil**	18000	Tr	210.0	20.00	Tr	Tr	Tr	Tr	Tr	Tr	Tr	Tr	Tr	0
378	**Corn oil**	0	Tr	0	17.24	Tr	Tr	Tr	Tr	Tr	0	Tr	Tr	Tr	0
379	**Evening primrose oil**	0	Tr	0	N	Tr	Tr	Tr	Tr	Tr	0	Tr	Tr	Tr	0
380	**Olive oil**	0	N	0	5.10	Tr	Tr	Tr	Tr	Tr	0	Tr	Tr	Tr	0
381	**Palm oil**	0	Tr[a]	0	33.12	Tr	Tr	Tr	Tr	Tr	0	Tr	Tr	Tr	0
382	**Peanut (Groundnut) oil**	0	Tr	0	15.16	Tr	Tr	Tr	Tr	Tr	0	Tr	Tr	Tr	0
383	**Rapeseed oil**	0	Tr	0	22.21	Tr	Tr	Tr	Tr	Tr	0	Tr	Tr	Tr	0
384	**Safflower oil**	0	Tr	0	40.68	Tr	Tr	Tr	Tr	Tr	0	Tr	Tr	Tr	0
385	**Sesame oil**	0	Tr	0	N	0.01	0.07	0.1	Tr	Tr	0	Tr	Tr	Tr	0
386	**Soya oil**	0	Tr	0	16.06	Tr	Tr	Tr	Tr	Tr	0	Tr	Tr	Tr	0
387	**Sunflower oil**	0	Tr	0	49.22	Tr	Tr	Tr	Tr	Tr	0	Tr	Tr	Tr	0
388	**Vegetable oil**, blended, average	0	Tr	0	N[b]	Tr	Tr	Tr	Tr	Tr	0	Tr	Tr	Tr	0
389	**Walnut oil**	0	Tr	0	N	Tr	Tr	Tr	Tr	Tr	0	Tr	Tr	Tr	0
390	**Wheatgerm oil**	0	Tr	0	136.65	Tr	Tr	Tr	Tr	Tr	0	Tr	Tr	Tr	0

[a] Unrefined palm oil contains approximately 30000mg β- and 24000mg α- carotene per 100g

[b] The vitamin E content will vary according to the type of oil

Section 2.5

Meat and meat products

This section of the Tables is largely based on the recent *Meat, Poultry and Game* (1995) and *Meat Products and Dishes* (1996) supplements. New analytical data for a few takeaway meat dishes have been incorporated. Foods for which new data are incorporated have been allocated a new food code and can thus easily be identified in the food index.

The nutrient values were constructed from the separable fat and lean in meat (and meat and skin for poultry) analysed following dissection into lean meat, separable fat and inedible matter (or meat, skin and inedible matter for poultry). Since it was not possible to analyse all samples for all nutrients, some values for minerals and vitamins were interpolated from analytical values from similar cuts and cooking methods, usually in proportion to the protein content of the samples.

The major source of variation in meat composition is the proportion of lean to fat, as a result of husbandry techniques and trimming practices, both at retail level and in the home. This affects levels of most other nutrients, which are distributed differently in the two fractions.

Users should note that all values are expressed per 100g edible portion. Guidance for calculating nutrient content 'as purchased' or 'as served' (e.g. including rind or bone) is given in Section 4.2. For weight loss on cooking and calculation of cooked edible proportion obtained from raw meat see Section 4.3.

Losses of labile vitamins assigned to cooked dishes and food were estimated using figures in Section 4.3.

Taxonomic names for foods in this part of the Tables can be found in Section 4.5.

Composition of food per 100g edible portion

No.	Food	Description and main data sources	Edible conversion factor	Water	Total nitrogen	Protein	Fat	Carbo-hydrate	Energy value	
				g	g	g	g	g	kcal	kJ
Bacon										
391	**Bacon rashers, back**, *raw*	10 samples; smoked and unsmoked, loose and prepacked British, Danish and Dutch bacon	0.97	63.9	2.64	16.5	16.5	0	215	891
392	*dry-fried*	10 samples; smoked and unsmoked, loose and prepacked British, Danish and Dutch bacon	1.00	49.7	3.87	24.2	22.0	0	295	1225
393	*grilled*	15 samples; smoked and unsmoked, loose and prepacked British, Danish and Dutch bacon	1.00	50.4	3.71	23.2	21.6	0	287	1194
394	*grilled crispy*	10 samples; smoked and unsmoked, loose and prepacked British, Danish and Dutch bacon	1.00	37.8	5.76	36.0	18.8	0	313	1308
395	*microwaved*	15 samples; smoked and unsmoked, loose and prepacked British, Danish and Dutch bacon	1.00	45.5	3.87	24.2	23.3	0	307	1274
396	fat trimmed, *raw*	24 samples, back fat removed. MLC data and calculation from No 395	1.00	69.5	3.01	18.8	6.7	0	136	568
397	fat trimmed, *grilled*	15 samples; smoked and unsmoked, loose and prepacked British, Danish & Dutch bacon, back fat removed	1.00	56.2	4.11	25.7	12.3	0	214	892
398	reduced salt, *grilled*	6 samples; smoked and unsmoked, loose and prepacked British and Danish bacon	1.00	51.6	3.86	24.1	20.6	0	282	1172
399	**middle**, *grilled*	9 samples; smoked and unsmoked, loose and prepacked British and Danish bacon	1.00	47.8	3.97	24.8	23.1	0	307	1276

Composition of food per 100g edible portion

No.	Food	Starch	Total sugars	Dietary fibre NSP	Fatty acids Satd	Mono-unsatd	Poly-unsatd	Trans	Cholest-erol
		g	g	g	g	g	g	g	mg
	Bacon								
391	**Bacon rashers, back**, *raw*	0	0	0	6.2	6.9	2.2	0.1	53
392	*dry-fried*	0	0	0	8.3	9.2	2.8	0.1	65
393	*grilled*	0	0	0	8.1	9.0	2.8	Tr	75
394	*grilled crispy*	0	0	0	7.1	7.9	2.4	0.1	68
395	*microwaved*	0	0	0	8.8	9.8	3.0	0.1	84
396	fat trimmed, *raw*	0	0	0	2.5	2.8	0.9	Tr	(31)
397	fat trimmed, *grilled*	0	0	0	4.6	5.2	1.6	0.1	44
398	reduced salt, *grilled*	0	0	0	7.8	8.7	2.7	0.1	74
399	**middle**, *grilled*	0	0	0	8.4	10.0	3.0	0.1	83

Inorganic constituents per 100g edible portion

No.	Food	mg										µg	
		Na	K	Ca	Mg	P	Fe	Cu	Zn	Cl	Mn	Se	I
	Bacon												
391	**Bacon rashers, back**, *raw*	1540	300	5	17	150	0.4	0.06	1.2	2350	0.01	8	5
392	*dry-fried*	1910	360	6	21	180	0.6	0.06	1.9	(3510)	0.01	18	7
393	*grilled*	1880	340	7	21	180	0.6	0.05	1.7	2780	0.01	12	7
394	*grilled crispy*	(2700)	510	10	32	300	1.1	0.10	3.1	(3510)	0.01	18	11
395	*microwaved*	2330	360	8	23	200	0.7	0.06	2.0	(2360)	0.01	12	7
396	fat trimmed, *raw*	(1350)	(250)	(6)	(16)	(150)	(0.5)	(0.05)	(1.5)	(1740)	(0.01)	(9)	(6)
397	fat trimmed, *grilled*	(1930)	360	8	23	210	0.7	0.07	2.2	(2500)	0.01	13	8
398	reduced salt, *grilled*	1130	340	7	22	200	0.7	0.09	2.1	1500	0.01	12	7
399	**middle**, *grilled*	1960	350	8	21	220	0.7	0.07	2.2	2050	0.01	11	8

No.	Food	Retinol µg	Carotene µg	Vitamin D µg	Vitamin E mg	Thiamin mg	Riboflavin mg	Niacin mg	Trypt/60 mg	Vitamin B_6 mg	Vitamin B_{12} µg	Folate µg	Pantothenate mg	Biotin µg	Vitamin C mg
	Bacon														
391	**Bacon rashers, back**, *raw*	Tr	Tr	0.3	0.02	0.63	0.11	5.6	2.6	0.46	Tr	3	1.00	2	1
392	*dry-fried*	Tr	Tr	0.6	0.07	0.86	0.14	6.8	4.4	0.53	1	2	1.26	5	Tr
393	*grilled*	Tr	Tr	0.6	0.07	1.16	0.15	7.2	3.8	0.52	1	5	1.24	3	Tr
394	*grilled crispy*	Tr	Tr	1.0	0.10	1.38	0.24	10.8	6.6	0.71	1	4	1.34	5	Tr
395	*microwaved*	Tr	Tr	0.6	0.07	1.10	0.16	7.9	4.4	0.55	1	2	1.26	5	Tr
396	fat trimmed, *raw*	Tr	Tr	(0.5)	(0.05)	(0.68)	(0.12)	(5.4)	(3.3)	(0.35)	(1)	(2)	(0.93)	(4)	Tr
397	fat trimmed, *grilled*	Tr	Tr	0.7	0.07	0.98	0.17	7.7	4.7	0.50	1	3	1.34	5	Tr
398	reduced salt, *grilled*	Tr	Tr	0.6	0.07	0.92	0.16	7.2	4.4	0.47	1	2	1.25	5	Tr
399	**middle**, *grilled*	Tr	Tr	0.6	0.13	0.77	0.17	7.5	5.1	0.42	1	3	1.27	6	Tr

Composition of food per 100g edible portion

No.	Food	Description and main data sources	Edible conversion factor	Water	Total nitrogen	Protein	Fat	Carbo-hydrate	Energy value	
				g	g	g	g	g	kcal	kJ
***Bacon** continued*										
400	**Bacon rashers**, **streaky**, *raw*	10 samples; smoked and unsmoked, loose and prepacked British and Danish bacon	(0.91)	57.3	2.53	15.8	23.6	0	276	1142
401	*grilled*	10 samples; smoked and unsmoked, loose and prepacked British and Danish bacon	1.00	44.0	3.81	23.8	26.9	0	337	1400
402	*fried*	10 samples; smoked and unsmoked, loose and prepacked British and Danish bacon	1.00	45.1	3.81	23.8	26.6	0	335	1389
403	**Bacon**, fat only, average, *raw*	Fat from five different cuts	1.00	12.8	0.76	4.8	80.9	0	747	3075
404	average, *cooked*	Fat from five different cuts	1.00	13.8	1.48	9.3	72.8	0	692	2852
405	**Ham**	10 samples, 9 brands; loose and prepacked including honey roast and smoked Ham. Added water 10–15%	1.00	73.2	2.94	18.4	3.3	1.0	107	451
406	gammon joint, *raw*	10 samples; smoked and unsmoked, prepacked British and Danish gammon	0.92	68.6	2.80	17.5	7.5	0	138	575
407	*boiled*	10 samples; smoked and unsmoked, prepacked British and Danish gammon	1.00	61.2	3.73	23.3	12.3	0	204	851
408	gammon rashers, *grilled*	5 samples; unsmoked British gammon	1.00	58.2	4.40	27.5	9.9	0	199	834
Beef										
409	**Beef**, average, trimmed lean, *raw*	LGC; average of 10 different cuts	1.00	71.9	3.60	22.5	4.3	0	129	542
410	trimmed fat, *raw*	LGC; average of 10 different cuts	1.00	35.0	3.02	18.9	53.6	0	558	2305
411	**Beef**, fat, *cooked*	LGC; average of 8 different cuts	1.00	33.6	2.48	15.5	52.3	0	533	2199

Composition of food per 100g edible portion

No.	Food	Starch	Total sugars	Dietary fibre NSP	Fatty acids Satd	Mono-unsatd	Poly-unsatd	Trans	Cholest-erol
		g	g	g	g	g	g	g	mg
	Bacon *continued*								
400	**Bacon rashers, streaky**, *raw*	0	0	0	8.2	10.2	3.5	0.1	65
401	*grilled*	0	0	0	9.8	11.5	3.7	0.1	90
402	*fried*	0	0	0	9.1	11.1	4.5	2.6	78
403	**Bacon**, fat only, average, *raw*	0	0	0	31.5	36.1	8.9	N	198
404	average, *cooked*	0	0	0	28.5	32.9	7.6	N	270
405	**Ham**	0	1.0	0	1.1	1.4	0.5	Tr	58
406	gammon joint, *raw*	0	0	0	2.5	3.3	1.2	Tr	23
407	*boiled*	0	0	0	4.1	5.4	1.9	Tr	83
408	gammon rashers, *grilled*	0	0	0	3.4	4.1	1.7	0.1	83
	Beef								
409	**Beef**, average, trimmed lean, *raw*	0	0	0	1.7	1.9	0.2	0.1	58
410	trimmed fat, *raw*	0	0	0	24.9	24.2	1.7	2.4	72
411	**Beef**, fat, *cooked*	0	0	0	24.3	23.4	1.8	2.4	97

Inorganic constituents per 100g edible portion

No.	Food	mg										µg	
		Na	K	Ca	Mg	P	Fe	Cu	Zn	Cl	Mn	Se	I
	***Bacon** continued*												
400	**Bacon rashers**, **streaky**, *raw*	1260	250	6	15	140	0.5	0.06	1.5	1500	0.01	7	7
401	*grilled*	1680	330	9	20	180	0.8	0.15	2.5	2630	0.01	11	6
402	*fried*	(1880)	350	7	21	200	0.7	0.07	2.1	(2320)	0.01	10	7
403	**Bacon**, fat only, average, *raw*	560	75	3	4	38[a]	0.7	0.06	0.6	810	Tr	(1)	(11)
404	average, *cooked*	990	130	7	10	90	0.8	0.09	0.8	1520	Tr	(2)	(11)
405	**Ham**	1200	340	7	24	340	0.7	0.12	1.8	1470	0.01	11	5
406	gammon joint, *raw*	(880)	190	7	17	130	0.6	0.08	1.5	(1980)	0.01	11	7
407	*boiled*	1180	250	9	18	170	0.8	0.10	2.1	(2640)	0.01	12	9
408	gammon rashers, *grilled*	1930	380	8	26	230	0.8	0.09	2.2	(2680)	0.02	14	8
	Beef												
409	**Beef**, average, trimmed lean, *raw*	63	350	5	22	200	2.7	0.03	4.1	51	0.01	7	10
410	trimmed fat, *raw*	26	140	5	9	79	0.7	0.02	1.1	28	Tr	2	10
411	**Beef**, fat, *cooked*	35	200	6	12	110	1.0	0.01	1.5	39	0.01	3	14

[a] Sweetcure bacon contains 140mg P per 100g fat

No.	Food	Retinol µg	Carotene µg	Vitamin D µg	Vitamin E mg	Thiamin mg	Ribo-flavin mg	Niacin mg	Trypt/60 mg	Vitamin B_6 mg	Vitamin B_{12} µg	Folate µg	Panto-thenate mg	Biotin µg	Vitamin C mg
	Bacon *continued*														
400	**Bacon rashers**, **streaky**, *raw*	Tr	Tr	0.9	0.07	0.45	0.14	4.7	2.7	0.30	1	3	0.92	2	Tr
401	*grilled*	Tr	Tr	0.7	0.07	0.70	0.17	6.3	4.3	0.40	1	3	1.22	4	Tr
402	*fried*	Tr	Tr	0.6	N	0.75	0.14	7.1	4.4	0.47	1	1	1.24	5	Tr
403	**Bacon**, fat only, average, *raw*	Tr	Tr	Tr	0.11	N	N	N	0.9	N	Tr	Tr	N	Tr	0
404	average, *cooked*	Tr	Tr	Tr	0.36	N	N	N	1.7	N	Tr	Tr	N	Tr	0
405	**Ham**	Tr	Tr	N	0.04	0.80	0.17	6.5	3.1	0.61	1	19	1.03	3	Tr
406	gammon joint, *raw*	Tr	Tr	0.6	0.06	0.44	0.13	5.3	2.9	0.43	Tr	4	1.07	2	Tr
407	*boiled*	Tr	Tr	0.8	0.08	0.58	0.16	5.4	3.9	0.42	Tr	3	1.43	2	Tr
408	gammon rashers, *grilled*	Tr	Tr	0.8	0.08	1.16	0.18	6.4	5.5	0.16	1	3	1.43	6	Tr
	Beef														
409	**Beef**, average, trimmed lean, *raw*	Tr	Tr	0.5	0.13	0.10	0.21	5.0	4.7	0.53	2	19	0.75	1	0
410	trimmed fat, *raw*	Tr	Tr	Tr	0.06	0.04	0.13	1.2	1.7	0.17	1	18	0.43	1	0
411	**Beef**, fat, *cooked*	Tr	Tr	Tr	0.08	0.05	0.18	1.6	1.8	0.23	2	26	0.60	2	0

Composition of food per 100g edible portion

No.	Food	Description and main data sources	Edible conversion factor	Water	Total nitrogen	Protein	Fat	Carbo-hydrate	Energy value	
				g	g	g	g	g	kcal	kJ
Beef *continued*										
412	**Braising steak**, braised, lean	10 samples	1.00	55.5	5.50	34.4	9.7	0	225	944
413	lean and fat	Calculated from 90% lean and 9% fat	1.00	53.1	5.26	32.9	12.7	0	246	1029
414	**Fore-rib/rib-roast**, *raw*, lean and fat	Calculated from 75% lean and 25% fat	0.85	61.4	3.01	18.8	19.8	0	253	1052
415	*roasted*, lean and fat	Calculated from 78% lean and 22% fat	0.84	49.8	4.64	29.1	20.4	0	300	1250
416	**Mince**, *raw*	10 samples	1.00	62.0[a]	3.15	19.7	16.2[b]	0	225	934
417	*microwaved*	10 samples	1.00	55.3	4.22	26.4	17.5	0	263	1096
418	*stewed*	10 samples	1.00	64.4	3.49	21.8	13.5	0	209	870
419	extra lean, *stewed*	17 samples	1.00	66.6	3.95	24.7	8.7	0	177	742
420	**Rump steak**, *raw*, lean and fat	Calculated from 88% lean and 11% fat	0.99	68.2	3.31	20.7	10.1	0	174	726
421	*barbecued*, lean	10 samples	1.00	62.4	4.99	31.2	5.7	0	176	741
422	*fried*, lean	10 samples	1.00	61.7	4.94	30.9	6.6	0	183	770
423	*fried*, lean and fat	Calculated from 87% lean and 12% fat	1.00	57.2	4.54	28.4	12.7	0	228	953
424	*grilled*, lean	10 samples	1.00	62.9	4.96	31.0	5.9	0	177	745
425	from steakhouse, lean	10 samples	1.00	63.0	4.77	29.8	4.7	0	162	681
426	strips, *stir-fried*, lean	10 samples	1.00	57.9	5.17	32.3	8.8	0	208	875
427	**Silverside**, salted, *boiled*, lean	Calculated from 88% lean and 11% fat	1.00	60.4	4.86	30.4	6.9	0	184	772
428	**Stewing steak**, *raw*, lean and fat	Calculated from 90% lean and 9% fat	1.00	70.1	3.54	22.1	6.4	0	146	613
429	*stewed*, lean and fat	Calculated from 84% lean and 14% fat	1.00	59.4	4.67	29.2	9.6	0	203	852
430	**Topside**, *raw*, lean and fat	Calculated from 84% lean and 15% fat	0.99	65.8	3.26	20.4	12.9	0	198	824
431	*roasted well-done*, lean	10 samples	1.00	56.9	5.79	36.2	6.3	0	202	849
432	lean and fat	Calculated from 88% lean and 11% fat	1.00	53.0	5.25	32.8	12.5	0	244	1020

[a] Water ranged from 57.3 to 70.0g per 100g

[b] Fat ranged from 7.8g to 26.5g per 100g

Composition of food per 100g edible portion

No.	Food	Starch	Total sugars	Dietary fibre NSP	Fatty acids Satd	Mono-unsatd	Poly-unsatd	Trans	Cholest-erol
		g	g	g	g	g	g	g	mg
	Beef *continued*								
412	**Braising steak**, braised, lean	0	0	0	4.1	4.1	0.6	0.4	100
413	lean and fat	0	0	0	5.3	5.2	0.8	0.5	100
414	**Fore-rib/rib-roast**, *raw*, lean and fat	0	0	0	8.9	8.8	0.7	0.8	59
415	*roasted*, lean and fat	0	0	0	9.2	9.1	0.7	0.8	83
416	**Mince**, *raw*	0	0	0	6.9	6.9	0.5	0.8	60
417	*microwaved*	0	0	0	7.6	7.7	0.7	0.8	80
418	*stewed*	0	0	0	5.7	5.7	0.6	0.7	79
419	extra lean, *stewed*	0	0	0	3.8	3.8	0.3	0.4	75
420	**Rump steak**, *raw*, lean and fat	0	0	0	4.3	4.4	0.6	0.3	60
421	*barbecued*, lean	0	0	0	2.4	2.4	0.4	0.2	76
422	*fried*, lean	0	0	0	2.4	2.5	0.9	0.2	86
423	*fried*, lean and fat	0	0	0	4.9	5.2	1.6	0.4	84
424	*grilled*, lean	0	0	0	2.5	2.5	0.5	0.2	76
425	from steakhouse, lean	0	0	0	2.0	2.0	0.3	0.1	73
426	strips, *stir-fried*, lean	0	0	0	3.3	3.5	1.2	0.3	92
427	**Silverside**, salted, *boiled*, lean	0	0	0	2.5	3.4	0.3	0.2	74
428	**Stewing steak**, *raw*, lean and fat	0	0	0	2.6	2.9	0.4	0.2	69
429	*stewed*, lean and fat	0	0	0	3.7	4.2	0.9	0.3	91
430	**Topside**, *raw*, lean and fat	0	0	0	5.4	5.8	0.8	0.4	48
431	*roasted well-done*, lean	0	0	0	2.6	2.8	0.3	0.2	88
432	lean and fat	0	0	0	5.2	5.7	0.6	0.5	83

Inorganic constituents per 100g edible portion

No.	Food	mg										µg	
		Na	K	Ca	Mg	P	Fe	Cu	Zn	Cl	Mn	Se	I
	***Beef** continued*												
412	**Braising steak**, braised, lean	62	340	8	23	220	2.7	Tr	9.5	62	Tr	11	15
413	lean and fat	60	330	8	22	210	2.6	Tr	8.7	61	Tr	10	15
414	**Fore-rib/rib-roast**, *raw*, lean and fat	52	290	5	19	170	1.5	0.03	2.9	44	Tr	6	10
415	*roasted*, lean and fat	54	320	8	20	190	1.8	Tr	6.1	58	Tr	10	12
416	**Mince**, *raw*	80	260	9	17	160	1.4	Tr	3.9	76	Tr	7	9
417	*microwaved*	91	290	12	20	190	2.0	Tr	5.2	110	0.02	9	16
418	*stewed*	73	210	20	15	150	2.7	0.10	5.0	63	0.02	7	14
419	extra lean, *stewed*	75	280	14	18	170	2.3	0.08	5.6	61	Tr	8	10
420	**Rump steak**, *raw*, lean and fat	56	350	4	22	200	2.7	0.04	3.5	38	Tr	7	11
421	*barbecued*, lean	78	460	8	29	270	3.2	0.10	5.1	61	0.04	10	11
422	*fried*, lean	78	390	5	25	240	3.0	0.02	5.2	50	0.02	10	9
423	*fried*, lean and fat	71	360	5	23	220	2.7	0.02	4.7	47	0.02	9	9
424	*grilled*, lean	74	430	7	29	260	3.6	0.04	5.6	62	0.02	10	12
425	from steakhouse, lean	72	410	7	28	250	2.4	0.04	5.4	60	0.02	10	11
426	strips, *stir-fried*, lean	78	450	7	30	270	2.6	0.04	5.8	64	0.02	11	12
427	**Silverside**, salted, *boiled*, lean	1020	190	10	17	150	2.0	Tr	5.3	1690	0.02	10	11
428	**Stewing steak**, *raw*, lean and fat	66	340	5	20	180	2.0	0.04	5.3	66	Tr	7	13
429	*stewed*, lean and fat	51	250	15	19	180	2.3	0.04	7.5	33	0.01	10	12
430	**Topside**, *raw*, lean and fat	67	340	5	22	190	1.7	0.07	3.5	44	0.02	7	9
431	*roasted well-done*, lean	62	410	8	27	230	2.9	0.04	6.5	36	0.01	12	13
432	lean and fat	57	370	7	24	210	2.6	0.04	5.8	34	0.01	11	12

No.	Food	Retinol µg	Carotene µg	Vitamin D µg	Vitamin E mg	Thiamin mg	Riboflavin mg	Niacin mg	Trypt/60 mg	Vitamin B_6 mg	Vitamin B_{12} µg	Folate µg	Pantothenate mg	Biotin µg	Vitamin C mg
	***Beef** continued*														
412	**Braising steak**, braised, lean	Tr	8	0.8	0.02	0.05	0.26	5.2	8.0	0.34	3	54	0.55	2	0
413	lean and fat	Tr	7	0.7	0.03	0.05	0.26	4.9	7.5	0.33	3	52	0.57	2	0
414	**Fore-rib/rib-roast**, *raw*, lean and fat	Tr	6	0.4	0.12	0.09	0.21	4.1	5.3	0.43	2	16	0.66	1	0
415	*roasted*, lean and fat	Tr	6	0.6	0.19	0.06	0.17	4.5	6.1	0.36	3	19	0.54	2	0
416	**Mince**, *raw*	Tr	Tr	0.7	0.17	0.06	0.13	5.8	3.6	0.37	2	14	0.49	1	0
417	*microwaved*	Tr	8	0.6	0.31	0.08	0.31	8.0	4.3	0.38	3	17	0.53	2	0
418	*stewed*	9	25	0.8	0.34	0.03	0.19	4.6	4.4	0.28	2	17	0.36	5	0
419	extra lean, *stewed*	Tr	8	0.6	0.30	0.03	0.13	4.8	4.5	0.16	3	20	0.36	2	0
420	**Rump steak**, *raw*, lean and fat	Tr	Tr	0.4	0.04	0.09	0.23	4.9	4.5	0.61	2	5	0.65	1	0
421	*barbecued*, lean	Tr	8	0.7	0.20	0.15	0.32	6.8	7.0	0.36	3	10	0.78	2	0
422	*fried*, lean	Tr	8	0.7	0.18	0.14	0.29	5.9	6.7	0.63	2	5	0.74	2	0
423	*fried*, lean and fat	Tr	7	0.6	N	0.13	0.27	5.3	6.0	0.57	2	5	0.70	2	0
424	*grilled*, lean	Tr	8	0.4	0.07	0.13	0.28	6.8	7.0	0.65	3	(5)	0.91	2	0
425	from steakhouse, lean	Tr	8	0.7	N	0.13	0.27	6.5	6.7	0.63	2	17	0.88	2	0
426	strips, *stir-fried*, lean	Tr	8	0.7	0.06	0.21	0.30	6.8	7.2	0.73	3	5	0.94	2	0
427	**Silverside**, salted, *boiled*, lean	Tr	8	0.7	0.10	0.05	0.27	2.6	6.8	0.39	2	12	0.40	2	0
428	**Stewing steak**, *raw*, lean and fat	Tr	Tr	0.7	0.20	0.07	0.26	4.0	4.3	0.42	2	6	0.65	1	0
429	*stewed*, lean and fat	Tr	7	0.6	0.17	0.03	0.15	2.4	6.2	0.23	2	11	0.30	2	0
430	**Topside**, *raw*, lean and fat	Tr	Tr	0.4	0.13	0.08	0.17	4.6	4.3	0.48	2	22	0.59	1	0
431	*roasted well-done*, lean	Tr	8	0.8	0.08	0.09	0.29	5.8	8.1	0.56	3	21	0.60	2	0
432	lean and fat	Tr	7	0.7	0.08	0.08	0.27	5.2	7.2	0.51	3	20	0.56	2	0

Composition of food per 100g edible portion

No.	Food	Description and main data sources	Edible conversion factor	Water	Total nitrogen	Protein	Fat	Carbo-hydrate	Energy value	
				g	g	g	g	g	kcal	kJ
Lamb										
433	**Lamb**, average, trimmed lean, *raw*	LGC; average of 8 different cuts	1.00	70.6	3.23	20.2	8.0	0	153	639
434	trimmed fat, *raw*	LGC; average of 8 different cuts	1.00	34.7	2.13	13.3	51.6	0	518	2135
435	fat, *cooked*	LGC; average of 8 different cuts	1.00	28.3	2.64	15.4	56.3	0	568	2345
436	**Best end neck cutlets**, *raw*, lean and fat	Calculated from 66% lean and 34% fat	0.74	53.9	2.61	16.3	27.9	0	316	1309
437	*grilled*, lean	33 samples	0.48	57.4	4.56	28.5	13.8	0	238	995
438	lean and fat	Calculated from 68% lean and 32% fat	0.72	46.1	3.91	24.5	29.9	0	367	1523
439	**Breast**, *roasted*, lean	10 samples	1.00	54.4	4.27	26.7	18.5	0	273	1138
440	lean and fat	Calculated from 62% lean and 36% fat	1.00	45.5	3.59	22.4	29.9	0	359	1487
441	**Leg**, average, *raw*, lean and fat	Calculated from 83% lean and 17% fat	1.00	67.4	3.05	19.0	12.3	0	187	778
442	**Leg, whole**, *roasted medium*, lean	10 samples	0.77	60.5	4.75	29.7	9.4	0	203	853
443	lean and fat	Calculated from 89% lean and 11% fat	1.00	57.3	4.50	28.1	14.2	0	240	1003
444	**Loin chops**, *raw*, lean and fat	Calculated from 72% lean and 28% fat	0.78	59.3	2.81	17.6	23.0	0	277	1150
445	*grilled*, lean	33 samples	0.61	59.6	4.67	29.2	10.7	0	213	892
446	lean and fat	Calculated from 76% lean and 24% fat	0.81	50.5	4.24	26.5	22.1	0	305	1268
447	*microwaved*, lean and fat	Calculated from 72% lean and 28% fat	0.82	45.3	4.39	27.5	26.9	0	352	1463
448	*roasted*, lean and fat	Calculated from 73% lean and 27% fat	0.88	43.8	4.66	29.1	26.9	0	359	1490
449	**Mince**, *raw*	10 samples	1.00	67.1[a]	3.06	19.1	13.3[b]	0	196	817
450	*stewed*	10 samples	1.00	62.8	3.90	24.4	12.3	0	208	870
451	**Neck fillet**, strips, *stir-fried*, lean	10 samples	1.00	55.3	3.90	24.4	20.0	0	278	1155

[a] Water ranged from 63.0g to 71.6g per 100g

[b] Fat ranged from 8.1g to 22.8g per 100g

Composition of food per 100g edible portion

No.	Food	Starch	Total sugars	Dietary fibre NSP	Fatty acids Satd	Fatty acids Mono-unsatd	Fatty acids Poly-unsatd	Fatty acids Trans	Cholest-erol
		g	g	g	g	g	g	g	mg
	Lamb								
433	**Lamb**, average, trimmed lean, *raw*	0	0	0	3.5	3.1	0.5	0.6	74
434	trimmed fat, *raw*	0	0	0	26.3	19.5	2.3	4.8	92
435	fat, *cooked*	0	0	0	28.4	21.6	2.4	5.3	100
436	**Best end neck cutlets**, *raw*, lean and fat	0	0	0	13.6	10.4	1.4	2.3	76
437	*grilled*, lean	0	0	0	6.5	5.1	0.7	1.1	100
438	lean and fat	0	0	0	14.5	11.2	1.5	2.5	105
439	**Breast**, *roasted*, lean	0	0	0	8.6	7.0	0.9	1.6	95
440	lean and fat	0	0	0	14.3	11.4	1.4	2.7	93
441	**Leg**, average, *raw*, lean and fat	0	0	0	5.4	4.9	0.7	0.9	78
442	**Leg, whole**, *roasted medium*, lean	0	0	0	3.8	3.9	0.6	0.7	100
443	lean and fat	0	0	0	5.7	6.1	0.8	1.2	100
444	**Loin chops**, *raw*, lean and fat	0	0	0	10.8	8.8	1.2	1.8	79
445	*grilled*, lean	0	0	0	4.9	4.0	0.6	0.9	96
446	lean and fat	0	0	0	10.5	8.4	1.3	1.9	100
447	*microwaved*, lean and fat	0	0	0	12.8	10.2	1.5	2.3	110
448	*roasted*, lean and fat	0	0	0	12.8	10.2	1.5	2.3	115
449	**Mince**, *raw*	0	0	0	6.2	5.3	0.6	1.1	77
450	*stewed*	0	0	0	5.8	4.8	0.6	0.9	96
451	**Neck fillet**, strips, *stir-fried*, lean	0	0	0	8.2	7.6	2.2	1.3	86

Inorganic constituents per 100g edible portion

No.	Food	mg										µg	
		Na	K	Ca	Mg	P	Fe	Cu	Zn	Cl	Mn	Se	I
	Lamb												
433	**Lamb**, average, trimmed lean, *raw*	70	330	12	22	190	1.4	0.08	3.3	74	0.01	4	6
434	trimmed fat, *raw*	36	140	9	9	86	0.7	0.03	0.9	43	0.01	2	6
435	fat, *cooked*	72	260	11	18	160	1.1	0.05	1.5	67	0.01	4	6
436	**Best end neck cutlets**, *raw*, lean and fat	58	250	11	17	150	1.0	0.05	1.9	62	0.01	3	8
437	*grilled*, lean	84	370	23	26	230	1.9	0.11	3.6	71	0.01	4	6
438	lean and fat	81	340	19	24	210	1.7	0.09	2.9	71	0.01	4	6
439	**Breast**, *roasted*, lean	93	330	8	22	200	1.6	0.07	5.1	67	0.01	4	6
440	lean and fat	85	300	9	21	180	1.4	0.06	3.7	67	0.01	4	6
441	**Leg**, average, *raw*, lean and fat	58	320	7	22	190	1.4	0.08	2.8	59	0.01	2	2
442	**Leg, whole**, *roasted medium*, lean	63	360	7	26	220	1.8	0.11	4.6	67	0.02	4	3
443	lean and fat	64	340	7	25	210	1.7	0.10	4.3	67	0.02	4	3
444	**Loin chops**, *raw*, lean and fat	63	280	13	19	170	1.3	0.07	2.0	65	0.01	3	7
445	*grilled*, lean	80	400	22	28	240	2.1	0.10	3.6	73	0.02	4	6
446	lean and fat	81	370	20	27	230	1.9	0.09	3.1	74	0.02	4	6
447	*microwaved*, lean and fat	74	310	17	24	200	1.8	0.09	3.3	76	0.01	4	6
448	*roasted*, lean and fat	85	370	20	27	230	2.1	0.11	4.6	80	0.01	4	6
449	**Mince**, *raw*	69	310	17	21	190	1.6	0.08	3.5	68	0.01	2	6
450	*stewed*	59	270	15	20	180	2.1	0.11	4.6	46	0.02	3	5
451	**Neck fillet**, strips, *stir-fried*, lean	68	360	7	23	210	1.8	0.08	5.2	61	0.02	4	6

Vitamins per 100g edible portion

No.	Food	Retinol µg	Carotene µg	Vitamin D µg	Vitamin E mg	Thiamin mg	Riboflavin mg	Niacin mg	Trypt/60 mg	Vitamin B_6 mg	Vitamin B_{12} µg	Folate µg	Pantothenate mg	Biotin µg	Vitamin C mg
Lamb															
433	**Lamb**, average, trimmed lean, *raw*	6	Tr	0.4	0.09	0.09	0.20	5.4	3.9	0.30	2	6	0.92	2	0
434	trimmed fat, *raw*	29	Tr	0.5	0.14	0.07	0.12	2.2	1.3	0.10	1	4	0.47	1	0
435	fat, *cooked*	29	Tr	0.5	0.28	0.09	0.17	3.6	2.0	0.20	1	4	0.74	1	0
436	**Best end neck cutlets**, *raw*, lean and fat	15	Tr	0.4	0.07	0.15	0.16	5.1	3.2	0.34	1	11	0.40	2	0
437	*grilled*, lean	Tr	Tr	0.6	0.10	0.16	0.19	7.0	5.9	0.40	3	4	1.40	2	0
438	lean and fat	9	Tr	0.6	0.16	0.14	0.19	6.0	4.7	0.32	2	4	1.21	2	0
439	**Breast**, *roasted*, lean	Tr	Tr	0.6	0.11	0.08	0.19	5.7	5.6	0.16	3	6	1.30	2	0
440	lean and fat	10	Tr	0.5	0.17	0.09	0.18	4.9	4.2	0.16	2	5	1.09	2	0
441	**Leg**, average, *raw*, lean and fat	9	Tr	0.7	0.05	0.14	0.23	5.1	3.7	0.33	1	11	1.25	2	0
442	**Leg, whole**, *roasted medium*, lean	Tr	Tr	0.7	0.03	0.12	0.29	6.2	5.8	0.34	2	2	1.50	3	0
443	lean and fat	Tr	Tr	0.6	0.05	0.12	0.28	5.9	5.4	0.32	2	2	1.41	2	0
444	**Loin chops**, *raw*, lean and fat	12	Tr	0.8	0.07	0.13	0.22	5.0	3.6	0.23	1	3	0.86	1	0
445	*grilled*, lean	Tr	Tr	0.6	0.02	0.17	0.26	8.3	6.1	0.52	3	6	1.40	3	0
446	lean and fat	7	Tr	(0.3)	0.09	0.16	0.25	7.3	5.2	0.44	3	6	1.28	2	0
447	*microwaved*, lean and fat	8	Tr	0.6	0.14	0.14	0.20	5.5	5.4	0.27	3	3	1.34	2	0
448	*roasted*, lean and fat	8	Tr	0.6	0.11	0.14	0.31	6.0	5.7	0.29	3	5	1.35	2	0
449	**Mince**, *raw*	5	Tr	0.8	0.18	0.12	0.18	4.8	3.7	0.20	2	2	0.90	2	0
450	*stewed*	5	Tr	0.5	0.11	0.09	0.21	5.2	5.3	0.21	2	9	0.90	4	0
451	**Neck fillet**, strips, *stir-fried*, lean	Tr	Tr	0.6	0.59	0.17	0.20	4.6	5.1	0.20	2	7	1.20	2	0

Composition of food per 100g edible portion

No.	Food	Description and main data sources	Edible conversion factor	Water	Total nitrogen	Protein	Fat	Carbo-hydrate	Energy value	
				g	g	g	g	g	kcal	kJ
***Lamb** continued*										
452	**Shoulder**, *raw*, lean and fat	Calculated from 76% lean and 24% fat	0.80	61.6	2.81	17.6	18.3	0	235	976
453	**diced, kebabs**, *grilled*, lean and fat	Calculated from 85% lean and 15% fat	1.00	52.1	4.56	28.5	19.3	0	288	1199
454	**whole**, *roasted*, lean	10 samples	1.00	56.9	4.35	27.2	12.1	0	218	910
455	lean and fat	Calculated from 78% lean and 22% fat	0.79	50.5	3.96	24.7	22.1	0	298	1238
456	**Stewing lamb**, *pressure cooked*, lean	10 samples	0.67	56.6	4.59	28.7	14.8	0	248	1036
457	*stewed*, lean	10 samples	0.68	58.9	4.26	26.6	14.8	0	240	1000
458	lean and fat	Calculated from 85% lean and 15% fat	1.00	56.1	3.91	24.4	20.1	0	279	1159
Pork										
459	**Pork**, average, trimmed lean, *raw*	LGC; average of 8 different cuts	1.00	74.0	3.49	21.8	4.0	0	123	519
460	trimmed fat, *raw*	LGC; average of 8 different cuts	1.00	33.6	1.62	10.1	56.4	0	548	2259
461	fat, *cooked*	LGC; average of 5 different cuts	1.00	33.1	2.27	14.2	50.9	0	515	2125
462	**Belly joint/slices**, *grilled*, lean and fat	25 samples, 58% lean and 42% fat	0.85	48.6	4.38	27.4	23.4	0	320	1332
463	**Diced**, *casseroled*, lean only	10 samples	1.00	62.2	5.07	31.7	6.4	0	184	776
464	**Fillet strips**, *stir-fried*, lean	10 samples	1.00	59.6	5.14	32.1	5.9	0	182	764
465	**Leg joint**, *raw*, lean and fat	Calculated from 79% lean and 21% fat	0.87	64.4	3.04	19.0	15.2	0	213	885
466	*roasted medium*, lean	10 samples	0.85	61.1	5.28	33.0	5.5	0	182	765
467	lean and fat	Calculated from 83% lean and 17% fat	1.00	58.3	4.94	30.9	10.2	0	215	903

No.	Food	Starch	Total sugars	Dietary fibre NSP	Fatty acids Satd	Mono-unsatd	Poly-unsatd	Trans	Cholest-erol
		g	g	g	g	g	g	g	mg
***Lamb** continued*									
452	**Shoulder**, *raw*, lean and fat	0	0	0	8.5	7.1	1.0	1.0	76
453	**diced, kebabs**, *grilled*, lean and fat	0	0	0	9.0	7.5	1.0	1.5	110
454	**whole**, *roasted*, lean	0	0	0	5.5	4.7	0.6	0.9	105
455	lean and fat	0	0	0	10.4	8.7	1.0	1.7	105
456	**Stewing lamb**, *pressure cooked*, lean	0	0	0	6.5	5.6	1.0	1.0	100
457	*stewed*, lean	0	0	0	6.5	5.6	1.0	1.0	94
458	lean and fat	0	0	0	9.2	7.7	1.3	1.5	92
Pork									
459	**Pork**, average, trimmed lean, *raw*	0	0	0	1.4	1.5	0.7	Tr	63
460	trimmed fat, *raw*	0	0	0	20.4	23.7	9.5	0.3	71
461	fat, *cooked*	0	0	0	17.9	21.5	8.9	0.3	98
462	**Belly joint/slices**, *grilled*, lean and fat	0	0	0	8.2	9.5	4.0	0.1	97
463	**Diced**, *casseroled*, lean only	0	0	0	1.9	2.3	1.6	Tr	99
464	**Fillet strips**, *stir-fried*, lean	0	0	0	1.3	1.8	2.2	Tr	90
465	**Leg joint**, *raw*, lean and fat	0	0	0	5.1	6.4	2.5	0.1	63
466	*roasted medium*, lean	0	0	0	1.9	2.3	0.7	Tr	100
467	lean and fat	0	0	0	3.6	4.4	1.4	Tr	100

Inorganic constituents per 100g edible portion

No.	Food	mg										µg	
		Na	K	Ca	Mg	P	Fe	Cu	Zn	Cl	Mn	Se	I
	Lamb *continued*												
452	**Shoulder**, *raw*, lean and fat	63	280	6	19	160	1.1	0.08	3.4	63	0.02	3	3
453	**diced, kebabs**, *grilled*, lean and fat	87	420	14	28	240	1.8	0.12	5.6	76	0.03	4	6
454	**whole**, *roasted*, lean	80	330	8	23	210	1.8	0.10	5.8	81	0.01	6	6
455	lean and fat	80	320	9	22	190	1.6	0.09	5.0	79	0.01	5	7
456	**Stewing lamb**, *pressure cooked*, lean	65	240	31	20	180	1.9	0.13	6.0	72	0.01	4	6
457	*stewed*, lean	49	160	37	16	140	1.9	0.09	6.1	67	0.01	4	6
458	lean and fat	50	170	33	16	140	1.7	0.08	5.4	65	0.01	4	6
	Pork												
459	**Pork**, average, trimmed lean, *raw*	63	380	7	24	190	0.7	0.05	2.1	51	0.01	13	5
460	trimmed fat, *raw*	47	160	9	9	91	0.4	0.04	0.6	51	Tr	7	5
461	fat, *cooked*	69	240	10	14	140	0.6	0.05	0.9	67	Tr	9	5
462	**Belly joint/slices**, *grilled*, lean and fat	97	350	20	23	220	0.9	0.12	2.9	96	0.02	17	5
463	**Diced**, *casseroled*, lean only	37	220	12	21	180	1.0	0.13	3.6	39	0.02	20	5
464	**Fillet strips**, *stir-fried*, lean	71	540	8	35	320	1.4	0.14	2.6	70	0.02	20	3
465	**Leg joint**, *raw*, lean and fat	60	330	6	21	180	0.7	0.02	1.9	50	Tr	12	5
466	*roasted medium*, lean	69	400	10	27	250	1.1	0.06	3.2	67	Tr	21	3
467	lean and fat	70	380	10	26	240	1.0	0.06	2.9	67	Tr	20	3

No.	Food	Retinol µg	Carotene µg	Vitamin D µg	Vitamin E mg	Thiamin mg	Riboflavin mg	Niacin mg	Trypt/60 mg	Vitamin B_6 mg	Vitamin B_{12} µg	Folate µg	Pantothenate mg	Biotin µg	Vitamin C mg
	Lamb *continued*														
452	**Shoulder**, *raw*, lean and fat	11	Tr	0.4	0.13	0.14	0.16	4.4	2.9	0.23	2	4	0.88	1	0
453	**diced, kebabs**, *grilled*, lean and fat	Tr	Tr	0.6	0.19	0.12	0.25	6.8	5.7	0.21	3	7	1.40	2	0
454	**whole**, *roasted*, lean	Tr	Tr	0.8	0.06	0.11	0.23	5.3	5.6	0.21	2	4	1.10	2	0
455	lean and fat	6	Tr	0.7	0.15	0.10	0.21	5.0	4.8	0.20	2	4	0.99	2	0
456	**Stewing lamb**, *pressure cooked*, lean	Tr	Tr	0.6	0.04	0.14	0.20	4.3	6.0	0.18	3	2	1.40	2	0
457	*stewed*, lean	Tr	Tr	0.6	0.20	0.04	0.12	2.3	5.5	0.11	3	2	1.30	2	0
458	lean and fat	Tr	Tr	0.6	0.20	0.05	0.12	2.4	4.9	0.11	2	2	1.19	2	0
	Pork														
459	**Pork**, average, trimmed lean, *raw*	Tr	Tr	0.5	0.05	0.98	0.24	6.9	4.5	0.54	1	3	1.46	2	0
460	trimmed fat, *raw*	Tr	Tr	1.3	0.03	0.20	0.13	2.1	1.1	0.11	1	2	0.61	5	0
461	fat, *cooked*	Tr	Tr	2.1	0.05	0.37	0.16	3.8	1.5	0.16	Tr	2	0.86	8	0
462	**Belly joint/slices**, *grilled*, lean and fat	Tr	Tr	1.1	0.03	0.60	0.18	7.0	4.9	0.38	1	8	1.77	4	0
463	**Diced**, *casseroled*, lean only	Tr	Tr	0.8	0.05	0.48	0.25	4.2	6.6	0.36	1	3	0.94	5	0
464	**Fillet strips**, *stir-fried*, lean	Tr	Tr	0.8	0.19	1.53	0.41	10.1	6.6	0.78	1	4	2.20	5	0
465	**Leg joint**, *raw*, lean and fat	Tr	Tr	0.9	0.07	0.68	0.18	5.8	3.9	0.42	1	1	1.32	3	0
466	*roasted medium*, lean	Tr	Tr	0.7	0.02	0.73	0.25	9.7	6.7	0.50	1	4	2.90	5	0
467	lean and fat	Tr	Tr	1.0	0.03	0.71	0.24	9.2	6.1	0.47	1	4	2.67	5	0

Composition of food per 100g edible portion

No.	Food	Description and main data sources	Edible conversion factor	Water	Total nitrogen	Protein	Fat	Carbo-hydrate	Energy value	
				g	g	g	g	g	kcal	kJ
***Pork** continued*										
468	**Loin chops**, *raw*, lean and fat	Calculated from 70% lean and 30% fat	0.84	59.8	2.98	18.6	21.7	0	270	1119
469	*barbecued*, lean and fat	Calculated from 82% lean and 18% fat	0.84	55.0	4.53	28.3	15.8	0	255	1066
470	*grilled*, lean	22 samples of a mixture of loin and pork chops	0.77	61.2	5.06	31.6	6.4	0	184	774
471	lean and fat	Calculated from 80% lean and 20% fat	1.00	54.6	4.64	29.0	15.7	0	257	1074
472	*microwaved*, lean and fat	Calculated from 82% lean and 18% fat	0.80	55.4	4.83	30.2	14.1	0	248	1035
473	*roasted*, lean and fat	Calculated from 78% lean and 22% fat	0.76	49.1	5.10	31.9	19.3	0	301	1256
474	**Steaks**, *raw*, lean and fat	Calculated from 89% lean and 11% fat	1.00	69.6	3.36	21.0	9.4	0	169	705
475	*grilled*, lean and fat	Calculated from 92% lean and 8% fat	1.00	59.1	5.19	32.4	7.6	0	198	832
Veal										
476	**Veal**, escalope, *raw*	9 samples	1.00	75.1	3.63	22.7	1.7	0	106	449
477	*fried*	9 samples	1.00	58.7	5.39	33.7	6.8	0	196	825
Chicken										
478	**Dark meat**, *raw*	31 samples	1.00	75.8	3.34	20.9	2.8	0	109	459
479	**Light meat**, *raw*	31 samples	1.00	74.2	3.84	24.0	1.1	0	106	449
480	**Meat**, average, *raw*	Calculated from 44% light meat and 56% dark meat	1.00	75.1	3.57	22.3	2.1	0	108	457
481	**Breast**, *casseroled*, meat only	Calculated from light meat from fresh and frozen chicken	1.00	67.7	4.54	28.4	5.2	0	160	675
482	*grilled without skin*, meat only	10 samples	1.00	66.6	5.11	32.0	2.2	0	148	626
483	strips, *stir-fried*	10 samples	1.00	65.9	4.76	29.7	4.6	0	161	677

No.	Food	Starch	Total sugars	Dietary fibre NSP	Fatty acids Satd	Mono-unsatd	Poly-unsatd	Trans	Cholest-erol
		g	g	g	g	g	g	g	mg
	Pork *continued*								
468	**Loin chops**, *raw*, lean and fat	0	0	0	8.0	8.5	3.6	0.1	61
469	*barbecued*, lean and fat	0	0	0	5.7	6.3	2.6	0.1	87
470	*grilled*, lean	0	0	0	2.2	2.6	1.0	Tr	75
471	lean and fat	0	0	0	5.6	6.5	2.5	0.1	86
472	*microwaved*, lean and fat	0	0	0	4.9	5.7	2.5	0.1	100
473	*roasted*, lean and fat	0	0	0	7.0	7.8	3.1	0.1	110
474	**Steaks**, *raw*, lean and fat	0	0	0	3.3	3.8	1.6	0.1	63
475	*grilled*, lean and fat	0	0	0	2.7	3.0	1.2	Tr	100
	Veal								
476	**Veal**, escalope, *raw*	0	0	0	0.6	0.7	0.3	Tr	52
477	*fried*	0	0	0	1.8	2.5	1.9	0.1	110
	Chicken								
478	**Dark meat**, *raw*	0	0	0	0.8	1.3	0.6	Tr	105
479	**Light meat**, *raw*	0	0	0	0.3	0.5	0.2	Tr	70
480	**Meat**, average, *raw*	0	0	0	0.6	1.0	0.4	Tr	90
481	**Breast**, *casseroled*, meat only	0	0	0	1.5	2.4	1.0	0.1	90
482	*grilled without skin*, meat only	0	0	0	0.6	1.0	0.4	Tr	94
483	strips, *stir-fried*	0	0	0	N	N	N	N	87

Inorganic constituents per 100g edible portion

No.	Food	mg										µg	
		Na	K	Ca	Mg	P	Fe	Cu	Zn	Cl	Mn	Se	I
***Pork** continued*													
468	**Loin chops**, *raw*, lean and fat	53	300	10	19	170	0.4	0.06	1.3	56	0.01	11	8
469	*barbecued*, lean and fat	68	400	21	26	240	0.8	0.08	2.3	64	0.02	17	5
470	*grilled*, lean	66	410	14	28	250	0.7	0.08	2.4	70	0.02	18	3
471	lean and fat	70	390	14	26	230	0.7	0.08	2.2	73	0.02	17	3
472	*microwaved*, lean and fat	58	330	19	24	220	0.7	0.07	2.4	65	0.02	19	5
473	*roasted*, lean and fat	68	360	19	25	230	0.8	0.09	2.4	70	0.02	20	5
474	**Steaks**, *raw*, lean and fat	58	360	6	22	210	0.7	0.02	1.7	51	Tr	13	5
475	*grilled*, lean and fat	76	460	8	32	280	1.1	0.10	2.7	68	0.02	20	5
Veal													
476	**Veal**, escalope, *raw*	59	350	4	24	230	0.6	Tr	2.4	54	0.02	9	9
477	*fried*	86	460	6	32	300	0.9	Tr	3.1	77	0.02	11	8
Chicken													
478	**Dark meat**, *raw*	90	390	7	24	110	0.8	0.02	1.7	110	0.01	14	6
479	**Light meat**, *raw*	60	370	5	29	220	0.5	0.05	0.7	77	0.01	12	6
480	**Meat**, average, *raw*	77	380	6	26	160	0.7	0.03	1.2	95	0.01	13	6
481	**Breast**, *casseroled*, meat only	60	270	9	25	210	0.5	0.06	1.1	60	0.01	13	8
482	*grilled without skin*, meat only	55	460	6	36	310	0.4	0.04	0.8	67	0.01	16	7
483	strips, *stir-fried*	61	420	6	33	280	0.5	0.08	0.8	63	0.01	15	7

No.	Food	Retinol µg	Carotene µg	Vitamin D µg	Vitamin E mg	Thiamin mg	Riboflavin mg	Niacin mg	Trypt/60 mg	Vitamin B_6 mg	Vitamin B_{12} µg	Folate µg	Pantothenate mg	Biotin µg	Vitamin C mg
	Pork *continued*														
468	**Loin chops**, *raw*, lean and fat	Tr	Tr	0.9	0.11	0.81	0.18	4.9	3.3	0.62	1	1	0.97	3	0
469	*barbecued*, lean and fat	Tr	Tr	1.0	0.02	1.03	0.17	8.6	5.6	0.32	1	1	1.82	6	0
470	*grilled*, lean	Tr	Tr	0.8	0.01	0.78	0.16	9.1	6.2	0.56	1	7	1.22	4	0
471	lean and fat	Tr	Tr	1.1	0.02	0.70	0.17	8.2	5.3	0.49	1	6	1.20	5	0
472	*microwaved*, lean and fat	Tr	Tr	1.0	0.03	0.92	0.17	7.0	6.0	0.37	1	4	1.93	5	0
473	*roasted*, lean and fat	Tr	Tr	1.1	0.03	0.77	0.16	8.4	6.3	0.42	1	2	2.05	6	0
474	**Steaks**, *raw*, lean and fat	Tr	Tr	0.6	0.05	0.85	0.22	6.8	4.2	0.54	1	1	1.42	2	0
475	*grilled*, lean and fat	Tr	Tr	0.9	0.02	1.45	0.27	9.1	6.6	0.68	1	8	2.09	5	0
	Veal														
476	**Veal**, escalope, *raw*	Tr	Tr	1.4	0.26	0.12	0.23	7.8	4.8	0.65	2	23	0.87	1	0
477	*fried*	6	Tr	1.3	0.39	0.08	0.25	7.8	7.6	0.70	4	17	1.02	5	0
	Chicken														
478	**Dark meat**, *raw*	20	Tr	0.1	0.17	0.14	0.22	5.6	4.1	0.28	1	9	1.09	3	0
479	**Light meat**, *raw*	Tr	Tr	0.2	0.13	0.14	0.14	10.7	4.7	0.51	Tr	14	1.26	2	0
480	**Meat**, average, *raw*	11	Tr	0.1	0.15	0.14	0.18	7.8	4.3	0.38	Tr	19	1.16	2	0
481	**Breast**, *casseroled*, meat only	Tr	Tr	0.1	0.07	0.06	0.13	8.8	5.6	0.36	Tr	6	1.34	2	0
482	*grilled without skin*, meat only	Tr	Tr	0.3	0.17	0.14	0.13	15.8	6.2	0.63	Tr	6	1.67	2	0
483	strips, *stir-fried*	Tr	Tr	0.2	N	0.11	0.16	14.4	5.8	0.44	Tr	5	1.56	2	0

Composition of food per 100g edible portion

No.	Food	Description and main data sources	Edible conversion factor	Water	Total nitrogen	Protein	Fat	Carbo-hydrate	Energy value	
				g	g	g	g	g	kcal	kJ
Chicken *continued*										
484	**Breast in crumbs**, *chilled, fried*	4 samples	1.00	53.2	2.88	18.0	12.7	14.8	242	1013
485	**Drumsticks**, *roasted*, meat and skin	Calculated from 89% dark meat and 11% skin from fresh and frozen chicken	0.63	63.0	4.14	25.8	9.1	0	185	775
486	**Roasted**, meat, average	Calculated from 46% light meat and 54% dark meat	1.00	65.3	4.37	27.3	7.5	0	177	742
487	dark meat	19 samples of a mixture of fresh and frozen chicken	1.00	63.9	3.90	24.4	10.9	0	196	819
488	light meat	19 samples of a mixture of fresh and frozen chicken	1.00	66.9	4.83	30.2	3.6	0	153	645
489	leg quarter, meat and skin	20 samples	0.51	60.9	3.34	20.9	16.9	0	236	981
490	wing quarter, meat and skin	20 samples	0.53	59.9	3.97	24.8	14.1	0	226	943
491	**Skin**, dry, *roasted/grilled*	34 samples; crisply roasted	1.00	31.1	3.45	21.5	46.1	0	501	2070
Turkey										
492	**Dark meat**, *raw*	20 samples	1.00	75.8	3.26	20.4	2.5	0	104	439
493	**Light meat**, *raw*	20 samples	1.00	74.9	3.90	24.4	0.8	0	105	444
494	**Meat**, average, *raw*	Calculated from 56% light meat and 44% dark meat	1.00	75.3	3.62	22.6	1.6	0	105	443
495	**Breast**, fillet, *grilled*, meat only	9 samples; skinless	1.00	63.0	5.60	35.0	1.7	0	155	658
496	strips, *stir-fried*	8 samples; skinless	1.00	64.4	4.96	31.0	4.5	0	164	692
497	**Roasted**, dark meat	27 samples including self-basting turkey	1.00	64.3	4.71	29.4	6.6	0	177	745
498	light meat	18 samples	1.00	65.1	5.39	33.7	2.0	0	153	648
499	meat, average	Calculated from 51% light meat and 49% dark meat from fresh, frozen and self-basting turkey	1.00	64.6	4.99	31.2	4.6	0	166	701
500	**Skin**, dry, *roasted*	10 samples	1.00	29.5	3.06	29.9	40.2	0	481	1995

Composition of food per 100g edible portion

No.	Food	Starch	Total sugars	Dietary fibre NSP	Fatty acids Satd	Mono-unsatd	Poly-unsatd	Trans	Cholest-erol
		g	g	g	g	g	g	g	mg
	Chicken *continued*								
484	**Breast in crumbs**, *chilled, fried*	14.0	0.8	(0.7)	2.1	5.3	4.6	0.4	(33)
485	**Drumsticks**, *roasted*, meat and skin	0	0	0	2.5	4.3	1.8	0.1	135
486	**Roasted**, meat, average	0	0	0	2.9	5.1	2.2	0.1	120
487	dark meat	0	0	0	1.0	1.6	0.7	0.1	82
488	light meat	0	0	0	2.1	3.4	1.5	0.1	105
489	leg quarter, meat and skin	0	0	0	4.6	7.8	3.2	0.2	115
490	wing quarter, meat and skin	0	0	0	3.9	6.4	2.7	0.2	100
491	**Skin**, dry, *roasted/grilled*	0	0	0	12.9	22.5	7.7	0.6	170
	Turkey								
492	**Dark meat**, *raw*	0	0	0	0.8	1.0	0.6	Tr	86
493	**Light meat**, *raw*	0	0	0	0.3	0.3	0.2	Tr	57
494	**Meat**, average, *raw*	0	0	0	0.5	0.6	0.4	Tr	70
495	**Breast**, fillet, *grilled*, meat only	0	0	0	0.6	0.6	0.3	Tr	74
496	strips, *stir-fried*	0	0	0	N	N	N	N	72
497	**Roasted**, dark meat	0	0	0	2.0	2.4	1.7	0.1	120
498	light meat	0	0	0	0.7	0.7	0.5	Tr	82
499	meat, average	0	0	0	1.4	1.7	1.1	0.1	100
500	**Skin**, dry, *roasted*	0	0	0	13.2	15.6	8.8	0.6	290

Inorganic constituents per 100g edible portion

No.	Food	mg										µg	
		Na	K	Ca	Mg	P	Fe	Cu	Zn	Cl	Mn	Se	I
	Chicken *continued*												
484	**Breast in crumbs**, *chilled, fried*	(420)	(280)	(21)	(24)	(180)	(0.1)	(0.06)	(0.5)	(620)	(0.12)	N	N
485	**Drumsticks**, *roasted*, meat and skin	130	280	15	25	210	1.0	0.09	2.3	90	0.02	17	7
486	**Roasted**, meat, average	100	300	17	23	200	0.8	0.08	2.2	88	0.02	17	6
487	dark meat	60	360	7	30	250	0.4	0.17	0.8	62	0.01	14	7
488	light meat	80	330	11	26	220	0.7	0.10	1.5	75	0.02	16	7
489	leg quarter, meat and skin	95	230	12	20	180	0.8	0.06	1.7	85	0.01	16	7
490	wing quarter, meat and skin	100	260	11	24	200	0.6	0.04	1.1	75	0.01	15	7
491	**Skin**, dry, *roasted/grilled*	80	260	16	26	210	1.3	0.05	1.2	N	0.03	N	N
	Turkey												
492	**Dark meat**, *raw*	90	310	7	22	200	1.0	0.04	3.1	73	Tr	15	5
493	**Light meat**, *raw*	50	360	4	27	230	0.3	0.01	1.0	39	Tr	10	6
494	**Meat**, average, *raw*	68	340	5	25	220	0.6	0.02	1.9	54	Tr	13	6
495	**Breast**, fillet, *grilled*, meat only	90	550	5	42	380	0.6	0.08	1.7	85	0.01	17	8
496	strips, *stir-fried*	60	420	5	32	280	0.4	0.04	1.3	75	0.01	15	7
497	**Roasted**, dark meat	110	330	17	25	260	1.2	0.11	3.4	86	0.02	17	8
498	light meat	50	400	6	30	260	0.5	0.05	1.4	52	0.01	14	8
499	meat, average	90	350	11	27	260	0.8	0.09	2.5	85	0.01	17	8
500	**Skin**, dry, *roasted*	110	330	20	33	250	1.6	0.07	1.8	N	0.03	N	N

No.	Food	Retinol µg	Carotene µg	Vitamin D µg	Vitamin E mg	Thiamin mg	Riboflavin mg	Niacin mg	Trypt/60 mg	Vitamin B_6 mg	Vitamin B_{12} µg	Folate µg	Pantothenate mg	Biotin µg	Vitamin C mg
	Chicken *continued*														
484	**Breast in crumbs**, *chilled, fried*	Tr	Tr	N	(0.61)	(0.11)	(0.06)	7.6	(4.1)	(0.49)	Tr	(6)	(1.11)	(2)	0
485	**Drumsticks**, *roasted*, meat and skin	24	Tr	0.2	0.21	0.09	0.14	5.5	4.9	0.19	1	12	1.31	3	0
486	**Roasted**, meat, average	24	Tr	0.1	0.23	0.07	0.11	6.2	5.3	0.27	1	10	1.34	4	0
487	dark meat	Tr	Tr	0.3	0.31	0.07	0.23	12.6	5.5	0.54	Tr	10	1.38	2	0
488	light meat	11	Tr	0.2	0.23	0.07	0.16	9.2	5.3	0.36	Tr	10	1.39	3	0
489	leg quarter, meat and skin	26	Tr	0.2	0.23	0.07	0.28	5.0	4.6	0.40	1	11	1.21	3	0
490	wing quarter, meat and skin	13	Tr	0.4	0.27	0.06	0.17	10.0	4.9	0.25	Tr	6	0.67	1	0
491	**Skin**, dry, *roasted/grilled*	N	Tr	1.0	N	N	N	N	N	N	N	N	N	N	0
	Turkey														
492	**Dark meat**, *raw*	Tr	Tr	0.4	Tr	0.08	0.31	4.6	4.0	0.35	2	28	0.75	2	0
493	**Light meat**, *raw*	Tr	Tr	0.3	Tr	0.06	0.15	10.7	4.3	0.81	1	9	0.66	1	0
494	**Meat**, average, *raw*	Tr	Tr	0.3	0.01	0.07	0.22	8.0	4.4	0.61	2	17	0.70	2	0
495	**Breast**, fillet, *grilled*, meat only	Tr	Tr	0.4	0.02	0.07	0.15	14.0	6.8	0.63	1	7	0.95	2	0
496	strips, *stir-fried*	Tr	Tr	0.3	N	0.07	0.12	13.5	6.1	0.69	1	8	0.84	2	0
497	**Roasted**, dark meat	Tr	Tr	0.3	Tr	0.05	0.25	7.2	5.7	0.44	2	20	1.06	3	0
498	light meat	Tr	Tr	0.1	0.02	0.05	0.16	12.9	6.8	0.47	1	18	0.97	2	0
499	meat, average	Tr	Tr	0.3	0.06	0.06	0.19	10.3	6.2	0.49	1	17	0.98	2	0
500	**Skin**, dry, *roasted*	N	Tr	N	N	N	N	N	N	N	N	N	N	N	0

Composition of food per 100g edible portion

No.	Food	Description and main data sources	Edible conversion factor	Water	Total nitrogen	Protein	Fat	Carbo-hydrate	Energy value	
				g	g	g	g	g	kcal	kJ
Duck										
501	**Duck**, *raw*, meat only	19 samples, meat from dressed carcase	0.28	74.8	3.15	19.7	6.5	0	137	575
502	*crispy, Chinese style*	10 samples from Chinese takeaways, seasoned roasted duck	1.00	44.0	4.46	27.9	24.2	0.3	331	1375
503	*roasted*, meat only	10 samples from dressed carcase	0.21	62.1	4.05	25.3	10.4	0	195	815
504	meat, fat and skin	20 samples; meat, fat and skin = 0.42 of dressed carcase	1.00	42.6	3.20	20.0	38.1	0	423	1750
Goose										
505	**Goose**, *roasted*, meat, fat and skin	5 samples; meat, fat and skin = 0.65 of dressed carcase	1.00	51.1	4.40	27.5	21.2	0	301	1252
Pheasant										
506	**Pheasant**, *roasted*, meat only	10 samples from dressed carcase	0.52	59.4	4.46	27.9	12.0	0	220	918
Rabbit										
507	**Rabbit**, *raw*, meat only	10 samples from leg and loin	1.00	71.5	3.50	21.9	5.5	0	137	576
508	*stewed*, meat only	30 samples of a mixture of fresh, wild, farmed and frozen imported	0.60	70.7	3.39	21.2	3.2	0	114	479
Venison										
509	**Venison**, *roast*	Haunch, meat only, calculated from raw	1.00	60.4	5.70	35.6	2.5	0	165	698

Composition of food per 100g edible portion

No.	Food	Starch	Total sugars	Dietary fibre NSP	Fatty acids Satd	Mono-unsatd	Poly-unsatd	Trans	Cholest-erol
		g	g	g	g	g	g	g	mg
Duck									
501	**Duck**, *raw*, meat only	0	0	0	2.0	3.2	1.0	0.1	110
502	*crispy, Chinese style*	0	0	0	(7.2)	(12.3)	(3.4)	(0.2)	(63)
503	*roasted*, meat only	0	0	0	3.3	5.2	1.3	0.1	115
504	meat, fat and skin	0	0	0	11.4	19.3	5.3	0.4	99
Goose									
505	**Goose**, *roasted*, meat, fat and skin	0	0	0	(6.6)	(9.9)	(2.4)	Tr	(91)
Pheasant									
506	**Pheasant**, *roasted*, meat only	0	0	0	4.1	5.6	1.6	0.1	(220)
Rabbit									
507	**Rabbit**, *raw*, meat only	0	0	0	2.1	1.3	1.8	0.1	53
508	*stewed*, meat only	0	0	0	1.7	0.7	0.6	0.1	49
Venison									
509	**Venison**, *roast*	0	0	0	N	N	N	Tr	N

Inorganic constituents per 100g edible portion

No.	Food	mg										µg	
		Na	K	Ca	Mg	P	Fe	Cu	Zn	Cl	Mn	Se	I
	Duck												
501	**Duck**, *raw*, meat only	110	290	12	19	200	2.4	0.34	1.9	98	Tr	22	N
502	*crispy, Chinese style*	453	292	(22)	25	257	4.0	0.30	2.8	(396)	0.10	(22)	N
503	*roasted*, meat only	96	270	13	20	200	2.7	0.31	2.6	96	Tr	(22)	N
504	meat, fat and skin	87	220	22	17	180	1.7	0.23	2.2	76	0.20	(22)	N
	Goose												
505	**Goose**, *roasted*, meat, fat and skin	80	320	10	23	220	3.3	0.15	2.6	80	0.01	N	N
	Pheasant												
506	**Pheasant**, *roasted*, meat only	66	360	28	26	220	2.2	0.10	1.3	170	0.02	(14)	N
	Rabbit												
507	**Rabbit**, *raw*, meat only	67	360	22	25	220	1.0	0.06	1.4	74	0.01	17	N
508	*stewed*, meat only	48	200	39	18	150	1.1	0.06	1.7	45	0.02	(16)	N
	Venison												
509	**Venison**, *roast*	52	290	6	27	240	5.1	0.36	3.9	59	0.04	(14)	N

Vitamins per 100g edible portion

No.	Food	Retinol µg	Carotene µg	Vitamin D µg	Vitamin E mg	Thiamin mg	Riboflavin mg	Niacin mg	Trypt/60 mg	Vitamin B_6 mg	Vitamin B_{12} µg	Folate µg	Pantothenate mg	Biotin µg	Vitamin C mg
Duck															
501	**Duck**, *raw*, meat only	(24)	Tr	N	0.02	0.36	0.45	5.3	4.2	0.34	3	25	1.60	6	0
502	*crispy, Chinese style*	9	Tr	1.0	2.17	0.09	0.39	3.5	(4.2)	0.15	3	(15)	(1.50)	Tr	(0)
503	*roasted*, meat only	N	N	N	0.02	0.26	0.47	5.1	5.4	0.25	3	10	1.50	4	0
504	meat, fat and skin	N	N	N	N	0.18	0.51	3.8	4.2	0.31	2	15	2.60	7	0
Goose															
505	**Goose**, *roasted*, meat, fat and skin	(21)	Tr	N	N	0.12	0.51	4.6	(5.5)	0.42	2	12	1.40	3	0
Pheasant															
506	**Pheasant**, *roasted*, meat only	N	N	N	N	0.02	0.29	9.2	6.0	0.57	3	20	(0.96)	N	0
Rabbit															
507	**Rabbit**, *raw*, meat only	N	N	N	0.13	0.10	0.19	8.4	4.1	0.50	10	5	0.80	1	0
508	*stewed*, meat only	N	N	N	N	0.02	0.16	6.2	5.1	0.29	3	5	0.80	1	0
Venison															
509	**Venison**, *roast*	N	N	N	N	0.16	0.69	5.5	6.5	0.65	1	6	N	N	0

Composition of food per 100g edible portion

No.	Food	Description and main data sources	Edible conversion factor	Water	Total nitrogen	Protein	Fat	Carbo-hydrate	Energy value	
				g	g	g	g	g	kcal	kJ
Offal										
510	**Heart**, lamb, *roasted*	10 samples, fat and valves removed	1.00	58.8	4.05	25.3	13.9	0	226	944
511	**Kidney**, lamb, *fried*	10 samples, skin and core removed	1.00	62.8	3.79	23.7	10.3	0	188	784
512	ox, *stewed*	10 samples, skin and core removed	1.00	69.2	3.92	24.5	4.4	0	138	579
513	pig, *stewed*	20 samples, core removed. Salt added	1.00	66.3	3.91	24.4	6.1	0	153	641
514	**Liver**, calf, *fried*	10 samples	1.00	64.5	3.57	22.3	9.6	Tr	176	734
515	chicken, *fried*	10 samples	1.00	65.9	3.54	22.1	8.9	Tr	169	705
516	lamb, *fried*	10 samples	1.00	53.9	4.82	30.1	12.9	Tr	237	989
517	ox, *stewed*	18 samples, coated in seasoned flour	1.00	62.6	3.96	24.8	9.5	3.6	198	831
518	pig, *stewed*	18 samples, coated in seasoned flour	1.00	62.1	4.09	25.6	8.1	3.6	189	793
519	**Oxtail**, *stewed*	12 samples, meat only. Salt added	1.00	53.9	4.88	30.5	13.4	0	243	1014
520	**Tongue**, sheep, *stewed*	Fat and skin removed	1.00	56.9	2.91	18.2	24.0	0	289	1197
521	**Tripe**, dressed, *raw*	6 samples	1.00	92.1	1.14	7.1	0.5	0	33	139
522	**Trotters and tails**, *boiled*	23% trotters and 77% pig tails. Salt added	0.54	53.5	3.17	19.8	22.3	0	280	1162
Burgers and grillsteaks										
523	**Beefburgers**, *chilled/frozen, raw*	8 samples, 3 brands. 98-99% meat	1.00	56.1	2.74	17.1	24.7	0.1	291	1206
524	*fried*	8 samples, 3 brands	1.00	46.2	4.56	28.5	23.9	0.1	329	1370
525	*grilled*	8 samples, 3 brands	1.00	47.9	4.24	26.5	24.4	0.1	326	1355
526	**Big Mac**	Manufacturer's data (McDonald's). Portion includes two beefburgers, bun, sauce, cheese, lettuce, onions and pickles	1.00	N	1.98	12.4	10.7	*22.0*	*228*	*959*

Composition of food per 100g edible portion

No.	Food	Starch	Total sugars	Dietary fibre NSP	Fatty acids Satd	Mono-unsatd	Poly-unsatd	Trans	Cholest-erol
		g	g	g	g	g	g	g	mg
	Offal								
510	**Heart**, lamb, *roasted*	0	0	0	N	N	N	Tr	260
511	**Kidney**, lamb, *fried*	0	0	0	N	N	N	Tr	610
512	ox, *stewed*	0	0	0	1.4	1.0	0.9	0.1	460
513	pig, *stewed*	0	0	0	2.0	1.6	0.9	Tr	700
514	**Liver**, calf, *fried*	0	0	0	N	N	N	Tr	330
515	chicken, *fried*	0	0	0	N	N	N	Tr	350
516	lamb, *fried*	0	0	0	N	N	N	Tr	400
517	ox, *stewed*	3.6	Tr	0	3.5	1.5	2.0	Tr	240
518	pig, *stewed*	3.6	0	0	2.5	1.3	2.2	Tr	290
519	**Oxtail**, *stewed*	0	0	0	N	N	N	Tr	110
520	**Tongue**, sheep, *stewed*	0	0	0	N	N	N	Tr	(270)
521	**Tripe**, dressed, *raw*	0	0	0	0.2	0.2	Tr	Tr	64
522	**Trotters and tails**, *boiled*	0	0	0	N	N	N	Tr	N
	Burgers and grillsteaks								
523	**Beefburgers**, *chilled/frozen, raw*	Tr	0.1	0	10.7	11.4	0.5	1.4	76
524	*fried*	Tr	0.1	0	10.7	10.8	0.8	0.8	96
525	*grilled*	Tr	0.1	0	10.9	11.2	0.7	1.4	(75)
526	**Big Mac**	16.6	5.4	N	4.6	4.4	1.6	0.1	23

Inorganic constituents per 100g edible portion

No.	Food	mg										µg	
		Na	K	Ca	Mg	P	Fe	Cu	Zn	Cl	Mn	Se	I
Offal													
510	**Heart**, lamb, *roasted*	84	210	7	21	240	6.0	0.66	2.8	100	0.03	N	N
511	**Kidney**, lamb, *fried*	230	280	14	21	350	11.2	0.58	3.6	410	0.13	(209)	N
512	ox, *stewed*	150	210	17	19	290	9.0	0.63	3.0	190	0.14	(210)	N
513	pig, *stewed*	370	190	13	21	330	6.4	0.84	4.7	480	0.18	(250)	N
514	**Liver**, calf, *fried*	70	350	8	24	380	12.2	23.86	15.9	110	0.29	(27)	N
515	chicken, *fried*	79	300	9	23	350	11.3	0.52	3.8	110	0.35	N	N
516	lamb, *fried*	82	340	8	25	500	7.7	13.54	5.9	140	0.45	(62)	N
517	ox, *stewed*	110	250	11	19	380	7.8	2.30	4.3	120	0.44	(50)	N
518	pig, *stewed*	130	250	11	22	390	17.0	2.50	8.2	150	0.40	(50)	N
519	**Oxtail**, *stewed*	190	170	14	18	140	3.8	0.27	8.8	270	N	N	N
520	**Tongue**, sheep, *stewed*	80	110	11	13	200	3.4	N	N	80	N	N	N
521	**Tripe**, dressed, *raw*	50	12	52	3	16	0.2	0.04	0.7	8	0.02	N	N
522	**Trotters and tails**, *boiled*	1620	30	130	8	110	0.7	0.07	2.4	2490	0.01	N	N
Burgers and grillsteaks													
523	**Beefburgers**, *chilled/frozen, raw*	290	290	7	16	150	1.7	0.12	3.8	350	0.02	8	(8)
524	*fried*	470	420	12	26	240	2.8	0.13	6.3	570	0.02	(10)	(13)
525	*grilled*	400	380	10	22	210	2.5	0.13	6.1	520	0.02	(9)	(12)
526	**Big Mac**	430	142	68	22	142	0.9	N	N	192	N	N	N

No.	Food	Retinol µg	Carotene µg	Vitamin D µg	Vitamin E mg	Thiamin mg	Riboflavin mg	Niacin mg	Trypt/60 mg	Vitamin B_6 mg	Vitamin B_{12} µg	Folate µg	Pantothenate mg	Biotin µg	Vitamin C mg
	Offal														
510	**Heart**, lamb, *roasted*	Tr	Tr	0.1	N	0.24	1.37	3.8	5.6	0.26	6	2	(3.80)	(8)	2
511	**Kidney**, lamb, *fried*	110	Tr	0.6	0.41	0.52	3.10	9.1	5.3	0.48	54	70	4.60	73	5
512	ox, *stewed*	45	N	N	0.42	0.24	3.29	6.2	(5.5)	0.57	38	130	3.10	79	5
513	pig, *stewed*	46	Tr	N	0.36	0.19	2.10	6.1	5.2	0.28	15	43	2.40	53	11
514	**Liver**, calf, *fried*	(25200)[a]	100	0.3	0.50	0.61	2.89	13.6	5.8	0.89	58	110	4.10	50	19
515	chicken, *fried*	(10500)[a]	Tr	N	0.34	0.63	2.72	12.9	4.4	0.55	45	1350	5.90	216	23
516	lamb, *fried*	(19700)[a]	60	0.9	0.32	0.38	5.65	19.9	4.9	0.53	83	207	8.00	33	19
517	ox, *stewed*	(17300)[a]	1500	1.1	0.44	0.18	3.60	10.3	5.3	0.52	110	290	5.70	50	15
518	pig, *stewed*	(22600)[a]	Tr	1.1	0.16	0.21	3.10	11.5	5.5	0.64	26	110	4.60	34	9
519	**Oxtail**, *stewed*	Tr	Tr	Tr	0.45	0.02	0.28	3.3	6.5	0.14	2	9	0.90	2	0
520	**Tongue**, sheep, *stewed*	Tr	Tr	Tr	(0.32)	(0.13)	(0.45)	(3.7)	3.9	(0.10)	(7)	(4)	(0.80)	(2)	(6)
521	**Tripe**, dressed, *raw*	Tr	Tr	Tr	0.08	Tr	Tr	Tr	1.2	Tr	Tr	7	Tr	1	3
522	**Trotters and tails**, *boiled*	Tr	Tr	Tr	N	0.06	0.20	0.9	3.7	N	1	3	N	N	0
	Burgers and grillsteaks														
523	**Beefburgers**, *chilled/frozen, raw*	Tr	Tr	1.2	0.28	0.01	0.15	3.5	2.5	0.28	2	9	0.78	1	0
524	*fried*	Tr	Tr	(1.9)	0.54	Tr	0.22	5.5	4.3	0.31	3	8	0.85	2	0
525	*grilled*	Tr	Tr	(1.8)	0.39	0.01	0.20	5.1	4.0	0.31	3	10	0.84	2	0
526	**Big Mac**	2	N	0.3	0.23	0.05	0.11	N	N	0.01	N	N	N	N	1

[a] Total retinol

Composition of food per 100g edible portion

No.	Food	Description and main data sources	Edible conversion factor	Water	Total nitrogen	Protein	Fat	Carbo-hydrate	Energy value	
				g	g	g	g	g	kcal	kJ
	Burgers and grillsteaks *continued*									
527	**Cheeseburger**, *takeaway*	Manufacturers' data and calculation from ingredient proportions. Includes beefburger, bun, cheese, mustard, ketchup, onions and pickles	1.00	(47.0)	2.18	13.6	11.8	*26.1*	*259*	*1086*
528	**Chicken burger**, *takeaway*	Manufacturers' data. Portion includes chicken burger, bun, lettuce and mayonnaise	1.00	N	2.00	12.5	10.8	*23.4*	*267*	*1118*
529	**Economy burgers**, *frozen, raw*	10 samples, 6 brands containing onion. 60% meat	1.00	57.1	2.19	13.7	21.2	4.0	261	1081
530	*grilled*	10 samples, 6 brands containing onion	1.00	50.8	2.53	15.8	19.3	9.7[a]	273	1138
531	**Grillsteaks**, beef, *chilled/frozen, grilled*	10 samples, 7 brands	1.00	50.1	3.54	22.1	23.9	0.5	305	1268
532	**Hamburger**, *takeaway*	Manufacturers' data and calculation from ingredient proportions. Portion includes bun, beefburger, mustard, ketchup, onions and pickles	1.00	(49.4)	2.22	13.9	9.6	*26.9*	*243*	*1022*
533	**Quarterpounder with cheese**, *takeaway*	Manufacturers' data (McDonald's). Portion includes a quarter pound beefburger, bun, ketchup, mustard,onions, pickles and slice of cheese	1.00	N	2.44	15.1	13.0	*19.5*	*250*	*1048*
534	**Whopper burger**	Manufacturers' data and calculation from ingredient proportions. Portion includes bun, beefburger, mayonnaise, lettuce, tomato, ketchup, onions and pickles	1.00	(52.5)	1.70	10.7	14.8	*17.4*	*241*	*1008*

[a] Includes 0.1g oligosaccharides per 100g food

Composition of food per 100g edible portion

No.	Food	Starch	Total sugars	Dietary fibre NSP	Fatty acids Satd	Fatty acids Mono-unsatd	Fatty acids Poly-unsatd	Fatty acids Trans	Cholest-erol
		g	g	g	g	g	g	g	mg
	Burgers and grillsteaks *continued*								
527	**Cheeseburger**, *takeaway*	21.2	4.9	(0.7)	(6.2)	(5.4)	(0.9)	(0.3)	(32)
528	**Chicken burger**, *takeaway*	N	N	Tr	N	N	N	N	N
529	**Economy burgers**, *frozen, raw*	3.1	0.9	0.9	8.0	9.4	2.1	0.8	(92)
530	*grilled*	8.9	0.7	0.8	7.3	8.6	1.9	0.7	84
531	**Grillsteaks**, beef, *chilled/frozen, grilled*	Tr	0.5	Tr	10.8	10.7	0.8	(0.7)	88
532	**Hamburger**, *takeaway*	21.2	5.7	(0.8)	(4.0)	(4.2)	(0.8)	(0.2)	(40)
533	**Quarterpounder with cheese**, *takeaway*	14.0	5.5	N	6.4	5.6	0.9	0.1	33
534	**Whopper burger**	14.4	3.0	(0.8)	4.4	(5.4)	(4.2)	(0.3)	31

Inorganic constituents per 100g edible portion

No.	Food	mg										µg	
		Na	K	Ca	Mg	P	Fe	Cu	Zn	Cl	Mn	Se	I
	Burgers and grillsteaks *continued*												
527	**Cheeseburger**, *takeaway*	678	210	85	(27)	(230)	1.1	(0.13)	(3.0)	(920)	(0.23)	(15)	(15)
528	**Chicken burger**, *takeaway*	560	190	19	N	N	0.4	N	N	N	Tr	(9)	N
529	**Economy burgers**, *frozen, raw*	590	210	32	18	170	2.1	0.11	2.5	830	0.13	N	N
530	*grilled*	800	270	110	25	200	2.5	0.15	1.2	(1200)	0.25	N	N
531	**Grillsteaks**, beef, *chilled/frozen, grilled*	710	360	18	19	190	2.4	0.10	4.7	980	0.02	(3)	N
532	**Hamburger**, *takeaway*	620	210	40	(28)	(170)	1.2	(0.12)	(3.0)	(900)	(0.25)	(16)	(13)
533	**Quarterpounder with cheese**, *takeaway*	511	168	110	22	141	1.0	N	N	391	N	N	N
534	**Whopper burger**	333	(230)	(50)	(20)	(130)	(1.8)	(0.09)	(2.2)	(670)	(0.21)	(12)	(13)

Vitamins per 100g edible portion

No.	Food	Retinol µg	Carotene µg	Vitamin D µg	Vitamin E mg	Thiamin mg	Riboflavin mg	Niacin mg	Trypt/60 mg	Vitamin B_6 mg	Vitamin B_{12} µg	Folate µg	Pantothenate mg	Biotin µg	Vitamin C mg
	Burgers and grillsteaks *continued*														
527	**Cheeseburger**, *takeaway*	(24)	(23)	(0.3)	(0.26)	0.17	0.18	2.2	(2.8)	0.19	2	(23)	(0.46)	(1)	N
528	**Chicken burger**, *takeaway*	N	N	N	N	0.26	0.07	4.3	N	0.28	N	N	N	N	N
529	**Economy burgers**, *frozen, raw*	(4)	Tr	N	N	0.07	0.23	2.6	N	0.17	2	12	N	N	Tr
530	*grilled*	5	Tr	N	N	0.07	0.07	3.8	N	0.17	2	24	N	N	Tr
531	**Grillsteaks**, beef, *chilled/frozen, grilled*	Tr	Tr	N	0.15	0.13	0.14	4.2	4.7	0.18	3	N	0.63	2	Tr
532	**Hamburger**, *takeaway*	N	Tr	(0.3)	(0.22)	0.19	0.12	2.5	(2.7)	(0.18)	(1)	(24)	(0.48)	(1)	N
533	**Quarterpounder with cheese**, *takeaway*	Tr	N	0.2	0.14	0.04	0.12	N	N	0.06	N	N	N	N	Tr
534	**Whopper burger**	(7)	(94)	(0.2)	(1.80)	(0.15)	(0.11)	(2.4)	(2.0)	(0.13)	(1)	(25)	(0.40)	(1)	(2)

Composition of food per 100g edible portion

No.	Food	Description and main data sources	Edible conversion factor	Water	Total nitrogen	Protein	Fat	Carbo-hydrate	Energy value	
				g	g	g	g	g	kcal	kJ
	Meat products									
535	**Black pudding**, *dry-fried*	8 samples, 6 brands	1.00	44.3	1.65	10.3	21.5	16.6	297	1236
536	**Chicken nuggets**, *takeaway*	2 samples	1.00	47.8	2.99	18.7	13.0	19.5	265	1111
537	**Chicken pie**, *individual, chilled/frozen, baked*	12 samples including chicken, chicken and ham, chicken and mushroom and chicken and vegetable pies. 10.5-25% meat	1.00	45.6	1.44	9.0	17.7	24.6	288	1202
538	**Chicken roll**	10 samples, 3 brands	1.00	71.3	2.74	17.1	4.8	5.2	131	552
539	**Corned beef**, canned	10 samples, 4 brands	1.00	59.5	4.14	25.9	10.9	1.0	205	860
540	**Cornish pastie**	10 samples, 5 brands	1.00	46.5	1.07	6.7	16.3	25.0	267	1117
541	**Frankfurter**	10 samples, 7 brands of continental style frankfurters. 75-90% meat	1.00	54.2	2.17	13.6	25.4	1.1	287	1189
542	**Game pie**	Recipe	1.00	29.3	1.95	12.2	22.5	34.7[a]	381	1595
543	**Haggis**, *boiled*	8 samples	1.00	46.2	1.71	10.7	21.7	19.2	310	1292
544	**Liver sausage**	10 samples, 4 brands	1.00	58.4	2.14	13.4	16.7	6.0	226	942
545	**Luncheon meat**, canned	10 samples, 9 brands	1.00	54.4	2.06	12.9	23.8	3.6	279	1158
546	**Meat spread**	10 samples of a mixture of beef and ham based spreads. 70-90% meat	1.00	64.6	2.51	15.7	13.4	2.3	192	800
547	**Pate**, liver	20 samples including canned	1.00	47.6	2.02	12.6	32.7	(0.8)	348	1437
548	meat, reduced fat	11 samples, assorted types; pork meat and liver based. 70-80% meat	1.00	65.0	2.88	18.0	12.0	3.0	191	798
549	**Polony**	24 samples	1.00	52.0	1.50	9.4	21.1	14.2	281	1168

[a] Includes 0.1g oligosaccharides per 100g food

No.	Food	Starch	Total sugars	Dietary fibre NSP	Fatty acids Satd	Fatty acids Mono-unsatd	Fatty acids Poly-unsatd	Fatty acids Trans	Cholest-erol
		g	g	g	g	g	g	g	mg
	Meat products								
535	**Black pudding**, *dry-fried*	16.4	0.2	(0.2)	(8.5)	(8.1)	(3.6)	N	68
536	**Chicken nuggets**, *takeaway*	18.4	1.1	0.2	3.3	6.8	2.2	1.5	55
537	**Chicken pie**, *individual, chilled/frozen, baked*	23.0	1.6	0.8	7.0	7.4	2.4	1.2	32
538	**Chicken roll**	5.2	0	Tr	1.5	2.1	0.9	0.08	40
539	**Corned beef**, canned	0	1.0	0	5.7	4.3	0.3	0.68	84
540	**Cornish pastie**	24.1	0.9	0.9	5.9	8.4	1.2	3.42	33
541	**Frankfurter**	Tr	1.1	0.1	9.2	11.5	3.0	0.12	76
542	**Game pie**	31.2	3.4	1.3	7.9	9.0	4.0	0.2	60
543	**Haggis**, *boiled*	(19.2)	Tr	(0.2)	7.6	6.9	1.4	N	91
544	**Liver sausage**	5.0	1.0	0.7	5.3	5.7	2.3	Tr	115
545	**Luncheon meat**, canned	3.6	Tr	0.2	8.7	11.0	3.0	0.4	64
546	**Meat spread**	2.1	0.2	Tr	5.5	5.8	1.2	0.2	62
547	**Pate**, liver	0.8	0.4	Tr	9.5	11.8	3.0	Tr	170
548	meat, reduced fat	1.7	1.3	Tr	3.5	3.9	1.5	0.1	160
549	**Polony**	(14.2)	Tr	N	N	N	N	N	40

Inorganic constituents per 100g edible portion

No.	Food	mg										µg	
		Na	K	Ca	Mg	P	Fe	Cu	Zn	Cl	Mn	Se	I
	Meat products												
535	**Black pudding**, *dry-fried*	(940)	(110)	(120)	(16)	(80)	(12.3)	(0.11)	(0.7)	(1560)	N	6	5
536	**Chicken nuggets**, *takeaway*	510	280	25	23	210	0.6	Tr	0.5	690	0.10	N	N
537	**Chicken pie**, *individual, chilled/frozen, baked*	430	140	60	15	90	0.8	0.06	0.6	710	0.23	N	N
538	**Chicken roll**	680	190	18	18	220	0.4	0.11	0.5	1050	0.02	N	N
539	**Corned beef**, canned	860	140	27	15	130	2.4	0.18	5.5	1560	0.02	(8)	14
540	**Cornish pastie**	400	140	60	14	75	1.1	0.31	0.6	720	0.20	(2)	3
541	**Frankfurter**	920	170	12	11	200	1.1	0.11	1.4	1280	0.02	(8)	18
542	**Game pie**	430	170	64	18	120	2.1	0.15	1.2	650	0.28	4	8
543	**Haggis**, *boiled*	770	170	29	36	160	4.8	0.44	1.9	1200	N	N	N
544	**Liver sausage**	810	180	20	14	260	6.0	0.91	2.6	1150	0.19	N	N
545	**Luncheon meat**, canned	920	120	39	10	200	1.0	0.10	1.5	1410	0.05	(7)	N[a]
546	**Meat spread**	810	220	15	14	140	4.7	0.13	3.3	1540	0.09	N	N
547	**Pate**, liver	750	150	16	11	450	5.9	0.46	2.8	880	0.16	N	N
548	meat, reduced fat	710	190	14	14	240	6.4	0.46	2.7	1180	0.16	N	N
549	**Polony**	870	120	42	13	130	1.3	0.32	1.2	1160	N	N	N[a]

[a] Iodine from erythrosine is present but largely unavailable

Vitamins per 100g edible portion

No.	Food	Retinol µg	Carotene µg	Vitamin D µg	Vitamin E mg	Thiamin mg	Ribo-flavin mg	Niacin mg	Trypt/60 mg	Vitamin B_6 mg	Vitamin B_{12} µg	Folate µg	Panto-thenate mg	Biotin µg	Vitamin C mg
	Meat products														
535	**Black pudding**, *dry-fried*	41	Tr	(0.7)	0.24	0.09	0.07	1.0	2.8	0.04	1	5	0.60	2	0
536	**Chicken nuggets**, *takeaway*	14	Tr	N	1.29	0.09	0.10	6.3	3.9	0.29	Tr	20	1.30	7	0
537	**Chicken pie**, *individual, chilled/frozen, baked*	Tr	Tr	N	N	0.41	0.09	1.5	1.6	0.12	Tr	8	0.64	4	N
538	**Chicken roll**	Tr	Tr	N	N	0.26	0.08	6.5	N	0.34	Tr	9	0.80	2	Tr
539	**Corned beef**, canned	Tr	Tr	1.3	0.78	Tr	0.20	2.6	6.5	0.18	2	5	0.40	2	0
540	**Cornish pastie**	Tr	N	N	1.30	0.09	0.06	1.3	1.7	0.19	Tr	5	0.60	1	Tr
541	**Frankfurter**	Tr	Tr	N	0.63	0.32	0.15	2.8	2.2	0.12	1	3	0.75	2	N
542	**Game pie**	610	77	(0.9)	(0.96)	0.21	0.24	3.0	2.4	0.21	3	75	0.62	12	1
543	**Haggis**, *boiled*	(1800)	Tr	(0.1)	0.41	0.16	0.35	1.5	2.0	0.07	2	8	0.50	12	Tr
544	**Liver sausage**	2600	N	(0.6)	0.10	0.36	1.16	3.7	2.4	0.25	10	36	1.50	7	Tr
545	**Luncheon meat**, canned	Tr	Tr	N	0.11	0.06	0.15	1.2	2.7	0.10	1	13	0.50	Tr	27[a]
546	**Meat spread**	Tr	Tr	N	0.49	0.07	0.19	3.4	1.8	0.13	3	6	0.75	4	0
547	**Pate**, liver	7300	130	1.2	N	0.10	1.17	1.9	2.8	0.25	8	99	2.10	14	N
548	meat, reduced fat	5930	N	N	0.77	0.46	1.12	7.1	2.2	0.35	12	31	2.68	27	18
549	**Polony**	Tr	Tr	N	0.09	0.17	0.10	1.5	1.8	0.08	Tr	4	0.50	Tr	N

[a] Some brands contain ascorbate, range 12-60mg per 100g

Composition of food per 100g edible portion

No.	Food	Description and main data sources	Edible conversion factor	Water g	Total nitrogen g	Protein g	Fat g	Carbo-hydrate g	Energy value kcal	Energy value kJ
	Meat products *continued*									
550	**Pork pie**, *individual*	8 samples of 8cm pies including Melton Mowbray. 28-39% meat	1.00	37.5	1.73	10.8	25.7	23.7	363	1514
551	**Salami**	22 samples including Danish, French, German and Italian. 90-100% meat	1.00	33.7	3.34	20.9[a]	39.2[a]	0.5[a]	438	1814
552	**Sausages**, beef, *chilled, grilled*	6 samples of thick sausages	1.00	48.0	2.13	13.3	19.5	13.1	278	1157
553	pork, *raw*, average	Average of frozen and chilled samples, thick and thin, 65-70% meat	1.00	49.4	1.91	11.9	25.0	9.6	309	1282
554	pork, *chilled, fried*	16 samples	1.00	46.4	2.22	13.9	23.9	9.9	308	1279
555	**Sausages**, pork, chilled, *grilled*	16 samples	1.00	45.9	2.32	14.5	22.1	9.8	294	1221
556	reduced fat, *chilled/frozen, grilled*	7 samples, 5 brands	1.00	50.1	2.59	16.2	13.8	10.8	230	959
557	**Sausages,** premium, *chilled, grilled*	Calculated from raw. 10 samples, 9 brands including Cumberland and Lincolnshire sausages. 65-90% meat	1.00	49.3	2.69	16.8	22.4	6.3	292	1215
558	**Sausage rolls**, puff pastry	Manufacturers' data	1.00	N	1.58	9.9	27.6	25.4	383	1596
559	**Saveloy**, unbattered, *takeaway*	20 samples	1.00	56.1	2.20	13.8	22.3	10.8[b]	296	1233
560	**Scotch eggs**, *retail*	10 samples, 8 brands	1.00	54.0	1.92	12.0	16.0	13.1	241	1006
561	**Steak and kidney pie**, single crust, *homemade*	Recipe	1.00	50.9	2.68	16.5	15.1	15.7	261	1091
562	**Steak and kidney/Beef pie**, *individual, chilled/frozen, baked*	16 samples including minced beef, minced beef and onion, minced beef and vegetable, steak, and steak and kidney pies. 12.5-30% meat	1.00	41.4	1.41	8.8	19.4	26.7	310	1295

[a] Danish salami contains 13.4g protein, 49.7g fat, 2.2g CHO, 509kcal, 2102kJ; German salami contains 20.7g protein, 31.5g fat, 2.6g CHO, 376kcal, 1559kj; French salami contains 21.0g protein, 37.4g fat, 1.9g CHO, 428kcal, 1771kJ; Italian salami contains 23.4g protein, 30.7g fat, 0.9g CHO 373kcal, 1548kj per 100g food

[b] Includes 0.64g oligosaccharides per 100g food

Composition of food per 100g edible portion

No.	Food	Starch	Total sugars	Dietary fibre NSP	Fatty acids Satd	Mono-unsatd	Poly-unsatd	Trans	Cholest-erol
		g	g	g	g	g	g	g	mg
	***Meat products** continued*								
550	**Pork pie**, *individual*	22.7	1.0	0.9	9.7	11.0	3.2	0.4	45
551	**Salami**	Tr	0.5	0.1	14.6	17.7	4.4	0.2	83
552	**Sausages**, beef, *chilled, grilled*	11.7	1.4	0.7	7.9	8.8	1.4	0.4	42
553	pork, *raw*, average	6.8	2.8	1.0	9.2	11.2	3.4	0.2	60
554	pork, *chilled, fried*	8.4	1.6	0.7	8.5	10.3	3.5	0.1	53
555	**Sausages**, pork, chilled, *grilled*	8.3	1.5	0.7	8.0	9.6	3.0	0.1	53
556	reduced fat, *chilled/frozen, grilled*	9.9	0.9	1.5	4.9	5.9	2.1	0.1	55
557	**Sausages,** premium, *chilled, grilled*	5.4	0.9	N	8.2	9.4	3.3	(0.1)	72
558	**Sausage rolls**, puff pastry	(24.5)	0.9	(1.0)	11.2	N	N	N	N
559	**Saveloy**, unbattered, *takeaway*	8.9	1.3	0.8	7.5	9.8	3.6	0.4	78
560	**Scotch eggs**, *retail*	13.1	Tr	N	4.3	6.8	2.8	0.2	165
561	**Steak and kidney pie**, single crust, *homemade*	15.4	0.3	0.6	4.9	5.8	3.5	0.8	112
562	**Steak and kidney/Beef pie**, *individual, chilled/frozen, baked*	25.2	1.5	0.5	8.4	7.8	1.9	1.2	39

Inorganic constituents per 100g edible portion

No.	Food	mg										µg	
		Na	K	Ca	Mg	P	Fe	Cu	Zn	Cl	Mn	Se	I
	***Meat products** continued*												
550	**Pork pie**, *individual*	650	160	68	17	100	1.1	0.08	1.0	760	0.23	6	7
551	**Salami**	1800[a]	320	11	18	170	1.3	0.12	3.0	3270	0.04	(7)	(15)[b]
552	**Sausages**, beef, *chilled, grilled*	1200	190	80	15	200	1.4	0.11	2.0	1640	0.21	4	6
553	pork, *raw*, average	860	160	103	13	175	0.9	0.07	0.9	1235	0.17	(5)	(7)
554	pork, *chilled, fried*	1070	180	110	15	220	1.1	0.12	1.1	1430	0.19	6	(8)
555	**Sausages**, pork, chilled, *grilled*	1080	190	110	15	220	1.1	0.11	1.4	1660	0.20	(6)	8
556	**Sausages**, reduced fat, *chilled/frozen, grilled*	1180	260	130	19	230	1.3	0.08	1.7	1580	0.24	(7)	(9)
557	premium, *chilled, grilled*	(840)	(220)	(180)	(16)	(180)	(1.2)	(0.07)	(1.4)	(920)	(0.16)	N	N
558	**Sausage rolls**, puff pastry	600	N	N	N	N	N	N	N	N	N	N	N
559	**Saveloy**, unbattered, *takeaway*	1150	180	81	18	230	4.5	0.11	1.2	(1770)	0.16	N	N
560	**Scotch eggs**, *retail*	670	130	50	15	170	1.8	0.23	1.2	980	0.20	N	17
561	**Steak and kidney pie**, single crust, *homemade*	669	264	33	19	176	2.7	0.20	3.4	1017	0.16	39	N
562	**Steak and kidney/Beef pie**, *individual, chilled/frozen, baked*	460	140	60	15	95	1.3	0.07	1.4	(690)	0.26	N	N

[a] Danish salami contains 1840mg Na; French 1700mg; German 1500mg; Italian 1335mg per 100g food

[b] Iodine from erythrosine is present but largely unavailable

No.	Food	Retinol µg	Carotene µg	Vitamin D µg	Vitamin E mg	Thiamin mg	Riboflavin mg	Niacin mg	Trypt/60 mg	Vitamin B_6 mg	Vitamin B_{12} µg	Folate µg	Pantothenate mg	Biotin µg	Vitamin C mg
	Meat products *continued*														
550	**Pork pie**, *individual*	Tr	Tr	N	0.21	0.19	0.06	2.3	1.0	0.15	Tr	5	0.74	2	4
551	**Salami**	Tr	Tr	N	0.23	0.60	0.23	5.6	2.8	0.36	2	3	1.66	7	N
552	**Sausages**, beef, *chilled, grilled*	Tr	Tr	N	0.67	Tr	0.11	2.5	1.8	0.11	1	7	0.64	3	N[a]
553	pork, *raw*, average	Tr	Tr	0.9	0.93	0.03	0.11	2.5	1.5	0.12	1	13	0.77	5	7
554	pork, *chilled, fried*	Tr	Tr	(1.1)	0.86	0.01	0.13	3.1	2.0	0.09	1	3	0.85	5	5
555	**Sausages**, pork, chilled, *grilled*	Tr	Tr	(1.1)	0.92	Tr	0.13	3.1	2.0	0.12	1	4	0.93	5	5
556	**Sausages**, reduced fat, *chilled/frozen, grilled*	Tr	Tr	N	0.30	Tr	0.13	2.8	2.0	0.11	1	32	1.04	3	37
557	premium, *chilled, grilled*	Tr	Tr	N	(0.80)	(0.05)	(0.10)	(2.7)	(1.5)	(0.14)	(1)	(8)	(0.76)	(3)	(8)
558	**Sausage rolls**, puff pastry	N	N	N	N	N	N	N	N	N	N	N	N	N	N
559	**Saveloy**, unbattered, *takeaway*	19	Tr	N	0.45	0.14	0.09	1.9	1.9	0.06	Tr	1	0.86	4	N
560	**Scotch eggs**, *retail*	30	Tr	0.7	N	0.08	0.21	1.0	2.9	0.13	1	42	(1.10)	(9)	N
561	**Steak and kidney pie**, single crust, *homemade*	69	24	N	2.05	0.15	0.56	3.7	3.3	0.31	4	6	1.12	(9)	2
562	**Steak and kidney/Beef pie**, *individual, chilled/frozen, baked*	5	20	0.7	1.04	0.40	(0.15)	(1.6)	(1.6)	(0.06)	(2)	(8)	0.61	3	Tr

[a] Ascorbic acid is added as an antioxidant. Measurable levels may be present

Composition of food per 100g edible portion

No.	Food	Description and main data sources	Edible conversion factor	Water	Total nitrogen	Protein	Fat	Carbo-hydrate	Energy value	
				g	g	g	g	g	kcal	kJ
***Meat products** continued*										
563	**Stewed steak with gravy**, canned	10 samples, 9 brands	1.00	71.0	2.59	16.2	10.1	0.6	158	659
564	**Tongue slices**	7 samples of a mixture of chilled, canned and delicatessen tongue. 90-100% meat	1.00	63.0	2.99	18.7	14.0	Tr	201	836
565	**Turkey roll**	10 samples, 8 brands	1.00	64.1	2.70	16.9	9.0	4.7	166	696
566	**White pudding**	6 samples	1.00	22.8	1.12	7.0	31.8	36.3	450	1876
Meat dishes										
567	**Beef bourguignonne**	Recipe	1.00	(71.2)	2.30	14.0	6.3	2.5[a]	122	511
568	made with lean beef	Recipe	1.00	(73.4)	2.35	14.3	4.3	2.5[b]	105	442
569	**Beef casserole**, made with canned cook-in sauce	Recipe	1.00	71.6	2.42	15.1	6.5	4.5	136	570
570	**Beef chow mein**, *retail, reheated*	12 samples from different shops. Noodles with beef and vegetables in sauce	1.00	71.7	1.07	6.7	6.0	14.7	136	571
571	**Beef curry**, *chilled/frozen, reheated*	6 samples, 3 brands, sauce only	1.00	69.5	2.16	13.5	6.6	6.3	137	575
572	with rice	Calculated from 57% beef curry and 43% boiled white rice	1.00	69.7	1.38	8.6	3.9	16.4	131	551
573	reduced fat	Recipe	1.00	70.2	3.00	18.8	7.1	1.0[c]	143	598
574	**Beef stew**	Recipe	1.00	76.3	1.92	12.0	4.6	4.7[b]	107	449
575	**Beef, stir-fried with green peppers**	Recipe	1.00	(71.2)	1.89	11.8	8.0	5.8	141	589

[a] Includes 0.3g oligosaccharides per 100g food
[b] Includes 0.4g oligosaccharides per 100g food
[c] Includes 0.2g oligosaccharides per 100g food

Composition of food per 100g edible portion

No.	Food	Starch	Total sugars	Dietary fibre NSP	Fatty acids Satd	Mono-unsatd	Poly-unsatd	Trans	Cholest-erol
		g	g	g	g	g	g	g	mg
	***Meat productts** continued*								
563	**Stewed steak with gravy**, canned	0.6	Tr	Tr	4.7	4.4	0.3	0.4	38
564	**Tongue slices**	0	Tr	Tr	6.0	6.4	0.9	0.7	115
565	**Turkey roll**	4.7	0	N	2.7	3.8	2.0	0.1	150
566	**White pudding**	(36.3)	Tr	N	N	N	N	N	22
	Meat dishes								
567	**Beef bourguignonne**	1.4	0.8	0.4	2.1	2.9	0.9	0.1	42
568	made with lean beef	1.4	0.8	0.4	1.3	2.0	0.7	0.1	40
569	**Beef casserole**, made with canned cook-in sauce	1.8	2.7	N	2.7	2.9	0.4	0.3	44
570	**Beef chow mein**, *retail, reheated*	12.3	2.4	N	1.3	3.1	1.4	N	N
571	**Beef curry**, *chilled/frozen, reheated*	1.8	4.5	1.2	3.1	2.5	0.6	N	32
572	with rice	13.8	2.6	0.8	1.8	1.4	0.3	N	18
573	reduced fat	0.2	0.6	0.2	2.2	3.3	0.9	0.3	53
574	**Beef stew**	2.3	2.0	0.7	1.5	2.1	0.6	0.1	35
575	**Beef, stir-fried with green peppers**	2.2	3.5	0.8	2.7	3.7	1.1	0.2	32

Inorganic constituents per 100g edible portion

No.	Food	mg										µg	
		Na	K	Ca	Mg	P	Fe	Cu	Zn	Cl	Mn	Se	I
	***Meat products** continued*												
563	**Stewed steak with gravy**, canned	340	200	11	15	120	2.1	0.18	3.9	510	0.02	N	N
564	**Tongue slices**	1000	140	10	16	260	2.6	0.18	3.0	1190	0.02	8	11
565	**Turkey roll**	690	180	15	17	200	0.8	0.11	1.5	1110	0.04	N	N
566	**White pudding**	370	190	38	61	230	2.1	0.43	1.6	600	N	N	N
	Meat dishes												
567	**Beef bourguignonne**	347	323	14	18	133	1.66	0.18	3.2	497	0.09	6	N
568	made with lean beef	365	337	14	19	139	1.71	0.18	3.4	545	0.09	6	N
569	**Beef casserole**, made with canned cook-in sauce	557	282	7	16	137	1.2	0.02	4.0	375	0.04	N	N
570	**Beef chow mein**, *retail, reheated*	590	N	N	N	N	1.3	N	N	910	N	N	N
571	**Beef curry**, *chilled/frozen, reheated*	540	340	N	N	N	N	N	N	690	N	N	N
572	with rice	260	210	Tr	2	15	0.1	0.03	0.2	400	0.13	2	2
573	reduced fat	224	339	25	27	174	1.93	0.04	5.2	310	0.10	6	15
574	**Beef stew**	357	234	15	13	105	1.18	0.05	2.7	513	0.05	4	8
575	**Beef, stir-fried with green peppers**	319	276	10	20	122	1.91	0.04	2.0	456	0.07	4	7

Vitamins per 100g edible portion

No.	Food	Retinol µg	Carotene µg	Vitamin D µg	Vitamin E mg	Thiamin mg	Riboflavin mg	Niacin mg	Trypt/60 mg	Vitamin B_6 mg	Vitamin B_{12} µg	Folate µg	Pantothenate mg	Biotin µg	Vitamin C mg
	***Meat products** continued*														
563	**Stewed steak with gravy**, canned	Tr	Tr	N	0.59	0.02	0.16	2.3	2.8	0.29	2	6	0.30	1	Tr
564	**Tongue slices**	Tr	Tr	N	N	0.03	0.18	2.0	N	0.12	5	4	N	N	0
565	**Turkey roll**	Tr	Tr	N	N	0.05	0.08	5.2	N	0.25	1	5	0.40	2	Tr
566	**White pudding**	Tr	Tr	N	1.00	0.26	0.08	0.5	1.3	0.06	1	6	0.80	18	0
	Meat dishes														
567	**Beef bourguignonne**	Tr	19	0.4	(0.16)	0.08	0.17	2.5	2.7	0.25	1	7	0.62	3	1
568	made with lean beef	Tr	19	0.5	(0.16)	0.09	0.17	2.7	2.8	0.27	1	7	0.63	3	1
569	**Beef casserole**, made with canned cook-in sauce	Tr	N	0.4	N	0.04	0.16	2.2	3.1	0.25	1	19	N	N	Tr
570	**Beef chow mein**, *retail, reheated*	Tr	Tr	N	(0.43)	0.03	0.03	N	1.1	N	Tr	N	N	N	Tr
571	**Beef curry**, *chilled/frozen, reheated*	Tr	Tr	N	0.62	0.05	0.16	2.4	1.6	0.20	N	N	0.71	3	Tr
572	with rice	Tr	Tr	N	0.35	0.03	0.10	1.5	1.1	0.14	N	N	0.49	2	Tr
573	reduced fat	Tr	110	0.4	0.08	0.07	0.20	3.0	2.9	0.32	1	22	0.42	1	1
574	**Beef stew**	0	1888	0.4	(0.19)	0.06	0.11	1.8	2.3	0.21	1	4	0.31	1	1
575	**Beef, stir-fried with green peppers**	Tr	165	0.2	(0.29)	0.05	0.11	2.3	2.5	0.38	1	11	0.32	N	38

Composition of food per 100g edible portion

No.	Food	Description and main data sources	Edible conversion factor	Water	Total nitrogen	Protein	Fat	Carbo-hydrate	Energy value	
				g	g	g	g	g	kcal	kJ
	***Meat dishes** continued*									
576	**Bolognese sauce** (with meat)	Recipe	1.00	70.8	1.88	11.8	11.6	2.5	161	670
577	**Chicken chasseur**	Recipe	1.00	76.9	2.07	12.8	4.1	2.3[a]	97	406
578	**Chicken chow mein**, *takeaway*	10 samples	1.00	69.0	1.36	8.5	7.2	12.7	147	614
579	**Chicken curry**, average, *takeaway*	50 samples, 10 each of Korma, Tikka Masala, Dhansak, Jalfrezi and Dopiaza. Meat and sauce only	1.00	70.2	1.88	11.7	9.8	2.5	145	603
580	*chilled/frozen, reheated,* with rice	Calculated from 55% chicken curry and 45% boiled white rice	1.00	69.2	1.22	7.6	5.0	16.3	137	575
581	**Chicken curry**, made with canned curry sauce	Recipe	1.00	67.0	2.99	18.7	6.5	4.4	150	628
582	**Chicken in white sauce**, canned	10 samples, 4 brands	1.00	73.6	2.29	14.3	8.3	2.5	141	590
583	**Chicken satay**	10 samples, takeaway	1.00	60.5	3.47	21.7	10.3	3.0	191	798
584	**Chicken tandoori**, *chilled, reheated*	7 samples, 6 brands. 95-96% meat	1.00	56.4	4.38	27.4	10.8	2.0	214	897
585	**Chicken tikka masala**, *retail*	21 samples, chilled, frozen and takeaway	1.00	68.2	2.08	12.9	10.6	2.6	157	656
586	**Chicken wings**, *marinated, chilled/frozen, barbecued*	4 samples including American and Chinese style and hot and spicy wings	0.65	50.5	4.38	27.4	16.6	4.1	274	1146
587	**Chicken**, stir-fried with rice and vegetables, *frozen, reheated*	6 samples. 10-13% meat	1.00	67.9	1.04	6.5	4.6	17.1[b]	132	554
588	**Chilli con carne**	Recipe	1.00	74.1	1.47	9.2	7.5	4.4	121	504
589	*chilled/frozen, reheated,* with rice	Calculated from 60% chilli con carne and 40% boiled white rice	1.00	73.8	0.88	5.5	2.7	16.1	107	451

[a] Includes 0.3g oligosaccharides per 100g food

[b] Includes 0.6g oligosaccharides per 100g food

No.	Food	Starch	Total sugars	Dietary fibre NSP	Fatty acids Satd	Mono-unsatd	Poly-unsatd	Trans	Cholest-erol
		g	g	g	g	g	g	g	mg
	Meat dishes *continued*								
576	**Bolognese sauce** (with meat)	0.2	2.1	0.6	4.2	5.2	1.0	0.5	33
577	**Chicken chasseur**	1.2	1.1	0.3	1.0	2.0	0.8	Tr	41
578	**Chicken chow mein**, *takeaway*	6.6	0.3	1.1	1.2	3.9	1.8	0	13
579	**Chicken curry**, average, *takeaway*	1.2	1.2	2.0	2.9	4.0	2.5	0.1	37
580	*chilled/frozen, reheated*, with rice	13.9	2.4	0.8	2.2	1.6	0.8	N	28
581	**Chicken curry**, made with canned curry sauce	2.1	2.3	N	N	N	N	Tr	72
582	**Chicken in white sauce**, canned	2.5	0	Tr	2.3	3.9	1.6	Tr	49
583	**Chicken satay**	1.4	1.6	2.2	3.0	4.3	2.5	0	57
584	**Chicken tandoori**, *chilled, reheated*	1.0	1.0	Tr	3.3	5.0	2.0	0.1	120
585	**Chicken tikka masala**, *retail*	0.6	1.8	1.6	3.6	4.3	2.3	0.2	46
586	**Chicken wings**, *marinated, chilled/frozen, barbecued*	0.5	3.6	Tr	4.6	7.5	3.3	(0.2)	(120)
587	**Chicken**, stir-fried with rice and vegetables, *frozen, reheated*	12.6	3.9	1.3	N	N	N	N	N
588	**Chilli con carne**	1.2	2.7	1.1	2.9	3.2	0.5	0.3	24
589	*chilled/frozen, reheated*, with rice	14.5	1.6	0.9	1.1	1.1	0.1	0.1	N

Inorganic constituents per 100g edible portion

No.	Food	mg										µg	
		Na	K	Ca	Mg	P	Fe	Cu	Zn	Cl	Mn	Se	I
	***Meat dishes** continued*												
576	**Bolognese sauce** (with meat)	306	305	16	16	105	1.06	0.05	2.3	449	0.07	4	7
577	**Chicken chasseur**	208	266	12	19	129	0.64	0.11	0.5	309	0.07	7	5
578	**Chicken chow mein**, *takeaway*	466	90	46	11	64	1.01	0.05	0.4	(720)	0.15	N	N
579	**Chicken curry**, average, *takeaway*	356	218	41	22	112	2.32	0.08	0.6	515	0.24	7	7
580	*chilled/frozen, reheated*, with rice	250	180	N	N	N	N	N	N	360	N	N	N
581	**Chicken curry**, made with canned curry sauce	663	413	23	32	147	1.2	0.05	1.1	542	0.13	N	N
582	**Chicken in white sauce**, canned	370	80	13	11	70	0.6	0.07	0.9	470	0.03	N	N
583	**Chicken satay**	613	363	30	43	223	1.0	0.14	0.9	690	0.39	12	23
584	**Chicken tandoori**, *chilled, reheated*	590	470	58	36	280	1.8	0.12	1.5	860	0.18	(16)	(7)
585	**Chicken tikka masala**, *retail*	424	289	55	25	145	1.2	0.11	0.7	(654)	0.23	(8)	N
586	**Chicken wings**, *marinated, chilled/frozen, barbecued*	390	350	42	27	200	1.3	0.09	1.6	610	0.15	(17)	(6)
587	**Chicken**, stir-fried with rice and vegetables, *frozen, reheated*	410	180	22	13	95	1.1	0.10	2.0	870	0.08	N	N
588	**Chilli con carne**	303	276	20	16	91	1.0	0.10	1.7	462	0.10	3	4
589	*chilled/frozen, reheated*, with rice	180	190	26	14	68	1.0	0.10	1.0	(290)	0.23	2	N

Vitamins per 100g edible portion

No.	Food	Retinol µg	Carotene µg	Vitamin D µg	Vitamin E mg	Thiamin mg	Ribo-flavin mg	Niacin mg	Trypt/60 mg	Vitamin B_6 mg	Vitamin B_{12} µg	Folate µg	Panto-thenate mg	Biotin µg	Vitamin C mg
	***Meat dishes** continued*														
576	**Bolognese sauce** (with meat)	Tr	738	0.4	0.60	0.06	0.07	2.9	2.1	0.22	1	9	0.32	1	3
577	**Chicken chasseur**	3	26	0.2	(0.14)	0.07	0.10	4.6	2.4	0.24	Tr	7	0.66	2	1
578	**Chicken chow mein**, *takeaway*	Tr	110	N	0.96	0.05	(0.03)	1.8	1.3	0.08	Tr	4	0.47	2	Tr
579	**Chicken curry**, average, *takeaway*	15	119	0.3	2.12	0.05	0.07	2.5	1.8	0.19	Tr	N	0.66	2	Tr
580	*chilled/frozen, reheated,* with rice	N	205	N	0.72	0.11	0.08	2.2	1.5	0.15	N	N	0.65	2	1
581	**Chicken curry**, made with canned curry sauce	9	N	0.1	N	0.09	0.13	5.0	3.6	0.25	Tr	N	N	N	Tr
582	**Chicken in white sauce**, canned	Tr	Tr	N	N	0.01	0.10	2.3	N	0.10	Tr	5	N	N	0
583	**Chicken satay**	5	23	N	1.41	0.07	0.07	11.0	3.2	0.22	Tr	27	0.93	6	Tr
584	**Chicken tandoori**, *chilled, reheated*	Tr	210	(0.2)	1.49	0.12	0.19	10.2	5.8	0.61	1	16	2.25	5	2
585	**Chicken tikka masala**, *retail*	43	104	0.3	1.72	0.08	0.14	4.2	1.9	0.29	1	21	0.98	6	Tr
586	**Chicken wings**, *marinated, chilled/frozen, barbecued*	(24)	N	(0.1)	(0.23)	(0.07)	(0.11)	(6.2)	(5.3)	(0.27)	(1)	(10)	(1.34)	(4)	Tr
587	**Chicken**, stir-fried with rice and vegetables, *frozen, reheated*	Tr	565	N	N	0.09	0.09	1.9	1.3	0.22	Tr	21	0.50	4	2
588	**Chilli con carne**	Tr	277	0.3	(0.56)	0.07	0.06	2.2	1.6	0.19	1	9	0.25	1	7
589	*chilled/frozen, reheated,* with rice	39	59	N	N	0.05	0.08	1.1	1.0	0.13	Tr	10	0.39	3	N

Composition of food per 100g edible portion

No.	Food	Description and main data sources	Edible conversion factor	Water	Total nitrogen	Protein	Fat	Carbo-hydrate	Energy value	
				g	g	g	g	g	kcal	kJ
	Meat dishes *continued*									
590	**Coq au vin**	Recipe	0.82	(68.7)	1.81	11.1	11.0	3.2	155	647
591	**Coronation chicken**	Recipe	1.00	46.7	2.67	16.6	31.7	3.2	364	1505
592	**Cottage/Shepherd's pie**, *chilled/frozen, reheated*	11 samples including beef and lamb. 11.5-25% meat	1.00	73.1	0.72	4.5	5.4	11.9	111	467
593	**Doner kebabs**, meat only	20 samples from assorted takeaways	1.00	42.0	3.76	23.5	31.4	0	377	1561
594	in pitta bread with salad	Calculated from 50% doner kebab, 22% pitta bread and 28% salad	1.00	53.7	2.27	14.2	16.2	14.0[a]	255	1065
595	**Faggots in gravy**, *chilled/frozen, reheated*	8 samples, 3 brands. 11-35% meat	1.00	69.6	1.31	8.2	7.5	12.6[b]	148	619
596	**Goulash**	Recipe	1.00	80.4	1.11	6.9	3.0	6.5[c]	79	332
597	**Irish stew**	Recipe	1.00	77.3	1.18	7.4	6.2	8.6[d]	118	493
598	made with lean lamb	Recipe	1.00	78.6	1.23	7.7	4.9	8.6[d]	107	450
599	canned	10 samples, 2 brands	1.00	82.5	0.75	4.7	5.1	6.8	91	379
600	**Lamb curry**, made with canned curry sauce	Recipe	1.00	59.6	2.49	15.6	18.9	4.4	249	1032
601	**Lamb kheema**	Recipe	1.00	69.2	1.76	11.0	13.4	3.6[e]	176	740
602	**Lamb/Beef hot pot with potatoes**, *chilled/frozen, retail, reheated*	10 samples, 6 brands of beef, lamb and Lancashire hot pot. 10-32% meat	1.00	74.4	1.15	7.2	4.4	10.6	108	455
603	**Lancashire hotpot**	Recipe	1.00	77.4	1.18	7.4	6.9	7.4[a]	119	500
604	**Lasagne**	Recipe	1.00	62.3	1.60	9.8	10.8	14.6[b]	191	800
605	*chilled/frozen, reheated*	12 samples, 11 brands. 10-20% meat	1.00	68.1	1.18	7.4	6.1	15.7	143	603

[a] Includes 0.2g oligosaccharides per 100g food
[b] Includes 0.1g oligosaccharides per 100g food
[c] Includes 0.4g oligosaccharides per 100g food
[d] Includes 0.3g oligosaccharides per 100g food
[e] Includes 0.6g oligosaccharides per 100g food

Composition of food per 100g edible portion

No.	Food	Starch	Total sugars	Dietary fibre NSP	Fatty acids Satd	Mono-unsatd	Poly-unsatd	Trans	Cholest-erol
		g	g	g	g	g	g	g	mg
	***Meat dishes** continued*								
590	**Coq au vin**	2.8	0.4	0.3	4.2	4.4	1.6	0.2	67
591	**Coronation chicken**	0.2	3.0	N	5.3	8.5	15.9	0.5	89
592	**Cottage/Shepherd's pie**, *chilled/frozen, reheated*	10.3	1.6	0.9	2.4	2.2	0.4	0.3	16
593	**Doner kebabs**, meat only	0	0	0	15.3	12.0	1.4	2.4	94
594	in pitta bread with salad	12.3	1.5	0.8	7.8	6.1	0.9	1.4	47
595	**Faggots in gravy**, *chilled/frozen, reheated*	10.8	1.7	0.2	2.5	2.9	1.0	0.1	45
596	**Goulash**	4.3	1.7	0.8	0.9	1.4	0.5	0.1	17
597	**Irish stew**	6.5	1.8	1.0	2.9	2.3	0.4	0.5	27
598	made with lean lamb	6.5	1.8	1.0	2.2	1.8	0.3	0.4	26
599	canned	5.6	1.2	N	2.5	2.0	0.3	0.4	15
600	**Lamb curry**, made with canned curry sauce	2.1	2.3	N	N	N	N	N	63
601	**Lamb kheema**	1.1	1.5	1.3	3.8	6.1	2.4	0.5	38
602	**Lamb/Beef hot pot with potatoes**, *chilled/frozen, retail, reheated*	9.6	1.0	0.9	1.7	1.9	0.5	0.4	N
603	**Lancashire hotpot**	5.8	1.4	0.9	3.0	2.7	0.9	0.5	27
604	**Lasagne**	11.7	2.7	0.8	4.5	3.8	1.4	0.3	26
605	*chilled/frozen, reheated*	12.7	3.0	0.7	2.8	2.2	0.7	0.3	18

Inorganic constituents per 100g edible portion

No.	Food	mg										µg	
		Na	K	Ca	Mg	P	Fe	Cu	Zn	Cl	Mn	Se	I
	***Meat dishes** continued*												
590	**Coq au vin**	250	281	16	19	86	1.1	0.11	0.9	383	0.10	8	N
591	**Coronation chicken**	236	213	12	17	142	0.8	0.07	0.9	377	0.03	N	17
592	**Cottage/Shepherd's pie**, *chilled/frozen, reheated*	420	240	20	14	65	0.7	0.04	0.9	710	0.08	N	N
593	**Doner kebabs**, meat only	860	350	23	25	210	2.1	0.11	4.0	N	0.06	6	4
594	in pitta bread with salad	550	260	37	20	130	1.6	0.11	2.2	N	0.17	3	3
595	**Faggots in gravy**, *chilled/frozen, reheated*	540	120	32	10	80	1.7	0.30	0.9	(830)	0.15	N	N
596	**Goulash**	360	278	12	15	70	0.9	0.07	1.5	547	0.08	2	5
597	**Irish stew**	94	275	13	14	78	0.7	0.06	1.5	157	0.08	1	4
598	made with lean lamb	95	281	12	14	85	0.7	0.06	1.6	69	0.08	2	6
599	canned	280	130	10	8	40	1.2	0.13	0.7	(430)	0.05	N	N
600	**Lamb curry**, made with canned curry sauce	649	358	22	26	155	1.6	0.09	3.2	520	0.13	N	N
601	**Lamb kheema**	235	284	29	23	124	1.7	0.08	2.0	347	0.15	1	5
602	**Lamb/Beef hot pot with potatoes**, *chilled/frozen, retail, reheated*	330	260	17	16	80	0.8	0.07	1.3	(510)	0.08	N	N
603	**Lancashire hotpot**	185	278	11	14	79	0.6	0.06	1.5	288	0.06	1	4
604	**Lasagne**	340	224	100	19	142	0.8	0.09	1.6	496	0.17	5	14
605	*chilled/frozen, reheated*	390	230	80	19	120	1.0	0.10	1.4	(600)	0.22	N	N

No.	Food	Retinol µg	Carotene µg	Vitamin D µg	Vitamin E mg	Thiamin mg	Riboflavin mg	Niacin mg	Trypt/60 mg	Vitamin B_6 mg	Vitamin B_{12} µg	Folate µg	Pantothenate mg	Biotin µg	Vitamin C mg
	Meat dishes *continued*														
590	**Coq au vin**	50	28	0.3	0.16	0.10	0.15	2.6	1.6	0.16	Tr	5	0.60	3	1
591	**Coronation chicken**	37	36	0.2	6.12	0.05	0.12	5.5	3.2	0.22	Tr	7	N	N	Tr
592	**Cottage/Shepherd's pie**, *chilled/frozen, reheated*	17	110	0.1[a]	0.28	0.15	0.10	1.3	0.8	0.19	1	14	0.39	1	1
593	**Doner kebabs**, meat only	Tr	Tr	0.6	0.56	0.11	0.25	5.8	4.9	0.20	2	7	1.10	2	0
594	in pitta bread with salad	Tr	91	0.3	0.47	0.14	0.14	3.4	2.9	0.14	1	16	0.60	1	2
595	**Faggots in gravy**, *chilled/frozen, reheated*	1100	55	0.5	0.33	0.10	0.56	2.0	N	0.13	6	19	N	N	Tr
596	**Goulash**	Tr	206	0.2	0.33	0.08	0.06	1.2	1.3	0.23	Tr	10	0.26	1	8
597	**Irish stew**	3	1730	0.1	0.12	0.12	0.05	1.3	1.4	0.21	1	8	0.38	1	3
598	made with lean lamb	2	1726	0.1	0.12	0.12	0.05	1.4	1.5	0.26	1	8	0.39	1	3
599	canned	Tr	N	N	N	0.02	0.06	0.9	N	0.14	Tr	3	N	N	N
600	**Lamb curry**, made with canned curry sauce	6	N	0.3	N	0.09	0.12	2.5	2.9	0.15	1	N	N	N	Tr
601	**Lamb kheema**	2	270	0.4	(0.33)	0.12	0.10	2.4	2.0	0.13	1	11	0.41	1	4
602	**Lamb/Beef hot pot with potatoes**, *chilled/frozen, retail, reheated*	N	N	N	N	0.43	0.09	1.5	N	0.29	1	25	N	N	Tr
603	**Lancashire hotpot**	3	867	0.1	(0.08)	0.11	0.05	1.3	1.4	0.20	1	8	0.37	1	3
604	**Lasagne**	50	519	0.4	0.81	0.11	0.11	1.7	1.9	0.13	1	7	0.41	2	2
605	*chilled/frozen, reheated*	Tr	N	N	N	0.33	0.12	1.4	1.3	0.14	1	11	0.38	4	N

[a] Contribution from 25-hydroxycholecalciferol not included

Composition of food per 100g edible portion

No.	Food	Description and main data sources	Edible conversion factor	Water	Total nitrogen	Protein	Fat	Carbo-hydrate	Energy value	
				g	g	g	g	g	kcal	kJ
	Meat dishes *continued*									
606	**Meat samosas**, *takeaway*	10 samples from Indian restaurants	1.00	44.5	1.82	11.4	17.3	18.9	272	1136
607	**Moussaka**, *chilled/frozen/ longlife, reheated*	8 samples, 4 brands of beef and lamb. 20-23% meat	1.00	70.6	1.33	8.3	8.3	8.6[a]	140	586
608	**Pasta with meat and tomato sauce**	Recipe	1.00	76.7	0.94	5.7	3.6	12.3	101	426
609	**Pork casserole**, made with canned cook-in sauce	Recipe	1.00	70.1	2.74	17.1	7.8	3.8	154	641
610	**Sausage casserole**	Recipe	1.00	68.5	1.90	11.9	10.9	5.1[b]	165	687
611	**Shish kebab**, meat only	20 samples from assorted takeaways	1.00	59.4	4.64	29.0	10.0	0	206	863
612	in pitta bread with salad	Calculated from 37% shish kebab, 27% pitta bread and 36% salad	1.00	64.1	2.16	13.5	4.1	17.2[b]	155	656
613	**Spaghetti bolognese**, *chilled/ frozen, reheated*	12 samples. Meat sauce portion only, 10-18% meat	1.00	76.9	1.49	9.3	5.7	5.3	108	454
614	*chilled/frozen, reheated*, with spaghetti	Recipe. Calculated using 56% spaghetti and 44% meat and sauce	1.00	75.2	1.01	6.1	2.9	14.8	106	447
615	**Spring rolls**, meat, *takeaway*	10 samples	1.00	54.9	1.04	6.5	16.4	18.2	242	1009
616	**Sweet and sour chicken**, *takeaway*	10 samples	1.00	59.2	1.21	7.6	10.0	19.7	194	814
617	**Sweet and sour pork**	Recipe	1.00	59.7	2.04	12.7	8.6	11.3[c]	177	741

[a] Includes 0.1g oligosaccharides per 100g food

[b] Includes 0.3g oligosaccharides per 100g food

[c] Includes 0.4g oligosaccharides per 100g food

Composition of food per 100g edible portion

No.	Food	Starch	Total sugars	Dietary fibre NSP	Fatty acids Satd	Mono-unsatd	Poly-unsatd	Trans	Cholest-erol
		g	g	g	g	g	g	g	mg
	Meat dishes *continued*								
606	**Meat samosas**, *takeaway*	16.8	1.9	(2.4)	4.5	7.0	4.8	0.2	20
607	**Moussaka**, *chilled/frozen/longlife, reheated*	6.5	2.0	0.8	2.9	3.6	1.1	0.4	26
608	**Pasta with meat and tomato sauce**	10.7	1.6	N	1.4	1.3	0.4	0.1	11
609	**Pork casserole**, made with canned cook-in sauce	1.5	2.3	Tr	2.7	3.2	1.4	0.1	50
610	**Sausage casserole**	2.7	2.1	0.9	3.5	4.5	2.0	0	40
611	**Shish kebab**, meat only	0	0	0	3.9	4.3	0.8	0.6	90
612	in pitta bread with salad	15.0	1.9	1.0	1.5	1.6	0.5	0.2	33
613	**Spaghetti bolognese**, *chilled/frozen, reheated*	2.3	3.0	0.9	2.3	2.5	0.5	0.2	N
614	*chilled/frozen, reheated*, with spaghetti	13.2	1.6	1.1	1.1	1.2	0.4	0.1	N
615	**Spring rolls**, meat, *takeaway*	12.0	1.8	1.9	3.8	7.1	4.8	Tr	7
616	**Sweet and sour chicken**, *takeaway*	8.8	10.7	N	1.3	5.2	3.0	Tr	24
617	**Sweet and sour pork**	3.4	7.5	0.6	2.0	3.9	2.0	0	51

Inorganic constituents per 100g edible portion

No.	Food	mg										µg	
		Na	K	Ca	Mg	P	Fe	Cu	Zn	Cl	Mn	Se	I
	***Meat dishes** continued*												
606	**Meat samosas**, *takeaway*	409	258	64	26	138	2.6	0.15	2.3	(631)	0.45	N	N
607	**Moussaka**, *chilled/frozen/longlife, reheated*	350	250	75	77	110	0.6	0.12	0.1	(540)	0.13	N	N
608	**Pasta with meat and tomato sauce**	117	183	11	16	63	0.7	0.09	1.0	224	0.20	N	N
609	**Pork casserole**, made with canned cook-in sauce	476	345	8	20	176	0.7	0.03	1.4	324	0.03	10	4
610	**Sausage casserole**	650	250	32	18	140	0.9	0.06	1.2	790	0.11	6	5
611	**Shish kebab**, meat only	510	420	7	29	250	2.6	0.14	6.1	N	0.03	4	6
612	in pitta bread with salad	330	260	34	19	130	1.6	0.12	2.5	N	0.19	2	3
613	**Spaghetti bolognese**, *chilled/frozen, reheated*	410	290	21	18	85	1.3	0.08	1.5	(630)	0.15	(1)	N
614	*chilled/frozen, reheated,* with spaghetti	180	141	13	16	62	0.9	0.09	0.9	(277)	0.23	(3)	N
615	**Spring rolls**, meat, *takeaway*	485	117	32	13	64	1.2	0.06	0.6	(749)	0.24	N	N
616	**Sweet and sour chicken**, *takeaway*	259	142	35	13	114	2.4	0.04	0.3	(400)	0.19	(6)	N
617	**Sweet and sour pork**	494	307	15	20	143	0.9	0.08	1.3	718	0.06	7	(6)

No.	Food	Retinol µg	Carotene µg	Vitamin D µg	Vitamin E mg	Thiamin mg	Riboflavin mg	Niacin mg	Trypt/60 mg	Vitamin B_6 mg	Vitamin B_{12} µg	Folate µg	Pantothenate mg	Biotin µg	Vitamin C mg
	***Meat dishes** continued*														
606	**Meat samosas**, takeaway	2	28	0.3	0.55	(0.21)	(0.09)	(2.3)	(2.2)	(0.15)	(1)	(6)	(0.75)	(2)	Tr
607	**Moussaka**, *chilled/frozen/ longlife, reheated*	40	235	0.3	N	0.05	0.19	1.5	1.5	0.15	1	8	0.48	2	N
608	**Pasta with meat and tomato sauce**	Tr	(25)	0.1	N	0.04	0.14	1.1	1.1	0.07	Tr	7	N	N	Tr
609	**Pork casserole**, made with canned cook-in sauce	Tr	Tr	0.5	0.03	0.54	0.14	4.4	3.4	0.36	1	1	0.90	1	Tr
610	**Sausage casserole**	Tr	19	0.5	N	0.32	0.10	2.7	2.1	0.21	Tr	5	0.60	2	Tr
611	**Shish kebab**, meat only	Tr	Tr	0.6	0.67	0.14	0.28	7.0	6.0	0.26	3	9	1.40	3	0
612	in pitta bread with salad	Tr	120	0.2	0.50	0.16	0.12	3.2	2.8	0.14	1	20	0.58	1	3
613	**Spaghetti bolognese**, *chilled/frozen, reheated*	Tr	N	N	N	N	N	N	N	N	N	N	N	N	Tr
614	*chilled/frozen, reheated,* with spaghetti	Tr	N	N	N	N	N	N	N	N	N	N	N	N	Tr
615	**Spring rolls**, meat, *takeaway*	Tr	175	N	1.47	0.10	0.07	1.0	1.0	0.06	Tr	3	0.33	3	Tr
616	**Sweet and sour chicken**, *takeaway*	2	135	0.6	2.14	0.04	0.05	2.1	1.3	0.12	Tr	(2)	0.39	2	Tr
617	**Sweet and sour pork**	7	474	0.4	(0.30)	0.47	0.14	3.4	2.7	0.30	1	10	0.73	2	14

Section 2.6

Fish and fish products

This section of the Tables is largely based on data in the *Fish and Fish Products* (1993) supplement although some new analytical data on takeaway fish dishes have been incorporated. Foods for which new data are incorporated have been allocated a new food code and can thus easily be identified in the food index.

Fish are mainly drawn from a wild population which means that their composition is probably more variable than that of foods drawn from domesticated inbred stock whose nutrition has been closely controlled. There is considerable variation in composition within one species and this variation is probably greater than that between species.

The fat content of many fish show considerable seasonal changes and it is difficult to assign definite values. The actual fat content of fish normally landed and consumed shows less variation because the fish tend to be caught during a limited part of the cycle; the values used are therefore based on the fat content of the fish during the period when the major landings of the species are made.

In fish with fine bones it is often difficult to remove the bones completely, whether before analysis or before consumption. The calcium and phosphorus content of these fish is more variable than in a fish which can be boned easily. The values in the Tables are based on samples which have been prepared for consumption in the normal way.

The crustaceans and molluscs tend to accumulate many cations from their environment, and the concentration of iron, copper and zinc reported in these fish shows very wide variation, depending on the source of the samples and the metallic contamination to which they have been exposed.

Users should note that all values are expressed per 100g edible portion. Guidance for calculating nutrient content 'as purchased' or 'as served' (e.g. including bone or shells) is given in Section 4.2. For weight loss on cooking and calculation of cooked edible portion obtainable from raw fish see Section 4.3. Losses of labile vitamins assigned to cooked dishes or foods were estimated from figures found in Section 4.3.

Taxonomic names for foods in this part of the Tables can be found in Section 4.5.

Composition of food per 100g edible portion

No.	Food	Description and main data sources	Edible conversion factor	Water g	Total nitrogen g	Protein g	Fat g	Carbo-hydrate g	Energy value kcal	Energy value kJ
White fish										
618	**Cod**, *raw*	11 samples from assorted outlets, fillets	0.86[a]	80.8	2.93	18.3	0.7	0	80	337
619	*baked*	Baked in the oven with added butter, fillets; flesh only	0.85	76.6	3.43	21.4	1.2	Tr	96	408
620	*poached*	Poached in milk, butter and salt added, fillets; flesh only	0.87	77.7	3.35	20.9	1.1	Tr	94	396
621	*frozen, raw*	11 samples from assorted supermarkets; steaks	1.00	82.4	2.67	16.7	0.6	0	72	306
622	*frozen, grilled*	12 samples, grilled with butter and salt added; steaks	1.00	78.0	3.32	20.8	1.3	Tr	95	402
623	in batter, *fried in blended oil*	Samples as fried in retail blend oil, fatty acids calculated	1.00	54.9	2.58	16.1	15.4	11.7	247	1031
624	in crumbs, *frozen, fried in blended oil*	10 samples, 7 brands; shallow fried in blended oil, 5 minutes per side[b]	0.98	55.9	1.98	12.4	14.3	15.2	235	983
625	in parsley sauce, *frozen, boiled*	10 samples, 4 brands; boiled in bag for 20 minutes	1.00	82.1	1.92	12.0	2.8	2.8	84	352
626	*dried, salted, boiled*	Soaked 24 hours and boiled	0.83	64.9	5.20	32.5	0.9	0	138	586
627	**Coley**, *raw*	Literature sources and estimation from frozen coley	0.47	80.2	2.93	18.3	1.0	0	82	348
628	*steamed*	Analytical and calculated values, pieces from tail end; flesh only	0.85	74.8	3.73	23.3	1.3	0	105	444
629	**Haddock**, *raw*	12 samples from assorted outlets; fillets	0.83[c]	79.4	3.04	19.0	0.6	0	81	345
630	*steamed*	12 samples steamed for 20 minutes, fillets; flesh only	0.84[c]	78.3	3.34	20.9	0.6	0	89	378
631	smoked, *steamed*	Analysis and calculation from raw, cutlets; flesh only	0.65	71.6	3.73	23.3	0.9	0	101	429
632	in crumbs, *frozen, fried in blended oil*	10 samples, 7 brands; shallow fried in blended oil for 10-15 minutes per side[b]	0.80[d]	59.9	2.35	14.7	10.0	12.6	196	822

[a] Some fillets contained skin and bones. Values ranged from 0.79 to 1.00

[b] Composition of oven baked fish in crumbs is very similar to fried in blended oil

[c] Some fillets contained skin and bones

[d] Levels ranged from 0.66 to 1.00

Composition of food per 100g edible portion

No.	Food	Starch	Total sugars	Individual sugars					Dietary fibre	Fatty acids				Cholest-erol
				Gluc	Fruct	Sucr	Malt	Lact	NSP	Satd	Mono-unsatd	Poly-unsatd	Trans	
		g	g	g	g	g	g	g	g	g	g	g	g	mg
	White fish													
618	**Cod**, *raw*	0	0	0	0	0	0	0	0	0.1	0.1	0.3	0	46
619	*baked*	0	Tr	0	0	0	0	Tr	0	(0.3)	(0.2)	(0.4)	0	(56)
620	*poached*	0	Tr	0	0	0	0	Tr	0	(0.3)	(0.1)	(0.3)	0	(53)
621	*frozen, raw*	0	0	0	0	0	0	0	0	0.1	0.1	0.2	0	39
622	*frozen, grilled*	0	0	0	0	0	0	Tr	0	0.4	0.2	0.3	0	(49)
623	in batter, *fried in blended oil*	11.7	Tr	Tr	Tr	Tr	Tr	Tr	0.5	1.6	5.5	7.5	N	N
624	in crumbs, *frozen, fried in blended oil*	15.0	0.2	0.1	Tr	Tr	Tr	0.1	(0.4)	(1.5)	(5.2)	(7.0)	N	N
625	in parsley sauce, *frozen, boiled*	2.8	Tr	Tr	Tr	Tr	0	Tr	(0.1)	N	N	N	N	N
626	*dried, salted, boiled*	0	0	0	0	0	0	0	0	0.2	0.1	0.4	N	59
627	**Coley**, *raw*	0	0	0	0	0	0	0	0	0.1	0.3	0.3	Tr	(40)
628	*steamed*	0	0	0	0	0	0	0	0	0.2	0.3	0.4	Tr	55
629	**Haddock**, *raw*	0	0	0	0	0	0	0	0	0.1	0.1	0.2	Tr	36
630	*steamed*	0	0	0	0	0	0	0	0	0.1	0.1	0.2	Tr	38
631	smoked, *steamed*	0	0	0	0	0	0	0	0	(0.2)	(0.1)	(0.3)	Tr	(47)
632	in crumbs, *frozen, fried in blended oil*	12.6	Tr	Tr	Tr	Tr	Tr	0	(0.6)	N	N	N	N	N

Inorganic constituents per 100g edible portion

No.	Food	mg										µg	
		Na	K	Ca	Mg	P	Fe	Cu	Zn	Cl	Mn	Se	I
	White fish												
618	**Cod**, *raw*	60	340	9	22	180	0.1	0.02	0.4	76	0.01	28	110
619	*baked*	340	350	11	26	190	0.1	0.02	0.5	520	0.01	34	(130)
620	*poached*	110	330	11	26	180	0.1	0.02	0.5	150	0.01	33	(120)
621	*frozen, raw*	71	340	8	22	180	0.1	0.06	0.4	120	0.01	27	(110)
622	*frozen, grilled*	91	380	10	26	200	0.1	0.07	0.5	140	0.01	33	(130)
623	in batter, *fried in blended oil*	160	290	67	25	200	0.5	0.04	0.5	160	0.12	N	N
624	in crumbs, *frozen, fried in blended oil*	480	230	43	19	190	0.4	0.08	0.4	650	0.12	17	N
625	in parsley sauce, *frozen, boiled*	260	270	51	19	170	0.1	0.04	0.4	N	0.02	N	N
626	*dried, salted, boiled*	400	31	22	35	160	1.8	N	N	670	0.01	52	N
627	**Coley**, *raw*	86	360	9	(25)	250	(0.3)	0.05	0.5	(84)	(0.01)	(18)	(36)
628	*steamed*	97	460	19	31	410	0.6	0.06	0.6	83	(0.01)	(23)	(46)
629	**Haddock**, *raw*	67	360	14	24	200	0.1	0.03	0.4	86	0.01	27	250
630	*steamed*	73	370	26	24	200	0.1	0.02	0.5	100	0.01	28	(260)
631	smoked, *steamed*	990	440	29	30	240	0.1	0.04	0.4	1570	0.01	36	(340)
632	in crumbs, *frozen, fried in blended oil*	290	230	120	21	200	0.8	0.05	0.4	400	0.21	18	250

Vitamins per 100g edible portion

No.	Food	Retinol	Carotene	Vitamin D	Vitamin E	Thiamin	Riboflavin	Niacin	Trypt/60	Vitamin B_6	Vitamin B_{12}	Folate	Pantothenate	Biotin	Vitamin C
		µg	µg	µg	mg	mg	mg	mg	mg	mg	µg	µg	mg	µg	mg
	White fish														
618	**Cod**, *raw*	2	Tr	Tr	0.44	0.04	0.05	2.4	3.4	0.18	1	12	0.27	1	Tr
619	*baked*	2	Tr	Tr	0.59	0.03	0.05	2.3	4.0	0.19	2	12	0.26	1	Tr
620	*poached*	(2)	Tr	Tr	0.61	0.04	0.06	2.8	3.9	0.21	2	14	0.31	1	Tr
621	*frozen, raw*	2	Tr	Tr	(0.44)	0.04	0.05	1.6	3.1	(0.18)	1	6	(0.27)	(1)	Tr
622	*frozen, grilled*	2	Tr	Tr	(1.00)	0.05	0.06	1.9	3.9	(0.22)	2	10	(0.34)	(1)	Tr
623	in batter, *fried in blended oil*	N	Tr	Tr	N	0.09	0.07	1.7	3.0	0.13	2	57	0.30	3	Tr
624	in crumbs, *frozen, fried in blended oil*	Tr	N	Tr	(3.45)	0.07	0.06	1.2	2.3	0.09	N	6	0.32	3	Tr
625	in parsley sauce, *frozen, boiled*	Tr	Tr	Tr	N	0.06	0.10	1.1	2.2	0.13	N	17	0.47	2	Tr
626	*dried, salted, boiled*	(2)	Tr	Tr	N	Tr	Tr	N	6.1	N	Tr	Tr	N	Tr	Tr
627	**Coley**, *raw*	4	Tr	Tr	0.36	(0.15)	(0.20)	(2.3)	3.4	(0.29)	(3)	N	(0.42)	(3)	Tr
628	*steamed*	5	Tr	0.3	0.46	(0.19)	(0.27)	(2.9)	4.4	(0.40)	(5)	N	(0.46)	(3)	Tr
629	**Haddock**, *raw*	Tr	Tr	Tr	0.39	0.04	0.07	4.4	3.6	0.39	1	9	0.26	2	Tr
630	*steamed*	Tr	Tr	Tr	0.41	0.04	0.11	4.1	3.9	0.41	2	9	0.25	1	Tr
631	smoked, *steamed*	Tr	Tr	Tr	N	0.05	0.16	4.3	4.4	0.46	2	N	0.26	1	Tr
632	in crumbs, *frozen, fried in blended oil*	Tr	Tr	Tr	N	0.08	0.08	2.8	2.7	0.24	1	N	0.26	2	Tr

Composition of food per 100g edible portion

No.	Food	Description and main data sources	Edible conversion factor	Water	Total nitrogen	Protein	Fat	Carbo-hydrate	Energy value	
				g	g	g	g	g	kcal	kJ
	White fish *continued*									
633	**Halibut**, *grilled*	Calculated from raw, cutlets and steaks; flesh only	0.78	72.1	4.05	25.3	2.2	0	121	513
634	**Lemon sole**, *raw*	Literature sources	N	81.2	2.78	17.4	1.5	0	83	351
635	*steamed*	Flesh only	0.71	77.2	3.29	20.6	0.9	0	91	384
636	goujons, *baked*	Calculated from manufacturers' proportions	1.00	54.1	2.58	16.0	14.6	14.7	187	775
637	goujons, *fried in blended oil*	Calculated from manufacturers' proportions	1.00	40.8	2.51	15.5	28.7	14.3	374	1553
638	**Plaice**, *raw*	8 fish purchased whole, and literature sources	0.42	79.5	2.67	16.7	1.4	0	79	336
639	*frozen, steamed*	12 samples, steamed for 15-20 minutes, fillets; flesh only	0.72[c]	78.0	3.14	19.6	1.5[a]	0	92	389
640	in batter, *fried in blended oil*	Samples as fried in retail blend oil, fatty acids calculated	1.00	52.4	2.43	15.2	16.8	12.0	257	1072
641	in crumbs, *fried in blended oil*	8 fillets, dipped in egg and breadcrumbs and fried; flesh only, light skin included	1.00	59.9	2.88	18.0	13.7	8.6	228	951
642	goujons, *baked*	Calculated from manufacturers' proportions	1.00	40.8	1.43	8.8	18.3	27.7	304	1270
643	goujons, *fried in blended oil*	Calculated from manufacturers' proportions	1.00	27.9	1.39	8.5	32.3	27.0	426	1771
644	**Rock Salmon/Dogfish**, in batter, *fried in blended oil*	Samples as fried in retail blend oil, fatty acids calculated	0.93	51.3	2.83	14.7[b]	21.9	10.3	295	1225
645	**Skate**, in batter, *fried in blended oil*	Samples as fried in retail blend oil, fatty acids calculated	0.85	50.7	3.67	14.7[b]	10.1	4.9	168	702
646	**Whiting**, *steamed*	Analysis and calculation from steamed, flesh only	0.93	76.9	3.35	20.9	0.9	0	92	389
647	in crumbs, *fried in blended oil*	Fillets coated in crumbs and fried	0.90	63.0	2.90	18.1	10.3	7.0	191	801

[a] Skin contains 7g fat per 100g

[b] (Total N - non-protein N) × 6.25

[c] Levels ranged from 0.49 to 0.81

Composition of food per 100g edible portion

No.	Food	Starch	Total sugars	Individual sugars					Dietary fibre	Fatty acids				Cholest-erol
				Gluc	Fruct	Sucr	Malt	Lact	NSP	Satd	Mono-unsatd	Poly-unsatd	Trans	
		g	g	g	g	g	g	g	g	g	g	g	g	mg
	***White fish** continued*													
633	**Halibut**, *grilled*	0	0	0	0	0	0	0	0	0.4	0.7	0.5	Tr	41
634	**Lemon sole**, *raw*	0	0	0	0	0	0	0	0	0.2	0.3	0.5	Tr	60
635	*steamed*	0	0	0	0	0	0	0	0	0.1	0.2	0.3	Tr	73
636	goujons, *baked*	13.9	0.9	Tr	Tr	Tr	Tr	0.4	N	N	N	N	N	55
637	goujons, *fried in blended oil*	13.4	0.9	0	0	Tr	Tr	0.4	N	(2.9)	(11.4)	(12.3)	N	53
638	**Plaice**, *raw*	0	0	0	0	0	0	0	0	0.2	0.4	0.3	Tr	42
639	*frozen, steamed*	0	0	0	0	0	0	0	0	0.2	0.4	0.4	Tr	54
640	in batter, *fried in blended oil*	12.0	Tr	Tr	Tr	Tr	0	Tr	(0.5)	1.8	6.1	8.2	N	N
641	in crumbs, *fried in blended oil*	(8.3)	(0.3)	Tr	Tr	Tr	(0.2)	Tr	(0.2)	1.5	4.9	6.7	N	N
642	goujons, *baked*	27.0	0.9	Tr	Tr	Tr	Tr	0.5	N	N	N	N	N	23
643	goujons, *fried in blended oil*	26.1	0.9	0	0	Tr	Tr	0.4	N	(3.4)	(12.9)	(13.3)	N	22
644	**Rock Salmon/Dogfish**, in batter, *fried in blended oil*	10.3	Tr	Tr	Tr	Tr	Tr	Tr	(0.4)	(2.9)	(8.0)	(9.9)	N	N
645	**Skate**, in batter, *fried in blended oil*	(4.8)	(0.1)	Tr	Tr	Tr	Tr	Tr	(0.2)	1.0	3.4	4.7	N	N
646	**Whiting**, *steamed*	0	0	0	0	0	0	0	0	0.1	0.3	0.2	Tr	55
647	in crumbs, *fried in blended oil*	6.8	0.2	Tr	Tr	Tr	Tr	0	0.2	1.1	3.7	5.0	N	N

Inorganic constituents per 100g edible portion

No.	Food	mg										µg	
		Na	K	Ca	Mg	P	Fe	Cu	Zn	Cl	Mn	Se	I
	White fish *continued*												
633	**Halibut**, *grilled*	71	490	34	29	240	0.6	0.05	0.5	71	0.01	N	47
634	**Lemon sole**, *raw*	(95)	(230)	(17)	(17)	(200)	(0.5)	0.01	0.4	(97)	N	60	N
635	*steamed*	120	280	21	20	250	0.6	0.01	0.5	120	N	73	N
636	goujons, *baked*	200	230	49	20	200	0.9	0.04	0.5	260	N	N	N
637	goujons, *fried in blended oil*	190	220	48	19	190	0.9	0.04	0.5	250	N	N	N
638	**Plaice**, *raw*	120	280	45	22	180	0.3	0.02	0.5	170	0.01	37	33
639	*frozen, steamed*	110	290	27[a]	22	160	0.2	0.02	0.6	140	0.01	40	(36)
640	in batter, *fried in blended oil*	210	210	73	20	170	0.5	0.06	0.7	240	0.14	6	N
641	in crumbs, *fried in blended oil*	220	280	67	24	180	0.8	(0.02)	0.7	310	(0.16)	(29)	(31)
642	goujons, *baked*	480	220	130	21	120	1.2	0.06	0.5	710	0.28	N	N
643	goujons, *fried in blended oil*	470	210	120	20	120	1.2	0.06	0.5	690	0.27	N	N
644	**Rock Salmon/Dogfish**, in batter, *fried in blended oil*	160	230	44	18	190	0.5	0.08	0.4	170	0.14	22	N
645	**Skate**, in batter, *fried in blended oil*	140	240	50	27	180	1.0	0.09	0.9	170	N	N	N
646	**Whiting**, *steamed*	110	400	21	28	190	0.1	0.01	0.4	140	0.01	25	80
647	in crumbs, *fried in blended oil*	200	320	48	33	260	0.7	N	N	190	N	N	N

[a] Skin contains 150mg Ca per 100g

Vitamins per 100g edible portion

No.	Food	Retinol µg	Carotene µg	Vitamin D µg	Vitamin E mg	Thiamin mg	Ribo-flavin mg	Niacin mg	Trypt/60 mg	Vitamin B_6 mg	Vitamin B_{12} µg	Folate µg	Panto-thenate mg	Biotin µg	Vitamin C mg
	White fish *continued*														
633	**Halibut**, *grilled*	N	Tr	N	1.00	0.07	0.07	6.1	4.7	0.40	1	11	0.35	4	Tr
634	**Lemon sole**, *raw*	Tr	Tr	Tr	N	0.09	0.08	3.5	0.2	N	1	11	0.30	(5)	Tr
635	*steamed*	Tr	Tr	Tr	N	0.10	0.10	3.8	3.8	N	1	13	0.29	5	Tr
636	goujons, *baked*	9	3	0	(3.07)	0.08	0.08	2.2	3.0	N	1	11	(0.24)	(4)	Tr
637	goujons, *fried in blended oil*	8	3	Tr	(6.49)	0.09	0.08	2.2	3.0	N	1	14	(0.24)	(3)	Tr
638	**Plaice**, *raw*	Tr	Tr	Tr	N	0.20	0.19	3.2	3.1	0.22	1	11	0.80	(47)	Tr
639	*frozen, steamed*	Tr	Tr	Tr	N	0.29	0.17	2.4	3.7	0.24	1	(14)	0.87	48	Tr
640	in batter, *fried in blended oil*	N	Tr	Tr	N	0.19	0.32	1.8	2.8	0.12	N	27	0.82	30	Tr
641	in crumbs, *fried in blended oil*	Tr	Tr	Tr	(3.31)	0.23	0.18	2.9	3.4	(0.15)	1	17	(0.52)	(30)	Tr
642	goujons, *baked*	9	3	0	(3.42)	0.07	0.11	1.2	1.7	0.20	1	5	(0.29)	N	Tr
643	goujons, *fried in blended oil*	9	3	Tr	6.83	0.08	0.10	1.2	1.7	0.16	1	6	(0.28)	N	Tr
644	**Rock Salmon/Dogfish**, in batter, *fried in blended oil*	94	Tr	N	N	0.07	0.08	3.2	2.8	0.21	N	4	0.57	12	Tr
645	**Skate**, in batter, *fried in blended oil*	9	Tr	N	1.20	0.03	0.10	2.4	2.7	N	N	N	N	N	Tr
646	**Whiting**, *steamed*	Tr	Tr	Tr	N	0.05	0.31	1.8	3.9	0.20	N	N	0.24	1	Tr
647	in crumbs, *fried in blended oil*	Tr	Tr	Tr	(2.48)	N	N	N	3.4	N	N	N	N	N	Tr

Composition of food per 100g edible portion

No.	Food	Description and main data sources	Edible conversion factor	Water	Total nitrogen	Protein	Fat	Carbo-hydrate	Energy value	
				g	g	g	g	g	kcal	kJ
	Fatty fish									
648	**Anchovies**, canned in oil, *drained*	10 samples, 4 brands.	0.74	46.4	4.03	25.2	10.0	0	191	798
649	**Eel**, *jellied*	10 samples from assorted outlets. Jelly included in analysis	0.95	82.7	1.35	8.4	7.1	Tr	98	406
650	**Herring**, *raw*	35 fish, purchased whole over the year	0.50	68.0[a]	2.85	17.8	13.2[b]	0	190	791
651	*grilled*	Samples gutted, grilled for 7 minutes per side; flesh only	0.68	63.9	3.22	20.1	11.2	0	181	756
652	**Kipper**, *raw*	10 samples from assorted outlets. Smoked cured herring	0.55	61.2	2.80	17.5	17.7[c]	0	229	952
653	*grilled*	10 samples, grilled 4-5 minutes; flesh only	0.63	55.9	3.22	20.1	19.4	0	255	1060
654	**Mackerel**, *raw*	10 samples from assorted outlets, purchased whole; flesh and skin	0.71	64.0[d]	2.99	18.7	16.1[e]	0	220	914
655	*grilled*	10 samples, grilled for 5 minutes per side; flesh and skin	0.92	58.6	3.33	20.8	17.3	0	239	994
656	*smoked*	10 samples, flesh and skin	0.99	47.1	3.02	18.9	30.9	0	354	1465
657	**Pilchards**, canned in tomato sauce	10 samples, 6 brands; whole contents	1.00	70.4	2.67	16.7	8.1	1.1	144	601
658	**Salmon**, *raw*	11 farmed and wild samples, whole fish and steaks	0.79[f]	67.2	3.23	20.2	11.0[g]	0	180	750
659	*grilled*	Calculated from raw, steaks; flesh only	0.82	60.7	3.87	24.2	13.1	0	215	896
660	*steamed*	Calculated from raw, steaks; flesh only	0.77	64.5	3.49	21.8	11.9	0	194	812
661	*smoked*	4 samples	1.00	64.9	4.06	25.4	4.5	0	142	598
662	pink, canned in brine, *flesh only, drained*	6 samples, 3 brands.	0.79	71.3	3.79	23.5	6.6	0	153	644

[a] Levels range from 57-79g per 100g being highest in spring and lowest in autumn/winter
[b] Levels range from 5g per 100g in spring to 20g per 100g in winter
[c] Levels range from 13.3g to 22.2g fat per 100g
[d] Levels range from 56 to 74g water per 100g
[e] Levels range from 6 to 23g fat per 100g
[f] This is an average value. The value for whole salmon is 0.62 and salmon steaks 0.81
[g] Wild salmon entering the river contain approximately 14.5g fat per 100g

Composition of food per 100g edible portion

No.	Food	Starch	Total sugars	Individual sugars: Gluc	Fruct	Sucr	Malt	Lact	Dietary fibre: NSP	Fatty acids: Satd	Mono-unsatd	Poly-unsatd	Trans	Cholest-erol
		g	g	g	g	g	g	g	g	g	g	g	g	mg
	Fatty fish													
648	**Anchovies**, canned in oil, *drained*	0	0	0	0	0	0	0	0	1.6	5.3	1.8	Tr	63
649	**Eel**, *jellied*	Tr	0	0	0	0	0	0	0	1.9	3.5	1.0	Tr	79
650	**Herring**, *raw*	0	0	0	0	0	0	0	0	3.3	5.5	2.7	Tr	50
651	*grilled*	0	0	0	0	0	0	0	0	2.8	4.7	2.3	Tr	43
652	**Kipper**, *raw*	0	0	0	0	0	0	0	0	2.8	9.3	3.9	0	64
653	*grilled*	0	0	0	0	0	0	0	0	3.1	10.2	4.2	0	70
654	**Mackerel**, *raw*	0	0	0	0	0	0	0	0	3.3	7.9	3.3	0	54
655	*grilled*	0	0	0	0	0	0	0	0	3.5	8.5	3.5	0	58
656	*smoked*	0	0	0	0	0	0	0	0	(6.3)	(15.1)	(6.3)	0	105
657	**Pilchards**, canned in tomato sauce	0.2	0.9	0.4	0.5	Tr	0	0	Tr	1.7	2.2	3.4	0	56
658	**Salmon**, *raw*	0	0	0	0	0	0	0	0	1.9	4.4	3.1	0	50
659	*grilled*	0	0	0	0	0	0	0	0	2.5	5.8	4.1	0	60
660	*steamed*	0	0	0	0	0	0	0	0	2.0	4.7	3.3	0	54
661	*smoked*	0	0	0	0	0	0	0	0	(0.8)	(1.8)	(1.3)	0	35
662	pink, canned in brine, *flesh only, drained*	0	0	0	0	0	0	0	0	1.3	2.4	1.9	0	28

Inorganic constituents per 100g edible portion

No.	Food	mg										µg	
		Na	K	Ca	Mg	P	Fe	Cu	Zn	Cl	Mn	Se	I
	Fatty fish												
648	**Anchovies**, canned in oil, *drained*	3930	230	300	56	300	4.1	0.17	3.0	6090	0.18	N	N
649	**Eel**, *jellied*	660	55	62	6	73	0.1	0.04	0.9	980	0.04	22	14
650	**Herring**, *raw*	120	320	60	32	230	1.2	0.14	0.9	170	0.04	35	29
651	*grilled*	160	430	79	42	310	1.6	0.19	1.2	220	0.05	46	38
652	**Kipper**, *raw*	830	340	53	27	230	1.6	0.12	1.0	1190	0.04	32	55
653	*grilled*	940	390	60	31	260	1.8	0.14	1.1	1350	0.05	36	63
654	**Mackerel**, *raw*	63	290	11	24	200	0.8	0.08	0.6	82	0.02	30	140
655	*grilled*	63	360	12	28	230	0.8	0.09	0.7	97	0.02	36	(170)
656	*smoked*	750	310	20	28	210	1.2	0.09	1.1	1130	0.02	33	(150)
657	**Pilchards**, canned in tomato sauce	290	310	250	29	280	2.5	0.16	1.3	520	0.11	30	64
658	**Salmon**, *raw*	45	360	21	27	250	0.4	0.03	0.6	58	0.02	(26)	37
659	*grilled*	54	430	25	32	300	0.5	0.04	0.7	69	0.02	(31)	44
660	*steamed*	49	390	23	29	270	0.4	0.03	0.7	63	0.02	(28)	4
661	*smoked*	1880	420	19	32	250	0.6	0.09	0.4	2850	0.02	(24)	N
662	pink, canned in brine, *flesh only, drained*	430	260	91	25	170	0.6	0.05	0.8	730	Tr	25	59

Vitamins per 100g edible portion

No.	Food	Retinol µg	Carotene µg	Vitamin D µg	Vitamin E mg	Thiamin mg	Riboflavin mg	Niacin mg	Trypt/60 mg	Vitamin B_6 mg	Vitamin B_{12} µg	Folate µg	Pantothenate mg	Biotin µg	Vitamin C mg
	Fatty fish														
648	**Anchovies**, canned in oil, *drained*	57	Tr	N	N	Tr	0.10	3.8	4.7	N	11	18	N	N	Tr
649	**Eel**, *jellied*	110	Tr	3.0	2.60	0.07	0.16	0.8	1.6	0.03	2	N	0.35	2	Tr
650	**Herring**, *raw*	44	Tr	19.0[a]	0.76	0.01	0:26	4.1	3.3	0.44	13	9	0.81	7	Tr
651	*grilled*	34	Tr	16.1	0.64	Tr	0.27	4.0	3.8	0.35	15	10	0.78	7	Tr
652	**Kipper**, *raw*	32	Tr	8.0	0.32	Tr	0.28	4.1	3.3	0.27	10	6	0.53	5	Tr
653	*grilled*	38	Tr	9.4	0.37	Tr	0.27	4.5	3.8	0.25	12	5	0.51	5	Tr
654	**Mackerel**, *raw*	45	Tr	8.2	0.43	0.14	0.29	8.6	3.5	0.41	8	N	0.81	5	Tr
655	*grilled*	48	Tr	8.8	0.46	0.15	0.32	9.4	3.9	0.45	1	N	0.93	6	Tr
656	*smoked*	31	Tr	8.0	0.25	0.26	0.52	9.5	3.5	0.50	6	N	1.03	3	Tr
657	**Pilchards**, canned in tomato sauce	7	(140)	14.0	2.56	0.01	0.33	5.9	3.1	0.27	13	N	0.85	11	Tr
658	**Salmon**, *raw*	13[b]	Tr	5.9[b]	1.91	0.23	0.13	7.2	3.8	0.75	4	16	1.02	7	Tr
659	*grilled*	16	Tr	7.1	2.29	0.25	0.14	7.7	4.5	0.81	5	19	1.16	9	Tr
660	*steamed*	14	Tr	8.7	2.07	0.22	0.14	7.0	4.1	0.81	4	17	0.88	7	Tr
661	*smoked*	N	Tr	N	N	0.16	0.17	8.8	4.7	(0.28)	(3)	(2)	(0.87)	N	Tr
662	pink, canned in brine, *flesh only, drained*	31	Tr	9.2	1.52	0.02	0.22	5.9	4.4	0.21	4	14	0.74	9	Tr

[a] Levels range from 7mg to 31mg vitamin D per 100g

[b] These are values for Atlantic salmon. Pacific salmon may contain 90 (20-150)mg retinol and 12.5 (5-20)mg vitamin D per 100g

No.	Food	Description and main data sources	Edible conversion factor	Water	Total nitrogen	Protein	Fat	Carbo-hydrate	Energy value	
				g	g	g	g	g	kcal	kJ
	Fatty fish *continued*									
663	**Sardines**, canned in brine, *drained*	10 samples, 4 brands.	0.79	66.2	3.44	21.5	9.6	0	172	721
664	canned in oil, *drained*	13 samples, 10 brands; canned in vegetable and olive oil	0.82	58.6	3.73	23.3	14.1[a]	0	220	918
665	canned in tomato sauce	10 samples, 8 brands; whole contents	1.00	69.3	2.72	17.0	9.9	1.4	162	678
666	**Swordfish**, *grilled*	Calculated from raw, steaks; flesh only	0.89	75.5	3.67	22.9	5.2	0	139	583
667	**Trout**, rainbow, *grilled*	11 samples, grilled 7 minutes per side; flesh only	0.73	73.3	3.44	21.5	5.4[b]	0	135	565
668	**Tuna**, canned in brine, *drained*	10 samples, 9 brands; skipjack tuna	0.81	74.6	3.76	23.5	0.6	0	99	422
669	canned in oil, *drained*	10 samples, 6 brands; skipjack tuna	0.79	63.3	4.34	27.1	9.0	0	189	794
670	**Whitebait**, in flour, *fried*	Whole fish; rolled in flour and fried	1.00	23.5	3.12	19.5	47.5	5.3	525	2174
	Crustacea									
671	**Crab**, *boiled*	12 samples; purchased boiled. Light and dark meat	0.35	71.0	3.12	19.5	5.5	Tr	128	535
672	canned in brine, *drained*	6 cans, 2 brands. White meat only	0.44	79.2	2.90	18.1	0.5	Tr	77	326
673	**Lobster**, *boiled*	Boiled in fresh water	0.36	74.3	3.54	22.1	1.6	Tr	103	435
674	**Prawns**, *boiled*	Samples cooked in sea or salt water	0.38	70.0	3.62	22.6	0.9	0	99	418
675	**Scampi**, in breadcrumbs, *frozen, fried in blended oil*	10 samples, 8 brands. Deep fried for 4 minutes	1.00	49.8	1.50	9.4	13.6	20.5	237	991
676	**Shrimps**, canned in brine, *drained*	10 cans, 3 brands	0.65	74.9	3.33	20.8	1.2	Tr	94	398
677	*frozen*	10 packets from Chinese supermarkets. Shrimps and prawns	1.00	81.2	2.64	16.5	0.8	Tr	73	310

[a] If not drained the fat content is approximately 24.4g per 100g

[b] Skin contains 17.3g fat per 100g

Composition of food per 100g edible portion

No.	Food	Starch	Total sugars	Individual sugars: Gluc	Fruct	Sucr	Malt	Lact	Dietary fibre NSP	Fatty acids: Satd	Mono-unsatd	Poly-unsatd	Trans	Cholest-erol
		g	g	g	g	g	g	g	g	g	g	g	g	mg
	***Fatty fish** continued*													
663	**Sardines**, canned in brine, *drained*	0	0	0	0	0	0	0	0	N	N	N	0	60
664	canned in oil, *drained*	0	0	0	0	0	0	0	0	2.9	4.8	5.0	0.1	65
665	canned in tomato sauce	Tr	1.4	0.6	0.8	Tr	0	0	Tr	2.8	2.9	3.2	0	76
666	**Swordfish**, *grilled*	0	0	0	0	0	0	0	0	1.2	2.1	1.4	0	52
667	**Trout**, rainbow, *grilled*	0	0	0	0	0	0	0	0	1.1	2.0	1.7	0	70[a]
668	**Tuna**, canned in brine, *drained*	0	0	0	0	0	0	0	0	0.2	0.1	0.2	0	51
669	canned in oil, *drained*	0	0	0	0	0	0	0	0	1.5	2.3	4.8	0.2	50
670	**Whitebait**, in flour, *fried*	(5.2)	(0.1)	Tr	Tr	Tr	Tr	0	0.2	N	N	N	N	N
	Crustacea													
671	**Crab**, *boiled*	Tr	Tr	0	0	0	0	0	0	0.7	1.5	1.6	0	72
672	canned in brine, *drained*	Tr	Tr	0	0	0	0	0	0	(0.1)	(0.1)	(0.1)	0	(72)
673	**Lobster**, *boiled*	Tr	Tr	0	0	0	0	0	0	0.2	0.3	0.6	0	110
674	**Prawns**, *boiled*	0	0	0	0	0	0	0	0	0.2	0.2	0.2	0	(280)
675	**Scampi**, in breadcrumbs, *frozen, fried in blended oil*	20.5	Tr	Tr	Tr	Tr	Tr	0	N	1.4	5.1	6.4	N	110
676	**Shrimps**, canned in brine, *drained*	Tr	Tr	0	0	0	0	0	0	0.2	0.3	0.4	0	(130)
677	*frozen*	Tr	Tr	0	0	0	0	0	0	(0.1)	(0.2)	(0.3)	0	(130)

[a] Skin contains 230mg cholesterol per 100g

Inorganic constituents per 100g edible portion

No.	Food	Na (mg)	K (mg)	Ca (mg)	Mg (mg)	P (mg)	Fe (mg)	Cu (mg)	Zn (mg)	Cl (mg)	Mn (mg)	Se (µg)	I (µg)
	Fatty fish *continued*												
663	**Sardines**, canned in brine, *drained*	530	320	540	45	510	2.3	0.16	2.3	810	0.20	41	23
664	canned in oil, *drained*	450	410	500	46	520	2.3	0.11	2.2	620	0.19	49	23
665	canned in tomato sauce	350	410	430	39	420	2.9	0.16	2.4	590	0.24	37	N
666	**Swordfish**, *grilled*	170	450	5	34	340	0.6	N	N	170	0.03	(57)	N
667	**Trout**, rainbow, *grilled*	55	410	21[a]	26	250[a]	0.4	0.05	0.5[a]	65	0.01	21	(15)
668	**Tuna**, canned in brine, *drained*	320	230	8	27	170	1.0	0.05	0.7	550	Tr	78	13
669	canned in oil, *drained*	290	260	12	33	200	1.6	0.20	1.1	530	0.05	90	14
670	**Whitebait**, in flour, *fried*	230	110	860	50	860	5.1	N	N	330	N	N	N
	Crustacea												
671	**Crab**, *boiled*	420	250	N	58	340	1.6	1.77	5.5	640	0.17	(84)	N
672	canned in brine, *drained*	550	100	120	32	140	2.8	0.42	5.7	830	N	N	N
673	**Lobster**, *boiled*	330	260	62	34	260	0.8	1.35	2.5	530	(0.03)	(54)	(100)
674	**Prawns**, *boiled*	1590	260	110	(49)	270	1.1	(0.20)	2.2	2550	0.01	(23)	(30)
675	**Scampi**, in breadcrumbs, *frozen, fried in blended oil*	660	130	210	24	310	1.7	0.16	0.6	610	0.35	17	41
676	**Shrimps**, canned in brine, *drained*	980	100	110	49	150	5.1	0.23	1.9	1510	0.04	(21)	81
677	*frozen*	380	75	130	47	150	2.6	0.15	1.1	520	0.15	(43)	(100)

[a] Skin contains 890mg Ca, 750mg P and 4.1mg Zn per 100g

Vitamins per 100g edible portion

No.	Food	Retinol µg	Carotene µg	Vitamin D µg	Vitamin E mg	Thiamin mg	Riboflavin mg	Niacin mg	Trypt/60 mg	Vitamin B_6 mg	Vitamin B_{12} µg	Folate µg	Pantothenate mg	Biotin µg	Vitamin C mg
	Fatty fish *continued*														
663	**Sardines**, canned in brine, *drained*	6	Tr	4.6	N	0.01	0.26	6.1	4.0	0.16	13	8	0.76	10	Tr
664	canned in oil, *drained*	7	Tr	5.0	0.31	0.01	0.29	6.9	4.4	0.18	15	8	0.86	5	Tr
665	canned in tomato sauce	9	140	8.0	3.08	0.02	0.28	5.5	3.2	0.35	14	13	0.50	5	Tr
666	**Swordfish**, *grilled*	N	Tr	N	N	0.19	0.20	9.5	4.3	0.59	5	N	0.50	N	Tr
667	**Trout**, rainbow, *grilled*	29	Tr	9.6[a]	1.01[a]	0.20	0.12	4.2	4.0	0.35	5	10	1.58	3	Tr
668	**Tuna**, canned in brine, *drained*	N	Tr	3.6	0.55	0.02	0.11	14.4	4.4	0.47	4	4	0.29	2	Tr
669	canned in oil, *drained*	N	Tr	3.0	1.94	0.02	0.12	16.1	5.1	0.51	5	5	0.32	3	Tr
670	**Whitebait**, in flour, *fried*	N	Tr	N	N	N	N	N	3.6	N	N	N	N	N	Tr
	Crustacea														
671	**Crab**, *boiled*	Tr	Tr	Tr	N	0.07	0.86	1.5	3.6	0.16	Tr	20	0.95	7	Tr
672	canned in brine, *drained*	Tr	Tr	Tr	N	Tr	0.05	1.1	3.4	N	Tr	N	N	Tr	Tr
673	**Lobster**, *boiled*	Tr	Tr	Tr	(1.47)	0.08	0.05	1.5	4.1	(0.08)	(3)	(9)	(1.00)	(7)	Tr
674	**Prawns**, *boiled*	Tr	Tr	Tr	N	(0.02)	(0.12)	(0.3)	4.8	(0.04)	(8)	N	(0.16)	(1)	Tr
675	**Scampi**, in breadcrumbs, *frozen, fried in blended oil*	Tr	Tr	Tr	N	0.11	0.04	1.2	1.8	0.09	1	N	0.26	1	Tr
676	**Shrimps**, canned in brine, *drained*	N	Tr	Tr	N	0.01	0.02	0.8	3.9	0.03	2	15	0.35	1	Tr
677	*frozen*	2	Tr	Tr	N	Tr	0.02	0.5	3.1	(0.08)	3	14	(0.24)	(1)	Tr

[a] Skin contains 2.9mg Vit E and 24ug Vit D per 100g

Composition of food per 100g edible portion

No.	Food	Description and main data sources	Edible conversion factor	Water	Total nitrogen	Protein	Fat	Carbo-hydrate	Energy value	
				g	g	g	g	g	kcal	kJ
	Molluscs									
678	**Cockles**, *boiled*	11 samples from assorted outlets; fresh and frozen	1.00	83.0	1.92	12.0	0.6	Tr	53	226
679	**Mussels**, *boiled*	11 fresh and frozen samples, boiled for 2 minutes	0.27	72.9	2.67	16.7	2.7	3.5[a]	104	440
680	**Squid**, *frozen, raw*	5 samples, 4 brands	0.59	84.2	2.10	13.1	1.5	(1.0)[a]	70	294
681	in batter, *fried in blended oil*	Calculated from dissection of shop bought samples	1.00	62.4	1.85	11.5	10.0	15.7	195	815
682	**Whelks**, *boiled*	10 samples from assorted outlets, boiled in salted water	0.34	73.9	3.12	19.5	1.2	Tr[a]	89	376
683	**Winkles**, *boiled*	11 samples,from stalls and fishmongers. Purchased cooked	0.19	73.0	2.46	15.4	1.2	Tr[a]	72	306
	Fish products and dishes									
684	**Crabsticks**	10 samples from assorted outlets. Crab flavoured minced fish sticks	1.00	76.6	1.60	10.0	0.4	6.6	68	290
685	**Curry**, fish, Bangladeshi	Recipe	1.00	70.8	1.97	12.3	8.0	1.5	127	529
686	prawn, *takeaway*	20 samples, average of 10 Bhuna and 10 Madras	1.00	76.4	1.31	8.2	8.5	2.2	117	488
687	**Fish balls**, *steamed*	7 varieties from different Chinese shops	1.00	82.2	1.89	11.8	0.5	5.5	74	313
688	**Fish cakes**, *fried in blended oil*	Samples as frozen; shallow fried 5 minutes each side	1.00	55.2	1.38	8.6	13.4	16.8	218	911
689	**Fish fingers**, cod, *fried in blended oil*	Samples as frozen; shallow fried 4 minutes each side	1.00	53.8	2.11	13.2	14.1	15.5	238	994
690	*grilled*	Samples as frozen; grilled 5 minutes each side	1.00	55.7	2.29	14.3	8.9	16.6	200	838
691	**Fish paste**	30 samples, sardine, crab, lobster and salmon	1.00	67.1	2.45	15.3	10.5	3.7	170	708
692	**Fisherman's pie**, *retail*	Calculated from manufacturers' proportions	1.00	75.3	1.42	8.9	5.4	8.9	118	493

[a] As glycogen

No.	Food	Starch	Total sugars	Individual sugars					Dietary fibre	Fatty acids				Cholest-erol
				Gluc	Fruct	Sucr	Malt	Lact	NSP	Satd	Mono-unsatd	Poly-unsatd	Trans	
		g	g	g	g	g	g	g	g	g	g	g	g	mg
Molluscs														
678	**Cockles**, *boiled*	Tr	Tr	0	0	0	0	0	0	0.2	0.1	0.2	0	53
679	**Mussels**, *boiled*	Tr	Tr	0	0	0	0	0	0	0.5	0.4	1.0	0	58
680	**Squid**, *frozen, raw*	Tr	Tr	0	0	0	0	0	0	0.3	0.2	0.5	0	200
681	in batter, *fried in blended oil*	12.9	2.2	0	0	0.1	Tr	2.0	0.5	2.1	3.3	3.7	N	145
682	**Whelks**, *boiled*	Tr	Tr	0	0	0	0	0	0	0.2	0.2	0.3	0	125
683	**Winkles**, *boiled*	Tr	Tr	0	0	0	0	0	0	0.2	0.2	0.4	0	105
Fish products and dishes														
684	**Crabsticks**	6.6	Tr	Tr	Tr	Tr	0	0	0	N	N	N	N	(39)
685	**Curry**, fish, Bangladeshi	Tr	1.0	0.4	0.3	0.3	0	0	0.3	N	N	N	Tr	N
686	prawn, *takeaway*	0.9	1.2	0.5	0.6	0.1	0.1	0.1	2.0	1.4	3.8	2.9	Tr	144
687	**Fish balls**, *steamed*	5.5	Tr	0	0	0	0	0	0	N	N	N	N	N
688	**Fish cakes**, *fried in blended oil*	16.8	Tr	Tr	Tr	Tr	Tr	Tr	N	1.8	5.4	5.6	N	21
689	**Fish fingers**, cod, *fried in blended oil*	15.5	Tr	Tr	Tr	Tr	Tr	0	0.6	(3.6)	(5.3)	(4.6)	N	32
690	*grilled*	16.6	Tr	Tr	Tr	Tr	Tr	0	0.7	2.8	3.4	2.3	N	35
691	**Fish paste**	3.2	0.5	Tr	Tr	0.5	Tr	0	(0.2)	N	N	N	N	N
692	**Fisherman's pie**, *retail*	7.2	1.7	0.1	Tr	0.1	Tr	1.4	0.5	1.8	1.3	0.7	N	N

Inorganic constituents per 100g edible portion

No.	Food	mg										µg	
		Na	K	Ca	Mg	P	Fe	Cu	Zn	Cl	Mn	Se	I
	Molluscs												
678	**Cockles**, *boiled*	490	110	91	46	140	28.0	0.38	2.1	750	0.84	(43)	(160)
679	**Mussels**, *boiled*	360	140	52	38	190	6.8	0.21	2.3	590	0.26	43	120
680	**Squid**, *frozen, raw*	190	150	13	36	170	0.2	0.68	1.2	280	0.02	(66)	(20)
681	in batter, *fried in blended oil*	88	230	81	23	160	0.7	0.52	0.9	65	0.11	(35)	(19)
682	**Whelks**, *boiled*	280	190	84	87	140	3.3	6.59[a]	12.1[b]	480	0.12	N	N
683	**Winkles**, *boiled*	750	220	130	310	160	10.2	1.70	3.3	1160	1.20	N	80
	Fish products and dishes												
684	**Crabsticks**	700	58	13	7	47	0.8	0.28	3.7	1010	0.08	N	N
685	**Curry**, fish, Bangladeshi	N	260	N	N	261	1.1	0.09	N	N	N	N	N
686	prawn, *takeaway*	311	198	97	26	95	4.2	0.12	0.8	440	0.31	14	34
687	**Fish balls**, *steamed*	750	110	26	13	100	0.2	0.07	0.3	1030	0.02	N	N
688	**Fish cakes**, *fried in blended oil*	510	230	110	18	110	0.8	0.03	0.4	(640)	0.20	N	N
689	**Fish fingers**, cod, *fried in blended oil*	450	260	85	22	220	0.8	0.04	0.3	550	0.19	(21)	(120)
690	*grilled*	440	290	92	22	220	0.8	0.02	0.4	590	0.21	(23)	110
691	**Fish paste**	600	300	280	33	310	18.3[c]	0.60	2.0	940	N	N	310[d]
692	**Fisherman's pie**, *retail*	130	260	71	21	120	0.4	0.04	0.5	(200)	0.05	13	N

[a] Levels ranged from 1.2 to 18.5mg Cu per 100g

[b] Levels ranged from 5.6 to 20.0mg Zn per 100g

[c] Iron oxides are often added as a colourant

[d] Crab paste contains 240µg I and salmon paste 370µg I per 100g

Vitamins per 100g edible portion

No.	Food	Retinol µg	Carotene µg	Vitamin D µg	Vitamin E mg	Thiamin mg	Riboflavin mg	Niacin mg	Trypt/60 mg	Vitamin B_6 mg	Vitamin B_{12} µg	Folate µg	Pantothenate mg	Biotin µg	Vitamin C mg
	Molluscs														
678	**Cockles**, *boiled*	40	Tr	Tr	N	0.05	0.11	1.2	2.6	0.04	47	N	0.27	9	Tr
679	**Mussels**, *boiled*	N	Tr	Tr	1.05	0.02	0.38	1.3	3.6	0.06	22	(37)	0.40	9	Tr
680	**Squid**, *frozen, raw*	(15)	0	Tr	(1.20)	0.05	0.02	2.1	2.8	(0.69)	3	2	(0.68)	N	0
681	in batter, *fried in blended oil*	46	13	0.1	2.35	0.10	0.13	1.6	2.5	0.32	2	15	0.52	N	Tr
682	**Whelks**, *boiled*	N	Tr	Tr	0.80	0.04	0.17	1.3	4.2	0.09	21	(6)	0.55	6	Tr
683	**Winkles**, *boiled*	N	Tr	Tr	3.90	0.29	0.38	1.7	3.3	0.10	36	N	0.38	3	Tr
	Fish products and dishes														
684	**Crabsticks**	Tr	Tr	Tr	N	0.01	0.06	0.2	1.9	0.02	1	N	N	N	Tr
685	**Curry**, fish, Bangladeshi	N	107	2.7	N	N	N	0.7	2.3	N	4	N	N	N	0
686	prawn, *takeaway*	7	279	Tr	3.16	0.02	0.05	0.6	1.2	0.08	0	4	0.30	2	2
687	**Fish balls**, *steamed*	N	Tr	N	N	0.02	0.03	0.8	2.2	N	1	4	N	N	0
688	**Fish cakes**, *fried in blended oil*	Tr	Tr	Tr	N	0.07	0.14	1.6	1.6	0.25	1	N	0.33	1	Tr
689	**Fish fingers**, cod, *fried in blended oil*	Tr	Tr	Tr	N	0.11	0.07	1.6	2.5	0.15	1	16	0.32	1	Tr
690	*grilled*	Tr	Tr	Tr	N	0.12	0.08	1.7	2.7	0.17	1	16	0.35	1	Tr
691	**Fish paste**	19[a]	Tr	N	0.87	0.02	0.20	4.1	2.9	N	N	N	N	N	Tr
692	**Fisherman's pie**, *retail*	57	32	0.2	(0.33)	0.06	0.07	0.9	1.8	(0.14)	(0)	(10)	(0.22)	(1)	2

[a] Salmon paste contains 49µg retinol per 100g

Composition of food per 100g edible portion

No.	Food	Description and main data sources	Edible conversion factor	Water	Total nitrogen	Protein	Fat	Carbo-hydrate	Energy value	
				g	g	g	g	g	kcal	kJ
	Fish products and dishes *continued*									
693	**Kedgeree**	Recipe	1.00	64.3	2.58	16.0	9.1	8.0	176	738
694	**Roe**, cod, hard, *fried in blended oil*	Parboiled, slices, coated in crumbs and fried in blended oil. Some nutrients calculated from raw	1.00	62.0	3.34	20.9	11.9	3.0	202	844
695	**Salmon en croute**, *retail*	Calculated from manufacturers' proportions	1.00	49.3	1.93	11.8	19.1	18.0	288	1202
696	**Seafood cocktail**	Recipe. Mussels, crabsticks, prawns, squid and cockles	1.00	75.3	2.50	15.6	1.5	2.9	87	369
697	**Seafood pasta**, *retail*	Calculated from manufacturers' proportions	1.00	77.1	1.45	8.9	4.8	7.6	110	460
698	**Szechuan prawns with vegetables**, *takeaway*	10 samples from different outlets	1.00	81.1	1.25	7.8	4.7	2.5	83	347
699	**Taramasalata**	10 assorted samples. Greek dish based on cod's roe	1.00	35.9	0.51	3.2	52.9	4.1	504	2077
700	**Tuna pate**	Calculated from manufacturers' proportions	1.00	61.8	2.71	17.0	18.6	0.4	236	982

Composition of food per 100g edible portion

No.	Food	Starch	Total sugars	Individual sugars					Dietary fibre	Fatty acids				Cholest-erol
				Gluc	Fruct	Sucr	Malt	Lact	NSP	Satd	Mono-unsatd	Poly-unsatd	Trans	
		g	g	g	g	g	g	g	g	g	g	g	g	mg
	Fish products and dishes *continued*													
693	**Kedgeree**	8.0	0	Tr	Tr	Tr	0	0	0	(2.1)	(3.0)	(3.1)	N	(126)
694	**Roe**, cod, hard, *fried in blended oil*	3.0	Tr	Tr	Tr	Tr	Tr	0	0.1	1.6	4.1	5.7	N	315
695	**Salmon en croute**, *retail*	17.1	0.9	0.1	Tr	0.1	0.5	0.2	N	3.1	2.9	1.4	N	31
696	**Seafood cocktail**	1.4	Tr	0	0	0	0	0	0	0.3	0.2	0.5	N	(115)
697	**Seafood pasta**, *retail*	5.9	1.6	0.2	0.2	0.1	0.1	1.0	0.4	2.8	1.2	0.3	N	41
698	**Szechuan prawns with vegetables**, *takeaway*	1.2	0.8	0.4	0.4	Tr	Tr	Tr	1.4	0.7	2.4	1.4	Tr	56
699	**Taramasalata**	4.1	Tr	Tr	Tr	Tr	Tr	Tr	Tr	4.1	29.3	16.7	0.9	25
700	**Tuna pate**	Tr	0.3	0.1	0.1	0.2	0	0	Tr	7.8	4.4	5.3	N	72

Inorganic constituents per 100g edible portion

No.	Food	mg										µg	
		Na	K	Ca	Mg	P	Fe	Cu	Zn	Cl	Mn	Se	I
	***Fish products and dishes** continued*												
693	**Kedgeree**	851	277	35	22	192	0.6	0.08	0.7	1314	0.06	23	192
694	**Roe**, cod, hard, *fried in blended oil*	120	170	13	9	300	1.0	0.22	3.3	230	0.12	N	N
695	**Salmon en croute**, *retail*	190	200	47	18	140	0.6	0.08	0.5	320	0.13	N	N
696	**Seafood cocktail**	620	160	52	(32)	170	5.6	(0.36)	2.4	940	0.20	N	N
697	**Seafood pasta**, *retail*	170	190	38	20	100	0.4	0.07	0.5	260	0.11	12	39
698	**Szechuan prawns with vegetables**, *takeaway*	536	102	40	15	65	1.1	0.11	0.4	(827)	Tr	N	N
699	**Taramasalata**	650	60	21	6	50	0.4	N	0.4	1040	0.12	N	N
700	**Tuna pate**	390	170	12	21	130	0.8	0.05	0.5	640	Tr	N	17

Vitamins per 100g edible portion

No.	Food	Retinol µg	Carotene µg	Vitamin D µg	Vitamin E mg	Thiamin mg	Riboflavin mg	Niacin mg	Trypt/60 mg	Vitamin B_6 mg	Vitamin B_{12} µg	Folate µg	Pantothenate mg	Biotin µg	Vitamin C mg
	Fish products and dishes *continued*														
693	**Kedgeree**	93	23	1.0	N	0.05	0.21	2.5	3.4	0.29	2	N	0.62	6	Tr
694	**Roe**, cod, hard, *fried in blended oil*	75	Tr	17.0	N	0.59	0.37	1.0	3.9	0.28	11	N	2.60	15	Tr
695	**Salmon en croute**, *retail*	30	13	(3.4)	N	0.13	0.07	2.5	2.3	0.31	2	12	N	N	Tr
696	**Seafood cocktail**	6	Tr	Tr	(0.56)	0.03	0.20	1.2	3.3	0.14	15	N	(0.30)	N	Tr
697	**Seafood pasta**, *retail*	52	52	0	0.40	0.03	0.06	1.0	1.7	0.09	1	6	0.21	1	1
698	**Szechuan prawns with vegetables,** *takeaway*	Tr	569	N	1.99	0.02	N	0.4	N	0.06	N	N	N	N	2
699	**Taramasalata**	N	N	N	N	0.08	0.10	0.3	0.6	N	3	4	N	N	1
700	**Tuna pate**	N	59	2.9	2.71	0.02	0.09	10.1	3.2	0.34	3	4	0.22	1	1

Section 2.7

Vegetables

The foods in this section of the Tables have largely been taken from the *Vegetables, Herbs and Spices* (1991) and *Vegetable Dishes* (1992) supplements. Some new analytical data on vegetable products have been incorporated. Foods for which new data are incorporated have been allocated a new food code and can thus easily be identified in the food index.

Because many of the vegetables and pulses eaten in this country are imported, a larger number of literature values from foreign sources have been used in this food group than many others in the Tables.

For most boiled vegetables, data is included for foods in boiled unsalted water. The amount of salt added to vegetables when boiled can vary considerably. Where foods are included as boiled in salted water, the water contained 0.5% salt. For fried foods the type of oil used for frying has been included in the name; this will determine the fatty acid profile of that particular food. Most values for cooked foods were obtained by analysis, but some were calculated from raw foods. For these any nutrient losses were estimated using the factors shown in Section 4.3. The changes in weight of beans and some other vegetables when soaked and cooked are shown in Section 4.3.

Samples of the same or similar foods always vary somewhat in composition. Some nutrients differ in a consistent way between varieties of a vegetable and with season as shown for potatoes. There are also differences with the length of storage, the depth of peeling or the number of outer leaves removed, and with cooking conditions (such as the extent to which a vegetable is cut up, the amount of water and the length of cooking, although there is little or no difference between vegetables cooked with microwaves or by more conventional methods). Any differences arising from the method of cultivation, for example 'organic' methods, appear to be small and inconsistent. It is not practical to give specific nutrient values for each of these factors, and the Tables therefore show average values for most products.

Users should note that all values are expressed per 100g edible portion. Guidance for calculating nutrient content 'as purchased' or 'as served' (e.g. including pods, tough skin and outer leaves) is given in Section 4.2.

Composition of food per 100g edible portion

No.	Food	Description and main data sources	Edible conversion factor	Water	Total nitrogen	Protein	Fat	Carbo-hydrate	Energy value	
				g	g	g	g	g	kcal	kJ
	Early potatoes									
701	**New potatoes**, average, *raw*	IFR; flesh only	0.89	81.7	0.28	1.7	0.3	16.1	70	298
702	*boiled in unsalted water*	IFR. Samples as raw; boiled 20 minutes	1.00	80.5	0.24	1.5	0.3	17.8	75	321
703	*in skins, boiled in unsalted water*	LGC; boiled 20 minutes	1.00	81.1	0.23	1.4	0.3	15.4	66	281
704	canned, *re-heated, drained*	LGC; 10 samples, 4 brands	0.65	81.3	0.23	1.5	0.1	15.1	63	271
	Main crop potatoes									
705	**Old potatoes**, average, *raw*	IFR; 4 varieties sampled over two years. Flesh only	0.80	79.0	0.33	2.1	0.2	17.2	75	318
706	*baked, flesh and skin*	Calculation from flesh only	1.00	62.6	0.62	3.9	0.2	31.7	136	581
707	*baked, flesh only*	IFR. Samples as raw; baked 90 minutes 200°C	0.67	78.9	0.35	2.2	0.1	18.0	77	329
708	*boiled in unsalted water*	IFR. Samples as raw; boiled 20 minutes	1.00	80.3	0.29	1.8	0.1	17.0	72	306
709	*mashed with butter*	Calculation from boiled (100g), butter (5g), milk (7g)	1.00	77.6	0.29	1.8	4.3	15.5	104	438
710	*roast in blended oil*	Calculation from roast in corn oil	1.00	64.7	0.46	2.9	4.5	25.9	149	630
711	*roast in corn oil*	IFR. Samples as raw; roasted in shallow oil 90 minutes 200°C	1.00	64.7	0.46	2.9	4.5	25.9	149	630
712	*roast in lard*	Calculation from roast in corn oil	1.00	64.7	0.46	2.9	4.5	25.9	149	630

Composition of food per 100g edible portion

No.	Food	Starch	Total sugars	Individual sugars					Dietary fibre	Fatty acids			Cholest-erol
				Gluc	Fruct	Sucr	Malt	Lact	NSP	Satd	Mono-unsatd	Poly-unsatd	
		g	g	g	g	g	g	g	g	g	g	g	mg
	Early potatoes												
701	**New potatoes**, average, *raw*	14.8	1.3	0.3	0.3	0.7	0	0	1.0	0.1	Tr	0.1	0
702	*boiled in unsalted water*	16.7	1.1	0.3	0.2	0.6	0	0	1.1	0.1	Tr	0.1	0
703	*in skins, boiled in unsalted water*	14.4	1.0	0.3	0.2	0.5	0	0	1.5	0.1	Tr	0.1	0
704	canned, *re-heated, drained*	14.4	0.7	0.1	0.1	0.5	0	0	0.8	Tr	Tr	0.1	0
	Main crop potatoes												
705	**Old potatoes**, average, *raw*	16.6	0.6	0.2	0.1	0.3	0	0	1.3	Tr	Tr	0.1	0
706	*baked, flesh and skin*	30.5	1.2	0.3	0.3	0.6	0	0	2.7	Tr	Tr	0.1	0
707	*baked, flesh only*	17.3	0.7	0.2	0.1	0.4	0	0	1.4	Tr	Tr	0.1	0
708	*boiled in unsalted water*	16.3	0.7	0.2	0.1	0.4	0	0	1.2	Tr	Tr	0.1	0
709	*mashed with butter*	14.5	1.0	0.2	0.1	0.4	0	0	1.1	2.8	1.0	0.2	12
710	*roast in blended oil*	25.3	0.6	0.2	0.1	0.3	0	0	1.8	0.4	2.2	1.6	0
711	*roast in corn oil*	25.3	0.6	0.2	0.1	0.3	0	0	1.8	0.6	1.1	2.6	0
712	*roast in lard*	25.3	0.6	0.2	0.1	0.4	0	0	1.8	1.8	2.0	0.4	4

Inorganic constituents per 100g edible portion

No.	Food	mg										µg	
		Na	K	Ca	Mg	P	Fe	Cu	Zn	Cl	Mn	Se	I
	Early potatoes												
701	**New potatoes**, average, *raw*	11	320	6	14	34	0.3	0.09	0.2	57	(0.1)	(1)	(3)
702	*boiled in unsalted water*	9	250	5	12	28	0.3	0.06	0.1	43	(0.1)	(1)	(3)
703	*in skins, boiled in unsalted water*	10	430	13	18	54	1.6	0.06	0.3	(43)	0.2	(1)	(3)
704	canned, *re-heated, drained*	250	220	24	11	27	0.9	0.04	Tr	430	0.1	N	N
	Main crop potatoes												
705	**Old potatoes**, average, *raw*	7	360	5	17	37	0.4	0.08	0.3	66	0.1	1	3
706	*baked, flesh and skin*	12	630	11	32	68	0.7	0.14	0.5	120	0.2	2	5
707	*baked, flesh only*	7	360	7	18	40	0.4	0.08	0.3	72	0.1	1	3
708	*boiled in unsalted water*	7	280	5	14	31	0.4	0.07	0.3	45	0.1	1	3
709	*mashed with butter*	43	260	13	13	35	0.4	0.06	0.3	98	0.1	1	5
710	*roast in blended oil*	9	570	8	25	55	0.7	0.11	0.4	99	0.1	1	4
711	*roast in corn oil*	9	570	8	25	55	0.7	0.11	0.4	99	0.1	1	4
712	*roast in lard*	9	570	8	25	55	0.7	0.11	0.4	99	0.1	1	4

Vitamins per 100g edible portion

No.	Food	Retinol µg	Carotene µg	Vitamin D µg	Vitamin E mg	Thiamin mg	Ribo-flavin mg	Niacin mg	Trypt/60 mg	Vitamin B_6 mg	Vitamin B_{12} µg	Folate µg	Panto-thenate mg	Biotin µg	Vitamin C mg
	Early potatoes														
701	**New potatoes**, average, *raw*	0	Tr	0	(0.06)	0.15	0.02	0.4	0.4	(0.44)	0	25	(0.37)	(0.3)	16
702	*boiled in unsalted water*	0	Tr	0	(0.06)	0.09	0.06	0.4	0.3	0.36	0	18	(0.38)	(0.3)	15
703	*in skins, boiled in unsalted water*	0	Tr	0	(0.06)	0.13	0.02	0.4	0.4	(0.33)	0	19	(0.38)	(0.3)	9
704	canned, *re-heated, drained*	0	Tr	0	(0.06)	(0.02)	(0.03)	(0.7)	0.3	(0.16)	0	(11)	N	Tr	5
	Main crop potatoes														
705	**Old potatoes**, average, *raw*	0	Tr	0	0.06	0.21	0.02	0.6	0.5	0.44	0	35	0.37	0.3	11[a]
706	*baked, flesh and skin*	0	Tr	0	0.11	0.37	0.02	1.1	0.9	0.54	0	44	0.46	0.5	14
707	*baked, flesh only*	0	Tr	0	0.06	0.21	0.01	0.6	0.5	0.31	0	25	0.26	0.3	8
708	*boiled in unsalted water*	0	Tr	0	0.06	0.18	0.01	0.5	0.4	0.33	0	19	0.38	0.3	6
709	*mashed with butter*	39	21	Tr	0.15	0.16	0.02	0.5	0.4	0.30	Tr	24	0.36	0.4	5
710	*roast in blended oil*	0	Tr	0	N	0.23	0.02	0.7	0.7	0.31	0	36	0.25	0.3	8
711	*roast in corn oil*	0	Tr	0	0.78	0.23	0.02	0.7	0.7	0.31	0	36	0.25	0.3	8
712	*roast in lard*	Tr	Tr	N	Tr	0.23	0.02	0.7	0.7	0.31	Tr	36	0.25	0.3	8

[a] Freshly dug potatoes contain 21mg vitamin C per 100g. This falls to 9mg per 100g after 3 months storage and to 7mg after 9 months

Composition of food per 100g edible portion

No.	Food	Description and main data sources	Edible conversion factor	Water g	Total nitrogen g	Protein g	Fat g	Carbo-hydrate g	Energy value kcal	Energy value kJ
	Chipped old potatoes									
713	**Chips**, *homemade, fried in blended oil*	Calculation from fried in corn oil	1.00	56.5	0.63	3.9	6.7[a]	30.1	189	796
714	*fried in corn oil*	IFR. Samples as raw potatoes; deep fried 6 minutes 190°C	1.00	56.5	0.63	3.9	6.7[a]	30.1	189	796
715	*fried in dripping*	Calculation from fried in corn oil	1.00	56.5	0.63	3.9	6.7[a]	30.1	189	796
716	*retail, fried in blended oil*	Calculation from fried in vegetable oil	1.00	52.3	0.51	3.2	12.4[a]	30.5	239	1001
717	*fried in dripping*	Calculation from fried in vegetable oil	1.00	52.3	0.51	3.2	12.4[a]	30.5	239	1001
718	*fried in vegetable oil*	5 samples from fish and chip shops	1.00	52.3	0.51	3.2	12.4[a]	30.5	239	1001
719	French fries, *retail*	5 samples from burger outlets. Manufacturers' data	1.00	43.8	0.54	3.3	15.5[a]	34.0	280	1174
720	straight cut, *frozen, fried in blended oil*	Calculation from fried in corn oil	1.00	40.3	0.66	4.1	13.5[a]	36.0	273	1145
721	*frozen, fried in corn oil*	LGC; 10 samples, 10 brands. Deep fried 3-5 minutes	1.00	40.3	0.66	4.1	13.5[a]	36.0	273	1145
722	*fried in dripping*	Calculation from fried in corn oil	1.00	40.3	0.66	4.1	13.5[a]	36.0	273	1145
723	fine cut, *frozen, fried in blended oil*	Calculation from fried in corn oil	1.00	26.0	0.72	4.5	21.3[a]	41.2	364	1524
724	*frozen, fried in corn oil*	LGC; 10 samples, 4 brands. Deep fried 1-4 minutes	1.00	26.0	0.72	4.5	21.3[a]	41.2	364	1524
725	*frozen, fried in dripping*	Calculation from fried in corn oil	1.00	26.0	0.72	4.5	21.3[a]	41.2	364	1524
726	**Microwave chips**, *cooked*	LGC; 10 samples, 2 brands; cooked as packet directions	1.00	50.4	0.58	3.6	9.6	32.1	221	930
727	**Oven chips**, *frozen, baked*	LGC; 10 samples, 7 brands. Oven baked 15-20 minutes	1.00	58.5	0.52	3.2	4.2	29.8	162	687

[a] The fat content of chips will be variable and dependent on a number of factors related to their preparation

No.	Food	Starch	Total sugars	Individual sugars					Dietary fibre	Fatty acids			Cholest-erol
				Gluc	Fruct	Sucr	Malt	Lact	NSP	Satd	Mono-unsatd	Poly-unsatd	
		g	g	g	g	g	g	g	g	g	g	g	mg
	Chipped old potatoes												
713	**Chips**, *homemade, fried in blended oil*	29.5	0.6	0.2	0.1	0.3	0	0	2.2	0.6	3.3	2.4	0
714	*fried in corn oil*	29.5	0.6	0.2	0.1	0.3	0	0	2.2	0.9	1.7	3.9	0
715	*fried in dripping*	29.5	0.6	0.2	0.1	0.3	0	0	2.2	3.7	2.5	0.2	6
716	*retail, fried in blended oil*	28.8	1.7	0.5	0.4	0.8	0	0	(2.2)	1.1	6.2	4.5	0
717	*fried in dripping*	28.8	1.7	0.5	0.4	0.8	0	0	(2.2)	6.8	4.6	0.3	11
718	*fried in vegetable oil*	28.8	1.7	0.5	0.4	0.8	0	0	(2.2)	3.6	5.3	3.1	0
719	French fries, *retail*	32.7	1.3	0.2	0.1	1.0	0	0	(2.1)	5.8	6.9	2.1	N
720	straight cut, *frozen, fried in blended oil*	35.3	0.7	0.2	0.2	0.3	0	0	2.4	1.2	6.7	4.9	0
721	*frozen, fried in corn oil*	35.3	0.7	0.2	0.2	0.3	0	0	2.4	2.5	3.4	7.0	0
722	*fried in dripping*	35.3	0.7	0.2	0.2	0.3	0	0	2.4	7.5	5.0	0.3	12
723	fine cut, *frozen, fried in blended oil*	40.6	0.6	0.2	0.2	0.2	0	0	(2.4)	1.8	10.6	7.8	0
724	*frozen, fried in corn oil*	40.6	0.6	0.2	0.2	0.2	0	0	2.7	4.0	5.4	11.0	0
725	*frozen, fried in dripping*	40.6	0.6	0.2	0.2	0.2	0	0	2.7	11.8	7.9	0.5	19
726	**Microwave chips**, *cooked*	31.5	0.6	0.2	0.1	0.3	0	0	2.9	N	N	N	0
727	**Oven chips**, *frozen, baked*	29.1	0.7	0.2	0.1	0.4	0	0	2.0	1.8	1.6	0.6	0

No.	Food	mg										µg	
		Na	K	Ca	Mg	P	Fe	Cu	Zn	Cl	Mn	Se	I
	Chipped old potatoes												
713	**Chips**, *homemade, fried in blended oil*	12	660	11	31	62	0.8	0.14	0.6	120	0.2	2	5
714	*fried in corn oil*	12	660	11	31	62	0.8	0.14	0.6	120	0.2	2	5
715	*fried in dripping*	12	660	11	31	63	0.8	0.14	0.6	120	0.2	2	5
716	*retail, fried in blended oil*	35	(660)	(11)	(31)	(62)	0.9	(0.14)	(0.6)	(120)	(0.2)	(2)	(5)
717	*fried in dripping*	35	(660)	(11)	(31)	(63)	0.9	(0.14)	(0.6)	(120)	(0.2)	(2)	(5)
718	*fried in vegetable oil*	35	(660)	(11)	(31)	(62)	0.9	(0.14)	(0.6)	(120)	(0.2)	(2)	(5)
719	French fries, *retail*	310[a]	650	14	(26)	(130)	1.0	(0.17)	(0.5)	480	(0.1)	(2)	N
720	straight cut, *frozen, fried in blended oil*	29	710	15	33	120	0.9	0.24	0.6	76	0.2	(2)	(7)
721	*frozen, fried in corn oil*	29	710	15	33	120	0.9	0.24	0.6	76	0.2	(2)	(7)
722	*fried in dripping*	30	710	15	33	120	0.9	0.24	0.6	76	0.2	(2)	(8)
723	fine cut, *frozen, fried in blended oil*	97	720	19	34	170	1.0	0.22	0.6	98	0.2	(3)	(8)
724	*frozen, fried in corn oil*	97	720	19	34	170	1.0	0.22	0.6	98	0.2	(3)	(8)
725	*frozen, fried in dripping*	98	720	19	34	170	1.0	0.22	0.6	98	0.2	(3)	(9)
726	**Microwave chips**, *cooked*	40	530	17	30	99	1.0	0.14	0.4	64	0.2	2	6
727	**Oven chips**, *frozen, baked*	53	530	(12)	27	120	0.8	0.22	0.4	74	0.2	N	N

[a] Unsalted French fries contain approximately 35mg Na per 100g

Vitamins per 100g edible portion

No.	Food	Retinol µg	Carotene µg	Vitamin D µg	Vitamin E mg	Thiamin mg	Ribo-flavin mg	Niacin mg	Trypt/60 mg	Vitamin B_6 mg	Vitamin B_{12} µg	Folate µg	Panto-thenate mg	Biotin µg	Vitamin C mg
	Chipped old potatoes														
713	**Chips**, *homemade, fried in blended oil*	0	Tr	0	N	0.24	0.02	0.7	0.9	0.32	0	43	0.25	0.4	9
714	*fried in corn oil*	0	Tr	0	4.90	0.24	0.02	0.7	0.9	0.32	0	43	0.25	0.4	9
715	*fried in dripping*	N	N	Tr	0.02	0.24	0.02	0.7	0.9	0.32	Tr	43	0.25	0.4	9
716	*retail, fried in blended oil*	Tr	Tr	0	N	0.08	0.01	(0.7)	0.8	(0.32)	0	N	(0.25)	(0.4)	(9)[a]
717	*fried in dripping*	N	N	Tr	0.04	0.08	0.01	(0.7)	0.8	(0.32)	Tr	N	(0.25)	(0.4)	(9)[a]
718	*fried in vegetable oil*	Tr	Tr	0	0.39	0.08	0.01	(0.7)	0.8	(0.32)	0	36	(0.25)	(0.4)	(9)[a]
719	French fries, *retail*	0	Tr	0	1.00	0.08	0.05	2.3	0.8	0.36	0	31	N	N	4
720	straight cut, *frozen, fried in blended oil*	0	Tr	0	N	0.16	0.08	2.1	1.0	0.46	0	30	N	N	16
721	*frozen, fried in corn oil*	0	Tr	0	3.27	0.16	0.08	2.1	1.0	0.46	0	30	N	N	16
722	*fried in dripping*	N	N	Tr	0.04	0.16	0.08	2.1	1.0	0.46	Tr	30	N	N	16
723	fine cut, *frozen, fried in blended oil*	0	Tr	0	N	0.18	0.09	2.4	1.1	0.52	0	34	N	N	12
724	*frozen, fried in corn oil*	0	Tr	0	(5.16)	0.18	0.09	2.4	1.1	0.52	0	34	N	N	12
725	*frozen, fried in dripping*	N	N	Tr	0.06	0.18	0.09	2.4	1.1	0.52	Tr	34	N	N	12
726	**Microwave chips**, *cooked*	0	Tr	0	N	0.12	0.07	2.1	0.9	0.29	0	20	N	N	11
727	**Oven chips**, *frozen, baked*	0	Tr	0	0.44	0.11	0.04	2.2	0.8	0.37	0	21	N	N	12

[a] Storage of uncooked chips under some conditions may significantly reduce vitamin C levels which could approach zero

Composition of food per 100g edible portion

No.	Food	Description and main data sources	Edible conversion factor	Water	Total nitrogen	Protein	Fat	Carbo-hydrate	Energy value	
				g	g	g	g	g	kcal	kJ
	Potato products									
728	**Instant potato powder**, *made up with water*	Calculated from ingredients; made up as packet directions	1.00	83.3	0.24	1.5	0.1	13.5	57	245
729	*made up with whole milk*	Calculated from ingredients; made up as packet directions	1.00	80.0	0.37	2.4	1.2	14.8	76	322
730	**Potato croquettes**, *fried in blended oil*	LGC; 10 samples, 5 brands. Shallow fried 5-7 minutes	1.00	58.2	0.59	3.7	13.1	21.6	214	893
731	**Potato fritters**, battered, *cooked*	2 samples of different brands, oven cooked	1.00	59.5	0.51	3.2	8.5	25.5	185	777
732	**Potato waffles**, *frozen, cooked*	IFR. 10 samples (Birds Eye); grilled, shallow and deep fried in corn oil, oven baked	1.00	52.7	0.51	3.2	8.2	30.3	200	842
	Beans and lentils									
733	**Aduki beans**, *dried, boiled in unsalted water*	LGC analysis and calculation from dried	1.00	59.4	1.48	9.3	0.2	22.5[a]	123	525
734	**Baked beans**, canned in tomato sauce, *re-heated*	LGC; 10 cans, 7 brands	1.00	71.5	0.83	5.2	0.6	15.3	84	355
735	reduced sugar, reduced salt	LGC; 5 cans, 2 own brands	1.00	73.6	0.85	5.4	0.6	12.5	73	311
736	**Beansprouts**, mung, *raw*	IFR; as purchased	1.00	90.4	0.47	2.9	0.5	4.0	31	131
737	*stir-fried in blended oil*	LGC. 6 samples; stir-fried 2 minutes. And calculated from raw	1.00	88.4	0.30	1.9	6.1	2.5	72	298
738	**Black gram**, urad gram, *dried, raw*	Whole beans. Literature sources	1.00	11.5	3.98	24.9	1.4	*40.8*[a]	*275*	*1169*
739	*dried, boiled in unsalted water*	As raw; soaked and boiled	1.00	71.3	1.25	7.8	0.4	*13.6*[a]	*89*	*379*

[a] Including oligosaccharides

No.	Food	Starch	Total sugars	Individual sugars					Dietary fibre	Fatty acids			Cholest-erol
				Gluc	Fruct	Sucr	Malt	Lact	NSP	Satd	Mono-unsatd	Poly-unsatd	
		g	g	g	g	g	g	g	g	g	g	g	mg
	Potato products												
728	**Instant potato powder**, *made up with water*	12.7	0.7	0.1	0.1	Tr	0	0.6	1.0	Tr	Tr	0.1	0
729	*made up with whole milk*	12.7	2.0	0.1	0.1	Tr	0	1.9	1.0	0.7	0.3	0.1	4
730	**Potato croquettes**, *fried in blended oil*	21.1	0.5	0.2	0.3	0.1	0	0	1.3	1.7	3.2	7.6	0
731	**Potato fritters**, battered, *cooked*	25.2	0.3	0.1	Tr	0.2	Tr	Tr	1.5	3.8	3.3	1.0	1
732	**Potato waffles**, *frozen, cooked*	29.8	0.6	0.1	0.1	0.3	0	0.1	2.3	1.0	2.0	4.7	0
	Beans and lentils												
733	**Aduki beans**, *dried, boiled in unsalted water*	20.8	0.5[a]	Tr	Tr	0.3	0.1	0	5.5	N	N	N	0
734	**Baked beans**, canned in tomato sauce, *re-heated*	9.4	5.9	0.9	1.0	3.9	0	0	3.7	0.1	0.1	0.3	0
735	reduced sugar, reduced salt	9.7	2.8	0.4	0.6	1.8	0	0	3.8	0.1	0.1	0.3	0
736	**Beansprouts**, mung, *raw*	1.8	2.2	1.1	1.1	Tr	0	0	1.5	0.1	0.1	0.2	0
737	*stir-fried in blended oil*	1.1	1.4	0.6	0.8	Tr	0	0	0.9	0.5	3.0	2.2	0
738	**Black gram**, urad gram, *dried, raw*	37.6	1.3[a]	0.1	0.1	1.1	0	0	N	0.2	0.2	0.7	0
739	*dried, boiled in unsalted water*	13.0	0.3[a]	Tr	Tr	0.3	0	0	N	0.1	0.1	0.2	0

[a] Not including oligosaccharides

Inorganic constituents per 100g edible portion

No.	Food	mg										µg	
		Na	K	Ca	Mg	P	Fe	Cu	Zn	Cl	Mn	Se	I
	Potato products												
728	**Instant potato powder**, *made up with water*	200	260	13	12	41	0.4	0.04	0.2	290	0.1	N	N
729	*made up with whole milk*	210	290	44	15	66	0.4	0.04	0.3	310	0.1	N	N
730	**Potato croquettes**, *fried in blended oil*	420	360	44	19	49	0.9	0.08	0.3	650	0.2	N	N
731	**Potato fritters**, battered, *cooked*	405	300	18	16	87	0.5	0.02	0.2	535	0.1	N	N
732	**Potato waffles**, *frozen, cooked*	430	480	32	21	120	0.5	Tr	0.3	630	0.1	N	N
	Beans and lentils												
733	**Aduki beans**, *dried, boiled in unsalted water*	2	570	39	60	180	1.9	0.51	2.3	N	0.8	1	N
734	**Baked beans**, canned in tomato sauce, *re-heated*	530	310	53	31	100	1.4	0.03	0.5	820	0.3	2	3
735	reduced sugar, reduced salt	330	270	45	29	90	1.2	0.10	0.5	480	0.3	2	3
736	**Beansprouts**, mung, *raw*	5	74	20	18	48	1.7	0.08	0.3	15	0.3	1	N
737	*stir-fried in blended oil*	3	45	12	11	29	1.0	0.05	0.2	9	0.2	(1)	N
738	**Black gram**, urad gram, *dried, raw*	40	800	150	160	370	6.3	0.72	2.8	N	1.2	N	N
739	*dried, boiled in unsalted water*	13	260	49	52	120	2.0	0.23	0.9	N	0.4	N	N

No.	Food	Retinol µg	Carotene µg	Vitamin D µg	Vitamin E mg	Thiamin mg	Riboflavin mg	Niacin mg	Trypt/60 mg	Vitamin B_6 mg	Vitamin B_{12} µg	Folate µg	Pantothenate mg	Biotin µg	Vitamin C mg
	Potato products														
728	**Instant potato powder**, *made up with water*	0	3	0	0.05	0.01	0.03	1.2	0.4	0.15	0	2	N	N	23
729	*made up with whole milk*	14	8	Tr	0.06	0.02	0.07	1.2	0.5	0.17	0.1	4	N	N	23
730	**Potato croquettes**, *fried in blended oil*	0	N	0	N	0.08	0.08	1.4	1.0	0.22	0	2	N	N	2
731	**Potato fritters**, battered, *cooked*	0	Tr	0	0.88	0.06	0.04	1.4	0.7	0.14	0.1	14	0.40	8.0	9
732	**Potato waffles**, *frozen, cooked*	0	Tr	0	N	N	N	N	0.7	N	0	N	N	N	36
	Beans and lentils														
733	**Aduki beans**, *dried, boiled in unsalted water*	0	6	0	N	0.14	0.08	0.9	1.5	N	0	N	N	N	Tr
734	**Baked beans**, canned in tomato sauce, *re-heated*	0	74	0	0.37	0.09	0.06	0.5	0.8	0.14	0	22	0.18	2.5	Tr
735	reduced sugar, reduced salt	0	77	0	0.39	0.09	0.06	0.5	0.9	0.14	0	23	0.19	2.6	Tr
736	**Beansprouts**, mung, *raw*	0	40	0	N	0.11	0.04	0.5	0.5	0.10	0	61	0.38	N	7
737	*stir-fried in blended oil*	0	24	0	N	0.06	0.02	0.3	0.3	0.07	0	43	0.23	N	7
738	**Black gram**, urad gram, *dried, raw*	0	38	0	N	0.42	0.37	2.0	4.6	N	0	132	N	N	Tr
739	*dried, boiled in unsalted water*	0	12	0	N	(0.11)	(0.09)	(0.5)	1.5	N	0	(33)	N	N	Tr

No.	Food	Description and main data sources	Edible conversion factor	Water g	Total nitrogen g	Protein g	Fat g	Carbo-hydrate g	Energy value kcal	Energy value kJ
	***Beans and lentils** continued*									
740	**Blackeye beans**, *dried, raw*	Whole beans. Analysis and literature sources	1.00	10.7	3.76	23.5	1.6	54.1[a]	311	1324
741	*dried, boiled in unsalted water*	As raw; soaked and boiled	1.00	66.2	1.41	8.8	0.7	19.9[a]	116	494
742	**Broad beans**, *frozen, boiled in unsalted water*	LGC; 10 samples, 7 brands. Boiled 3-10 minutes	1.00	73.8	1.27	7.9	0.6	11.7[a]	81	344
743	**Butter beans**, canned, *re-heated, drained*	LGC; 10 cans, 5 brands	0.57	74.0	0.95	5.9	0.5	13.0[a]	77	327
744	**Chick pea flour/besan flour**	Literature sources and estimation from whole peas	1.00	(10.0)	3.15	19.7	(5.4)	(49.6)[a]	313	1328
745	**Chick peas**, whole, *dried, raw*	Analytical and literature sources. Kabuli variety	1.00	10.0	3.42	21.3	5.4	49.6[a]	320	1355
746	*dried, boiled in unsalted water*	As raw. Soaked and boiled	1.00	65.8	1.35	8.4	2.1	18.2[a]	121	512
747	canned, *re-heated, drained*	LGC. Whole peas; 10 samples, 5 brands	0.60	67.5	1.15	7.2	2.9	16.1[a]	115	487
748	**Green beans/French beans**, *raw*	IFR; pods and beans, ends trimmed	0.83	90.7	0.31	1.9	0.5	3.2	24	99
749	*frozen, boiled in unsalted water*	LGC; 10 samples, 8 brands. Boiled 3-8 minutes	1.00	90.0	0.28	1.7	0.1	4.7	25	108
750	**Hummus**	LGC. Chick pea spread; 10 samples, retail and homemade	1.00	61.4	1.22	7.6	12.6	11.6[a]	187	781
751	**Lentils**, green and brown, whole, *dried, raw*	LGC; 10 samples, 6 brands. Continental type	1.00	10.8	3.90	24.3	1.9	48.8[a]	297	1264
752	*dried, boiled in salted water*	LGC; as raw. Boiled 10 minutes, simmered 25 minutes	1.00	66.7	1.41	8.8	0.7	16.9[a]	105	446
753	**Lentils**, red, split, *dried, raw*	LGC; as purchased	1.00	11.1	3.80	23.8	1.3	56.3[a]	318	1353
754	*dried, boiled in unsalted water*	LGC. As purchased; boiled 20 minutes	1.00	72.1	1.22	7.6	0.4	17.5[a]	100	424

[a] Including oligosaccharides

No.	Food	Starch	Total sugars	Individual sugars					Dietary fibre	Fatty acids			Cholest-erol
				Gluc	Fruct	Sucr	Malt	Lact	NSP	Satd	Mono-unsatd	Poly-unsatd	
		g	g	g	g	g	g	g	g	g	g	g	mg
	***Beans and lentils** continued*												
740	**Blackeye beans**, *dried, raw*	47.5	2.9[a]	0.2	0.1	2.5	0.1	0	8.2	0.5	0.1	0.7	0
741	*dried, boiled in unsalted water*	18.0	1.1[a]	0.1	Tr	0.9	Tr	0	3.5	0.2	0.1	0.3	0
742	**Broad beans**, *frozen, boiled in unsalted water*	10.0	1.3[a]	Tr	Tr	1.3	0	0	6.5	0.1	0.1	0.3	0
743	**Butter beans**, canned, *re-heated, drained*	10.9	1.1[a]	Tr	Tr	1.1	0	0	4.6	0.1	Tr	0.2	0
744	**Chick pea flour/besan flour**	(43.8)	(2.6)[a]	Tr	(0.2)	(2.4)	0	0	(10.7)	(0.5)	(1.1)	(2.7)	0
745	**Chick peas**, whole, *dried, raw*	43.8	2.6[a]	Tr	0.2	2.4	0	0	10.7	0.5	1.1	2.7	0
746	*dried, boiled in unsalted water*	16.6	1.0[a]	Tr	0.1	0.9	0	0	4.3	0.2	0.4	1.0	0
747	canned, *re-heated, drained*	15.1	0.4[a]	Tr	Tr	0.4	0	0	4.1	0.3	0.7	1.3	0
748	**Green beans/French beans**, *raw*	0.9	2.3	0.8	1.1	0.4	0	0	2.2	(0.1)	Tr	(0.3)	0
749	*frozen, boiled in unsalted water*	2.6	2.1	0.7	1.0	0.4	0	0	4.1	Tr	Tr	Tr	0
750	**Hummus**	9.3	1.9[a]	Tr	0.2	1.7	0	0	2.4	N	N	N	0
751	**Lentils**, green and brown, whole, *dried, raw*	44.5	1.2[a]	Tr	0.1	1.1	0	0	8.9	0.2	0.3	0.8	0
752	*dried, boiled in salted water*	15.9	0.4[a]	Tr	Tr	0.4	0	0	3.8	0.1	0.1	0.3	0
753	**Lentils**, red, split, *dried, raw*	50.8	2.4[a]	Tr	0.2	2.2	0	0	4.9	0.2	0.2	0.5	0
754	*dried, boiled in unsalted water*	16.2	0.8[a]	Tr	0.1	0.7	0	0	1.9	Tr	0.1	0.2	0

[a] Not including oligosaccharides

Inorganic constituents per 100g edible portion

No.	Food	mg										µg	
		Na	K	Ca	Mg	P	Fe	Cu	Zn	Cl	Mn	Se	I
	Beans and lentils *continued*												
740	**Blackeye beans**, *dried, raw*	16	1170	81	140	410	7.6	0.75	3.2	N	1.3	7	N
741	*dried, boiled in unsalted water*	5	320	21	45	140	1.9	0.22	1.1	N	0.5	3	N
742	**Broad beans**, *frozen, boiled in unsalted water*	8	280	56	36	150	1.6	0.32	1.0	15	0.3	1	(6)
743	**Butter beans**, canned, *re-heated, drained*	420	290	15	27	68	1.5	0.14	0.6	660	0.3	(3)	N
744	**Chick pea flour/besan flour**	(39)	(1000)	180	120	340	8.3	0.62	3.1	(60)	2.1	2	N
745	**Chick peas**, whole, *dried, raw*	39	1000	160	130	310	5.5	0.95	3.0	60	2.4	2	N
746	*dried, boiled in unsalted water*	5	270	46	37	83	2.1	0.28	1.2	7	0.7	1	N
747	canned, *re-heated, drained*	220	110	43	24	81	1.5	0.05	0.8	280	0.8	1	N
748	**Green beans/French beans**, *raw*	Tr	230	36	17	38	1.2	0.01	0.2	9	N	Tr	N
749	*frozen, boiled in unsalted water*	8	160	56	17	33	0.6	0.05	0.2	21	0.2	Tr	N
750	**Hummus**	670	190	41	62	160	1.9	0.30	1.4	670	0.5	4	N
751	**Lentils**, green and brown, whole, *dried, raw*	12	940	71	110	350	11.1	1.02	3.9	87	1.4	105	N
752	*dried, boiled in salted water*	3	310	22	34	130	3.5	0.33	1.4	26	0.5	40	N
753	**Lentils**, red, split, *dried, raw*	36	710	51	83	320	7.6	0.58	3.1	64	N	(6)	N
754	*dried, boiled in unsalted water*	12	220	16	26	100	2.4	0.19	1.0	20	N	(2)	N

No.	Food	Retinol µg	Carotene µg	Vitamin D µg	Vitamin E mg	Thiamin mg	Riboflavin mg	Niacin mg	Trypt/60 mg	Vitamin B_6 mg	Vitamin B_{12} µg	Folate µg	Pantothenate mg	Biotin µg	Vitamin C mg
	***Beans and lentils** continued*														
740	**Blackeye beans**, *dried, raw*	0	35	0	N	0.87	0.19	2.1	5.0	0.36	0	630	1.50	18.4	1
741	*dried, boiled in unsalted water*	0	13	0	N	0.19	0.05	0.5	1.9	0.10	0	210	0.30	7.0	Tr
742	**Broad beans**, *frozen, boiled in unsalted water*	0	225	0	0.61	(0.03)	(0.06)	(3.0)	1.3	(0.08)	0	(32)	(3.80)	(2.1)	8
743	**Butter beans**, canned, *re-heated, drained*	0	Tr	0	0.33	0.05	0.03	0.2	0.9	0.05	0	12	N	N	Tr
744	**Chick pea flour/besan flour**	0	(60)	0	(2.88)	(0.39)	(0.24)	(1.9)	2.6	(0.53)	0	(180)	(1.59)	N	Tr
745	**Chick peas**, whole, *dried, raw*	0	60	0	2.88	0.39	0.24	1.9	2.9	0.53	0	180	1.59	N	Tr
746	*dried, boiled in unsalted water*	0	23	0	1.10	0.10	0.07	0.7	1.1	0.14	0	66	0.29	N	Tr
747	canned, *re-heated, drained*	0	21	0	1.55	0.05	0.03	0.2	1.0	0.04	0	11	N	N	Tr
748	**Green beans/French beans**, *raw*	0	(330)	0	0.20	0.05	0.07	0.9	0.5	0.05	0	64	0.09	1.0	12
749	*frozen, boiled in unsalted water*	0	520	0	0.12	0.05	0.09	0.4	0.4	0.06	0	48	N	N	7
750	**Hummus**	0	N	0	N	0.16	0.05	1.1	1.0	N	0	N	N	N	1
751	**Lentils**, green and brown, whole, *dried, raw*	0	N	0	N	0.41	0.27	2.2	3.3	0.93	0	110	N	N	Tr
752	*dried, boiled in salted water*	0	N	0	N	0.14	0.08	0.6	1.2	0.28	0	30	N	N	Tr
753	**Lentils**, red, split, *dried, raw*	0	(60)	0	N	0.50	0.20	2.0	3.2	0.60	0	35	1.36	N	Tr
754	*dried, boiled in unsalted water*	0	(20)	0	N	0.11	0.04	0.4	1.0	0.11	0	33	0.31	N	Tr

Composition of food per 100g edible portion

No.	Food	Description and main data sources	Edible conversion factor	Water	Total nitrogen	Protein	Fat	Carbo-hydrate	Energy value	
				g	g	g	g	g	kcal	kJ
	***Beans and lentils** continued*									
755	**Mung beans**, whole, *dried, raw*	Literature sources	1.00	11.0	3.80	23.9	1.1	46.3[a]	279	1188
756	*dried, boiled in unsalted water*	As raw, soaked and boiled	1.00	69.3	1.21	7.6	0.4	15.3[a]	91	389
757	**Red kidney beans**, *dried, raw*	Whole beans. Analytical and literature sources	1.00	11.2	3.54	22.1	1.4	44.1[a]	266	1133
758	*dried, boiled in unsalted water*	As raw, soaked and boiled	1.00	66.0	1.35	8.4	0.5	17.4[a]	103	440
759	canned, *re-heated, drained*	LGC; 10 cans, 6 brands	0.64	67.5	1.11	6.9	0.6	17.8[a]	100	424
760	**Runner beans**, *raw*	IFR; ends and sides trimmed	0.86	91.2	0.26	1.6	0.4	3.2	22	93
761	*boiled in unsalted water*	IFR. Sliced and boiled 20 minutes	1.00	92.8	0.19	1.2	0.5	2.3	18	76
762	**Soya beans**, *dried, raw*	Whole beans. Analysis and literature sources	1.00	8.5	5.74	35.9	18.6	15.8[a]	370	1551
763	*dried, boiled in unsalted water*	As raw	1.00	64.3	2.24	14.0	7.3	5.1[a]	141	590
764	**Tofu**, soya bean, *steamed*	LGC. Soya bean curd; 7 assorted samples	1.00	85.0	1.29	8.1	4.2	0.7[a]	73	304
765	*steamed, fried*	Calcd. from steamed and ref. Haytowitz and Matthews (1986)	1.00	51.0	3.76	23.5	17.7	2.0[a]	261	1086
	Peas									
766	**Mange-tout peas**, *raw*	LGC. Whole pods, ends trimmed; 10 samples	0.92	88.7	0.58	3.6	0.2	4.2	32	136
767	*boiled in salted water*	LGC. As raw; boiled 3 minutes. And calculated from raw	1.00	89.2	0.51	3.2	0.1	3.3	26	111
768	*stir-fried in blended oil*	LGC. As raw; stir-fried 5 minutes. And calculated from raw	1.00	83.6	0.61	3.8	4.8	3.5	71	298
769	**Mushy peas**, canned, *re-heated*	LGC; 10 samples, 3 brands	1.00	76.5	0.92	5.8	0.7	13.8[a]	81	345

[a] Including oligosaccharides

Composition of food per 100g edible portion

No.	Food	Starch	Total sugars	Individual sugars					Dietary fibre	Fatty acids			Cholest-erol
				Gluc	Fruct	Sucr	Malt	Lact	NSP	Satd	Mono-unsatd	Poly-unsatd	
		g	g	g	g	g	g	g	g	g	g	g	mg
	Beans and lentils *continued*												
755	**Mung beans**, whole, *dried, raw*	40.9	1.5[a]	0.2	0.3	1.0	0	0	10.0	0.3	0.1	0.5	0
756	*dried, boiled in unsalted water*	14.1	0.5[a]	0.1	0.1	0.3	0	0	3.0	0.1	Tr	0.2	0
757	**Red kidney beans**, *dried, raw*	38.0	2.5[a]	0.2	0.1	2.2	0	0	15.7	0.2	0.1	0.8	0
758	*dried, boiled in unsalted water*	14.5	1.0[a]	0.1	Tr	0.8	0	0	6.7	0.1	Tr	0.3	0
759	canned, *re-heated, drained*	12.8	3.6[a]	0.1	0.1	3.3	0	0	6.2	0.1	0.1	0.3	0
760	**Runner beans**, *raw*	0.4	2.8	0.9	1.3	0.6	0	0	2.0	0.1	Tr	0.2	0
761	*boiled in unsalted water*	0.3	2.0	0.6	0.9	0.5	0	0	1.9	0.1	Tr	0.3	0
762	**Soya beans**, *dried, raw*	4.8	5.5[a]	0.2	0.5	4.8	0	0	15.7	2.3	3.5	9.1	0
763	*dried, boiled in unsalted water*	1.9	2.1[a]	0.1	0.2	1.9	0	0	6.1	0.9	1.4	3.5	0
764	**Tofu**, soya bean, *steamed*	0.3	0.3[a]	Tr	Tr	0.2	0	0	N	0.5	0.8	2.0	0
765	*steamed, fried*	0.9	0.9[a]	0.1	0.1	0.6	0	0	N	N	N	N	0
	Peas												
766	**Mange-tout peas**, *raw*	0.8	3.4	2.6	0.3	0.5	0	0	2.3	Tr	Tr	0.1	0
767	*boiled in salted water*	0.5	2.8	2.1	0.1	0.6	0	0	2.2	Tr	Tr	Tr	0
768	*stir-fried in blended oil*	0.2	3.3	2.4	0.2	0.7	0	0	2.4	0.4	2.4	1.8	0
769	**Mushy peas**, canned, *re-heated*	10.7	1.7[a]	Tr	Tr	1.6	0	0	1.8	0.1	0.1	0.3	0

[a] Not including oligosaccharides

Inorganic constituents per 100g edible portion

No.	Food	mg										µg	
		Na	K	Ca	Mg	P	Fe	Cu	Zn	Cl	Mn	Se	I
	Beans and lentils *continued*												
755	**Mung beans**, whole, *dried, raw*	12	1250	89	150	360	6.0	0.47	2.7	12	0.8	16	N
756	*dried, boiled in unsalted water*	2	270	24	43	81	1.4	0.19	0.9	4	0.3	5	N
757	**Red kidney beans**, *dried, raw*	18	1370	100	150	410	6.4	0.68	3.0	(2)	1.2	16	N
758	*dried, boiled in unsalted water*	2	420	37	45	130	2.5	0.23	1.0	(1)	0.5	6	N
759	canned, *re-heated, drained*	390	280	71	30	130	2.0	Tr	0.7	640	0.3	6	N
760	**Runner beans**, *raw*	Tr	220	33	19	34	1.2	0.02	0.2	21	0.2	N	2
761	*boiled in unsalted water*	1	130	22	14	21	1.0	0.01	0.2	5	0.2	N	Tr
762	**Soya beans**, *dried, raw*	5	1730	240	250	660	9.7	1.55	4.3	7	2.6	14	6
763	*dried, boiled in unsalted water*	1	510	83	63	250	3.0	0.32	0.9	3	0.7	5	2
764	**Tofu**, soya bean, *steamed*	4	63	510[a]	23[a]	95	1.2	0.20	0.7	16	0.4	N	N
765	*steamed, fried*	12	180	1480	67	270	3.5	0.58	2.0	46	1.2	N	N
	Peas												
766	**Mange-tout peas**, *raw*	2	200	44	28	62	0.8	0.06	0.5	28	0.3	Tr	N
767	*boiled in salted water*	42	170	35	22	55	0.8	0.06	0.4	72	0.3	Tr	N
768	*stir-fried in blended oil*	2	210	46	29	65	0.8	0.06	0.5	29	0.3	Tr	N
769	**Mushy peas**, canned, *re-heated*	340	170	14	22	100	1.3	0.11	0.7	490	0.2	N	N

[a] If nigari is used as a coagulant Ca and Mg are 150 and 59mg per 100g respectively

Vitamins per 100g edible portion

No.	Food	Retinol µg	Carotene µg	Vitamin D µg	Vitamin E mg	Thiamin mg	Riboflavin mg	Niacin mg	Trypt/60 mg	Vitamin B_6 mg	Vitamin B_{12} µg	Folate µg	Pantothenate mg	Biotin µg	Vitamin C mg
	Beans and lentils *continued*														
755	**Mung beans**, whole, *dried, raw*	0	24	0	N	0.36	0.26	2.1	3.8	0.38	0	140	1.91	N	Tr
756	*dried, boiled in unsalted water*	0	12	0	N	0.09	0.07	0.5	1.2	0.07	0	35	0.41	N	Tr
757	**Red kidney beans**, *dried, raw*	0	11	0	0.52	0.65	0.19	2.1	3.5	0.40	0	130	0.78	N	4
758	*dried, boiled in unsalted water*	0	4	0	0.20	0.17	0.05	0.6	1.3	0.12	0	42	0.22	N	1
759	canned, *re-heated, drained*	0	4	0	0.19	0.21	0.06	0.6	1.1	0.11	0	15	0.15	N	Tr
760	**Runner beans**, *raw*	0	145	0	0.23	0.06	0.03	Tr	0.4	0.08	0	60	0.05	0.7	18
761	*boiled in unsalted water*	0	120	0	0.23	0.05	0.02	Tr	0.3	0.04	0	42	0.04	0.5	10
762	**Soya beans**, *dried, raw*	0	12	0	2.90	0.61	0.27	2.2	5.7	0.38	0	370	0.79	65.0	Tr
763	*dried, boiled in unsalted water*	0	6	0	1.13	0.12	0.09	0.5	2.2	0.23	0	54	0.18	25.0	Tr
764	**Tofu**, soya bean, *steamed*	0	2	0	0.95	0.06	0.02	0.1	1.3	0.07	0	15	0.05	N	0
765	*steamed, fried*	0	2	0	N	0.09	0.02	0.1	3.8	0.10	0	27	0.14	N	0
	Peas														
766	**Mange-tout peas**, *raw*	0	695	0	0.39	0.22	0.15	0.6	0.6	0.18	0	10	0.72	5.3	54
767	*boiled in salted water*	0	665	0	0.37	0.14	0.16	0.4	0.5	0.14	0	8	0.67	3.7	28
768	*stir-fried in blended oil*	0	725	0	N	0.17	0.14	0.6	0.6	0.17	0	9	0.68	5.0	51
769	**Mushy peas**, canned, *re-heated*	0	Tr	0	(0.30)	N	N	N	0.9	N	0	N	N	Tr	Tr

Composition of food per 100g edible portion

No.	Food	Description and main data sources	Edible conversion factor	Water	Total nitrogen	Protein	Fat	Carbo-hydrate	Energy value	
				g	g	g	g	g	kcal	kJ
	Peas *continued*									
770	**Peas**, raw	IFR; whole peas, no pods	0.37	74.6	1.10	6.9	1.5	11.3[a]	83	344
771	*boiled in unsalted water*	IFR. As raw; boiled 20 minutes	1.00	75.6	1.07	6.7	1.6	10.0[a]	79	329
772	*frozen, boiled in salted water*	Based on frozen, boiled in unsalted water	1.00	78.3	0.95	6.0	0.9	9.7[a]	69	291
773	*boiled in unsalted water*	LGC; 10 samples, 8 brands. Boiled 2-5 minutes	1.00	78.3	0.95	6.0	0.9	9.7[a]	69	291
774	canned, *re-heated, drained*	LGC; 10 samples, 9 brands	0.67	77.9	0.85	5.3	0.9	13.5[a]	80	339
775	**Petit pois**, *frozen, boiled in unsalted water*	Based on boiled in salted water	1.00	81.1	0.80	5.0	0.9	5.5[a]	49	206
776	**Processed peas**, canned, *re-heated, drained*	LGC; 10 samples, 7 brands	0.65	69.6	1.10	6.9	0.7	17.5[a]	99	423
	Vegetables, general									
777	**Asparagus**, *raw*	IFR; tough base of stems removed	0.75	91.4	0.47	2.9	0.6	2.0	25	103
778	*boiled in salted water*	IFR. Soft tips only; boiled 15 minutes	0.48	91.5	0.55	3.4	0.8	1.4	26	110
779	**Aubergine**, *raw*	IFR; ends trimmed	0.96[b]	92.9	0.14	0.9	0.4	2.2	15	64
780	*fried in corn oil*	IFR. Sliced; shallow fried 10 minutes	1.00	59.5	0.19	1.2	31.9	2.8	302	1246
781	**Beetroot**, *raw*	IFR; top and root trimmed, peeled	0.80	87.1	0.27	1.7	0.1	7.6	36	154
782	*boiled in salted water*	IFR. As raw; boiled 45 minutes	0.80	82.4	0.37	2.3	0.1	9.5	46	195
783	pickled, *drained*	LGC; 10 samples, 5 brands. Whole and sliced	0.65	88.6	0.19	1.2	0.2	5.6	28[c]	117[c]
784	**Broccoli**, green, *raw*	IFR; tough stems removed	0.61	88.2	0.71	4.4	0.9	1.8[a]	33	138
785	*boiled in unsalted water*	IFR. As raw, cut into florets; boiled 15 minutes	1.00	91.1	0.50	3.1	0.8	1.1[a]	24	100

[a] Including oligosaccharides
[b] If peeled = 0.77
[c] Acetic acid from vinegar will contribute to the energy value

Composition of food per 100g edible portion

No.	Food	Starch	Total sugars	Individual sugars					Dietary fibre	Fatty acids			Cholesterol
				Gluc	Fruct	Sucr	Malt	Lact	NSP	Satd	Mono-unsatd	Poly-unsatd	
		g	g	g	g	g	g	g	g	g	g	g	mg
	Peas *continued*												
770	**Peas**, *raw*	7.0	2.3[a]	0.1	0.1	2.1	0	0	4.7	0.3	0.2	0.7	0
771	*boiled in unsalted water*	7.6	1.2[a]	Tr	Tr	1.2	0	0	4.5	0.3	0.2	0.8	0
772	*frozen, boiled in salted water*	4.7	2.7[a]	0.1	0.1	2.5	0	0	5.1	0.2	0.1	0.5	0
773	*boiled in unsalted water*	4.7	2.7[a]	0.1	0.1	2.5	0	0	5.1	0.2	0.1	0.5	0
774	canned, *re-heated, drained*	6.3	3.9[a]	Tr	0.1	3.8	0	0	5.1	0.2	0.1	0.4	0
775	**Petit pois**, *frozen, boiled in unsalted water*	Tr	3.0[a]	1.0	0.5	1.5	0	0	4.5	0.2	0.1	0.5	0
776	**Processed peas**, canned, *re-heated, drained*	14.7	1.5[a]	Tr	Tr	1.5	0	0	4.8	0.1	0.1	0.3	0
	Vegetables, general												
777	**Asparagus**, *raw*	0.1	1.9	0.7	1.1	0.1	0	0	1.7	0.1	0.1	0.2	0
778	*boiled in salted water*	Tr	1.4	0.5	0.7	0.2	0	0	1.4	0.1	0.2	0.3	0
779	**Aubergine**, *raw*	0.2	2.0	1.1	0.8	0.1	0	0	2.0	0.1	Tr	0.2	0
780	*fried in corn oil*	0.2	2.6	1.4	1.1	0.1	0	0	2.3	4.1	7.9	18.5	0
781	**Beetroot**, *raw*	0.6	7.0	0.2	0.1	6.7	0	0	1.9	Tr	Tr	0.1	0
782	*boiled in salted water*	0.7	8.8	0.2	0.1	8.5	0	0	1.9	Tr	Tr	0.1	0
783	pickled, *drained*	Tr	5.6	0.6	0.6	4.4	0	0	1.7	Tr	Tr	0.1	0
784	**Broccoli**, green, *raw*	0.1	1.5[a]	0.5	0.8	0.2	0	0	2.6	0.2	0.1	0.5	0
785	*boiled in unsalted water*	Tr	0.9[a]	0.3	0.5	0.1	0	0	2.3	0.2	0.1	0.4	0

[a] Not including oligosaccharides

Inorganic constituents per 100g edible portion

No.	Food	mg										µg	
		Na	K	Ca	Mg	P	Fe	Cu	Zn	Cl	Mn	Se	I
	***Peas** continued*												
770	**Peas**, *raw*	1	330	21	34	130	2.8	0.05	1.1	39	0.4	Tr	2
771	*boiled in unsalted water*	Tr	230	19	29	130	1.5	0.03	1.0	8	0.4	Tr	2
772	*frozen, boiled in salted water*	(94)	150	35	21	99	1.6	(0.03)	0.7	(120)	0.3	(1)	(2)
773	*boiled in unsalted water*	2	150	35	21	99	1.6	(0.03)	0.7	9	0.3	Tr	(2)
774	canned, *re-heated, drained*	250	130	30	20	81	1.9	0.02	0.6	360	0.2	Tr	N
775	**Petit pois**, *frozen, boiled in unsalted water*	(2)	130	42	24	95	1.6	0.11	0.9	(9)	0.3	(1)	(2)
776	**Processed peas**, canned, *re-heated, drained*	380	150	33	25	89	1.8	0.09	0.7	520	0.3	N	13
	Vegetables, general												
777	**Asparagus**, *raw*	1	260	27	13	72	0.7	0.08	0.7	60	0.2	(1)	Tr
778	*boiled in salted water*	60	220	25	13	50	0.6	0.08	0.7	110	0.2	(1)	Tr
779	**Aubergine**, *raw*	2	210	10	11	16	0.3	0.01	0.2	14	0.1	(1)	1
780	*fried in corn oil*	2	170	8	8	25	0.5	0.03	0.1	16	0.2	(1)	(1)
781	**Beetroot**, *raw*	66	380	20	11	51	1.0	0.02	0.4	59	0.7	Tr	N
782	*boiled in salted water*	110	510	29	16	87	0.8	0.03	0.5	N	0.9	Tr	N
783	pickled, *drained*	120	190	19	13	17	0.5	0.04	0.3	210	0.2	Tr	N
784	**Broccoli**, green, *raw*	8	370	56	22	87	1.7	0.02	0.6	100	0.2	Tr	2
785	*boiled in unsalted water*	(13)	170	40	13	57	1.0	0.02	0.4	(23)	0.2	Tr	2

Vitamins per 100g edible portion

No.	Food	Retinol µg	Carotene µg	Vitamin D µg	Vitamin E mg	Thiamin mg	Riboflavin mg	Niacin mg	Trypt/60 mg	Vitamin B_6 mg	Vitamin B_{12} µg	Folate µg	Pantothenate mg	Biotin µg	Vitamin C mg
	Peas *continued*														
770	**Peas**, *raw*	0	300	0	0.21	0.74	0.02	2.5	1.1	0.12	0	62	(0.15)	0.5	24
771	*boiled in unsalted water*	0	250	0	0.21	0.70	0.03	1.8	1.1	0.09	0	27	0.15	0.4	16
772	*frozen, boiled in salted water*	0	405	0	0.18	0.26	0.09	1.6	0.9	0.09	0	47	0.14	0.4	12
773	*boiled in unsalted water*	0	571	0	0.18	0.26	0.09	1.6	0.9	0.09	0	33	0.14	0.4	12
774	canned, *re-heated, drained*	0	534	0	0.22	0.09	0.07	1.2	0.9	0.06	0	25	(0.04)	Tr	1
775	**Petit pois**, *frozen, boiled in unsalted water*	0	(405)	0	(0.18)	0.13	0.12	1.5	0.9	0.09	0	50	(0.14)	(0.4)	8
776	**Processed peas**, canned, *re-heated, drained*	0	60	0	0.30	0.10	0.04	0.4	1.1	0.10	0	11	(0.04)	Tr	Tr
	Vegetables, general														
777	**Asparagus**, *raw*	0	315	0	1.16	0.16	0.06	1.0	0.5	0.09	0	175	0.17	(0.4)	12
778	*boiled in salted water*	0	389	0	(1.16)	0.12	0.06	0.8	0.6	0.07	0	173	0.16	0.4	10
779	**Aubergine**, *raw*	0	70	0	0.03	0.02	0.01	0.1	0.2	0.08	0	18	0.08	N	4
780	*fried in corn oil*	0	125	0	5.50	0	Tr	Tr	0.2	0.07	0	(5)	(0.07)	N	1
781	**Beetroot**, *raw*	0	20	0	Tr	0.01	0.01	0.1	0.3	0.03	0	150	0.12	Tr	5
782	*boiled in salted water*	0	27	0	Tr	0.01	0.01	0.1	0.3	0.04	0	110	0.10	Tr	5
783	pickled, *drained*	0	Tr	0	Tr	0.02	0.03	0.1	0.2	0.04	0	2	(0.10)	Tr	N
784	**Broccoli**, green, *raw*	0	575	0	(1.30)	0.10	0.06	0.9	0.8	0.14	0	90	N	N	87
785	*boiled in unsalted water*	0	475	0	(1.10)	0.05	0.05	0.7	0.6	0.11	0	64	N	N	44

Composition of food per 100g edible portion

No.	Food	Description and main data sources	Edible conversion factor	Water	Total nitrogen	Protein	Fat	Carbo-hydrate	Energy value	
				g	g	g	g	g	kcal	kJ
Vegetables, general *continued*										
786	**Brussels sprouts**, *raw*	IFR; base trimmed, outer leaves removed	0.69	84.3	0.56	3.5	1.4	4.1[a]	42	177
787	*boiled in unsalted water*	IFR. As raw; boiled 15 minutes	1.00	86.9	0.46	2.9	1.3	3.5[a]	35	153
788	*frozen, boiled in unsalted water*	LGC. 10 samples, 8 brands; boiled 5-10 minutes	1.00	86.8	0.56	3.5	(1.3)	2.5[a]	35	148
789	**Cabbage**, *raw, average*	Average of January King, Savoy, summer and white	0.77	90.1	0.28	1.7	0.4	4.1	26	109
790	*boiled in unsalted water, average*	As raw	1.00	93.1	0.16	1.0	0.4	2.2	16	67
791	white, *raw*	IFR; outer leaves and stem removed	0.91	90.7	0.23	1.4	0.2	5.0	27	113
792	**Carrots**, old, *raw*	IFR; ends trimmed, peeled	0.70	89.8	0.10	0.6	0.3	7.9[a]	35	146
793	*boiled in unsalted water*	IFR. As raw; sliced and boiled 12.5 minutes	1.00	90.5	0.10	0.6	0.4	4.9[a]	24	100
794	young, *raw*	IFR; ends trimmed, scrubbed	0.87	88.8	0.11	0.7	0.5	6.0[a]	30	125
795	young, *boiled in unsalted water*	IFR. As raw; sliced and boiled 15 minutes	1.00	90.7	0.09	0.6	0.4	4.4[a]	22	93
796	canned, *re-heated, drained*	LGC; 10 cans, 5 brands	0.61	91.9	0.09	0.5	0.3	4.2[a]	20	87
797	**Cauliflower**, *raw*	IFR; florets only	0.45	88.4	0.58	3.6	0.9	3.0[a]	34	142
798	*boiled in unsalted water*	IFR. As raw; boiled 13 minutes	1.00	90.6	0.47	2.9	0.9	2.1[a]	28	117
799	**Celery**, *raw*	IFR; stem only	0.91	95.1	0.08	0.5	0.2	0.9	7	30
800	*boiled in salted water*	IFR. Stem only; boiled 20 minutes	1.00	95.2	0.08	0.5	0.3	0.8	8	34
801	**Chicory**, *raw*	IFR. Stem and inner leaves; pale variety	0.80	94.3	0.09	0.5	0.6	2.8	11[b]	45[b]
802	**Courgette**, *raw*	IFR; ends trimmed	0.88	93.7	0.29	1.8	0.4	1.8	18	74
803	*boiled in unsalted water*	Analysis and calculation from raw	1.00	93.0	0.32	2.0	0.4	2.0	19	81
804	*fried in corn oil*	IFR. As raw; sliced and shallow fried 5 minutes	1.00	86.8	0.41	2.6	4.8	2.6	63	265
805	**Cucumber**, *raw*	IFR; ends trimmed, not peeled	0.97[c]	96.4	0.11	0.7	0.1	1.5	10	40

[a] Including oligosaccharides

[b] Contains inulin; 32 per cent total carbohydrate taken to be available for energy purposes

[c] If peeled = 0.77

Composition of food per 100g edible portion

No.	Food	Starch	Total sugars	Individual sugars					Dietary fibre	Fatty acids			Cholest-erol
				Gluc	Fruct	Sucr	Malt	Lact	NSP	Satd	Mono-unsatd	Poly-unsatd	
		g	g	g	g	g	g	g	g	g	g	g	mg
	Vegetables, general *continued*												
786	**Brussels sprouts**, *raw*	0.8	3.1[a]	1.1	1.3	0.7	0	0	4.1	0.3	0.1	0.7	0
787	*boiled in unsalted water*	0.3	3.0[a]	1.3	1.1	0.6	0	0	3.1	0.3	0.1	0.7	0
788	*frozen, boiled in unsalted water*	0.4	2.4[a]	1.0	0.9	0.5	0	0	4.3	0.3	0.1	0.7	0
789	**Cabbage**, *raw, average*	0.1	4.0	2.0	1.8	0.3	0	0	2.4	0.1	Tr	0.3	0
790	*boiled in unsalted water, average*	0.1	2.0	1.0	0.9	0.1	0	0	1.8	0.1	Tr	0.3	0
791	white, *raw*	0.1	4.9	2.3	2.1	0.5	0	0	2.1	Tr	Tr	0.1	0
792	**Carrots**, old, *raw*	0.3	7.4[a]	2.3	1.9	3.2	0	0	2.4	0.1	Tr	0.2	0
793	*boiled in unsalted water*	0.2	4.6[a]	1.4	1.2	2.0	0	0	2.5	0.1	Tr	0.2	0
794	young, *raw*	0.2	5.6[a]	(1.7)	(1.5)	(2.4)	0	0	2.4	0.1	Tr	0.3	0
795	young, *boiled in unsalted water*	0.2	4.2[a]	(1.3)	(1.1)	(1.8)	0	0	2.3	0.1	Tr	0.2	0
796	canned, *re-heated, drained*	0.4	3.7[a]	0.8	0.7	2.2	0	0	1.9	(0.1)	Tr	(0.2)	0
797	**Cauliflower**, *raw*	0.4	2.5[a]	1.3	1.1	0.1	0	0	1.8	0.2	0.1	0.5	0
798	*boiled in unsalted water*	0.2	1.8[a]	0.9	0.8	0.1	0	0	1.6	0.2	0.1	0.5	0
799	**Celery**, *raw*	Tr	0.9	0.4	0.3	0.2	0	0	1.1	Tr	Tr	0.1	0
800	*boiled in salted water*	Tr	0.8	0.3	0.3	0.2	0	0	1.2	0.1	0.1	0.1	0
801	**Chicory**, *raw*	0.2	0.7	0.3	0.4	Tr	0	0	0.9	0.2	Tr	0.3	0
802	**Courgette**, *raw*	0.1	1.7	0.7	0.8	0.2	0	0	0.9	0.1	Tr	0.2	0
803	*boiled in unsalted water*	0.1	1.9	0.8	0.9	0.2	0	0	1.2	0.1	Tr	0.2	0
804	*fried in corn oil*	0.1	2.5	1.0	1.2	0.3	0	0	1.2	0.6	1.2	2.8	0
805	**Cucumber**, *raw*	0.1	1.4[b]	0.7	0.7	Tr	0	0	0.6	Tr	Tr	Tr	0

[a] Not including oligosaccharides

[b] Peeled cucumbers contain approximately 2.0g total sugars per 100g as equal quantities of glucose and fructose

Inorganic constituents per 100g edible portion

No.	Food	Na	K	Ca	Mg	P	Fe	Cu	Zn	Cl	Mn	Se	I
		mg										µg	
	Vegetables, general *continued*												
786	**Brussels sprouts**, *raw*	6	450	26	8	77	0.7	0.02	0.5	38	0.2	N	1
787	*boiled in unsalted water*	2	310	20	13	61	0.5	0.03	0.3	16	0.2	N	1
788	*frozen, boiled in unsalted water*	8	340	29	17	66	0.6	0.04	0.4	15	0.2	N	(1)
789	**Cabbage**, *raw, average*	5	270	52	8	41	0.7	0.02	0.3	37	0.2	(1)	2
790	*boiled in unsalted water, average*	8	120	33	4	25	0.3	0.01	0.1	9	0.2	(2)	2
791	white, *raw*	7	240	49	6	29	0.5	0.01	0.2	40	0.2	Tr	2
792	**Carrots**, old, *raw*	25	170	25	3	15	0.3	0.02	0.1	33	0.1	1	2
793	*boiled in unsalted water*	50	120	24	3	17	0.4	0.01	0.1	31	0.1	1	2
794	young, *raw*	40	240	34	9	25	0.4	0.02	0.2	39	(0.1)	(1)	(2)
795	young, *boiled in unsalted water*	23	160	30	6	15	0.4	0.02	0.2	28	(0.1)	(1)	(2)
796	canned, *re-heated, drained*	370	110	25	5	14	0.6	0.04	0.1	490	0.1	(1)	N
797	**Cauliflower**, *raw*	9	380	21	17	64	0.7	0.03	0.6	28	0.3	Tr	Tr
798	*boiled in unsalted water*	4	120	17	12	52	0.4	0.02	0.4	14	0.2	Tr	Tr
799	**Celery**, *raw*	60	320	41	5	21	0.4	0.01	0.1	130	0.1	(3)	N
800	*boiled in salted water*	160	230	45	4	20	0.3	0.01	0.1	250	0.1	(3)	N
801	**Chicory**, *raw*	1	170	21	6	27	0.4	0.05	0.2	25	0.3	N	N
802	**Courgette**, *raw*	1	360	25	22	45	0.8	0.02	0.3	45	0.1	(1)	N
803	*boiled in unsalted water*	1	210	19	17	36	0.6	0.01	0.2	26	0.1	(1)	N
804	*fried in corn oil*	1	490	38	32	61	1.4	0.05	0.5	65	0.1	(1)	N
805	**Cucumber**, *raw*	3	140	18	8	49	0.3	0.01	0.1	17	0.1	Tr	3

No.	Food	Retinol µg	Carotene µg	Vitamin D µg	Vitamin E mg	Thiamin mg	Riboflavin mg	Niacin mg	Trypt/60 mg	Vitamin B_6 mg	Vitamin B_{12} µg	Folate µg	Pantothenate mg	Biotin µg	Vitamin C mg
	Vegetables, general *continued*														
786	**Brussels sprouts**, *raw*	0	215	0	1.00	0.15	0.11	0.2	0.7	0.37	0	135	1.00	0.4	115
787	*boiled in unsalted water*	0	320	0	0.90	0.07	0.09	Tr	0.5	0.19	0	110	0.28	0.3	60
788	*frozen, boiled in unsalted water*	0	320	0	(0.90)	0.09	0.08	Tr	0.7	0.28	0	67	(0.28)	(0.3)	69
789	**Cabbage**, *raw, average*	0	1150[a]	0	0.20[b]	0.15	0.02	0.5	0.3	0.17	0	75	0.21	0.1	49
790	*boiled in unsalted water, average*	0	805[a]	0	0.20[b]	0.08	0.01	0.3	0.2	0.08	0	39	0.15	Tr	20
791	white, *raw*	0	19	0	0.20	0.12	0.01	0.3	0.2	0.18	0	34	0.21	0.1	35
792	**Carrots**, old, *raw*	0	12472	0	0.56	0.10	0.01	0.2	0.1	0.14	0	12	0.25	0.6	6
793	*boiled in unsalted water*	0	13402	0	0.56	0.09	Tr	Tr	0.1	0.10	0	16	0.18	0.4	2
794	young, *raw*	0	7807	0	(0.56)	0.04	0.02	0.2	0.1	0.07	0	28	(0.25)	(0.6)	4
795	young, *boiled in unsalted water*	0	7703	0	(0.56)	0.05	0.01	0.1	0.1	0.05	0	17	0.18	0.4	2
796	canned, *re-heated, drained*	0	2070	0	0.64	0.01	0.02	0.2	0.1	0.07	0	8	0.10	0.4	1
797	**Cauliflower**, *raw*	0	50	0	0.22	0.17	0.05	0.6	0.9	0.28	0	66	0.60	1.5	43
798	*boiled in unsalted water*	0	60	0	0.11	0.07	0.04	0.4	0.7	0.15	0	51	0.42	1.0	27
799	**Celery**, *raw*	0	50	0	0.20	0.06	0.01	0.3	0.1	0.03	0	16	0.40	0.1	8
800	*boiled in salted water*	0	50	0	0.20	0.06	0.01	Tr	0.1	0.03	0	10	0.28	Tr	4
801	**Chicory**, *raw*	0	120	0	N	0.14	Tr	0.1	0.2	0.01	0	14	N	N	5
802	**Courgette**, *raw*	0	610	0	N	0.12	0.02	0.3	0.3	0.15	0	52	0.08	N	21
803	*boiled in unsalted water*	0	(440)	0	N	0.08	0.02	0.2	0.3	0.09	0	31	0.11	N	11
804	*fried in corn oil*	0	500	0	0.83	0.10	0.01	0.4	0.4	0.09	0	42	N	N	15
805	**Cucumber**, *raw*	0	60[c]	0	0.07	0.03	0.01	0.2	0.1	0.04	0	9	0.30	0.9	2

[a] Average figures. The amount of carotene in leafy vegetables depends on the amount of chlorophyll, and the outer green leaves may contain 50 times as much as inner white ones

[b] The value for inner leaves. Outer leaves contain 7.0mg α-tocopherol per 100g

[c] Carotene can be as high as 260µg per 100g. In peeled cucumbers the carotene ranges from 0 to 35µg per 100g

No.	Food	Description and main data sources	Edible conversion factor	Water	Total nitrogen	Protein	Fat	Carbo-hydrate	Energy value	
				g	g	g	g	g	kcal	kJ
	Vegetables, general *continued*									
806	**Curly kale**, *raw*	IFR; main ribs and stalks removed	0.85	88.4	0.55	3.4	1.6	1.4	33	140
807	*boiled in salted water*	IFR. As raw; shredded and boiled 7 minutes	1.00	90.9	0.39	2.4	1.1	1.0	24	100
808	**Fennel**, Florence, *raw*	IFR; inner leaves and bulb only	0.80	94.2	0.15	0.9	0.2	1.8	12	50
809	*boiled in salted water*	IFR. As raw; boiled 14 minutes	1.00	94.4	0.14	0.9	0.2	1.5	11	47
810	**Garlic**, *raw*	IFR; peeled cloves	0.79	64.3	1.27	7.9	0.6	16.3	98	411
811	**Gherkins**, pickled, *drained*	LGC; 10 samples, 5 brands	0.67	92.8	0.14	0.9	0.1	2.6	14[a]	61[a]
812	**Gourd**, karela, *raw*	LGC; 5 samples. Ends trimmed	0.93	93.3	0.26	1.6	0.2	0.8	11	47
813	**Leeks**, *raw*	IFR; trimmed and outer leaves removed	0.57[b]	90.8	0.26	1.6	0.5	2.9[c]	22	93
814	*boiled in unsalted water*	IFR. As raw; chopped and boiled 22 minutes	1.00	92.2	0.20	1.2	0.7	2.6[c]	21	87
815	**Lettuce**, average, *raw*	Average of 4 varieties	0.74	95.1	0.13	0.8	0.5	1.7	14	59
816	Iceberg, *raw*	IFR; outer leaves removed	0.83	95.6	0.11	0.7	0.3	1.9	13	53
817	**Marrow**, *raw*	IFR; flesh only, seeds removed	0.54	95.6	0.08	0.5	0.2	2.2	12	51
818	*boiled in unsalted water*	IFR. As raw; cut and boiled 19 minutes	1.00	95.9	0.07	0.4	0.2	1.6	9	38
819	**Mixed vegetables**, *frozen, boiled in salted water*	LGC; 10 samples. Assorted varieties. Simmered 3-7 minutes	1.00	85.8	0.53	3.3	0.5	6.6	42	180
820	**Mushrooms**, common, *raw*	IFR; stalks trimmed where necessary	0.97[d]	92.6	0.64[e]	1.8[f]	0.5	0.4	13	55
821	*fried in butter*	Calculation from fried in corn oil	1.00	74.8	0.95[e]	2.4[f]	16.2	0.3	157	645
822	*fried in corn oil*	IFR. As raw; sliced and fried 8 minutes	1.00	74.8	0.95[e]	2.4[f]	16.2	0.3	157	645
823	**Mustard and cress**, *raw*	IFR; leaves and cut stems	1.00[g]	95.3	0.26	1.6	0.6	0.4	13	56
824	**Okra**, *raw*	IFR and literature sources. Ends trimmed	0.74	86.6	0.40	2.8	1.0	3.0	31	130
825	*boiled in unsalted water*	Calculated from raw	1.00	87.9	0.40	2.5	0.9	2.7	28	119
826	*stir-fried in corn oil*	IFR. As raw; sliced and fried 5 minutes	1.00	54.5	0.69	4.3	26.1	4.4	269	1122

[a] Acetic acid from vinegar will contribute to the energy value
[b] Bulb only = 0.36
[c] Including oligosaccharides
[d] If peeled = 0.75
[e] 60 per cent of this nitrogen is non-protein nitrogen
[f] (Total N - non-protein N) × 6.25
[g] If purchased on soil block = 0.27

Composition of food per 100g edible portion

No.	Food	Starch	Total sugars	Individual sugars					Dietary fibre	Fatty acids			Cholest-erol
				Gluc	Fruct	Sucr	Malt	Lact	NSP	Satd	Mono-unsatd	Poly-unsatd	
		g	g	g	g	g	g	g	g	g	g	g	mg
	Vegetables, general *continued*												
806	**Curly kale**, *raw*	0.1	1.3	(0.6)	(0.6)	(0.1)	0	0	3.1	0.2	0.1	0.9	0
807	*boiled in salted water*	0.1	0.9	(0.4)	(0.4)	(0.1)	0	0	2.8	0.2	0.1	0.6	0
808	**Fennel**, Florence, *raw*	0.1	1.7	0.9	0.7	0.1	0	0	2.4	Tr	Tr	Tr	0
809	*boiled in salted water*	0.1	1.4	0.7	0.6	0.1	0	0	2.3	Tr	Tr	Tr	0
810	**Garlic**, *raw*	14.7	1.6	0.4	0.6	0.6	0	0	4.1	0.1	Tr	0.3	0
811	**Gherkins**, pickled, *drained*	0.2	2.4	0.5	0.8	1.1	0	0	(1.2)	Tr	Tr	Tr	0
812	**Gourd**, karela, *raw*	0.8	Tr	Tr	Tr	Tr	0	0	2.6	N	N	N	0
813	**Leeks**, *raw*	0.3	2.2[a]	0.8	0.9	0.5	0	0	2.2	0.1	Tr	0.3	0
814	*boiled in unsalted water*	0.2	2.0[a]	0.7	0.8	0.5	0	0	1.7	0.1	Tr	0.4	0
815	**Lettuce**, average, *raw*	Tr	1.7	0.7	0.8	0.2	0	0	0.9	0.1	Tr	0.3	0
816	Iceberg, *raw*	Tr	1.9	0.7	0.9	0.2	0	0	0.6	Tr	Tr	0.2	0
817	**Marrow**, *raw*	0.1	2.1	0.8	1.1	0.2	0	0	0.5	Tr	Tr	Tr	0
818	*boiled in unsalted water*	0.2	1.4	0.5	0.7	0.1	0	0	0.6	Tr	Tr	Tr	0
819	**Mixed vegetables**, *frozen, boiled in salted water*	3.0	3.6	0.4	0.4	2.8	0	0	N	N	N	N	0
820	**Mushrooms**, common, *raw*	0.2	0.2	0.1	0.1	0.1	0	0	1.1	0.1	Tr	0.3	0
821	*fried in butter*	0.2	0.1	Tr	Tr	Tr	0	0	1.5	10.7	3.9	0.5	37
822	*fried in corn oil*	0.2	0.1	Tr	Tr	Tr	0	0	1.5	2.1	4.0	9.4	0
823	**Mustard and cress**, *raw*	Tr	0.4	N	N	N	0	0	1.1	Tr	0.2	0.2	0
824	**Okra**, *raw*	0.5	2.5	0.6	0.9	0.9	0	0	4.0	0.3	0.1	0.3	0
825	*boiled in unsalted water*	0.5	2.3	0.6	0.8	0.9	0	0	3.6	0.3	0.1	0.3	0
826	*stir-fried in corn oil*	0.8	3.6	0.9	1.3	1.4	0	0	6.3	3.3	6.5	15.1	0

[a] Not including oligosaccharides

Inorganic constituents per 100g edible portion

No.	Food	mg										µg	
		Na	K	Ca	Mg	P	Fe	Cu	Zn	Cl	Mn	Se	I
	Vegetables, general *continued*												
806	**Curly kale**, *raw*	43	450	130	34	61	1.7	0.03	0.4	68	0.8	(2)	N
807	*boiled in salted water*	100	160	150	8	39	2.0	0.02	0.2	N	0.4	(2)	N
808	**Fennel**, Florence, *raw*	11	440	24	8	26	0.3	0.02	0.5	27	N	N	N
809	*boiled in salted water*	96	300	20	7	21	0.2	0.01	0.4	120	N	N	N
810	**Garlic**, *raw*	4	620	19	25	170	1.9	0.06	1.0	73	0.5	2	3
811	**Gherkins**, pickled, *drained*	690	110	20	11	22	0.7	0.10	0.3	1060	0.1	N	N
812	**Gourd**, karela, *raw*	1	330	19	31	48	1.4	0.27	0.4	21	0.3	N	N
813	**Leeks**, *raw*	2	260	24	3	44	1.1	0.02	0.2	59	0.2	(1)	N
814	*boiled in unsalted water*	6	150	20	2	32	0.7	0.02	0.2	43	0.2	(1)	N
815	**Lettuce**, average, *raw*	3	220	28	6	28	0.7	0.01	0.2	47	0.3	(1)	2
816	Iceberg, *raw*	2	160	19	5	18	0.4	0.01	0.1	42	0.3	(1)	2
817	**Marrow**, *raw*	1	140	18	10	17	0.2	0.02	0.2	30	N	N	N
818	*boiled in unsalted water*	1	110	14	7	18	0.1	0.01	0.2	14	N	N	N
819	**Mixed vegetables**, *frozen, boiled in salted water*	96	130	35	16	57	0.8	0.02	0.4	140	0.2	N	N
820	**Mushrooms**, common, *raw*	5	320	6	9	80	0.6	0.72	0.4	69	0.1	9	3
821	*fried in butter*	150	340	11	19	110	1.0	0.40	0.5	320	0.1	12	11
822	*fried in corn oil*	4	340	8	19	100	1.0	0.40	0.5	89	0.1	12	4
823	**Mustard and cress**, *raw*	19	110	50	22	33	1.0	0.01	0.3	39	N	N	N
824	**Okra**, *raw*	8	330	160	71	59	1.1	0.13	0.6	41	N	(1)	N
825	*boiled in unsalted water*	5	310	120	57	54	0.6	0.09	0.5	N	N	(1)	N
826	*stir-fried in corn oil*	13	480	220	110	89	1.5	0.19	1.0	64	N	(2)	N

No.	Food	Retinol µg	Carotene µg	Vitamin D µg	Vitamin E mg	Thiamin mg	Riboflavin mg	Niacin mg	Trypt/60 mg	Vitamin B_6 mg	Vitamin B_{12} µg	Folate µg	Pantothenate mg	Biotin µg	Vitamin C mg
	Vegetables, general *continued*														
806	**Curly kale**, *raw*	0	3145	0	(1.70)	0.08	0.09	1.0	0.7	0.26	0	120	0.09	0.5	110
807	*boiled in salted water*	0	3375	0	(1.33)	0.02	0.06	0.8	0.5	0.13	0	86	0.05	0.4	71
808	**Fennel**, Florence, *raw*	0	140	0	N	0.06	0.01	0.6	N	0.06	0	42	N	N	5
809	*boiled in salted water*	0	60	0	N	0.05	0.01	0.4	N	0.08	0	26	N	N	2
810	**Garlic**, *raw*	0	Tr	0	0.01	0.13	0.03	0.3	1.9	0.38	0	5	N	N	17
811	**Gherkins**, pickled, *drained*	0	2	0	N	Tr	0.02	0.1	0.1	N	0	6	N	N	1
812	**Gourd**, karela, *raw*	0	345	0	N	0.09	0.05	0.4	0.3	N	0	45	N	N	185
813	**Leeks**, *raw*	0	177	0	0.92	0.29	0.05	0.4	0.2	0.48	0	56	0.12	1.4	17
814	*boiled in unsalted water*	0	150	0	0.78	0.02	0.02	0.4	0.2	0.05	0	40	0.10	1.0	7
815	**Lettuce**, average, *raw*	0	1023[a]	0	0.57	0.12	0.02	0.4	0.1	0.04	0	55	(0.18)	0.7	5
816	Iceberg, *raw*	0	50[a]	0	0.57	0.11	0.01	0.3	0.1	0.03	0	53	(0.18)	0.7	3
817	**Marrow**, *raw*	0	110	0	Tr	0.08	Tr	0.2	0.1	0.03	0	13	0.10	0.4	11
818	*boiled in unsalted water*	0	110	0	Tr	0.08	Tr	0.2	0.1	0.01	0	15	0.07	0.4	3
819	**Mixed vegetables**, *frozen, boiled in salted water*	0	2520	0	N	0.12	0.09	0.8	0.5	0.11	0	52	N	N	13
820	**Mushrooms**, common, *raw*	0	0	0	0.12	0.09	0.31	3.2	0.3	0.18	0	44	2.00	12.0	1
821	*fried in butter*	160	85	0.1	0.40	0.09	0.34	2.3	0.4	0.19	Tr	11	1.40	8.0	1
822	*fried in corn oil*	0	Tr	0	2.84	0.09	0.34	2.3	0.4	0.19	0	11	1.40	8.0	1
823	**Mustard and cress**, *raw*	0	(1280)	0	0.70	0.04	0.04	1.0	0.3	0.15	0	60	N	N	33
824	**Okra**, *raw*	0	515	0	N	0.20	0.06	1.0	0.4	0.21	0	88	0.25	N	21
825	*boiled in unsalted water*	0	465	0	N	0.13	0.05	0.9	0.3	0.19	0	46	0.21	N	16
826	*stir-fried in corn oil*	0	560	0	4.50	0.17	0.06	0.9	0.6	0.20	0	83	0.23	N	21

[a] Average figures. The outer green leaves may contain 50 times as much as the inner white ones

Composition of food per 100g edible portion

No.	Food	Description and main data sources	Edible conversion factor	Water g	Total nitrogen g	Protein g	Fat g	Carbo-hydrate g	Energy value kcal	Energy value kJ
	Vegetables, general *continued*									
827	**Onions**, *raw*	IFR; flesh only	0.91	89.0	0.20	1.2	0.2	7.9[a]	36	150
828	*fried in corn oil*	IFR. As raw: sliced into rings and fried 15 minutes[b]	1.00	65.7	0.37	2.3	11.2	14.1[a]	164	684
829	pickled, *drained*	LGC; 10 samples, 7 brands	0.58	90.6	0.14	0.9	0.2	4.9[a]	24[c]	101[c]
830	pickled, cocktail/silverskin, *drained*	LGC; 10 samples, 8 brands	0.59	91.8	0.10	0.6	0.1	3.1[a]	15[c]	63[c]
831	**Parsnip**, *raw*	LGC; ends trimmed and peeled	0.72	79.3	0.29	1.8	1.1	12.5[a]	64	271
832	*boiled in unsalted water*	LGC. As raw; sliced and boiled 12 minutes[d]	1.00	78.7	0.26	1.6	1.2	12.9[a]	66	278
833	**Peppers**, capsicum, chilli, green, *raw*	Refs. Cashel et al. (1989), Gopalan et al. (1980)	0.90	85.7	0.46	2.9	0.6	0.7	20	83
834	capsicum, green, *raw*	IFR; stalk and seeds removed	0.84	93.3	0.13	0.8	0.3	2.6[a]	15	65
835	green, *boiled in salted water*	IFR. As raw; sliced and boiled 15 minutes	1.00	92.6	0.16	1.0	0.5	2.6[a]	18	76
836	capsicum, red, *raw*	IFR; stalk and seeds removed	0.83	90.4	0.17	1.0	0.4	6.4[a]	32	134
837	red, *boiled in salted water*	Calculation from raw	1.00	89.5	0.19	1.1	0.4	7.0[a]	34	145
838	**Plantain**, *boiled in unsalted water*	10 samples. Flesh only; boiled 30 minutes. And literature sources	1.00	68.5	0.13	0.8	0.2	28.5	112	477
839	ripe, *fried in vegetable oil*	8 samples	1.00	34.7	0.24	1.5	9.2	47.5	267	1126
840	**Pumpkin**, *raw*	IFR; flesh only, peeled thickly, seeds removed	0.67	95.0	0.12	0.7	0.2	2.2	13	55
841	*boiled in salted water*	IFR. As raw; boiled 15 minutes	1.00	94.9	0.10	0.6	0.3	2.1	13	56
842	**Quorn**, *pieces, as purchased*	Manufacturer's information (Marlow Foods); pieces	1.00	74.2	2.25[e]	14.1	3.2	1.9[a]	92	389
843	**Radish**, red, *raw*	IFR; ends trimmed, flesh and skin	0.81	95.4	0.11	0.7	0.2	1.9	12	49
844	**Shallots**, *raw*	Literature sources	0.72	92.8	0.24	1.5	0.2	3.3	20	86

[a] Including oligosaccharides
[b] Onion rings in batter, frozen, oven baked contain 3.6g protein, 15.3g fat, 31.3g carbohydrate (26.4g starch, 4.9g sugars), 269kcal and 1128kJ energy per 100g
[c] Acetic acid from vinegar will contribute to the energy value
[d] Roast parsnip, frozen, contains 2.3g protein, 5.7g fat, 21.0g carbohydrate (11.9g starch, 9.1g sugar), 139kcal and 585kJ energy per 100g
[e] Additional non-protein nitrogen from chitin is present in variable amounts

Composition of food per 100g edible portion

No.	Food	Starch	Total sugars	Individual sugars					Dietary fibre	Fatty acids			Cholest-erol
				Gluc	Fruct	Sucr	Malt	Lact	NSP	Satd	Mono-unsatd	Poly-unsatd	
		g	g	g	g	g	g	g	g	g	g	g	mg
	Vegetables, general *continued*												
827	**Onions**, *raw*	Tr	5.6[a]	2.1	1.6	1.9	0	0	1.4	Tr	Tr	0.1	0
828	*fried in corn oil*	0.1	10.0[a]	3.8	2.8	3.4	0	0	3.1	1.4	2.8	6.5	0
829	pickled, *drained*	Tr	3.5[a]	0.9	1.4	1.2	0	0	1.2	Tr	Tr	0.1	0
830	pickled, cocktail/silverskin, *drained*	Tr	2.2[a]	0.7	1.3	0.2	0	0	N	Tr	Tr	Tr	0
831	**Parsnip**, *raw*	6.2	5.7[a]	0.8	0.5	4.3	0	0	4.6	0.2	0.5	0.2	0
832	*boiled in unsalted water*	6.4	5.9[a]	0.8	0.5	4.5	0	0	4.7	0.2	0.5	0.2	0
833	**Peppers**, capsicum, chilli, green, *raw*	Tr	0.7	0.4	0.2	0.1	0	0	N	N	N	N	0
834	capsicum, green, *raw*	0.1	2.4[a]	1.0	1.4	Tr	0	0	1.6	0.1	Tr	0.2	0
835	green, *boiled in salted water*	0.2	2.3[a]	0.9	1.4	Tr	0	0	1.8	0.1	Tr	0.3	0
836	capsicum, red, *raw*	0.1	6.1[a]	2.5	3.6	Tr	0	0	1.6	0.1	Tr	0.2	0
837	red, *boiled in salted water*	0.1	6.7[a]	2.7	3.9	Tr	0	0	1.7	0.1	Tr	0.2	0
838	**Plantain**, *boiled in unsalted water*	23.0	5.5	0.8	0.9	3.9	0	0	1.2	0.1	Tr	0.1	0
839	ripe, *fried in vegetable oil*	36.0	11.5	2.3	2.3	6.9	0	0	2.3	1.0	3.3	4.5	0
840	**Pumpkin**, *raw*	0.3	1.7	0.7	0.6	0.4	0	0	1.0	0.1	Tr	Tr	0
841	*boiled in salted water*	0.1	1.8	0.7	0.6	0.4	0	0	1.1	0.1	Tr	Tr	0
842	**Quorn**, *pieces, as purchased*	1.1	0.8[a]	0.8	Tr	Tr	Tr	Tr	4.8	0.6	0.7	1.9	0
843	**Radish**, red, *raw*	Tr	1.9	1.2	0.7	Tr	0	0	0.9	0.1	Tr	0.1	0
844	**Shallots**, *raw*	Tr	3.3	1.2	1.0	1.1	0	0	(1.4)	Tr	Tr	0.1	0

[a] Not including oligosaccharides

Inorganic constituents per 100g edible portion

No.	Food	mg										µg	
		Na	K	Ca	Mg	P	Fe	Cu	Zn	Cl	Mn	Se	I
	Vegetables, general *continued*												
827	**Onions**, *raw*	3	160	25	4	30	0.3	0.05	0.2	25	0.1	(1)	3
828	*fried in corn oil*	4	370	47	8	44	0.8	0.04	0.3	53	0.2	(2)	6
829	pickled, *drained*	450	93	22	5	23	0.2	0.04	0.1	730	0.1	(1)	(3)
830	pickled, cocktail/silverskin, *drained*	620	60	29	5	16	0.5	0.04	0.1	990	0.1	N	N
831	**Parsnip**, *raw*	10	450	41	23	74	0.6	0.05	0.3	49	0.5	2	N
832	*boiled in unsalted water*	4	350	50	23	76	0.6	0.04	0.3	33	0.3	N	N
833	**Peppers**, capsicum, chilli, green, *raw*	7	220	30	24	80	1.2	N	0.4	15	N	N	N
834	capsicum, green, *raw*	4	120	8	10	19	0.4	0.02	0.1	19	0.1	Tr	1
835	green, *boiled in salted water*	70	140	9	10	23	0.4	0.03	0.2	100	0.1	Tr	1
836	capsicum, red, *raw*	4	160	8	14	22	0.3	0.01	0.1	24	0.1	Tr	(1)
837	red, *boiled in salted water*	70	180	9	14	26	0.3	0.01	0.2	100	0.1	Tr	(1)
838	**Plantain**, *boiled in unsalted water*	4	400	5	33	31	0.5	0.08	0.2	50	N	(2)	N
839	ripe, *fried in vegetable oil*	3	610	6	54	66	0.8	0.20	0.4	110	N	(3)	N
840	**Pumpkin**, *raw*	Tr	130	29	10	19	0.4	0.02	0.2	37	(0.1)	N	N
841	*boiled in salted water*	76	84	23	7	15	0.1	0.02	0.2	N	(0.1)	N	N
842	**Quorn**, *pieces, as purchased*	348	120	29	37	237	0.6	0.10	7.0	N	2.8	N	N
843	**Radish**, red, *raw*	11	240	19	5	20	0.6	0.01	0.2	37	0.1	(2)	(1)
844	**Shallots**, *raw*	10	180	24	(4)	50	0.8	(0.05)	0.4	(25)	(0.1)	(1)	(3)

Vitamins per 100g edible portion

No.	Food	Retinol µg	Carotene µg	Vitamin D µg	Vitamin E mg	Thiamin mg	Riboflavin mg	Niacin mg	Trypt/60 mg	Vitamin B_6 mg	Vitamin B_{12} µg	Folate µg	Pantothenate mg	Biotin µg	Vitamin C mg
	Vegetables, general *continued*														
827	**Onions**, *raw*	0	10	0	0.31	0.13	Tr	0.7	0.3	0.20	0	17	0.11	0.9	5
828	*fried in corn oil*	0	40	0	1.93	0.08	0.01	Tr	0.5	0.10	0	38	0.12	1.3	3
829	pickled, *drained*	0	(10)	0	(0.31)	0.02	Tr	0.1	0.2	0.10	0	14	N	N	Tr
830	pickled, cocktail/silverskin, *drained*	0	Tr	0	N	N	Tr	N	0.1	N	0	N	N	N	Tr
831	**Parsnip**, *raw*	0	30	0	1.00	0.23	0.01	1.0	0.5	0.11	0	87	0.50	0.1	17
832	*boiled in unsalted water*	0	30	0	1.00	0.07	0.01	0.7	0.4	0.09	0	41	0.35	Tr	10
833	**Peppers**, capsicum, chilli, green, *raw*	0	175	0	N	0.07	0.08	1.1	0.5	N	0	29	N	N	120
834	capsicum, green, *raw*	0	265	0	0.80	0.01	0.01	0.1	0.1	0.30	0	36	0.08	N	120
835	green, *boiled in salted water*	0	240	0	0.80	0.01	0.02	Tr	0.2	0.26	0	19	0.06	N	69
836	capsicum, red, *raw*	0	3840	0	0.80	0.01	0.03	1.3	0.2	0.36	0	21	0.08	N	140
837	red, *boiled in salted water*	0	3780	0	0.90	0.01	0.03	0.9	0.2	0.31	0	11	0.06	N	81
838	**Plantain**, *boiled in unsalted water*	0	(350)	0	(0.20)	0.03	0.04	0.5	0.1	0.24	0	22	0.25	N	9
839	ripe, *fried in vegetable oil*	0	N	0	N	0.11	0.02	0.6	0.2	(1.00)	0	37	0.73	N	12
840	**Pumpkin**, *raw*	0	450	0	1.06	0.16	Tr	0.1	0.1	0.02	0	10	0.40	(0.4)	14
841	*boiled in salted water*	0	955	0	(1.06)	0.14	Tr	0.1	0.1	0.03	0	10	0.30	(0.4)	7
842	**Quorn**, *pieces, as purchased*	0	0	0	0	0.10	0.39	0.3	(2.7)	0.08	0.3	21	0.36	5.9	0
843	**Radish**, red, *raw*	0	Tr	0	0	0.03	Tr	0.4	0.1	0.07	0	38	0.18	N	17
844	**Shallots**, *raw*	0	N	0	(0.31)	0.04	0.06	0.6	0.4	(0.20)	0	(17)	(0.11)	(0.9)	13

Composition of food per 100g edible portion

No.	Food	Description and main data sources	Edible conversion factor	Water	Total nitrogen	Protein	Fat	Carbo-hydrate	Energy value	
				g	g	g	g	g	kcal	kJ
	Vegetables, general *continued*									
845	**Spinach**, raw	IFR; ribs and stems removed	0.81	89.7	0.45	2.8	0.8	1.6	25	103
846	*boiled in unsalted water*	IFR. As raw; shredded	1.00	91.8	0.35	2.2	0.8	0.8	19	79
847	*frozen, boiled in unsalted water*	LGC; 10 samples, 8 brands. Boiled 2-10 minutes	1.00	91.6	0.50	3.1	(0.8)	0.5	21	90
848	**Spring greens**, *raw*	IFR; main ribs and stems removed	0.84	86.2	0.48	3.0	1.0	3.1	33	136
849	*boiled in unsalted water*	IFR. As raw; boiled 12 minutes	1.00	92.2	0.30	1.9	0.7	1.6	20	82
850	**Spring onions**, *bulbs and tops, raw*	IFR; peeled bulb and leaves	0.69	92.2	0.32	2.0	0.5	3.0	23	98
851	**Swede**, *raw*	IFR; flesh only, peeled thinly	0.73	91.2	0.11	0.7	0.3	5.0	24	101
852	*boiled in unsalted water*	IFR. As raw; diced and boiled 22 minutes	1.00	95.8	0.05	0.3	0.1	2.3	11	46
853	**Sweet potato**, *raw*	IFR; flesh only, yellow variety	0.84	73.7	0.19	1.2	0.3	21.3	87	372
854	*boiled in salted water*	IFR. As raw; boiled 27 minutes	1.00	74.7	0.18	1.1	0.3	20.5	84	358
855	**Sweetcorn**, baby, canned, *drained*	Ref. Wu Leung et al. (1972)	0.53	92.5	0.46	2.9	0.4	2.0	23	96
856	kernels, canned, *re-heated, drained*	LGC; 10 samples, 5 brands	0.82	72.3	0.47	2.9	1.2	26.6[a]	122	519
857	on-the-cob, whole, *boiled in unsalted water*	IFR; boiled 19 minutes	0.59	69.9	0.68	4.2	2.3	19.6[a]	111	470
858	**Tomatoes**, *raw*	IFR; flesh, skin and seeds	1.00	93.1	0.11	0.7	0.3	3.1	17	73
859	*fried in corn oil*	IFR. As raw; sliced and fried 10 minutes	1.00	84.4	0.13	0.7	7.7	5.0	91	377
860	*grilled*	Calculated from raw using water loss of 13%	1.00	92.1	0.13	0.08	0.3	3.5	20	83
861	canned, *whole contents*	LGC; 10 samples, 10 brands. Tomatoes and juice	1.00[b]	94.0	0.16	1.0	0.1	3.0	16	69
862	**Turnip**, *raw*	IFR; flesh only, peeled thinly	0.75	91.2	0.14	0.9	0.3	4.7	23	98
863	*boiled in unsalted water*	IFR. As raw; diced and boiled 19 minutes	1.00	93.1	0.10	0.6	0.2	2.0	12	51

[a] Including oligosaccharides

[b] Drained = 0.6

Composition of food per 100g edible portion

No.	Food	Starch	Total sugars	Individual sugars					Dietary fibre	Fatty acids			Cholest-erol
				Gluc	Fruct	Sucr	Malt	Lact	NSP	Satd	Mono-unsatd	Poly-unsatd	
		g	g	g	g	g	g	g	g	g	g	g	mg
	Vegetables, general *continued*												
845	**Spinach**, *raw*	0.1	1.5	0.5	0.5	0.5	0	0	2.1	0.1	0.1	0.5	0
846	*boiled in unsalted water*	Tr	0.8	0.3	0.3	0.3	0	0	2.1	0.1	0.1	0.5	0
847	*frozen, boiled in unsalted water*	0.2	0.3	0.1	Tr	0.1	0	0	(2.1)	0.1	0.1	0.5	0
848	**Spring greens**, *raw*	0.4	2.7	(1.3)	(1.2)	(0.2)	0	0	3.4	0.1	0.1	0.6	0
849	*boiled in unsalted water*	0.2	1.4	(0.7)	(0.6)	(0.1)	0	0	2.6	0.1	0.1	0.4	0
850	**Spring onions**, *bulbs and tops, raw*	0.2	2.8	1.2	1.4	0.2	0	0	1.5	0.1	0.1	0.2	0
851	**Swede**, *raw*	0.1	4.9	2.7	2.0	0.2	0	0	1.9	Tr	Tr	0.2	0
852	*boiled in unsalted water*	0.1	2.2	1.2	0.9	0.1	0	0	0.7	Tr	Tr	0.1	0
853	**Sweet potato**, *raw*	15.6	5.7	0.7	0.6	4.4	Tr	0	2.4	0.1	Tr	0.1	0
854	*boiled in salted water*	8.9	11.6	N	N	N	N	0	2.3	0.1	Tr	0.1	0
855	**Sweetcorn**, baby, canned, *drained*	0.6	1.4	1.0	0.4	Tr	0	0	1.5	N	N	N	0
856	kernels, canned, *re-heated, drained*	16.6	9.6[a]	0.7	0.5	8.4	0.1	0	1.4	0.2	0.3	0.5	0
857	on-the-cob, whole, *boiled in unsalted water*	16.9	2.3[a]	0.5	0.3	1.3	0.1	0	2.2	0.2	0.3	0.5	0
858	**Tomatoes**, *raw*	Tr	3.1	1.4	1.5	0.1	0	0	1.0	0.1	0.1	0.2	0
859	*fried in corn oil*	0.1	4.9	2.3	2.5	0.2	0	0	1.3	1.0	1.9	4.5	0
860	*grilled*	Tr	3.5	1.6	1.7	0.1	0	0	1.5	1.1	0.1	0.2	0
861	canned, *whole contents*	0.2	2.8	1.3	1.4	Tr	0	0	0.7	Tr	Tr	Tr	0
862	**Turnip**, *raw*	0.2	4.5	2.3	1.7	0.6	0	0	2.4	Tr	Tr	0.2	0
863	*boiled in unsalted water*	0.1	1.9	0.9	0.7	0.3	0	0	1.9	Tr	Tr	0.1	0

[a] Not including oligosaccharides

Inorganic constituents per 100g edible portion

No.	Food	mg										µg	
		Na	K	Ca	Mg	P	Fe	Cu	Zn	Cl	Mn	Se	I
	Vegetables, general *continued*												
845	**Spinach**, *raw*	140	500	170	54	45	2.1	0.04	0.7	98	0.6	(1)	2
846	*boiled in unsalted water*	120	230	160	34	28	1.6	0.01	0.5	56	0.5	(1)	2
847	*frozen, boiled in unsalted water*	16	340	150	31	48	1.7	0.09	0.6	31	0.2	(1)	(2)
848	**Spring greens**, *raw*	20	370	210	19	91	3.0	0.02	0.4	78	N	N	N
849	*boiled in unsalted water*	10	160	75	8	29	1.4	0.02	0.3	16	N	N	N
850	**Spring onions**, *bulbs and tops, raw*	7	260	39	12	29	1.9	0.06	0.4	31	0.2	N	N
851	**Swede**, *raw*	15	170	53	9	40	0.1	0.01	0.3	31	(0.1)	(1)	N
852	*boiled in unsalted water*	14	86	26	4	11	0.1	Tr	0.1	9	(0.1)	(1)	N
853	**Sweet potato**, *raw*	40	370	24	18	50	0.7	0.14	0.3	65	0.4	(1)	2
854	*boiled in salted water*	32	300	23	45	50	0.7	0.14	0.3	52	0.4	(1)	2
855	**Sweetcorn**, baby, canned, *drained*	1140	180	8	N	N	1.2	N	N	1760	N	N	N
856	kernels, canned, *re-heated, drained*	270	220	4	23	79	0.5	Tr	0.5	390	0.1	Tr	N
857	on-the-cob, whole, *boiled in unsalted water*	1	240	3	34	81	0.6	0.04	0.4	14	0.2	Tr	N
858	**Tomatoes**, *raw*	9	250	7	7	24	0.5	0.01	0.1	55	0.1	Tr	2
859	*fried in corn oil*	10	300	12	13	24	0.5	0.02	0.1	57	0.1	Tr	2
860	*grilled*	10	286	8	8	27	0.6	0.01	0.1	63	0.1	Tr	2
861	canned, *whole contents*	39	250	12	11	19	0.4	0.07	0.1	93	0.1	Tr	(3)
862	**Turnip**, *raw*	15	280	48	8	41	0.2	0.01	0.1	39	0.1	(1)	N
863	*boiled in unsalted water*	28	200	45	6	31	0.2	0.01	0.1	31	0.1	(1)	N

Vitamins per 100g edible portion

No.	Food	Retinol µg	Carotene µg	Vitamin D µg	Vitamin E mg	Thiamin mg	Riboflavin mg	Niacin mg	Trypt/60 mg	Vitamin B_6 mg	Vitamin B_{12} µg	Folate µg	Pantothenate mg	Biotin µg	Vitamin C mg
	Vegetables, general *continued*														
845	**Spinach**, *raw*	0	3535	0	1.71	0.07	0.09	1.2	0.7	0.17	0	114	(0.27)	(0.1)	26
846	*boiled in unsalted water*	0	6604	0	(1.71)	0.06	0.05	0.9	0.6	0.09	0	81	0.21	0.1	8
847	*frozen, boiled in unsalted water*	0	(6604)	0	(1.71)	(0.06)	(0.05)	(0.9)	0.8	(0.09)	0	52	(0.21)	(0.1)	6
848	**Spring greens**, *raw*	0	8295	0	N	0.07	0.11	1.5	0.5	0.23	0	92	0.39	(0.4)	180
849	*boiled in unsalted water*	0	2270	0	N	0.05	0.06	1.2	0.3	0.18	0	66	0.30	(0.4)	77
850	**Spring onions**, *bulbs and tops, raw*	0	620	0	N	0.05	0.03	0.5	0.5	0.13	0	54	0.07	N	26
851	**Swede**, *raw*	0	350	0	Tr	0.15	Tr	1.2	0.1	0.21	0	31	0.11	0.1	31
852	*boiled in unsalted water*	0	165	0	Tr	0.13	0.01	1.0	0.1	0.04	0	18	0.07	Tr	15
853	**Sweet potato**, *raw*	0	3930[a]	0	(0.28)	0.17	Tr	0.5	0.3	0.09	0	17	0.59	N	23
854	*boiled in salted water*	0	3960[b]	0	(0.28)	0.07	0.01	0.5	0.3	0.05	0	8	0.53	N	17
855	**Sweetcorn**, baby, canned, *drained*	0	(140)	0	N	0.02	0.04	0.1	0.3	N	0	N	N	N	14
856	kernels, canned, *re-heated, drained*	0	110	0	0.46	0.04	0.06	1.5	0.5	0.13	0	20	0.22	N	1
857	on-the-cob, whole, *boiled in unsalted water*	0	120	0	0.88	0.18	0.05	2.0	0.5	0.15	0	34	0.63	N	7
858	**Tomatoes**, *raw*	0	564	0	1.22	0.09	0.01	1.0	0.1	0.14	0	22	0.25	1.5	17
859	*fried in corn oil*	0	765	0	N	0.09	0.01	0.5	0.1	0.10	0	17	(0.25)	(1.5)	16
860	*grilled*	0	646	0	1.40	0.11	0.01	1.1	0.1	0.16	0	25	0.29	1.7	19
861	canned, *whole contents*	0	362	0	1.22	0.05	0.02	0.7	0.1	0.11	0	18	0.20	1.5	12
862	**Turnip**, *raw*	0	20	0	Tr	0.05	0.01	0.4	0.2	0.08	0	14	0.20	0.1	17
863	*boiled in unsalted water*	0	20	0	Tr	0.05	0.02	0.2	0.1	0.04	0	8	0.14	Tr	10

[a] Value for orange fleshed varieties. Carotene can range from 1820 to 16000µg per 100g. White fleshed varieties contain approximately 69µg per 100g

[b] White fleshed varieties contain approximately 66µg per 100g

Composition of food per 100g edible portion

No.	Food	Description and main data sources	Edible conversion factor	Water	Total nitrogen	Protein	Fat	Carbo-hydrate	Energy value	
				g	g	g	g	g	kcal	kJ
	Vegetables, general *continued*									
864	**Watercress**, *raw*	IFR; large stalks removed	0.62	92.5	0.48	3.0	1.0	0.4	22	94
865	**Yam**, *raw*	IFR; flesh only	0.81	67.2	0.25	1.5	0.3	28.2	114	488
866	*boiled in unsalted water*	IFR. As raw; boiled 25 minutes	1.00	64.4	0.27	1.7	0.3	33.0	133	568
	Vegetable dishes									
867	**Beanburger**, soya, *fried in vegetable oil*	Recipe from review of recipe collection	1.00	57.4	1.71	10.6	11.0	13.7	193	807
868	**Bubble and squeak**, *fried in vegetable oil*	Fried cabbage and potato. Recipe from Wiles et al, 1980	1.00	77.5	0.23	1.4	9.1	9.8	125	519
869	**Cannelloni**, vegetable	Pasta tubes with mixed vegetable filling. Recipe from dietary survey records	1.00	68.7	0.81	5.1	9.1	15.5	161	675
870	**Casserole**, vegetable	Recipe from dietary survey records	1.00	84.2	0.35	2.2	0.4	10.6	52	221
871	**Cauliflower cheese**, *made with semi-skimmed milk*	Recipe	1.00	79.0	0.96	6.0	6.5	5.1	102	423
872	**Chilli**, vegetable	Recipe from dietary survey records	1.00	82.7	0.42	2.6	0.6	10.7	56	238
873	**Coleslaw**, with mayonnaise, *retail*	Recipe from average of manufacturers' proportions	1.00	65.6	0.20	1.2	26.4	4.2	258	939
874	with reduced calorie dressing, *retail*	Recipe from average of manufacturers' proportions	1.00	86.2	0.14	0.9	4.5	6.1	67	280
875	**Curry**, chick pea dahl	Punjabi dish. Split chick peas and tomato. Recipe from dietary survey records	1.00	63.7	1.27	7.9	6.1	17.9	154	645
876	vegetable, *retail*, with rice	4 samples, 2 brands; cooked in conventional and microwave ovens according to packet directions	1.00	74.5	0.53	3.3	3.0	16.4	102	429

No.	Food	Starch	Total sugars	Individual sugars					Dietary fibre	Fatty acids			Cholest-erol
				Gluc	Fruct	Sucr	Malt	Lact	NSP	Satd	Mono-unsatd	Poly-unsatd	
		g	g	g	g	g	g	g	g	g	g	g	mg
	Vegetables, general *continued*												
864	**Watercress**, *raw*	Tr	0.4	0.2	0.1	0.1	0	0	1.5	0.3	0.1	0.4	0
865	**Yam**, *raw*	27.5	0.7	0.2	0.1	0.4	0	0	1.3	0.1	Tr	0.1	0
866	*boiled in unsalted water*	32.3	0.7	0.2	0.1	0.4	0	0	1.4	0.1	Tr	0.1	0
	Vegetable dishes												
867	**Beanburger**, soya, *fried in vegetable oil*	9.1	3.5	0.9	0.9	1.6	0	0	4.7	1.52	4.18	3.97	33
868	**Bubble and squeak**, *fried in vegetable oil*	8.4	1.4	0.6	0.5	0.3	0	0	1.5	1.1	4.7	2.9	0
869	**Cannelloni**, vegetable	13.0	2.4	0.2	0.2	0.1	0.2	1.7	0.8	3.5	3.4	1.5	15
870	**Casserole**, vegetable	5.5	4.6	1.4	1.3	1.8	0	0	2.1	0.1	0.1	0.2	0
871	**Cauliflower cheese**, *made with semi-skimmed milk*	2.1	2.9	0.9	0.7	0.1	0	1.2	1.3	2.9	1.7	1.4	11
872	**Chilli**, vegetable	5.6	4.5	0.9	0.8	2.8	0	0	2.5	0.1	0.1	0.2	Tr
873	**Coleslaw**, with mayonnaise, *retail*	0.2	3.9	1.5	1.4	1.0	0	0	1.4	3.9	6.0	15.3	26
874	with reduced calorie dressing, *retail*	0.1	6.0	1.5	1.4	3.1	0	0	1.4	0.5	1.5	2.2	0
875	**Curry**, chick pea dahl	14.7	1.8	0.4	0.4	1.0	0	0	N	0.7	2.6	2.2	0
876	vegetable, *retail*, with rice	14.0	2.4	0.7	0.8	0.7	Tr	0.2	N	N	N	N	Tr

Inorganic constituents per 100g edible portion

No.	Food	mg										µg	
		Na	K	Ca	Mg	P	Fe	Cu	Zn	Cl	Mn	Se	I
	***Vegetables, general** continued*												
864	**Watercress**, *raw*	49	230	170	15	52	2.2	0.01	0.7	170	0.6	N	N
865	**Yam**, *raw*	2	380	15	15	27	0.7	0.01	0.3	10	0.1	N	N
866	*boiled in unsalted water*	17	260	12	12	21	0.4	0.03	0.4	40	Tr	N	N
	Vegetable dishes												
867	**Beanburger**, soya, *fried in vegetable oil*	265	449	69	53	211	2.7	0.28	1.1	412	0.9	4	N
868	**Bubble and squeak**, *fried in vegetable oil*	8	204	19	9	29	0.4	0.04	0.2	28	0.2	(2)	4
869	**Cannelloni**, vegetable	266	117	104	14	93	0.5	0.05	0.6	N	0.2	N	N
870	**Casserole**, vegetable	73	318	23	16	51	0.6	0.06	0.3	143	0.2	1	(3)
871	**Cauliflower cheese**, *made with semi-skimmed milk*	199	309	120	18	119	0.6	0.03	0.9	314	0.2	1	11
872	**Chilli**, vegetable	307	251	29	17	58	0.8	0.03	0.3	481	0.2	2	N
873	**Coleslaw**, with mayonnaise, *retail*	160	150	32	3	26	0.4	0.01	0.2	290	0.1	N	13
874	with reduced calorie dressing, *retail*	200	160	31	6	19	0.3	0.01	0.1	330	0.1	Tr	2
875	**Curry**, chick pea dahl	19	398	23	40	120	2.0	0.35	1.3	38	0.6	1	N
876	vegetable, *retail*, with rice	250	150	37	16	54	0.8	0.10	0.4	380	0.3	N	N

No.	Food	Retinol µg	Carotene µg	Vitamin D µg	Vitamin E mg	Thiamin mg	Riboflavin mg	Niacin mg	Trypt/60 mg	Vitamin B_6 mg	Vitamin B_{12} µg	Folate µg	Pantothenate mg	Biotin µg	Vitamin C mg
	***Vegetables, general** continued*														
864	**Watercress**, *raw*	0	2520	0	1.46	0.16	0.06	0.3	0.5	0.23	0	45	0.10	0.4	62
865	**Yam**, *raw*	0	Tr[a]	0	N	0.16	0.01	0.2	0.3	0.16	0	8	0.31	N	4
866	*boiled in unsalted water*	0	Tr	0	N	0.14	0.01	0.2	0.4	0.12	0	6	0.31	N	4
	Vegetable dishes														
867	**Beanburger**, soya, *fried in vegetable oil*	16	190	0.2	1.28	0.19	0.11	0.9	2.0	0.17	0.2	22	0.48	18.1	4
868	**Bubble and squeak**, *fried in vegetable oil*	0	411	0	N	0.13	0.01	0.4	0.3	0.21	0	30	0.27	0.2	13
869	**Cannelloni**, vegetable	56	493	N	0.26	0.06	0.11	0.4	1.0	0.06	0.4	(4)	(0.13)	(0.7)	3
870	**Casserole**, vegetable	0	1464	0	0.78	0.15	0.09	1.2	0.4	0.18	0	22	0.24	0.7	7
871	**Cauliflower cheese**, *made with semi-skimmed milk*	56	58	0.2	1.01	0.11	0.11	0.4	1.5	0.18	0.3	25	0.43	1.8	15
872	**Chilli**, vegetable	Tr	1559	0	0.54	0.09	0.03	0.6	0.4	0.10	0	13	0.15	N	6
873	**Coleslaw**, with mayonnaise, *retail*	29	870	0.1	6.74	0.08	0.03	0.2	0.2	0.12	0.2	21	N	N	20
874	with reduced calorie dressing, *retail*	0	755	0	1.22	0.08	0.01	0.2	0.1	0.12	0	20	0.14	0.1	20
875	**Curry**, chick pea dahl	0	198	0	1.16	0.13	0.07	0.7	1.1	0.17	0	32	0.46	N	2
876	vegetable, *retail*, with rice	Tr	N	Tr	N	0.06	0.08	0.6	0.7	0.13	Tr	8	N	N	N

[a] Yellow fleshed varieties contain 400 to 1440µg carotene per 100g

Composition of food per 100g edible portion

No.	Food	Description and main data sources	Edible conversion factor	Water	Total nitrogen	Protein	Fat	Carbo-hydrate	Energy value	
				g	g	g	g	g	kcal	kJ
	***Vegetable dishes** continued*									
877	**Flan**, vegetable	Recipe from dietary survey records and dissection of shop-bought samples	1.00	58.1	0.87	5.3	14.5	19.9	226	945
878	**Garlic mushrooms** (not coated)	Recipe from review of recipe collection[a]	1.00	77.5	0.70	2.1	14.4	0.7	140	579
879	**Lasagne**, vegetable, *retail*	8 assorted samples; cooked in conventional and microwave ovens according to packet directions	1.00	72.6	0.77	4.8	5.3	13.4	117	492
880	**Moussaka**, vegetable, *retail*	7 samples, 3 brands; cooked in conventional and microwave ovens according to packet directions	1.00	75.4	0.94	5.9	4.9	8.0	98	410
881	**Nut roast**	Recipe from review of recipe collection. Mixed nuts	1.00	38.8	2.42	13.2	23.5	18.4	333	1386
882	**Pakora/bhajia**, vegetable, *retail*	Recipe from manufacturer	1.00	50.5	1.04	6.4	14.7	21.4	235	975
883	**Pancakes**, stuffed with vegetables	Tomato, mushroom and onion stuffing. Recipe adapted from a recipe from Leeds Polytechnic	1.00	72.7	0.67	3.9	7.6	14.1	137	573
884	**Pasty**, vegetable	Recipe from dietary survey records	1.00	45.8	0.69	4.0	16.5	33.1	289	1208
885	**Pie**, vegetable	Recipe from review of recipe collection	1.00	68.0	0.52	3.0	8.4	18.8	159	663
886	**Ratatouille**, *retail*	9 frozen samples, 3 brands; shallow fried then simmered for 35-40 minutes	1.00	85.5	0.20	1.2	6.6	3.7	78	324
887	**Risotto**, vegetable	White rice, vegetables, red kidney beans and cashew nuts. Recipe from Leeds Polytechnic	1.00	66.9	0.77	4.1	6.5	19.2	147	619
888	**Salad**, green	Lettuce, cucumber, pepper and celery. Recipe from review of recipe collection	1.00	95.1	0.12	0.7	0.3	1.8	12	51
889	potato, with mayonnaise, *retail*	Recipe from manufacturers	1.00	59.0	0.25	1.5	26.5	11.4	287	1063

[a]Garlic mushrooms in breadcrumbs contain 4.3g protein, 17.3g fat, 19.4g carbohydrate (18.3g starch, 1.1g sugars), 246kcal and 1024kJ energy per 100g

No.	Food	Starch	Total sugars	Individual sugars					Dietary fibre	Fatty acids			Cholest-erol
				Gluc	Fruct	Sucr	Malt	Lact	NSP	Satd	Mono-unsatd	Poly-unsatd	
		g	g	g	g	g	g	g	g	g	g	g	mg
	***Vegetable dishes** continued*												
877	**Flan**, vegetable	17.3	2.4	0.4	0.4	0.5	Tr	0.7	1.5	5.4	6.4	3.2	15
878	**Garlic mushrooms** (not coated)	0.3	0.3	0.1	0.1	0.1	0	0.1	1.2	8.9	3.5	0.8	36
879	**Lasagne**, vegetable, *retail*	9.0	4.4	0.7	0.7	0.2	1.6	1.2	N	N	N	N	N
880	**Moussaka**, vegetable, *retail*	3.5	4.5	1.0	1.0	0.5	0.9	1.1	N	N	N	N	N
881	**Nut roast**	15.3	4.0	0.4	0.4	2.6	0.5	0	4.1	3.6	12.8	5.8	0
882	**Pakora/bhajia**, vegetable, *retail*	17.5	2.4	0.7	0.6	1.1	Tr	0	(3.6)	1.0	7.6	4.8	0
883	**Pancakes**, stuffed with vegetables	9.9	3.8	0.9	0.8	0.4	Tr	1.6	1.0	2.2	2.3	2.6	27
884	**Pasty**, vegetable	31.3	1.6	0.3	0.2	0.6	0.1	0	2.0	5.1	6.2	4.3	8
885	**Pie**, vegetable	15.7	2.8	1.0	1.0	0.6	0	0	1.5	2.6	3.1	2.2	4
886	**Ratatouille**, *retail*	0.1	3.6	1.7	1.8	Tr	0	0	(1.0)	0.8	1.6	3.6	0
887	**Risotto**, vegetable	16.3	2.4	0.6	0.7	1.2	0	0	2.2	1.0	3.3	1.8	0
888	**Salad**, green	0.1	1.7	0.8	0.9	0.1	0	0	1.0	0.1	Tr	0.1	0
889	potato, with mayonnaise, *retail*	10.2	1.1	0.2	0.1	0.7	0	0	0.8	3.9	6.1	15.4	26

Inorganic constituents per 100g edible portion

No.	Food	mg										µg	
		Na	K	Ca	Mg	P	Fe	Cu	Zn	Cl	Mn	Se	I
	***Vegetable dishes** continued*												
877	**Flan**, vegetable	268	124	121	12	94	0.8	0.05	0.6	417	0.2	1	N
878	**Garlic mushrooms** (not coated)	108	348	10	10	90	0.7	0.76	0.4	242	0.1	10	10
879	**Lasagne**, vegetable, *retail*	390	190	73	18	87	0.8	0.02	0.4	620	0.2	N	N
880	**Moussaka**, vegetable, *retail*	450	330	76	25	100	1.0	0.06	0.5	690	0.2	N	N
881	**Nut roast**	189	421	77	113	261	2.1	0.54	2.0	297	1.4	4	9
882	**Pakora/bhajia**, vegetable, *retail*	61	490	99	47	130	3.7	0.21	1.1	71	0.7	(1)	N
883	**Pancakes**, stuffed with vegetables	139	238	76	13	80	0.8	0.13	0.4	247	0.2	2	N
884	**Pasty**, vegetable	237	141	62	12	55	0.9	0.08	0.3	392	0.3	1	N
885	**Pie**, vegetable	170	246	41	14	52	0.8	0.10	0.3	298	0.2	1	N
886	**Ratatouille**, *retail*	19	220	22	15	27	0.6	0.06	0.2	55	0.1	N	N
887	**Risotto**, vegetable	348	242	32	28	100	1.1	0.30	0.8	557	0.4	7	N
888	**Salad**, green	6	165	19	8	34	0.4	0.01	0.1	31	0.2	Tr	2
889	potato, with mayonnaise, *retail*	160	180	6	9	29	0.4	0.05	0.2	290	0.1	N	14

No.	Food	Retinol µg	Carotene µg	Vitamin D µg	Vitamin E mg	Thiamin mg	Ribo-flavin mg	Niacin mg	Trypt/60 mg	Vitamin B_6 mg	Vitamin B_{12} µg	Folate µg	Panto-thenate mg	Biotin µg	Vitamin C mg
	Vegetable dishes *continued*														
877	**Flan**, vegetable	77	2172	0.6	2.57	0.09	0.08	0.6	1.2	0.10	0.3	19	N	N	8
878	**Garlic mushrooms** (not coated)	162	103	0.2	0.44	0.08	0.27	2.7	0.4	0.16	0.1	23	1.70	10.2	1
879	**Lasagne**, vegetable, *retail*	N	N	N	N	0.25	0.29	2.0	1.1	0.44	N	6	N	N	N
880	**Moussaka**, vegetable, *retail*	N	N	N	N	0.06	0.07	0.7	1.3	0.18	Tr	12	N	N	N
881	**Nut roast**	0	17	0	5.53	0.43	0.31	6.3	3.3	0.24	Tr	48	0.83	23.6	0
882	**Pakora/bhajia**, vegetable, *retail*	0	965	0	3.66	(0.17)	(0.10)	(1.2)	1.0	(0.23)	0	(30)	(0.53)	N	7
883	**Pancakes**, stuffed with vegetables	66	184	0.6	2.81	0.08	0.13	0.8	0.8	0.11	0.4	13	0.52	3.6	3
884	**Pasty**, vegetable	59	927	0.7	3.17	0.14	0.01	0.8	0.8	0.12	Tr	14	0.19	0.5	3
885	**Pie**, vegetable	30	1674	0.3	1.94	0.13	0.13	1.3	0.5	0.15	Tr	23	0.24	1.0	9
886	**Ratatouille**, *retail*	0	185	0	2.66	0.04	0.05	0.6	0.2	0.10	0	41	N	N	12
887	**Risotto**, vegetable	0	518	0	1.14	0.13	0.08	1.3	0.8	0.16	0	13	0.50	N	10
888	**Salad**, green	0	370	0	0.41	0.05	0.01	0.2	0.1	0.11	0	29	(0.21)	N	36
889	potato, with mayonnaise, *retail*	30	35	0.1	6.68	0.12	0.03	0.3	0.4	0.21	0.2	18	N	N	4

Composition of food per 100g edible portion

No.	Food	Description and main data sources	Edible conversion factor	Water g	Total nitrogen g	Protein g	Fat g	Carbo-hydrate g	Energy value kcal	Energy value kJ
	Vegetable dishes *continued*									
890	**Salad**, rice	Rice, vegetables, nut and raisin. Recipe from dietary survey records	1.00	65.8	0.51	3.0	7.5	23.1	165	696
891	**Samosas**, vegetable, *retail*	5 samples, 3 brands; mixed vegetable filling	1.00	51.3	0.82	5.1	9.3	30.0	217	911
892	**Sauerkraut**	IFR; 10 samples, 3 brands. Bottled and canned, drained	0.71	91.0	0.17	1.1	Tr	1.1	9	36
893	**Shepherd's pie**, vegetable, *retail*	Vegetable, lentil and barley base with potato topping. Recipe from manufacturer	1.00	77.5	0.30	1.9	4.9	13.3	101	425
894	**Tagliatelle**, with vegetables, *retail*	Recipe from manufacturer	1.00	83.3	0.28	1.6	3.0	11.0	74	315
895	**Vegeburger**, *retail, grilled*	6 samples, 3 brands; soya protein based. Grilled 6-10 minutes	1.00	50.3	2.91	16.6	11.1	8.0	196	821
896	**Vegetable and cheese grill/burger**, in crumbs, *baked/grilled*	9 samples, 6 brands including cheese grills and cheese and onion crispbakes	1.00	53.6	1.12	7.0	14.0	23.0	240	1005
897	**Vegetable bake**	Assorted vegetables topped with cheese sauce and breadcrumbs. Recipe from dietary survey records	1.00	73.9	0.69	4.3	7.2	13.1	131	548
898	**Vegetable kiev**, *baked*	4 samples, 2 brands including cordon bleu and traditional style	1.00	56.8	1.59	9.9	13.7	17.6	229	957
899	**Vegetable stir fry mix**, *fried in vegetable oil*	8 assorted frozen samples; stir-fried 4-7 minutes	1.00	83.8	0.32	2.0	3.6	6.4	64	270
900	**Vegetables**, *stir-fried, takeaway*	10 samples from different outlets	1.00	88.1	0.29	1.8	4.1	2.1	52	216
901	**Vegetarian sausages**, *baked/ grilled*	4 samples, 3 brands, baked or grilled	1.00	57.8	2.38	14.9	9.4	9.2	179	748

No.	Food	Starch	Total sugars	Individual sugars					Dietary fibre	Fatty acids			Cholest-erol
				Gluc	Fruct	Sucr	Malt	Lact	NSP	Satd	Mono-unsatd	Poly-unsatd	
		g	g	g	g	g	g	g	g	g	g	g	mg
Vegetable dishes *continued*													
890	**Salad**, rice	18.3	4.8	1.9	1.9	0.9	0	0	0.7	1.1	3.8	2.1	0
891	**Samosas**, vegetable, *retail*	27.3	2.7	0.3	0.2	1.1	1.1	0	2.5	N	N	N	0
892	**Sauerkraut**	Tr	1.1	0.7	0.3	0.1	0	0	2.2	Tr	Tr	Tr	0
893	**Shepherd's pie**, vegetable, *retail*	11.4	1.7	0.6	0.5	0.5	Tr	0.1	1.2	2.0	1.2	1.4	7
894	**Tagliatelle**, with vegetables, *retail*	8.6	2.2	0.5	0.5	0.9	0.1	0.3	0.7	0.8	0.7	1.3	2
895	**Vegeburger**, *retail, grilled*	4.4	3.6	0.5	0.4	1.8	0.9	0	4.2	N	N	N	N
896	**Vegetable and cheese grill/burger**, in crumbs, *baked/grilled*	18.6	1.3	0.2	0.2	0.2	0.3	0.4	1.6	4.6	4.4	3.0	N
897	**Vegetable bake**	8.6	4.3	0.8	0.7	0.9	Tr	1.8	1.2	2.9	2.1	1.7	11
898	**Vegetable kiev**, *baked*	13.4	1.3	0.2	0.1	0.3	0.5	0.2	1.2	5.0	5.2	2.8	13
899	**Vegetable stir fry mix**, *fried in vegetable oil*	2.5	3.9	1.4	1.2	1.1	0.2	0	N	0.3	1.8	1.3	0
900	**Vegetables**, *stir-fried, takeaway*	1.6	0.2	Tr	0.2	Tr	Tr	0	1.8	0.8	2.2	0.6	1
901	**Vegetarian sausages**, *baked/grilled*	2.9	1.3	0.5	0.5	0.2	0.1	0	2.6	2.3	4.1	1.9	0

Inorganic constituents per 100g edible portion

No.	Food	mg										µg	
		Na	K	Ca	Mg	P	Fe	Cu	Zn	Cl	Mn	Se	I
	***Vegetable dishes** continued*												
890	**Salad**, rice	234	179	21	25	70	1.0	0.19	0.8	351	0.3	(4)	(4)
891	**Samosas**, vegetable, *retail*	390	150	65	19	65	1.5	0.11	0.5	590	0.3	N	N
892	**Sauerkraut**	590	180	50	10	23	1.2	0.05	0.3	860	0.2	Tr	(1)
893	**Shepherd's pie**, vegetable, *retail*	340	240	12	14	36	0.6	0.07	0.3	560	(0.1)	(7)	N
894	**Tagliatelle**, with vegetables, *retail*	6	100	13	8	29	0.4	0.04	0.2	22	0.1	Tr	(1)
895	**Vegeburger**, *retail, grilled*	490	610	100	80	240	4.5	0.40	1.6	660	1.1	8	N
896	**Vegetable and cheese grill/burger**, in crumbs, *baked/grilled*	587	260	154	20	147	0.9	0.10	0.8	921	0.2	4	32
897	**Vegetable bake**	118	238	105	15	91	0.5	0.04	0.5	(204)	(0.1)	(2)	(16)
898	**Vegetable kiev**, *baked*	521	209	105	22	115	1.5	0.12	0.8	(804)	0.3	N	N
899	**Vegetable stir fry mix**, *fried in vegetable oil*	11	230	30	16	46	0.5	0.11	0.3	27	0.1	Tr	N
900	**Vegetables**, *stir-fried, takeaway*	399	119	13	9	32	1.0	0.05	0.2	(616)	0.1	Tr	N
901	**Vegetarian sausages**, *baked/grilled*	895	351	136	54	193	3.1	0.24	1.0	1347	0.8	4	Tr

No.	Food	Retinol µg	Carotene µg	Vitamin D µg	Vitamin E mg	Thiamin mg	Riboflavin mg	Niacin mg	Trypt/60 mg	Vitamin B_6 mg	Vitamin B_{12} µg	Folate µg	Pantothenate mg	Biotin µg	Vitamin C mg
	Vegetable dishes *continued*														
890	**Salad**, rice	0	134	0	(1.19)	0.05	0.02	0.9	0.7	0.12	0	21	(0.14)	(1.2)	12
891	**Samosas**, vegetable, *retail*	0	N	0	N	0.12	0.08	1.1	0.7	0.15	0	44	N	N	N
892	**Sauerkraut**	0	18	0	N	0.04	0.01	0.2	0.2	0.15	0	16	0.23	N	10
893	**Shepherd's pie**, vegetable, *retail*	29	585	Tr	(0.75)	0.11	0.01	0.4	0.4	0.18	Tr	10	0.24	(0.5)	4
894	**Tagliatelle**, with vegetables, *retail*	10	170	Tr	0.63	0.03	0.01	0.3	0.3	0.05	Tr	4	0.07	0.4	2
895	**Vegeburger**, *retail, grilled*	N	Tr	N	N	2.40[a]	0.42[a]	2.8[b]	3.9	0.30[b]	N	95	N	N	N
896	**Vegetable and cheese, grill/burger,** in crumbs, *baked/grilled*	17	265	1.0	1.12	0.07	0.11	1.0	1.2	0.05	0.4	8	0.30	2.0	2
897	**Vegetable bake**	59	2512	(0.3)	1.58	0.10	0.10	0.4	0.9	0.15	0.4	(11)	0.33	1.2	4
898	**Vegetable kiev**, *baked*	10	513	N	1.03	2.78	0.11	0.7	N	0.04	N	10	N	N	N
899	**Vegetable stir fry mix**, *fried in vegetable oil*	0	N	0	N	0.07	0.13	1.0	0.3	0.25	0	16	N	N	8
900	**Vegetables**, *stir-fried, takeaway*	Tr	575	0	1.26	0.03	0.13	0.3	0.3	0.05	0	3	0.30	3.3	2
901	**Vegetarian sausages**, *baked/grilled*	0	N	0	N	1.72	0.14	1.1	2.5	0.05	N	34	0.30	5.5	Tr

[a] 2 samples contained added thiamin and riboflavin
[b] 1 sample contained added niacin and vitamin B6

Section 2.8

Herbs and spices

The foods in this section of the Tables have been taken from the *Vegetables, Herbs, and Spices* (1991) supplement. The majority of values are derived from literature.

Many of the values for carbohydrate are not analysed but are calculated by ‘difference’. For spices this carbohydrate is likely to include much woody material and aromatic oils, resulting in an overestimate of both the carbohydrate and energy value. For some spices energy values and carbohydrate have therefore been given as ‘unknown’, i.e. as ‘N’. Variation in the nutrient content of spices may arise due to the processing methods used, e.g. contamination from processing machinery can result in variation of the iron content of ground spices.

Taxonomic names for foods in this part of the Tables can be found in Section 4.5.

Composition of food per 100g edible portion

No.	Food	Description and main data sources	Edible conversion factor	Water	Total nitrogen	Protein	Fat	Carbo-hydrate	Energy value	
				g	g	g	g	g	kcal	kJ
	Herbs and spices									
902	**Chilli powder**	Ref. Marsh *et al.* (1977)[a]	1.00	7.8	1.96	12.3	16.8	N	N	N
903	**Chinese 5 spice**	Calculated from recipe	1.00	9.1	1.53	9.5	8.7	N	N	N
904	**Cinnamon**, *ground*	Ref. Marsh *et al.* (1977)	1.00	9.5	0.62	3.9	3.2	N	N	N
905	**Curry powder**	2 samples[b]	1.00	8.5	1.52	9.5	10.8	26.1	233	979
906	**Garam masala**	Ref. Wharton *et al.* (1983)	1.00	10.1	2.50	15.6	15.1	*45.2*	*379*	*1592*
907	**Mint**, *fresh*	Literature sources	1.00	86.4	0.61	3.8	0.7	*5.3*	*43*	*181*
908	**Mixed herbs**, *dried*	Recipe; Equal quantities of Marjoram, Parsley, Sage and Thyme	1.00	8.1	1.93	12.1	8.5	36.3	261	1092
909	**Mustard powder**	2 brands	1.00	(8.0)	4.62	28.9	28.7	20.7	452	1884
910	**Nutmeg**, *ground*	Ref. Marsh *et al.* (1977)	1.00	6.2	1.10	5.8	36.3	N	N	N
911	**Paprika**	Ref. Marsh *et al.* (1977)	1.00	9.5	2.36	14.8	13.0	*34.9*	*289*	*1209*
912	**Parsley**, *fresh*	IFR; tough stalks removed	0.80	83.1	0.47	3.0	1.3	2.7	34	141
913	**Pepper**, black	Ref. Marsh *et al.* (1977)	1.00	10.5	2.05	10.9	3.3	N	N	N
914	white	Ref. Marsh *et al.* (1977)	1.00	11.4	1.95	10.4	2.1	N	N	N
915	**Rosemary**, *dried*	Ref. Marsh *et al.* (1977)	1.00	9.3	0.78	4.9	15.2	*46.4*	*331*	*1387*
916	**Sage**, *dried, ground*	Ref. Marsh *et al.* (1977)	1.00	8.0	1.70	10.6	12.7	*42.7*	*315*	*1317*
917	**Thyme**, *dried, ground*	Ref. Marsh *et al.* (1977)	1.00	7.8	1.46	9.1	7.4	*45.3*	*276*	*1156*

[a] Mix of chilli pepper 83%, cumin 9%, oregano 4%, salt 2.5% and garlic powder 1.5%

[b] Composition will vary according to variety

Composition of food per 100g edible portion

No.	Food	Starch	Total sugars	Individual sugars					Dietary fibre	Fatty acids			Cholest-erol
				Gluc	Fruct	Sucr	Malt	Lact	NSP	Satd	Mono-unsatd	Poly-unsatd	
		g	g	g	g	g	g	g	g	g	g	g	mg
	Herbs and spices												
902	**Chilli powder**	N	N	N	N	N	0	0	N	N	N	N	0
903	**Chinese 5 spice**	N	N	N	N	N	0	0	N	N	N	N	0
904	**Cinnamon**, *ground*	N	N	N	N	N	0	0	N	0.7	0.5	0.5	0
905	**Curry powder**	N	N	N	N	N	0	0	23.0	N	N	N	0
906	**Garam masala**	N	N	N	N	N	0	0	N	N	N	N	0
907	**Mint**, *fresh*	N	N	N	N	N	0	0	N	N	N	N	0
908	**Mixed herbs**, *dried*	N	N	N	N	N	0	0	N	N	N	N	0
909	**Mustard powder**	N	N	N	N	N	0	0	N	1.5	19.8	5.4	0
910	**Nutmeg**, *ground*	N	N	N	N	N	0	0	N	25.9	3.2	0.3	0
911	**Paprika**	N	N	N	N	N	0	0	N	1.9	1.4	7.1	0
912	**Parsley**, *fresh*	0.4	2.3	1.4	0.9	Tr	0	0	5.0	N	N	N	0
913	**Pepper**, black	Tr	N	N	N	N	0	0	N	N	N	N	0
914	white	Tr	N	N	N	N	0	0	N	N	N	N	0
915	**Rosemary**, *dried*	N	N	N	N	N	0	0	N	N	N	N	0
916	**Sage**, *dried, ground*	N	N	N	N	N	0	0	N	7.0	1.9	1.8	0
917	**Thyme**, *dried, ground*	N	N	N	N	N	0	0	N	2.7	0.5	1.2	0

Inorganic constituents per 100g edible portion

No.	Food	mg										µg	
		Na	K	Ca	Mg	P	Fe	Cu	Zn	Cl	Mn	Se	I
	Herbs and spices												
902	**Chilli powder**	1010	1920	280	170	300	14.3	0.43	2.7	1510	2.2	N	N
903	**Chinese 5 spice**	63	1070	1040	210	260	25.6	0.74	2.9	N	8.4	N	N
904	**Cinnamon**, *ground*	26	500	1230	56	61	38.1[a]	0.46	2.0	N	5.7	(15)	N
905	**Curry powder**	450	1830	640	280	270	58.3	1.04	4.1	470	4.7	N	N
906	**Garam masala**	97	1450	760	330	390	32.6	1.62	3.8	N	6.0	N	N
907	**Mint**, *fresh*	15	260	210	N	75	9.5	N	N	34	1.4	N	N
908	**Mixed herbs**, *dried*	81	1873	1653	280	235	69.0	0.73	4.6	N	9.8	N	N
909	**Mustard powder**	5	940	330	260	180	9.5	0.20	(6.5)	62	1.7	N	N
910	**Nutmeg**, *ground*	16	350	180	180	210	3.0	1.03	2.2	N	2.9	N	N
911	**Paprika**	34	2340	180	190	350	23.6	0.61	4.1	N	0.8	N	N
912	**Parsley**, *fresh*	33	760	200	23	64	7.7	0.03	0.7	160	0.2	(1)	N
913	**Pepper**, black	44	1260	430	190	170	11.2	1.13	1.4	60	6.5	(3)	N
914	white	5	73	270	90	180	14.3	1.13	1.1	60	4.5	(3)	N
915	**Rosemary**, *dried*	50	950	1280	220	70	29.3	0.55	3.2	N	0.5	N	N
916	**Sage**, *dried, ground*	11	1070	1650	430	91	28.1	0.76	4.7	N	25.0	N	N
917	**Thyme**, *dried, ground*	55	810	1890	220	200	123.6	0.86	6.2	N	7.6	N	N

[a] Whole unground cinnamon contains 4mg Fe per 100g

Vitamins per 100g edible portion

No.	Food	Retinol µg	Carotene µg	Vitamin D µg	Vitamin E mg	Thiamin mg	Riboflavin mg	Niacin mg	Trypt/60 mg	Vitamin B_6 mg	Vitamin B_{12} µg	Folate µg	Pantothenate mg	Biotin µg	Vitamin C mg
	Herbs and spices														
902	**Chilli powder**	0	(21000)	0	N	0.35	0.79	7.9	2.6	N	0	0	N	N	0
903	**Chinese 5 spice**	0	138	0	N	0.21	0.24	5.3	N	N	0	0	N	N	0
904	**Cinnamon**, *ground*	0	(155)	0	N	0.08	0.14	1.3	N	N	0	0	N	N	0
905	**Curry powder**	0	(100)	0	N	0.25	0.28	3.5	N	N	0	0	N	N	0
906	**Garam masala**	0	(340)	0	N	0.35	0.33	2.5	N	N	0	0	N	N	0
907	**Mint**, *fresh*	0	(740)	0	5.00	0.12	0.33	1.1	N	N	0	110	N	N	31
908	**Mixed herbs**, *dried*	0	(8103)	0	N	N	0.34	5.0	N	N	N	N	N	N	30
909	**Mustard powder**	0	N	0	N	N	N	N	8.5	N	0	0	N	N	0
910	**Nutmeg**, *ground*	0	(60)	0	N	0.35	0.06	1.3	N	N	0	0	N	N	0
911	**Paprika**	0	36250	0	N	0.65	1.74	15.3	3.1	N	0	0	N	N	0
912	**Parsley**, *fresh*	0	4040	0	1.70	0.23	0.06	1.0	0.5	0.09	0	170	0.30	0.4	190
913	**Pepper**, black	0	(115)	0	N	0.11	0.24	1.1	N	N	0	0	N	N	0
914	white	0	Tr	0	N	0.02	0.13	0.2	N	N	0	0	N	N	0
915	**Rosemary**, *dried*	0	(1880)	0	N	N	N	1.0	N	N	0	0	N	N	0
916	**Sage**, *dried, ground*	0	(3540)	0	N	N	0.34	5.7	N	N	0	0	N	N	0
917	**Thyme**, *dried, ground*	0	(2280)	0	N	N	0.40	4.9	3.1	N	0	0	N	N	0

Section 2.9

Fruit

The data in this section of the Tables have been taken from the *Fruit and Nuts* (1992) supplement.

Because much of the fruit eaten in this country is imported, a larger number of literature values from foreign sources have been used in this food group than many others in the Tables.

In general the word 'raw' has not been included in the food name unless there is a processed or cooked version of the same food. The description 'whole' refers to fruit with skin and pips, but excluding any inedible stone.

The nutrient content of fruit samples can vary widely, the variation often being greater within the same fruit type than between different varieties of fruit.

During the process of stewing fruit, sucrose becomes inverted into glucose and fructose, the extent depending on the length of cooking time and level of acidity. A factor of 10% hydrolysis of sucrose has been applied to all stewed fruit. The nutrient values for stewed fruits have been derived from both analyses and calculation. The proportions of sugar used for cooking and the method of calculation of the data have been included in the description of the food and are those for average consumption of sugar. However, for fruit cooked with a different proportion of sugar, the values for fruit 'stewed without sugar' can be used, with the appropriate quantity of sugar added. Corrections have been made for both vitamin losses (see Section 4.3 for the factors used) and evaporative losses of 10% during stewing.

Values for canned fruit include either syrup or juice, unless it is stated that the contents have been drained. It has been found by analysis that sugar diffuses between the syrup or juice and the fruit until it reaches an equilibrium, so that there are no significant differences between the levels of sugars in the fruit and the syrup or juice. One study found that the only significant differences between the fruit and its canning liquid were that the fruit contained higher levels of carotenoids and fibre.

Users should note that all values are expressed per 100g edible portion. Guidance for calculating nutrient content 'as purchased' or 'as served' (e.g. including citrus rind and inedible stones) is given in Section 4.2.

Taxonomic names for foods included in this part of the Tables can be found in Section 4.5.

Composition of food per 100g edible portion

No.	Food	Description and main data sources	Edible conversion factor	Water g	Total nitrogen g	Protein g	Fat g	Carbo-hydrate g	Energy value kcal	Energy value kJ
	Fruit, general									
918	**Apples**, cooking, *raw, peeled*	Bramley variety; flesh only	0.73	87.7	0.05	0.3	0.1	8.9	35	151
919	*stewed with sugar*	Samples as raw. 1000g fruit, 100g water, 120g sugar	1.00	77.7	0.05	0.3	0.1	19.1	74	314
920	*stewed without sugar*	Samples as *raw*. 1000g fruit, 100g water and calculation from No. 920	1.00	87.5	0.04	0.3	0.1	8.1	33	138
921	eating, average, *raw*	15 varieties; flesh and skin	0.89	84.5	0.06	0.4	0.1	11.8	47	199
922	*raw, peeled*	Literature sources and calculation from No. 921; flesh only	0.76	85.4	0.06	0.4	0.1	11.2	45	190
923	**Apricots**, *raw*	18 samples; flesh and skin	0.92	87.2	0.14	0.9	0.1	7.2	31	134
924	ready-to-eat	10 samples, no stones; semi-dried	1.00	29.7	0.63	4.0	0.6	36.5	158	674
925	canned in juice	10 samples, 5 brands. Drained proportion = 0.64	1.00	87.5	0.08	0.5	0.1	8.4	34	147
926	canned in syrup	10 samples, 9 brands. Drained proportion = 0.64	1.00	80.0	0.07	0.4	0.1	16.1	63	268
927	**Avocado**, average	Average of Fuerte and Hass varieties	0.71	72.5[a]	0.30	1.9	19.5[b]	1.9[c]	190	784
928	**Bananas**	10 samples; flesh only	0.66	75.1	0.19	1.2	0.3	23.2	95	403
929	**Blackberries**, *raw*	Cultivated and wild berries; whole fruit	1.00	85.0	0.14	0.9	0.2	5.1	25	104
930	*stewed with sugar*	Calculated from 700g fruit, 210g water, 84g sugar	1.00	78.9	0.11	0.7	0.2	13.8	56	239
931	**Blackcurrants**, *raw*	Whole fruit, stalks removed	0.98	77.4	0.15	0.9	Tr	6.6	28	121
932	*stewed with sugar*	Calculated from 700g fruit, 210g water, 84g sugar	1.00	72.9	0.12	0.7	Tr	15.0	58	252

[a] Water can range from 50 to 80g per 100g
[b] Fat can range from 10 to 40g per 100g
[c] Including mannoheptulose

Composition of food per 100g edible portion

No.	Food	Starch	Total sugars	Individual sugars					Dietary fibre	Fatty acids			Cholest-erol
				Gluc	Fruct	Sucr	Malt	Lact	NSP	Satd	Mono-unsatd	Poly-unsatd	
		g	g	g	g	g	g	g	g	g	g	g	mg
	Fruit, general												
918	**Apples**, cooking, *raw, peeled*	Tr	8.9	2.0	5.9	1.0	0	0	1.6	Tr	Tr	0.1	0
919	*stewed with sugar*	Tr	19.1	2.8	6.3	10.1	0	0	1.2	Tr	Tr	0.1	0
920	*stewed without sugar*	Tr	8.1	1.8	5.5	0.8	0	0	1.5	Tr	Tr	0.1	0
921	eating, average, *raw*	Tr	11.8	1.7	6.2	3.9	0	0	1.8	Tr	Tr	0.1	0
922	*raw, peeled*	Tr	11.2	1.6	5.9	3.7	0	0	1.6	Tr	Tr	0.1	0
923	**Apricots**, *raw*	0	7.2	1.6	0.9	4.6	0	0	1.7	Tr	Tr	Tr	0
924	ready-to-eat	Tr	36.5	17.5	8.4	10.6	0	0	6.3	N	N	N	0
925	canned in juice	0	8.4	3.0	4.1	1.4	0	0	0.9	Tr	Tr	Tr	0
926	canned in syrup	0	16.1	6.7	5.8	3.7	0	0	0.9	Tr	Tr	Tr	0
927	**Avocado**, average	Tr	0.5[a]	0.3	0.1	0.1	0	0	3.4	4.1	12.1	2.2	0
928	**Bananas**	2.3[b]	20.9[b]	4.8	5.0	11.1	0	0	1.1	0.1	Tr	0.1	0
929	**Blackberries**, *raw*	0	5.1	2.5	2.6	Tr	0	0	3.1	Tr	0.1	0.1	0
930	*stewed with sugar*	0	13.8	2.5	2.5	8.9	0	0	2.4	Tr	0.1	0.1	0
931	**Blackcurrants**, *raw*	0	6.6	3.0	3.4	0.3	0	0	3.6	Tr	Tr	Tr	0
932	*stewed with sugar*	0	15.0	2.8	3.2	9.0	0	0	2.8	Tr	Tr	Tr	0

[a] Not including mannoheptulose

[b] These are proportions for yellow ripe bananas. The starch content falls and the sugar content rises on ripening

Inorganic constituents per 100g edible portion

No.	Food	mg										µg	
		Na	K	Ca	Mg	P	Fe	Cu	Zn	Cl	Mn	Se	I
	Fruit, general												
918	**Apples**, cooking, *raw, peeled*	2	88	4	3	7	0.1	0.02	Tr	2	Tr	Tr	Tr
919	*stewed with sugar*	4	140	4	3	7	0.1	0.02	Tr	2	Tr	Tr	Tr
920	*stewed without sugar*	4	150	4	3	8	0.1	0.02	Tr	2	Tr	Tr	Tr
921	eating, average, *raw*	3	120	4	5	11	0.1	0.02	0.1	Tr	0.1	Tr	Tr
922	*raw, peeled*	3	100	3	3	8	0.1	0.02	0.1	Tr	0.1	Tr	Tr
923	**Apricots**, *raw*	2	270	15	11	20	0.5	0.06	0.1	3	0.1	(1)	N
924	ready-to-eat	14	1380	73	43	82	3.4	0.35	0.5	29	0.3	(5)	N
925	canned in juice	5	170	21	7	12	0.4	0.03	0.1	2	Tr	Tr	7
926	canned in syrup	10	150	19	5	8	0.2	Tr	0.1	2	Tr	Tr	7
927	**Avocado**, average	6	450	11	25	39	0.4	0.19	0.4	6	0.2	Tr	2
928	**Bananas**	1	400	6	34	28	0.3	0.10	0.2	79	0.4	(1)	8
929	**Blackberries**, *raw*	2	160	41	23	31	0.7	0.11	0.2	22	1.4	Tr	N
930	*stewed with sugar*	1	130	32	17	24	0.5	0.09	0.2	17	1.1	Tr	N
931	**Blackcurrants**, *raw*	3	370	60	17	43	1.3	0.14	0.3	15	0.3	N	N
932	*stewed with sugar*	2	290	47	13	33	1.0	0.11	0.3	11	0.2	N	N

No.	Food	Retinol µg	Carotene µg	Vitamin D µg	Vitamin E mg	Thiamin mg	Riboflavin mg	Niacin mg	Trypt/60 mg	Vitamin B_6 mg	Vitamin B_{12} µg	Folate µg	Pantothenate mg	Biotin µg	Vitamin C mg
	***Fruit**, general*														
918	**Apples**, cooking, *raw, peeled*	0	(17)	0	0.27	0.04	0.02	0.1	0.1	0.06	0	5	Tr	1.2	14[a]
919	*stewed with sugar*	0	(14)	0	0.22	0.01	0.01	0.1	0.1	0.05	0	Tr	Tr	0.8	10[b]
920	*stewed without sugar*	0	(15)	0	0.25	0.01	0.01	0.1	Tr	0.05	0	Tr	Tr	0.9	11
921	eating, average, *raw*	0	18	0	0.59	0.03	0.02	0.1	0.1	0.06	0	1	Tr	1.2	6[c]
922	*raw, peeled*	0	17	0	0.27	0.03	0.02	0.1	0.1	0.06	0	1	Tr	1.1	4
923	**Apricots**, *raw*	0	405[d]	0	N	0.04	0.05	0.5	0.1	0.08	0	5	0.24	N	6
924	ready-to-eat	0	545	0	N	Tr	0.16	2.3	0.5	0.14	0	11	0.58	N	1
925	canned in juice	0	210	0	N	0.02	0.01	0.3	0.1	0.06	0	2	0.06	0.4	14
926	canned in syrup	0	810	0	N	0.01	0.01	0.3	0.1	(0.06)	0	2	(0.06)	(0.4)	5
927	**Avocado**, average	0	16	0	3.20	0.10	0.18	1.1	0.3	0.36	0	11	1.10	3.6	6
928	**Bananas**	0	21	0	0.27	0.04	0.06	0.7	0.2	0.29	0	14	0.36	2.6	11
929	**Blackberries**, *raw*	0	80	0	2.37	0.02	0.05	0.5	0.1	0.05	0	34	0.25	0.4	15
930	*stewed with sugar*	0	62	0	1.85	0.01	0.03	0.3	0.1	0.03	0	5	0.15	0.2	9
931	**Blackcurrants**, *raw*	0	100	0	1.00	0.03	0.06	0.3	0.1	0.08	0	N	0.40	2.4	200[e]
932	*stewed with sugar*	0	78	0	0.78	0.02	0.04	0.2	0.1	0.05	0	N	0.23	1.4	115

[a] Unpeeled cooking apples contain 20mg vitamin C per 100g
[b] Frozen apple slices, stewed with sugar, contain 12mg vitamin C per 100g
[c] Levels ranged from 3 to 20mg vitamin C per 100g
[d] Levels ranged from 200 to 3370µg carotene per 100g
[e] Levels ranged from 150 to 230mg vitamin C per 100g

Composition of food per 100g edible portion

No.	Food	Description and main data sources	Edible conversion factor	Water	Total nitrogen	Protein	Fat	Carbo-hydrate	Energy value	
				g	g	g	g	g	kcal	kJ
Fruit, general *continued*										
933	**Cherries**, *raw*	10 samples of black and red cherries; flesh and skin	0.83	82.8	0.14	0.9	0.1	11.5	48	203
934	canned in syrup	10 samples, red and black. Drained proportion = 0.61 with stones and 0.47 without stones	1.00	77.8	0.09	0.5	Tr	18.5	71	305
935	glace	10 samples, 8 brands; red and multicoloured	1.00	23.6	0.07	0.4	Tr	66.4	251	1069
936	**Cherry pie filling**	10 samples, 7 brands	1.00	75.8	0.07	0.4	Tr	21.5	82	351
937	**Clementines**	10 samples; flesh only	0.75	87.5	0.14	0.9	0.1	8.7	37	158
938	**Currants**	10 samples, 9 brands	1.00	15.7	0.37	2.3	0.4	67.8	267	1139
939	**Damsons**, *raw*	Flesh and skin	0.90	77.5	0.08	0.5	Tr	9.6	38	162
940	*stewed with sugar*	Calculated from 1050g fruit, 210g water, 126g sugar	1.00	70.6	0.07	0.4	Tr	19.3	74	316
941	**Dates**, *raw*	5 samples; flesh and skin	0.86	60.7	0.24	1.5	0.1	31.3	124	530
942	*dried*	Flesh and skin	0.84	14.6	0.53	3.3	0.2	68.0	270	1151
943	**Dried mixed fruit**	Calculated from recipe proportions[a]	1.00	15.5	0.37	2.3	0.4	68.1	268	1144
944	**Figs**, *dried*	Analysis and literature sources; whole fruit	1.00	16.8	0.57	3.6	1.6	52.9	227	967
945	ready-to-eat	6 samples; semi-dried	1.00	23.6	0.52	3.3	1.5	48.6	209	889
946	**Fruit cocktail**, canned in juice	10 samples, 6 brands. Drained proportion = 0.65	1.00	86.9	0.07	0.4	Tr	7.2	29	122
947	canned in syrup	Analysis and calculation from recipe proportions. Drained proportion = 0.66[b]	1.00	81.8	0.06	0.4	Tr	14.8	57	244
948	**Fruit pie filling**	10 samples, 7 brands. Assorted flavours	1.00	79.5	0.06	0.4	Tr	20.1	77	328
949	**Fruit salad**, *homemade*	Recipe	1.00	81.8	0.11	0.7	0.1	14.8	60	253

[a] Calculated as sultanas 49%, currants 24%, raisins 18% and peel 9%

[b] Calculated as pears 42%, peaches 41%, pineapple 8%, grapes 5% and cherries 4%

No.	Food	Starch	Total sugars	Individual sugars					Dietary fibre	Fatty acids			Cholest-erol
				Gluc	Fruct	Sucr	Malt	Lact	NSP	Satd	Mono-unsatd	Poly-unsatd	
		g	g	g	g	g	g	g	g	g	g	g	mg
	***Fruit, general** continued*												
933	**Cherries**, *raw*	0	11.5	5.9	5.3	0.2	0	0	0.9	Tr	Tr	Tr	0
934	canned in syrup	0	18.5	7.3	6.6	4.3	0.3	0	0.6	Tr	Tr	Tr	0
935	glace	0	66.4	23.6	12.7	9.5	20.7	0	0.9	Tr	Tr	Tr	0
936	**Cherry pie filling**	3.9	17.6	7.2	6.4	3.9	0.1	0	0.4	Tr	Tr	Tr	0
937	**Clementines**	0	8.7	1.5	1.7	5.6	0	0	1.2	Tr	Tr	Tr	0
938	**Currants**	0	67.8	34.4	33.3	Tr	0	0	1.9	N	N	N	0
939	**Damsons**, *raw*	0	9.6	5.2	3.4	1.0	0	0	(1.8)	Tr	Tr	Tr	0
940	*stewed with sugar*	0	19.3	4.9	3.4	11.1	0	0	(1.5)	Tr	Tr	Tr	0
941	**Dates**, *raw*	0	31.3	16.2	15.1	Tr	0	0	1.8	Tr	Tr	Tr	0
942	*dried*	0	68.0	(35.4)	(32.6)	Tr	0	0	4.0	0.1	0.1	Tr	0
943	**Dried mixed fruit**	0	68.1	33.3	31.6	0.8	2.3	0	2.2	N	N	N	0
944	**Figs**, *dried*	0	52.9	28.6	22.7	1.6	0	0	7.5	N	N	N	0
945	ready-to-eat	0	48.6	26.2	20.8	1.5	0	0	6.9	N	N	N	0
946	**Fruit cocktail**, canned in juice	0	7.2	3.2	3.5	0.5	0	0	1.0	Tr	Tr	Tr	0
947	canned in syrup	0	14.8	6.1	6.4	1.9	0.3	0	1.0	Tr	Tr	Tr	0
948	**Fruit pie filling**	5.5	14.6	5.2	5.5	3.9	0	0	1.0	N	N	N	0
949	**Fruit salad**, *homemade*	0.3	14.4	2.8	4.1	7.6	0	0	1.3	Tr	Tr	Tr	0

Inorganic constituents per 100g edible portion

No.	Food	mg										µg	
		Na	K	Ca	Mg	P	Fe	Cu	Zn	Cl	Mn	Se	I
	Fruit, general *continued*												
933	**Cherries**, *raw*	1	210	13	10	21	0.2	0.07	0.1	Tr	0.1	(1)	Tr
934	canned in syrup	8	120	15	7	13	2.9	Tr	Tr	N	0.1	(1)	N[a]
935	glace	27	24	56	5	9	0.9	0.08	0.1	N	Tr	Tr	N[a]
936	**Cherry pie filling**	30	75	28	5	17	2.6	Tr	Tr	N	0.1	N	N
937	**Clementines**	4	130	31	10	18	0.1	0.01	0.1	(2)	Tr	N	N
938	**Currants**	14	720	93	30	71	1.3	0.81	0.3	16	0.7	N	N
939	**Damsons**, *raw*	2	290	24	11	16	0.4	0.08	(0.1)	Tr	N	Tr	N
940	*stewed with sugar*	1	240	19	9	13	0.3	0.07	(0.1)	Tr	N	Tr	N
941	**Dates**, *raw*	7	410	24	24	28	0.3	0.12	0.2	210	0.2	(1)	N
942	*dried*	10	700	45	41	60	1.3	0.26	0.4	370	0.3	(3)	N
943	**Dried mixed fruit**	48	880	73	29	73	2.2	0.47	0.4	13	0.4	N	N
944	**Figs**, *dried*	62	970	250	80	89	4.2	0.30	0.7	170	0.5	Tr	N
945	ready-to-eat	57	890	230	73	82	3.9	0.27	0.6	160	0.5	Tr	N
946	**Fruit cocktail**, canned in juice	3	95	9	7	14	0.4	0.04	0.1	2	0.1	Tr	N[a]
947	canned in syrup	3	95	5	5	9	0.3	0.02	0.1	3	0.1	Tr	N[a]
948	**Fruit pie filling**	43	84	30	5	15	1.0	0.02	Tr	45	0.1	Tr	N
949	**Fruit salad**, *homemade*	3	175	17	11	17	0.2	0.06	0.1	16	0.1	(1)	N

[a] Iodine from erythrosine is present but largely unavailable

No.	Food	Retinol µg	Carotene µg	Vitamin D µg	Vitamin E mg	Thiamin mg	Riboflavin mg	Niacin mg	Trypt/60 mg	Vitamin B_6 mg	Vitamin B_{12} µg	Folate µg	Pantothenate mg	Biotin µg	Vitamin C mg
	Fruit, general *continued*														
933	**Cherries**, *raw*	0	25	0	0.13	0.03	0.03	0.2	0.1	0.05	0	5	0.26	0.4	11
934	canned in syrup	0	17	0	(0.06)	0.02	0.01	0.1	Tr	(0.22)	0	5	(0.08)	(0.1)	1
935	glace	0	7	0	Tr	Tr	Tr	Tr	Tr	Tr	0	Tr	Tr	Tr	Tr
936	**Cherry pie filling**	0	18	0	N	0.02	0.01	0.2	0.1	N	0	2	N	N	1
937	**Clementines**	0	75	0	N	0.09	0.04	0.3	0.1	(0.07)	0	18	(0.20)	N	54
938	**Currants**	0	6	0	N	0.16	0.05	0.9	0.2	0.23	0	4	0.07	4.8	Tr
939	**Damsons**, *raw*	0	(295)	0	0.70	0.10	0.03	0.3	0.1	(0.05)	0	(3)	0.27	0.1	(5)
940	*stewed with sugar*	0	(240)	0	0.57	0.06	0.02	0.2	0.1	(0.03)	0	Tr	0.17	0.1	(3)
941	**Dates**, *raw*	0	(18)	0	N	0.06	0.07	0.7	0.7	0.12	0	25	0.21	N	14
942	*dried*	0	(40)	0	N	0.07	0.09	1.8	1.5	0.19	0	13	0.78	N	Tr
943	**Dried mixed fruit**	0	9	0	N	0.10	0.05	0.7	0.2	0.22	0	15	0.09	3.9	Tr
944	**Figs**, *dried*	0	(64)	0	N	0.08	0.10	0.8	0.5	0.26	0	9	0.51	N	1
945	ready-to-eat	0	(59)	0	N	0.07	0.09	0.7	0.4	0.24	0	8	0.47	N	1
946	**Fruit cocktail**, canned in juice	0	54	0	N	0.01	0.01	0.3	0.1	0.04	0	6	0.05	0.3	14
947	canned in syrup	0	(54)	0	N	0.02	0.01	0.4	0.1	0.03	0	5	0.05	0.1	4
948	**Fruit pie filling**	0	17	0	N	0.01	0.01	0.2	0.1	N	0	3	N	N	7
949	**Fruit salad**, *homemade*	0	23	0	N	0.05	0.03	0.3	0.1	0.10	0	N	N	N	27

Composition of food per 100g edible portion

No.	Food	Description and main data sources	Edible conversion factor	Water	Total nitrogen	Protein	Fat	Carbo-hydrate	Energy value	
				g	g	g	g	g	kcal	kJ
Fruit, general *continued*										
950	**Gooseberries**, cooking, *raw*	Tops and tails removed	0.91	90.1	0.18	1.1	0.4	3.0	19	81
951	*stewed with sugar*	1000g fruit, 150g water, 120g sugar	1.00	82.1	0.11	0.7	0.3	12.9	54	229
952	**Grapefruit**, *raw*	10 samples; flesh only	0.68	89.0	0.13	0.8	0.1	6.8	30	126
953	canned in juice	10 samples, 8 brands. Drained proportion = 0.52	1.00	88.6	0.09	0.6	Tr	7.3	30	120
954	canned in syrup	10 samples. Drained proportion = 0.52	1.00	81.8	0.08	0.5	Tr	15.5	60	257
955	**Grapes**, average	10 samples, white, black and seedless[a]	0.95	81.8	0.06	0.4	0.1	15.4	60	257
956	**Guava**, *raw*	Literature sources	0.90	84.7	0.13	0.8	0.5	5.0	26	112
957	canned in syrup	10 samples. Drained proportion = 0.62	1.00	77.6	0.06	0.4	Tr	15.7	60	258
958	**Kiwi fruit**	Analysis and literature sources, flesh and seeds	0.86	84.0	0.18	1.1	0.5	10.6	49	207
959	**Lemon peel**	Ref. 3	1.00	81.6	0.24	1.5	0.3	N	N	N
960	**Lemons**, *whole, without pips*	Analysis and literature sources; includes peel but no pips	0.99	86.3	0.16	1.0	0.3	3.2	19	79
961	**Lychees**, *raw*	Analysis and literature sources; flesh only	0.62	81.1	0.14	0.9	0.1	14.3	58	248
962	canned in syrup	Analysis and literature sources. Drained proportion = 0.50	1.00	79.3	0.06	0.4	Tr	17.7	68	290
963	**Mandarin oranges**, canned in juice	10 samples, 4 brands. Drained proportion = 0.56	1.00	89.6	0.11	0.7	Tr	7.7	32	135
964	canned in syrup	10 samples, 10 brands. Drained proportion = 0.56	1.00	84.8	0.08	0.5	Tr	13.4	52	223
965	**Mangoes**, *ripe, raw*	Literature sources; flesh only	0.68	82.4	0.11	0.7	0.2	14.1	57	245

[a] Few significant differences reported between varieties

Composition of food per 100g edible portion

No.	Food	Starch	Total sugars	Individual sugars					Dietary fibre	Fatty acids			Cholest-erol
				Gluc	Fruct	Sucr	Malt	Lact	NSP	Satd	Mono-unsatd	Poly-unsatd	
		g	g	g	g	g	g	g	g	g	g	g	mg
	Fruit, general *continued*												
950	**Gooseberries**, cooking, *raw*	0	3.0	1.3	1.6	Tr	0	0	2.4	N	N	N	0
951	*stewed with sugar*	0	12.9	2.4	2.6	7.8	0	0	1.9	N	N	N	0
952	**Grapefruit**, *raw*	0	6.8	2.1	2.3	2.4	0	0	1.3	Tr	Tr	Tr	0
953	canned in juice	0	7.3	3.6	3.4	0.3	0	0	0.4	Tr	Tr	Tr	0
954	canned in syrup	0	15.5	6.7	6.9	1.9	0	0	0.6	Tr	Tr	Tr	0
955	**Grapes**, average	0	15.4	7.6	7.8	0.1	0	0	0.7	Tr	Tr	Tr	0
956	**Guava**, *raw*	0.1	4.9	2.1	2.3	0.5	0	0	3.7	N	N	N	0
957	canned in syrup	Tr	15.7	5.5	6.5	3.7	0	0	3.0	Tr	Tr	Tr	0
958	**Kiwi fruit**	0.3	10.3	4.6	4.3	1.3	0	0	1.9	N	N	N	0
959	**Lemon peel**	0	N	N	N	N	0	0	N	0.1	Tr	0.1	0
960	**Lemons**, *whole, without pips*	0	3.2	1.4	1.4	0.4	0	0	N	0.1	Tr	0.1	0
961	**Lychees**, *raw*	0	14.3	7.0	7.3	Tr	0	0	0.7	Tr	Tr	Tr	0
962	canned in syrup	0	17.7	8.5	8.6	0.6	0	0	0.5	Tr	Tr	Tr	0
963	**Mandarin oranges**, canned in juice	0	7.7	2.8	3.1	1.8	0	0	0.3	Tr	Tr	Tr	0
964	canned in syrup	0	13.4	4.1	4.2	5.1	0	0	0.2	Tr	Tr	Tr	0
965	**Mangoes**, *ripe, raw*	0.3	13.8	0.7	3.0	10.1	0	0	2.6	0.1	Tr	Tr	0

No.	Food	mg										µg	
		Na	K	Ca	Mg	P	Fe	Cu	Zn	Cl	Mn	Se	I
	Fruit, general *continued*												
950	**Gooseberries**, cooking, *raw*	2	210	28	7	34	0.3	0.06	0.1	7	0.1	Tr	Tr
951	*stewed with sugar*	7	140	19	6	22	0.3	0.07	0.1	5	0.3	Tr	Tr
952	**Grapefruit**, *raw*	3	200	23	9	20	0.1	0.02	Tr	3	Tr	(1)	N
953	canned in juice	10	72	22	8	16	0.3	0.01	Tr	(5)	Tr	Tr	N
954	canned in syrup	10	79	17	7	13	0.7	(0.01)	0.4	(5)	Tr	Tr	N
955	**Grapes**, average	2	210	13	7	18	0.3	0.12	0.1	Tr	0.1	(1)	1
956	**Guava**, *raw*	5	230	13	12	25	0.4	0.10	0.2	4	0.1	N	N
957	canned in syrup	7	120	8	6	11	0.5	0.10	0.4	10	N	N	N
958	**Kiwi fruit**	4	290	25	15	32	0.4	0.13	0.1	39	0.1	N	N
959	**Lemon peel**	6	160	130	15	12	0.8	N	N	N	N	N	N
960	**Lemons**, *whole, without pips*	5	150	85	12	18	0.5	0.26	0.1	5	N	(1)	N
961	**Lychees**, *raw*	2	75	4	6	12	0.7	0.11	0.2	(5)	N	N	N
962	canned in syrup	1	160	6	9	30	0.5	0.15	0.3	3	0.1	N	N
963	**Mandarin oranges**, canned in juice	6	85	17	9	13	0.5	Tr	0.1	2	Tr	Tr	Tr
964	canned in syrup	6	49	17	7	8	0.2	Tr	Tr	2	Tr	Tr	Tr
965	**Mangoes**, *ripe, raw*	2	180	12	13	16	0.7	0.12	0.1	N	0.3	N	N

No.	Food	Retinol µg	Carotene µg	Vitamin D µg	Vitamin E mg	Thiamin mg	Ribo-flavin mg	Niacin mg	Trypt/60 mg	Vitamin B_6 mg	Vitamin B_{12} µg	Folate µg	Panto-thenate mg	Biotin µg	Vitamin C mg
	Fruit, general *continued*														
950	**Gooseberries**, cooking, *raw*	0	110	0	0.37	0.03	0.03	0.3	0.2	0.02	0	(8)	0.29	0.5	14
951	*stewed with sugar*	0	41	0	0.29	0.01	0.02	0.2	0.1	0.01	0	6	0.17	0.3	11
952	**Grapefruit**, *raw*	0	17[a]	0	(0.19)	0.05	0.02	0.3	0.1	0.03	0	26	0.28	(1.0)	36
953	canned in juice	0	Tr	0	(0.10)	0.04	0.01	0.3	0.1	(0.02)	0	6	(0.12)	(1.0)	33
954	canned in syrup	0	Tr	0	(0.11)	0.04	0.01	0.2	0.1	0.02	0	4	0.12	1.0	30
955	**Grapes**, average	0	17	0	Tr	0.05	0.01	0.2	Tr	0.10	0	2	0.05	0.3	3
956	**Guava**, *raw*	0	435[b]	0	N	0.04	0.04	1.0	0.1	0.14	0	N	0.15	N	230[c]
957	canned in syrup	0	(145)	0	N	(0.02)	(0.02)	(0.6)	0.1	(0.09)	0	N	(0.09)	N	180
958	**Kiwi fruit**	0	40	0	N	0.01	0.03	0.3	0.3	0.15	0	N	N	N	59
959	**Lemon peel**	0	30	0	N	0.06	0.08	0.4	0.2	0.17	0	N	0.32	N	130
960	**Lemons**, *whole, without pips*	0	18	0	N	0.05	0.04	0.2	0.1	0.11	0	N	0.23	0.5	58
961	**Lychees**, *raw*	0	0	0	N	Tr	0.04	Tr	0.1	N	0	N	N	N	8
962	canned in syrup	0	0	0	N	0.04	0.06	0.5	0.1	N	0	N	N	N	45
963	**Mandarin oranges**, canned in juice	0	95	0	Tr	0.08	0.01	0.2	0.1	(0.03)	0	12	(0.15)	(0.8)	20
964	canned in syrup	0	105	0	Tr	0.06	0.01	0.2	Tr	0.03	0	12	(0.15)	(0.8)	15
965	**Mangoes**, *ripe, raw*	0	696	0	1.05	0.04	0.05	0.5	1.3	0.13	0	N	0.16	N	37

[a] Pink varieties contain approximately 770µg carotene per 100g

[b] Peel included on analysis

[c] Levels ranged from 9 to 410mg vitamin C per 100g

Composition of food per 100g edible portion

No.	Food	Description and main data sources	Edible conversion factor	Water	Total nitrogen	Protein	Fat	Carbo-hydrate	Energy value	
				g	g	g	g	g	kcal	kJ
	***Fruit, general** continued*									
966	**Melon**, Canteloupe-type	10 samples, Canteloupe, Charantais and Rock; flesh only	0.59	92.1	0.10	0.6	0.1	4.2	19	81
967	Galia	11 samples; flesh only	0.64	91.7	0.08	0.5	0.1	5.6	24	102
968	Honeydew	10 samples; flesh only	0.63	92.2	0.10	0.6	0.1	6.6	28	119
969	watermelon	Literature sources; flesh only	0.57	92.3	0.07	0.5	0.3	7.1	31	133
970	**Mixed peel**	10 samples, 9 brands	1.00	20.9	0.05	0.3	0.9	59.1	231	984
971	**Nectarines**	10 samples; flesh and skin	0.89	88.9	0.22	1.4	0.1	9.0	40	171
972	**Olives**, in brine	Bottled, drained; flesh and skin, green	0.80	76.5	0.14	0.9	11.0	Tr	103	422
973	**Oranges**	Assorted varieties; flesh only	0.70[a]	86.1	0.18	1.1	0.1	8.5	37	158
974	**Passion fruit**	Analysis and literature sources; flesh and pips	0.61	74.9	0.45	2.6	0.4	5.8	36	152
975	**Paw-paw**, *raw*	Literature sources; flesh only	0.75	88.5	0.08	0.5	0.1	8.8	36	153
976	canned in juice	10 samples. Drained proportion = 0.59	1.00	80.4	0.03	0.2	Tr	17.0	65	275
977	**Peaches**, *raw*	10 samples; flesh and skin	0.90	88.9	0.16	1.0	0.1	7.6	33	142
978	canned in juice	10 samples, 7 brands; halves and slices. Drained proportion = 0.68	1.00	86.7	0.09	0.6	Tr	9.7	39	165
979	canned in syrup	10 samples, 9 brands; halves and slices. Drained proportion = 0.62	1.00	81.1	0.08	0.5	Tr	14.0	55	233

[a] Levels ranged from 0.60 to 0.74

No.	Food	Starch	Total sugars	Individual sugars					Dietary fibre	Fatty acids			Cholest-erol
				Gluc	Fruct	Sucr	Malt	Lact	NSP	Satd	Mono-unsatd	Poly-unsatd	
		g	g	g	g	g	g	g	g	g	g	g	mg
	***Fruit, general** continued*												
966	**Melon**, Canteloupe-type	0	4.2	1.8	2.2	0.1	0	0	1.0	Tr	Tr	Tr	0
967	Galia	0	5.6	1.6	2.0	2.0	0	0	0.4	Tr	Tr	Tr	0
968	Honeydew	0	6.6	2.8	3.2	0.6	0	0	0.6	Tr	Tr	Tr	0
969	watermelon	0	7.1	1.3	2.3	3.4	0	0	0.1	0.1	0.1	0.1	0
970	**Mixed peel**	0	59.1	19.9	4.5	9.1	25.6	0	4.8	N	N	N	0
971	**Nectarines**	0	9.0	1.3	1.3	6.3	0	0	1.2	Tr	Tr	Tr	0
972	**Olives**, in brine	0	Tr	Tr	Tr	Tr	0	0	2.9	1.7	5.7	1.3	0
973	**Oranges**	0	8.5	2.2	2.4	3.9	0	0	1.7	Tr	Tr	Tr	0
974	**Passion fruit**	0	5.8	2.2	1.9	1.7	0	0	3.3	0.1	0.1	0.1	0
975	**Paw-paw**, *raw*	0	8.8	2.8	2.8	3.1	0	0	2.2	Tr	Tr	Tr	0
976	canned in juice	0	17.0	7.8	7.0	2.2	0	0	0.7	Tr	Tr	Tr	0
977	**Peaches**, *raw*	0	7.6	1.1	1.1	5.2	0	0	1.5	Tr	Tr	Tr	0
978	canned in juice	0	9.7	2.4	3.7	3.6	0	0	0.8	Tr	Tr	Tr	0
979	canned in syrup	0	14.0	3.7	3.6	6.7	0	0	0.9	Tr	Tr	Tr	0

Inorganic constituents per 100g edible portion

No.	Food	mg										µg	
		Na	K	Ca	Mg	P	Fe	Cu	Zn	Cl	Mn	Se	I
	Fruit, general *continued*												
966	**Melon**, Canteloupe-type	8	210	20	11	13	0.3	Tr	Tr	44	Tr	Tr	(4)
967	Galia	31	150	13	12	10	0.2	Tr	0.1	75	Tr	Tr	N
968	Honeydew	32	210	9	10	16	0.1	Tr	Tr	45	Tr	Tr	N
969	watermelon	2	100	7	8	9	0.3	0.03	0.2	N	Tr	Tr	Tr
970	**Mixed peel**	280	21	130	12	6	1.3	0.15	0.2	N	0.1	N	N
971	**Nectarines**	1	170	7	10	22	0.4	0.06	0.1	5	0.1	(1)	3
972	**Olives**, in brine	2250	91	61	22	17	1.0	0.23	N	3750	N	N	N
973	**Oranges**	5	150	47	10	21	0.1	0.05	0.1	3	Tr	(1)	2
974	**Passion fruit**	19	200	11	29	64	1.3	N	0.8	N	N	N	N
975	**Paw-paw**, *raw*	5	200	23	11	13	0.5	0.08	0.2	11	0.1	N	N
976	canned in juice	8	110	23	8	6	0.4	0.10	0.3	40	N	N	N
977	**Peaches**, *raw*	1	160	7	9	22	0.4	0.06	0.1	Tr	0.1	(1)	3
978	canned in juice	12	170	4	7	19	0.4	0.04	0.1	(4)	0.1	Tr	N
979	canned in syrup	4	110	3	5	11	0.2	Tr	Tr	4	Tr	Tr	N

Vitamins per 100g edible portion

No.	Food	Retinol µg	Carotene µg	Vitamin D µg	Vitamin E mg	Thiamin mg	Riboflavin mg	Niacin mg	Trypt/60 mg	Vitamin B_6 mg	Vitamin B_{12} µg	Folate µg	Pantothenate mg	Biotin µg	Vitamin C mg
	Fruit, general *continued*														
966	**Melon**, Canteloupe-type	0	1765	0	0.10	0.04	0.02	0.6	Tr	0.11	0	5	0.13	N	26
967	Galia	0	N	0	(0.10)	(0.03)	(0.01)	(0.4)	Tr	(0.09)	0	(3)	(0.17)	N	15
968	Honeydew	0	48	0	0.10	0.03	0.01	0.3	Tr	0.06	0	2	0.21	N	9
969	watermelon	0	116	0	(0.10)	0.05	0.01	0.1	Tr	0.14	0	2	0.21	1.0	8
970	**Mixed peel**	0	Tr	0	N	N	N	N	0.1	N	0	N	N	N	Tr
971	**Nectarines**	0	114	0	N	0.02	0.04	0.6	0.3	0.03	0	Tr	0.16	(0.2)	37
972	**Olives**, in brine	0	180[a]	0	1.99	Tr	Tr	Tr	0.1	0.02	0	Tr	0.02	Tr	0
973	**Oranges**	0	47[b]	0	0.24	0.11	0.04	0.4	0.1	0.10	0	31	0.37	1.0	54[c]
974	**Passion fruit**	0	750	0	N	0.03	0.12	1.5	0.4	N	0	N	N	N	23
975	**Paw-paw**, *raw*	0	810	0	N	0.03	0.04	0.3	0.1	0.03	0	1	0.22	N	60
976	canned in juice	0	(255)	0	N	0.02	0.02	0.2	Tr	(0.01)	0	Tr	(0.20)	N	15
977	**Peaches**, *raw*	0	114	0	N	0.02	0.04	0.6	0.2	0.02	0	3	0.17	(0.2)	31
978	canned in juice	0	67	0	N	0.01	0.01	0.6	0.1	0.02	0	2	0.06	0.2	6
979	canned in syrup	0	75	0	N	0.01	0.01	0.6	0.1	0.02	0	7	0.05	0.1	5

[a] Values for green olives. Ripe black olives contain 40µg carotene per 100g

[b] Blood oranges have been found to contain 155µg carotene per 100g

[c] Levels ranged from 44 to 79mg vitamin C per 100g

Composition of food per 100g edible portion

No.	Food	Description and main data sources	Edible conversion factor	Water	Total nitrogen	Protein	Fat	Carbo-hydrate	Energy value	
				g	g	g	g	g	kcal	kJ
Fruit, general *continued*										
980	**Pears**, average, *raw*	Average of Comice, Conference and Williams varieties; flesh and skin	0.91	83.8	0.05	0.3	0.1	10.0	40	169
981	*raw, peeled*	Literature sources and calculation from No.980; flesh only	0.70	83.8	0.05	0.3	0.1	10.4	41	175
982	canned in juice	10 samples, 7 brands. Drained proportion = 0.60	1.00	86.8	0.04	0.3	Tr	8.5	33	141
983	canned in syrup	10 samples, 8 brands. Drained proportion = 0.61	1.00	82.6	0.04	0.2	Tr	13.2	50	215
984	**Pineapple**, *raw*	10 samples; flesh only	0.53	86.5	0.06	0.4	0.2	10.1	41	176
985	canned in juice	10 samples, 10 brands; cubes and slices. Drained proportion = 0.54	1.00	86.8	0.05	0.3	Tr	12.2	47	200
986	canned in syrup	10 samples, 10 brands; cubes and slices. Drained proportion = 0.56	1.00	82.2	0.08	0.5	Tr	16.5	64	273
987	**Plums**, average, *raw*	Assorted varieties; flesh and skin	0.94	83.9	0.09	0.6	0.1	8.8	36	155
988	average, *stewed with sugar*	1350g fruit, 100g water, 162g sugar; stones removed	0.95	74.2	0.08	0.5	0.1	20.2	79	335
989	canned in syrup	10 samples, 7 brands; Red, Golden and Victoria. Drained proportion = 0.45 (without stones)	1.00	81.4	0.05	0.3	Tr	15.5	59	253
990	**Prunes**, canned in juice	10 samples; stones removed.	0.93	74.1	0.12	0.7	0.2	19.7	79	335
991	canned in syrup	11 samples, 6 brands; stones removed.	0.92	69.9	0.10	0.6	0.2	23.0	90	386
992	ready-to-eat	4 samples; semi-dried	0.86	31.1	0.40	2.5	0.4	34.0	141	601
993	**Raisins**	10 samples, 8 brands. Large stoned variety	1.00	13.2	0.34	2.1	0.4	69.3	272	1159

No.	Food	Starch	Total sugars	Individual sugars					Dietary fibre	Fatty acids			Cholest-erol
				Gluc	Fruct	Sucr	Malt	Lact	NSP	Satd	Mono-unsatd	Poly-unsatd	
		g	g	g	g	g	g	g	g	g	g	g	mg
Fruit, general *continued*													
980	**Pears**, average, *raw*	0	10.0	2.3	7.1	0.7	0	0	2.2	Tr	Tr	Tr	0
981	*raw, peeled*	0	10.4	2.4	7.4	0.7	0	0	1.7	Tr	Tr	Tr	0
982	canned in juice	0	8.5	2.3	5.7	0.6	0	0	1.4	Tr	Tr	Tr	0
983	canned in syrup	0	13.2	3.4	6.1	3.4	0.2	0	1.1	Tr	Tr	Tr	0
984	**Pineapple**, *raw*	0	10.1	2.0	2.5	5.5	0	0	1.2	Tr	0.1	0.1	0
985	canned in juice	0	12.2	4.0	4.0	4.2	0	0	0.5	Tr	Tr	Tr	0
986	canned in syrup	0	16.5	6.0	4.8	5.8	0	0	0.7	Tr	Tr	Tr	0
987	**Plums**, average, *raw*	0	8.8	4.3	2.0	2.5	0	0	1.6	Tr	Tr	Tr	0
988	average, *stewed with sugar*	0	20.2	4.5	2.5	13.3	0	0	1.3	Tr	Tr	Tr	0
989	canned in syrup	0	15.5	7.1	6.2	2.2	0	0	0.8	Tr	Tr	Tr	0
990	**Prunes**, canned in juice	0	19.7	10.2	8.4	1.1	0	0	2.4	Tr	0.1	0.1	0
991	canned in syrup	0	23.0	11.0	5.5	6.5	0	0	2.8	Tr	0.1	0.1	0
992	ready-to-eat	0	34.0	17.9	12.1	4.1	0	0	5.7	N	N	N	0
993	**Raisins**	0	69.3	34.5	34.8	Tr	0	0	2.0	N	N	N	0

Inorganic constituents per 100g edible portion

No.	Food	mg										µg	
		Na	K	Ca	Mg	P	Fe	Cu	Zn	Cl	Mn	Se	I
	Fruit, general *continued*												
980	**Pears**, average, *raw*	3	150	11	7	13	0.2	0.06	0.1	1	Tr	Tr	1
981	*raw, peeled*	3	150	11	7	13	0.2	0.06	0.1	1	Tr	Tr	1
982	canned in juice	3	81	6	5	10	0.2	Tr	0.1	(3)	Tr	Tr	Tr
983	canned in syrup	3	68	6	4	7	0.2	0.02	0.1	3	Tr	Tr	Tr
984	**Pineapple**, *raw*	2	160	18	16	10	0.2	0.11	0.1	29	0.5	Tr	Tr
985	canned in juice	1	71	8	13	5	0.5	0.08	0.1	(4)	0.9	Tr	Tr
986	canned in syrup	2	79	6	11	5	0.2	0.02	0.1	4	0.9	Tr	Tr
987	**Plums**, average, *raw*	2	240	13	8	23	0.4	0.10	0.1	Tr	0.1	Tr	Tr
988	average, *stewed with sugar*	2	200	11	7	19	0.3	0.08	0.1	Tr	0.1	Tr	Tr
989	canned in syrup	6	79	9	4	10	N	Tr	Tr	N	Tr	Tr	Tr
990	**Prunes**, canned in juice	18	340	26	15	30	2.2	0.09	1.0	N	0.1	Tr	N
991	canned in syrup	(18)	(340)	(26)	(15)	(30)	(2.2)	(0.09)	(1.0)	N	(0.1)	Tr	N
992	ready-to-eat	11	760	34	24	73	2.6	0.14	0.4	3	0.3	3	N
993	**Raisins**	60	1020	46	35	76	3.8	0.39	0.7	9	0.3	(8)	N

No.	Food	Retinol µg	Carotene µg	Vitamin D µg	Vitamin E mg	Thiamin mg	Ribo-flavin mg	Niacin mg	Trypt/60 mg	Vitamin B_6 mg	Vitamin B_{12} µg	Folate µg	Panto-thenate mg	Biotin µg	Vitamin C mg
	***Fruit, general** continued*														
980	**Pears**, average, *raw*	0	18	0	0.50	0.02	0.03	0.2	Tr	0.02	0	2	0.07	0.2	6
981	*raw, peeled*	0	19	0	Tr	0.02	0.03	0.2	Tr	0.02	0	2	0.07	0.2	6
982	canned in juice	0	Tr	0	Tr	0.01	0.01	0.2	Tr	0.03	0	4	0.04	0.2	3
983	canned in syrup	0	Tr	0	Tr	0.01	0.01	0.2	Tr	0.03	0	3	0.04	0.2	2
984	**Pineapple**, *raw*	0	18	0	0.10	0.08	0.03	0.3	0.1	0.09	0	5	0.16	0.3	12
985	canned in juice	0	12	0	(0.05)	0.09	0.01	0.2	0.1	0.09	0	1	0.11	0.1	11
986	canned in syrup	0	11	0	0.06	0.07	0.01	0.2	0.1	0.07	0	(1)	0.07	0.1	13
987	**Plums**, average, *raw*	0	376	0	0.61	0.05	0.03	1.1	0.1	0.05	0	3	0.15	Tr	4
988	average, *stewed with sugar*	0	65	0	0.51	0.03	0.02	0.7	0.1	0.03	0	Tr	0.09	Tr	3
989	canned in syrup	0	29	0	0.25	0.01	0.01	0.3	Tr	(0.02)	0	Tr	(0.04)	Tr	1
990	**Prunes**, canned in juice	0	140	0	N	0.02	0.02	0.5	0.1	(0.06)	0	5	(0.07)	Tr	Tr
991	canned in syrup	0	(140)	0	N	(0.02)	(0.02)	(0.5)	0.1	(0.05)	0	(5)	(0.06)	Tr	Tr
992	ready-to-eat	0	140	0	N	0.09	0.18	1.3	0.4	0.21	0	3	0.41	Tr	Tr
993	**Raisins**	0	12	0	N	0.12	0.05	0.6	0.2	0.25	0	10	0.15	2.0	1

Composition of food per 100g edible portion

No.	Food	Description and main data sources	Edible conversion factor	Water	Total nitrogen	Protein	Fat	Carbo-hydrate	Energy value	
				g	g	g	g	g	kcal	kJ
	***Fruit, general** continued*									
994	**Raspberries**, *raw*	9 samples; whole fruit	1.00	87.0	0.22	1.4	0.3	4.6	25	109
995	canned in syrup	Mixed sample. Drained proportion = 0.52	1.00	74.0	0.10	0.6	0.1	22.5	88	374
996	**Rhubarb**, *raw*	Stems only	0.87	94.2	0.14	0.9	0.1	0.8	7	32
997	*stewed with sugar*	1000g fruit, 100g water, 120g sugar	1.00	84.6	0.14	0.9	0.1	11.5	48	203
998	canned in syrup	10 samples, 6 brands. Drained proportion = 0.56	1.00	90.6	0.08	0.5	Tr	7.6	31	130
999	**Satsumas**	10 samples; flesh only	0.71	87.4	0.14	0.9	0.1	8.5	36	155
1000	**Strawberries**, *raw*	9 samples; flesh and pips	0.95	89.5	0.13	0.8	0.1	6.0	27	113
1001	canned in syrup	10 samples. Drained proportion = 0.38	1.00	81.7	0.07	0.5	Tr	16.9	65	279
1002	**Sultanas**	10 samples, 9 brands; whole fruit	1.00	15.2	0.43	2.7	0.4	69.4	275	1171
1003	**Tangerines**	Flesh only	0.73	86.7	0.14	0.9	0.1	8.0	35	147

No.	Food	Starch	Total sugars	Individual sugars					Dietary fibre	Fatty acids			Cholest-erol
				Gluc	Fruct	Sucr	Malt	Lact	NSP	Satd	Mono-unsatd	Poly-unsatd	
		g	g	g	g	g	g	g	g	g	g	g	mg
	***Fruit, general** continued*												
994	**Raspberries**, *raw*	0	4.6	1.9	2.4	0.2	0	0	2.5	0.1	0.1	0.1	0
995	canned in syrup	0	22.5	N	N	N	0	0	1.5	Tr	Tr	Tr	0
996	**Rhubarb**, *raw*	0	0.8	0.4	0.4	0.1	0	0	1.4	Tr	Tr	Tr	0
997	*stewed with sugar*	0	11.5	1.2	1.2	9.1	0	0	1.2	Tr	Tr	Tr	0
998	canned in syrup	0	7.6	2.9	2.6	2.1	0	0	0.8	Tr	Tr	Tr	0
999	**Satsumas**	0	8.5	1.5	1.8	5.1	0	0	1.3	Tr	Tr	Tr	0
1000	**Strawberries**, *raw*	0	6.0	2.6	3.0	0.3	0	0	1.1	Tr	Tr	Tr	0
1001	canned in syrup	0	16.9	4.7	4.9	7.3	0	0	0.7	Tr	Tr	Tr	0
1002	**Sultanas**	0	69.4	34.8	34.6	Tr	0	0	2.0	N	N	N	0
1003	**Tangerines**	0	8.0	1.4	1.6	5.1	0	0	1.3	Tr	Tr	Tr	0

Inorganic constituents per 100g edible portion

No.	Food	Na (mg)	K (mg)	Ca (mg)	Mg (mg)	P (mg)	Fe (mg)	Cu (mg)	Zn (mg)	Cl (mg)	Mn (mg)	Se (µg)	I (µg)
	***Fruit, general** continued*												
994	**Raspberries**, *raw*	3	170	25	19	31	0.7	0.10	0.3	22	0.4	N	N
995	canned in syrup	4	100	14	11	14	1.7	0.10	N	5	0.3	N	N
996	**Rhubarb**, *raw*	3	290	93	13	17	0.3	0.07	0.1	87	0.2	Tr	N
997	*stewed with sugar*	1	210	33	6	18	0.1	0.02	Tr	75	0.3	Tr	N
998	canned in syrup	4	89	36	5	8	0.8	Tr	0.1	15	0.1	Tr	N
999	**Satsumas**	4	130	31	10	18	0.1	0.01	0.1	(2)	Tr	N	N
1000	**Strawberries**, *raw*	6	160	16	10	24	0.4	0.07	0.1	18	0.3	Tr	9
1001	canned in syrup	9	87	11	7	15	1.1	Tr	0.1	(5)	0.2	Tr	Na
1002	**Sultanas**	19	1060	64	31	86	2.2	0.40	0.3	16	0.3	N	N
1003	**Tangerines**	2	160	42	11	17	0.3	0.01	0.1	2	Tr	N	N

[a] Iodine from erythrosine is present but largely unavailable

Vitamins per 100g edible portion

No.	Food	Retinol µg	Carotene µg	Vitamin D µg	Vitamin E mg	Thiamin mg	Ribo-flavin mg	Niacin mg	Trypt/60 mg	Vitamin B_6 mg	Vitamin B_{12} µg	Folate µg	Panto-thenate mg	Biotin µg	Vitamin C mg
	Fruit, general *continued*														
994	**Raspberries**, *raw*	0	6	0	0.48	0.03	0.05	0.5	0.3	0.06	0	33	0.24	1.9	32
995	canned in syrup	0	3	0	0.15	0.01	0.03	0.3	0.1	0.04	0	(10)	0.17	(0.7)	7
996	**Rhubarb**, *raw*	0	60	0	0.20	0.03	0.03	0.3	0.1	0.02	0	7	0.09	N	6
997	*stewed with sugar*	0	28	0	0.17	0.03	0.02	0.2	0.1	0.02	0	4	0.08	N	5
998	canned in syrup	0	(18)	0	(0.11)	0.02	0.01	0.1	0.1	0.01	0	3	0.05	N	3
999	**Satsumas**	0	75	0	N	0.09	0.04	0.3	0.1	(0.07)	0	33	(0.20)	N	27
1000	**Strawberries**, *raw*	0	8	0	0.20	0.03	0.03	0.6	0.1	0.06	0	20	0.34	1.1	77
1001	canned in syrup	0	4	0	N	0.01	0.02	0.3	0.1	0.03	0	6	0.21	(1.0)	29
1002	**Sultanas**	0	12	0	0.70	0.09	0.05	0.8	0.2	0.25	0	27	0.09	4.8	Tr
1003	**Tangerines**	0	97	0	N	0.07	0.02	0.2	0.1	0.07	0	21	0.20	N	30

Section 2.10

Nuts

The data in this section of the Tables have been taken from the *Fruit and Nuts* (1992) supplement.

Users should note that all values are expressed per 100g edible portion. Guidance for calculating nutrient content 'as purchased' or 'as served' (e.g. including shells) is given in Section 4.2.

Cooked foods including nuts are not included in this Section. Nut roast is included under Vegetable dishes.

Taxonomic names for foods in this part of the Tables can be found in Section 4.5.

Composition of food per 100g edible portion

No.	Food	Description and main data sources	Edible conversion factor	Water	Total nitrogen	Protein	Fat	Carbo-hydrate	Energy value	
				g	g	g	g	g	kcal	kJ
	Nuts and seeds, general									
1004	**Almonds**	10 blanched samples, flaked and ground	0.37	4.2	4.07	21.1	55.8	6.9	612	2534
1005	**Brazil nuts**	10 samples, kernel only	0.46	2.8	2.61	14.1	68.2	3.1	682	2813
1006	**Cashew nuts**, *roasted and salted*	10 samples, kernel only	1.00	2.4	3.87	20.5	50.9	18.8	611	2533
1007	**Chestnuts**	Analysis and literature sources; kernel only	0.83	51.7	0.37	2.0	2.7	36.6	170	719
1008	**Coconut**, *creamed block*	7 samples, 2 brands; block of dried kernel	1.00	2.5	1.14	6.0	68.8	7.0	669	2760
1009	*desiccated*	Analytical and literature sources	1.00	2.3	1.05	5.6	62.0	6.4	604	2492
1010	**Coconut milk**	Analysis and literature sources; drained fluid from fresh coconut	1.00	92.2	0.06	0.3	0.3	4.9	22	95
1011	**Hazelnuts**	10 samples, kernel only	0.38	4.6	2.66	14.1	63.5	6.0	650	2685
1012	**Macadamia nuts**, *salted*	8 samples	1.00	1.3	1.49	7.9	77.6	4.8	748	3082
1013	**Marzipan**, *home-made*	Recipe	1.00	10.2	1.97	10.4	25.8	50.1	462	1934
1014	*retail*	10 samples, white and yellow	1.00	8.1	1.02	5.3	12.7	67.6	389	1642
1015	**Mixed nuts**	Calculated from recipe proportions[a]	1.00	2.5	4.27	22.9	54.1	7.9	607	2515
1016	**Peanut butter**, smooth	10 samples, 3 brands	1.00	1.1	4.17	22.6	51.8	13.1	606	2510
1017	**Peanuts and raisins**	Calculated from recipe proportions	1.00	9.3	2.80	15.3	25.9	37.5	435	1819
1018	**Peanuts**, *plain*	10 samples, kernel only	0.69	6.3	4.73	25.6	46.0	12.5	563	2337
1019	*dry roasted*	10 samples, 5 brands	1.00	1.8	4.71	25.5	49.8	10.3	589	2441
1020	*roasted and salted*	20 samples	1.00	1.9	4.53	24.5	53.0	7.1	602	2491

[a] Calculated as peanuts 67%, almonds 17%, cashews 8% and hazelnuts 7%

No.	Food	Starch	Total sugars	Individual sugars: Gluc	Fruct	Sucr	Malt	Lact	Dietary fibre NSP	Fatty acids: Satd	Mono-unsatd	Poly-unsatd	Trans	Cholest-erol
		g	g	g	g	g	g	g	g	g	g	g	g	mg
	Nuts and seeds, general													
1004	**Almonds**	2.7	4.2	Tr	Tr	4.2	0	0	(7.4)	4.4	38.2	10.5	0	0
1005	**Brazil nuts**	0.7	2.4	0	0	2.4	0	0	4.3	16.4	25.8	23.0	0	0
1006	**Cashew nuts**, *roasted and salted*	13.2	5.6	0	0	5.6	0	0	3.2	10.1	29.4	9.1	0	0
1007	**Chestnuts**	29.6	7.0	Tr	Tr	7.0	0	0	4.1	0.5	1.0	1.1	0	0
1008	**Coconut**, *creamed block*	0	7.0	Tr	0.1	6.9	0	0	N	59.3	3.9	1.6	0	0
1009	*desiccated*	0	6.4	Tr	0.8	5.6	0	0	13.7	53.4	3.5	1.5	0	0
1010	**Coconut milk**	0	4.9	0.3	Tr	4.6	0	0	Tr	0.2	Tr	Tr	0	0
1011	**Hazelnuts**	2.0	4.0	0.2	0.1	3.7	0	0	6.5	4.7	50.0	5.9	0	0
1012	**Macadamia nuts**, *salted*	0.8	4.0	0.1	0.1	3.8	0	0	5.3	11.2	60.8	1.6	0	0
1013	**Marzipan**, *home-made*	1.5	48.7	0	0	48.6	0	0	(3.3)	2.2	17.4	4.8	Tr	29
1014	*retail*	0	67.6	2.7	1.1	62.2	1.6	0	(1.9)	1.0	8.0	3.1	0	0
1015	**Mixed nuts**	3.9	4.0	Tr	Tr	4.0	0	0	6.0	8.4	28.2	14.8	0	0
1016	**Peanut butter**, smooth	6.4	6.7	0	0	6.7	0	0	5.4	12.8	19.9	16.8	0.9	0
1017	**Peanuts and raisins**	3.5	34.0	15.2	15.3	3.5	0	0	4.4	4.9	12.3	7.3	N	0
1018	**Peanuts**, *plain*	6.3	6.2	0	0	6.2	0	0	6.2	8.7	22.0	13.1	0	0
1019	*dry roasted*	6.5	3.8	0	0	3.8	0	0	6.4	8.9	22.8	15.5	0	0
1020	*roasted and salted*	3.3	3.8	0	0	3.8	0	0	6.0	9.5	24.2	16.5	0	0

Inorganic constituents per 100g edible portion

No.	Food	mg										µg	
		Na	K	Ca	Mg	P	Fe	Cu	Zn	Cl	Mn	Se	I
	Nuts and seeds, general												
1004	**Almonds**	14	780	240	270	550	3.0	1.00	3.2	18	1.7	2	2
1005	**Brazil nuts**	3	660	170	410	590	2.5	1.76	4.2	57	1.2	254[a]	20
1006	**Cashew nuts**, *roasted and salted*	290	730	35	250	510	6.2	2.04	5.7	490	1.8	34	11
1007	**Chestnuts**	11	500	46	33	74	0.9	0.23	0.5	15	0.5	Tr	N
1008	**Coconut**, *creamed block*	30	650	23	73	170	3.7	0.56	0.9	190	1.8	(12)	(2)
1009	*desiccated*	28	660	23	90	160	3.6	0.55	0.9	200	1.8	(12)	(3)
1010	**Coconut milk**	110	280	29	30	30	0.1	0.04	0.1	180	N	N	N
1011	**Hazelnuts**	6	730	140	160	300	3.2	1.23	2.1	18	4.9	2	17
1012	**Macadamia nuts**, *salted*	280	300	47	100	200	1.6	0.43	1.1	390	5.5	7	N
1013	**Marzipan**, *home-made*	21	366	115	122	262	1.6	0.49	1.6	20	0.8	2	5
1014	*retail*	20	160	66	68	130	0.9	0.24	0.8	23	0.4	1	Tr
1015	**Mixed nuts**	300	790	78	200	430	2.1	0.79	3.1	490	2.1	5	12
1016	**Peanut butter**, smooth	350	700	37	180	330	2.1	0.70	3.0	500	1.7	3	N
1017	**Peanuts and raisins**	28	824	54	133	274	3.1	0.74	2.3	8	1.3	5	N
1018	**Peanuts**, *plain*	2	670	60	210	430	2.5	1.02	3.5	7	2.1	3	20
1019	*dry roasted*	790	730	52	190	420	2.1	0.64	3.3	1140	2.2	3	19
1020	*roasted and salted*	400	810	37	180	410	1.3	0.54	2.9	660	1.9	4	19

[a] Selenium can range from 85 to 690mg per 100g

No.	Food	Retinol µg	Carotene µg	Vitamin D µg	Vitamin E mg	Thiamin mg	Riboflavin mg	Niacin mg	Trypt/60 mg	Vitamin B_6 mg	Vitamin B_{12} µg	Folate µg	Pantothenate mg	Biotin µg	Vitamin C mg
	Nuts and seeds, general														
1004	**Almonds**	0	0	0	23.98	0.21	0.75	3.1	3.4	0.15	0	48	0.44	64.0	0
1005	**Brazil nuts**	0	0	0	7.18	0.67	0.03	0.3	3.0	0.31	0	21	0.41	11.0	0
1006	**Cashew nuts**, *roasted and salted*	0	6	0	1.30	0.41	0.16	1.3	5.2	0.43	0	68	1.08	13.0	0
1007	**Chestnuts**	0	0	0	1.20	0.14	0.02	0.5	0.4	0.34	0	N	0.49	1.4	Tr
1008	**Coconut**, *creamed block*	0	0	0	1.40	(0.03)	(0.05)	(0.9)	1.2	N	0	(9)	(0.50)	N	0
1009	*desiccated*	0	0	0	1.26	0.03	0.05	0.9	1.1	(0.09)	0	9	0.50	N	0
1010	**Coconut milk**	0	0	0	Tr	0.03	0.06	0.1	0.1	0.03	0	N	0.04	N	2
1011	**Hazelnuts**	0	0	0	24.98	0.43	0.16	1.1	4.0	0.59	0	72	1.51	76.0	0
1012	**Macadamia nuts**, *salted*	0	0	0	1.49	0.28	0.06	1.6	1.7	0.28	0	N	0.61	6.0	0
1013	**Marzipan**, *home-made*	14	0	0.1	10.82	0.10	0.37	1.4	1.8	0.08	0.2	26	0.33	30.2	1
1014	*retail*	0	0	0	6.18	(0.05)	(0.19)	(0.7)	0.9	(0.04)	0	(12)	(0.11)	(16.0)	0
1015	**Mixed nuts**	0	Tr	0	6.44	0.22	0.22	9.9	4.9	0.53	0	54	1.42	86.4	0
1016	**Peanut butter**, smooth	0	0	0	4.99	0.17	0.09	12.5	4.9	0.58	0	53	1.56	94.0	0
1017	**Peanuts and raisins**	0	5	0	5.65	0.69	0.08	8.0	3.2	0.44	0	66	1.56	41.2	0
1018	**Peanuts**, *plain*	0	0	0	10.09	1.14	0.10	13.8	5.5	0.59	0	110	2.66	72.0	0
1019	*dry roasted*	0	0	0	1.11	0.18	0.13	13.1	5.5	0.54	0	44	1.59	130.0	0
1020	*roasted and salted*	0	0	0	0.66	0.18	0.10	13.6	5.3	0.63	0	52	1.70	102.0	0

Composition of food per 100g edible portion

No.	Food	Description and main data sources	Edible conversion factor	Water	Total nitrogen	Protein	Fat	Carbo-hydrate	Energy value	
				g	g	g	g	g	kcal	kJ
	Nuts and seeds, general *continued*									
1021	**Pecan nuts**	9 samples, kernel only	0.49	3.7	1.74	9.2	70.1	5.8	689	2843
1022	**Pine nuts**	20 samples, pine kernels	1.00	2.7	2.64	14.0	68.6	4.0	688	2840
1023	**Pistachio nuts**, *roasted and salted*	10 samples, kernel only	0.55	2.1	3.38	17.9	55.4	8.2	601	2485
1024	**Sesame seeds**	10 samples, with and without hulls	1.00	4.6	3.44	18.2	58.0	0.9	598	2470
1025	**Sunflower seeds**	Analysis and literature sources	1.00	4.4	3.74	19.8	47.5	*18.6*[a]	*581*	*2410*
1026	**Tahini paste**	Ref. McCarthy and Matthews (1984) and calculation from No. 1028	1.00	3.1	3.49	18.5	58.9	0.9	607	2508
1027	**Trail Mix**	10 samples; mix of nuts and dried fruit	1.00	8.9	1.45	9.1	28.5	37.2	432	1804
1028	**Walnuts**	10 samples, kernel only	0.43	2.8	2.77	14.7	68.5	3.3	688	2837

[a] Including oligosaccharides

No.	Food	Starch	Total sugars	Individual sugars					Dietary fibre	Fatty acids				Cholest-erol
				Gluc	Fruct	Sucr	Malt	Lact	NSP	Satd	Mono-unsatd	Poly-unsatd	Trans	
		g	g	g	g	g	g	g	g	g	g	g	g	mg
	Nuts and seeds, general *continued*													
1021	**Pecan nuts**	1.5	4.3	0.3	0.3	3.7	0	0	4.7	5.7	42.5	18.7	0	0
1022	**Pine nuts**	0.1	3.9	0.1	0.1	3.7	0	0	1.9	4.6	19.9	41.1	0	0
1023	**Pistachio nuts**, *roasted and salted*	2.5	5.7	Tr	Tr	5.7	0	0	6.1	7.4	27.6	17.9	0	0
1024	**Sesame seeds**	0.5	0.4	0.1	0.1	0.2	0	0	7.9	8.3	21.7	25.5	0	0
1025	**Sunflower seeds**	16.3	1.7[a]	0	0	1.7	0	0	6.0	4.5	9.8	31.0	0	0
1026	**Tahini paste**	0.5	0.4	0.1	0.1	0.2	0	0	8.0	8.4	22.0	25.8	N	0
1027	**Trail Mix**	0.1	37.1	17.3	16.1	3.4	0.4	0	4.3	N	N	N	N	0
1028	**Walnuts**	0.7	2.6	0.2	0.2	2.2	0	0	3.5	5.6	12.4	47.5	0	0

[a] Not including oligosaccharides

Inorganic constituents per 100g edible portion

No.	Food	mg										µg	
		Na	K	Ca	Mg	P	Fe	Cu	Zn	Cl	Mn	Se	I
	Nuts and seeds, general *continued*												
1021	**Pecan nuts**	1	520	61	130	310	2.2	1.07	5.3	15	4.6	12	N
1022	**Pine nuts**	1	780	11	270	650	5.6	1.32	6.5	41	7.9	N	N
1023	**Pistachio nuts**, *roasted and salted*	530	1040	110	130	420	3.0	0.83	2.2	810	0.9	(6)	N
1024	**Sesame seeds**	20	570	670	370	720	10.4	1.46	5.3	10	1.5	N	N
1025	**Sunflower seeds**	3	710	110	390	640	6.4	2.27	5.1	N	2.2	(49)	N
1026	**Tahini paste**	20	580	680	380	730	10.6	1.48	5.4	10	1.5	N	N
1027	**Trail Mix**	27	620	69	110	210	3.7	0.55	1.5	N	1.6	N	N
1028	**Walnuts**	7	450	94	160	380	2.9	1.34	2.7	24	3.4	3	9

No.	Food	Retinol µg	Carotene µg	Vitamin D µg	Vitamin E mg	Thiamin mg	Ribo-flavin mg	Niacin mg	Trypt/60 mg	Vitamin B_6 mg	Vitamin B_{12} µg	Folate µg	Panto-thenate mg	Biotin µg	Vitamin C mg
	Nuts and seeds, general *continued*														
1021	**Pecan nuts**	0	50	0	4.34	0.71	0.15	1.4	4.1	0.19	0	39	1.71	N	0
1022	**Pine nuts**	0	10	0	13.65	0.73	0.19	3.8	3.1	N	0	N	N	N	Tr
1023	**Pistachio nuts**, *roasted and salted*	0	130	0	4.16	0.70	0.23	1.7	3.9	N	0	58	N	N	0
1024	**Sesame seeds**	0	6	0	2.53	0.93	0.17	5.0	5.4	0.75	0	97	2.14	11.0	0
1025	**Sunflower seeds**	0	15	0	37.77	1.60	0.19	4.1	5.0	N	0	N	N	N	0
1026	**Tahini paste**	0	6	0	2.57	0.94	0.17	5.1	4.1	0.76	0	99	2.17	11.0	0
1027	**Trail Mix**	0	47	0	4.53	0.23	0.09	2.0	1.5	N	0	25	N	N	Tr
1028	**Walnuts**	0	0	0	3.83	0.40	0.14	1.2	2.8	0.67	0	66	1.60	19.0	0[a]

[a] Value for ripe dried walnuts. Unripe walnuts contain 1300 to 3000mg vitamin C per 100g

Section 2.11

Sugars, preserves and snacks

The data in this section of the Tables have been taken from the *Miscellaneous Foods* (1994) supplement. New analytical data have been incorporated for a few foods. Foods for which new data are incorporated have been allocated a new food code and can thus easily be identified in the food index.

Composition of food per 100g edible portion

No.	Food	Description and main data sources	Edible conversion factor	Water	Total nitrogen	Protein	Fat	Carbo-hydrate	Energy value	
				g	g	g	g	g	kcal	kJ
	Sugars, syrups and preserves									
1029	**Chocolate spread**	6 samples, 3 brands	1.00	0.2	0.66	4.1	37.6	57.1	569	2375
1030	**Chocolate nut spread**	8 samples, 5 brands	1.00	Tr	0.99	6.2	33.0	60.5	549	2294
1031	**Fruit spread**	8 samples, 4 brands; assorted flavours	1.00	64.0	0.11	0.7	0.1	31.4	121	518
1032	**Glucose liquid**, BP	1 sample	1.00	20.4	Tr	Tr	0	84.7[a]	318	1355
1033	**Honey**	8 samples; assorted types	1.00	17.5	0.06	0.4	0	76.4	288	1229
1034	**Honeycomb**	2 samples, honey and comb together	1.00	20.2	0.09	0.6	4.6[b]	74.4	281	1201
1035	**Ice-cream sauce**, topping	8 samples, 3 brands; strawberry and chocolate flavours	1.00	40.6	0.13	0.8	0.2	53.9	207	883
1036	**Jaggery**	5 assorted samples	1.00	3.4	0.08	0.5	0	97.2	367	1564
1037	**Jam**, fruit with edible seeds	10 samples, 5 flavours	1.00	29.8	0.10	0.6	0	69.0	261	1114
1038	reduced sugar	9 samples, 5 brands; assorted flavours	1.00	65.3	0.08	0.5	0.1	31.9	123	523
1039	stone fruit	8 samples, 4 flavours	1.00	29.6	0.06	0.4	0	69.3	261	1116
1040	**Lemon curd**	10 jars, 4 brands	1.00	30.7	0.09	0.6	4.9	62.7	282	1195
1041	**Marmalade**	4 brands	1.00	28.0	0.01	0.1	0	69.5[c]	261	1114
1042	**Mincemeat**	10 samples of the same brand	1.00	27.5	0.10	0.6	4.3	62.1	274	1163
1043	**Sugar**, Demerara	5 samples	1.00	Tr	0.08	0.5	0	104.5	394[d]	1681[d]
1044	white	Granulated and loaf sugar	1.00	Tr	Tr	Tr	0	105.0	394	1680
1045	**Syrup**, golden	3 samples of the same brand	1.00	20.0	0.05	0.3	0	79.0	298	1269
1046	**Treacle**, black	3 samples	1.00	28.5	0.19	1.2	0	67.2	257	1096

[a] Includes oligosaccharides
[b] Waxy material, probably not available as fat; disregarded in calculating energy values
[c] Reduced sugar marmalade contains about 33.0g carbohydrate
[d] Light muscovado sugar provides 376kcal, 1705kJ per 100g. Dark muscovado sugar provides 355kcal, 1607kJ per 100g

Composition of food per 100g edible portion

No.	Food	Starch	Total sugars	Individual sugars					Dietary fibre	Fatty acids				Cholest-erol
				Gluc	Fruct	Sucr	Malt	Lact	NSP	Satd	Mono-unsatd	Poly-unsatd	Trans	
		g	g	g	g	g	g	g	g	g	g	g	g	mg
	Sugars, syrups and preserves													
1029	**Chocolate spread**	Tr	57.1	Tr	Tr	53.1	Tr	4.0	N	N	N	N	N	(2)
1030	**Chocolate nut spread**	0.8	59.7	Tr	Tr	56.7	0	3.0	0.8	10.1	16.8	4.6	0.3	2
1031	**Fruit spread**	0.7	30.7	13.6	16.7	0.4	0	0	N	Tr	Tr	Tr	Tr	0
1032	**Glucose liquid**, BP	N	40.2	N	0	0	N	0	0	0	0	0	0	0
1033	**Honey**	0	76.4	34.6	41.8	Tr	Tr	0	0	0	0	0	0	0
1034	**Honeycomb**	0	74.4	34.2	40.2	0	0	0	0	0	0	0	0	0
1035	**Ice-cream sauce**, topping	2.9	51.0	17.1	15.9	18.0	0	0	0	Tr	Tr	Tr	Tr	0
1036	**Jaggery**	7.9	89.3	N	N	N	N	0	0	0	0	0	0	0
1037	**Jam**, fruit with edible seeds	0	69.0	27.4	14.9	18.7	8.0	0	N	0	0	0	0	0
1038	reduced sugar	0	31.9	10.4	15.0	6.5	0	0	(0.8)	Tr	Tr	Tr	Tr	0
1039	stone fruit	0	69.3	27.5	14.9	18.8	8.0	0	N	0	0	0	0	0
1040	**Lemon curd**	22.3	40.4	16.5	7.6	12.0	4.3	0	(0.2)	1.5	2.0	1.2	0.4	21
1041	**Marmalade**	0	69.5	27.6	15.0	18.8	8.0	0	(0.3)	0	0	0	0	0
1042	**Mincemeat**	Tr	62.1	30.7	30.8	0.6	Tr	0	1.3	N	N	N	N	4
1043	**Sugar**, Demerara	0	104.5	0	0	104.5	0	0	0	0	0	0	0	0
1044	white	0	105.0	0	0	105.0	0	0	0	0	0	0	0	0
1045	**Syrup**, golden	0	79.0	23.1	23.0	32.8	0	0	0	0	0	0	0	0
1046	**Treacle**, black	0	66.8	17.4	16.7	32.7	0	0	Tr	0	0	0	0	0

Inorganic constituents per 100g edible portion

No.	Food	mg										µg	
		Na	K	Ca	Mg	P	Fe	Cu	Zn	Cl	Mn	Se	I
	Sugars, syrups and preserves												
1029	**Chocolate spread**	N	N	N	N	N	N	N	N	N	N	N	N
1030	**Chocolate nut spread**	50	390	130	65	180	2.2	0.48	1.0	60	1.10	N	N
1031	**Fruit spread**	10	190	11	10	20	0.6	Tr	0.1	10	0.20	Tr	N
1032	**Glucose liquid**, BP	150	3	8	2	11	0.5	0.09	N	190	Tr	Tr	Tr
1033	**Honey**	11	51	5	2	17	0.4	0.05	0.9	18	0.30	(1)	Tr
1034	**Honeycomb**	7	35	8	2	32	0.2	0.04	N	26	N	(1)	Tr
1035	**Ice-cream sauce**, topping	140	68	9	15	26	0.8	0.09	0.1	40	0.14	N	N
1036	**Jaggery**	79	290	92	120	72	1.6	0.75	0.1	250	0.50	Tr	Tr
1037	**Jam**, fruit with edible seeds	29	43	12	5	10	0.2	0.01	(0.1)	9	0.13	Tr	7
1038	reduced sugar	20	120	19	7	15	0.4	0.05	Tr	Tr	0.10	0	(2)
1039	stone fruit	46	67	10	3	6	0.2	0.02	Tr	4	0.02	Tr	7
1040	**Lemon curd**	65	11	9	2	15	0.5	(0.30)	1.3	150	N	N	N
1041	**Marmalade**	64	35	26	3	6	0.2	0.03	(0.1)	7	0.01	(1)	(7)
1042	**Mincemeat**	18	44	35	4	13	0.6	0.12	0.2	7	N	(1)	(7)
1043	**Sugar**, Demerara	5	48	29	9	3	0.9	0.11	(0.1)	35	Tr	Tr	Tr
1044	white	5	5	(10)	(2)	(1)	(0.2)	0.12	(0.1)	Tr	Tr	Tr	Tr
1045	**Syrup**, golden	270	58	16	3	(1)	0.4	0.06	(0.1)	42	0.01	Tr	Tr
1046	**Treacle**, black	180	1760	550	180	29	21.3	0.78	0.8	820	2.67	N	Tr

Vitamins per 100g edible portion

No.	Food	Retinol µg	Carotene µg	Vitamin D µg	Vitamin E mg	Thiamin mg	Ribo-flavin mg	Niacin mg	Trypt/60 mg	Vitamin B_6 mg	Vitamin B_{12} µg	Folate µg	Panto-thenate mg	Biotin µg	Vitamin C mg
	Sugars, syrups and preserves														
1029	**Chocolate spread**	Tr	Tr	Tr	N	N	N	N	N	N	Tr	N	N	N	0
1030	**Chocolate nut spread**	Tr	Tr	Tr	N	0.03	0.10	0.5	1.5	0.1	Tr	N	N	N	Tr
1031	**Fruit spread**	0	Tr	0	Tr	Tr	Tr	Tr	Tr	Tr	0	Tr	Tr	Tr	6
1032	**Glucose liquid**, BP	0	0	0	0	0	0	0	0	0	0	0	0	0	0
1033	**Honey**	0	0	0	0	Tr	0.05	0.2	Tr	N	0	N	N	N	0
1034	**Honeycomb**	0	0	0	0	Tr	0.05	0.2	Tr	N	0	N	N	N	0
1035	**Ice-cream sauce**, topping	0	Tr	0	N	Tr	Tr	Tr	Tr	Tr	0	Tr	Tr	Tr	0
1036	**Jaggery**	0	0	0	N	Tr	0.04	Tr	Tr	Tr	0	Tr	Tr	Tr	0
1037	**Jam**, fruit with edible seeds	0	Tr	0	0	Tr	Tr	Tr	Tr	Tr	0	Tr	Tr	Tr	10[a]
1038	reduced sugar	0	(26)	0	(0.14)	Tr	Tr	Tr	Tr	Tr	0	Tr	Tr	Tr	26
1039	stone fruit	0	Tr	0	0	Tr	Tr	Tr	Tr	Tr	0	Tr	Tr	Tr	0
1040	**Lemon curd**	(10)	Tr	(0.1)	N	Tr	(0.02)	Tr	(0.1)	Tr	Tr	Tr	(0.10)	(1.0)	Tr
1041	**Marmalade**	0	50	0	Tr	Tr	Tr	Tr	Tr	Tr	0	5	Tr	Tr	10
1042	**Mincemeat**	0	9	Tr	N	0.04	0.02	0.4	0.1	(0.10)	Tr	8	0.03	Tr	Tr
1043	**Sugar**, Demerara	0	0	0	0	Tr	Tr	Tr	Tr	Tr	0	Tr	Tr	Tr	0
1044	white	0	0	0	0	0	0	0	0	0	0	0	0	0	0
1045	**Syrup**, golden	0	0	0	0	Tr	Tr	Tr	Tr	Tr	0	Tr	Tr	Tr	0
1046	**Treacle**, black	0	0	0	0	Tr	Tr	Tr	Tr	Tr	0	Tr	Tr	Tr	0

[a] Blackcurrant jam contains 24mg vitamin C per 100g

Composition of food per 100g edible portion

No.	Food	Description and main data sources	Edible conversion factor	Water	Total nitrogen	Protein	Fat	Carbo-hydrate	Energy value	
				g	g	g	g	g	kcal	kJ
	Chocolate confectionery									
1047	**Bounty bar**	Analysis and manufacturer's data (Mars). Milk chocolate[a]	1.00	8.6	0.59	3.7	26.3	58.1	469	1966
1048	**Chocolate covered caramels**	18 samples, 4 brands including Rolo, Caramel	1.00	5.6	0.80	5.0	21.7	66.5[b]	465	1952
1049	**Chocolate**, fancy and filled	10 samples of different brands	1.00	6.1	0.78	4.9	21.3	62.9[c]	447	1878
1050	milk	12 bars, 5 brands including Dairy Milk, Galaxy, Chocolate buttons	1.00	1.3	1.23	7.7	30.7	56.9	520	2177
1051	plain	6 bars, 3 brands	1.00	0.6	0.80	5.0	28.0	63.5	510	2137
1052	white	14 samples, 5 brands; buttons and bars	1.00	0.6	1.28	8.0	30.9	58.3	529	2212
1053	**Creme egg**	10 samples and manufacturer's data (Cadbury)	1.00	6.7	0.64	4.0	15.9	*71.0*[d]	*425*	*1792*
1054	**Kit Kat**	Analysis and manufacturer's data	1.00	2.0	1.20	7.5	26.0	63.0	500	2098
1055	**Mars bar**	Analysis and manufacturer's data (Mars)	1.00	6.9	0.69	4.5	18.3	77.3	473	1990
1056	**Milky Way**	Analysis and manufacturer's data (Mars)	1.00	6.4	0.56	3.5	16.7	74.8	445	1874
1057	**Smartie-type sweets**	10 samples including Smarties and M & M's	1.00	1.5	0.86	5.4	17.5	73.9	456	1922
1058	**Snickers**	Manufacturer's data (Mars) and literature (Cutrufelli and Pehrsson 1991)	1.00	5.6	1.50	9.4	27.8	55.8	497	2081
1059	**Twix**	Analysis and manufacturer's data (Mars)	1.00	3.6	0.74	4.6	24.1	68.5	492	2066

[a] Bounty bar made with plain chocolate contains 3.2g protein, 26.8 g fat, 56.3g carbohydrate (11.0g starch, 45.3g sugar), 465kcal and 1947kJ of energy per 100g

[b] Includes 10.7g maltodextrins [c] Includes 2.7g maltodextrins [d] Includes 16.0g maltodextrins

Composition of food per 100g edible portion

No.	Food	Starch	Total sugars	Individual sugars					Dietary fibre	Fatty acids				Cholest-erol
				Gluc	Fruct	Sucr	Malt	Lact	NSP	Satd	Mono-unsatd	Poly-unsatd	Trans	
		g	g	g	g	g	g	g	g	g	g	g	g	mg
	Chocolate confectionery													
1047	**Bounty bar**	10.6	47.5	3.0	0.1	39.3	2.8	2.3	3.2	20.5	4.8	0.8	0.1	4
1048	**Chocolate covered caramels**	Tr	55.8	5.2	2.9	36.8	4.3	7.0	N	10.7	9.1	0.7	2.5	23
1049	**Chocolate**, fancy and filled	0.2	60.0	5.4	3.1	45.7	2.2	3.6	1.3	11.3	8.0	1.0	1.1	11
1050	milk	Tr	56.9	0.1	0.1	46.6	Tr	10.1	0.8	18.3	9.9	1.2	0.4	23
1051	plain	0.9	62.6	Tr	Tr	62.4	Tr	0.2	2.5	16.8	9.0	1.0	0.1	6
1052	white	Tr	58.3	Tr	Tr	47.6	Tr	10.7	N	18.4	10.0	1.1	N	N
1053	**Creme egg**	Tr	58.0	3.6	1.8	45.7	2.0	4.9	N	2.0	N	N	N	10
1054	**Kit Kat**	12.9	50.1	0.1	0.1	42.2	0.1	7.6	N	16.2	7.5	0.7	0.2	12
1055	**Mars bar**	11.1	66.2	9.7	0.1	43.1	6.8	6.6	0.4	10.3	6.7	1.0	0.5	8
1056	**Milky Way**	6.7	68.1	8.7	Tr	47.4	5.4	6.7	0.2	9.5	6.1	0.9	0.4	7
1057	**Smartie-type sweets**	3.1	70.8	0.3	Tr	65.6	0.1	4.8	N	10.4	5.7	0.6	N	(17)
1058	**Snickers**	8.3	47.5	6.6	0.5	29.0	6.7	4.7	1.7	10.9	10.9	4.3	0.4	4
1059	**Twix**	16.8	51.6	6.8	0.1	34.0	4.2	6.4	0.8	11.7	11.5	0.9	5.0	4

Inorganic constituents per 100g edible portion

No.	Food	mg										µg	
		Na	K	Ca	Mg	P	Fe	Cu	Zn	Cl	Mn	Se	I
	Chocolate confectionery												
1047	**Bounty bar**	180	320	57	39	102	1.5	0.47	0.6	400	N	N	N
1048	**Chocolate covered caramels**	180	270	145	31	150	1.1	0.02	0.6	260	0.12	N	N
1049	**Chocolate**, fancy and filled	88	270	110	48	150	1.2	0.30	0.8	140	0.39	(2)	120
1050	milk	85	390	220	50	220	1.4	0.24	1.1	190	0.22	(4)	30
1051	plain	6	300	33	89	140	2.3	0.71	1.3	9	0.63	4	3
1052	white	110	350	270	26	230	0.2	Tr	0.9	250	0.02	N	N
1053	**Creme egg**	63	145	85	27	130	0.8	0.10	0.6	110	0.10	Tr	N
1054	**Kit Kat**	120	330	200	52	200	1.5	0.28	1.1	210	0.34	N	N
1055	**Mars bar**	150	250	95	32	110	1.2	0.31	(0.7)	300	N	(2)	N
1056	**Milky Way**	100	240	90	21	80	1.1	0.13	0.8	160	0.25	N	N
1057	**Smartie-type sweets**	58	280	150	48	160	1.5	0.25	0.9	120	0.25	N	N
1058	**Snickers**	270	330	98	32	110	1.2	(0.40)	(1.3)	N	0.49	N	N
1059	**Twix**	190	190	110	28	130	1.1	0.08	0.7	250	0.22	N	N

No.	Food	Retinol µg	Carotene µg	Vitamin D µg	Vitamin E mg	Thiamin mg	Ribo-flavin mg	Niacin mg	Trypt/60 mg	Vitamin B_6 mg	Vitamin B_{12} µg	Folate µg	Panto-thenate mg	Biotin µg	Vitamin C mg
	Chocolate confectionery														
1047	**Bounty bar**	15	(40)	0.3	0.32	(0.04)	(0.10)	(0.3)	0.8	(0.03)	Tr	(4)	(0.25)	(1)	0
1048	**Chocolate covered caramels**	33	20	Tr	2.37	0.06	0.34	0.3	1.1	0.02	Tr	4	0.60	3	0
1049	**Chocolate**, fancy and filled	81	120	Tr	1.65	0.05	0.20	0.4	0.9	0.03	Tr	17	(0.73)	(3)	0
1050	milk	25	11	0	0.45	0.07	0.49	0.4	2.3	0.04	1.0	11	0.70	4	0
1051	plain	15	15	0	1.44	0.04	0.06	0.4	0.7	0.03	0	12	0.30	3	0
1052	white	13	75	Tr	1.14	0.08	0.49	0.2	2.6	0.07	Tr	(10)	(0.59)	3	0
1053	**Creme eggs**	47	(55)	0.6	1.07	0.06	0.34	0.2	1.3	0.03	1.0	12	(0.59)	(3)	0
1054	**Kit Kat**	8	47	Tr	1.00	0.11	0.44	0.5	2.6	0.06	Tr	N	0.70	4	0
1055	**Mars bar**	(31)	(40)	0.3	(0.47)	(0.05)	(0.20)	(0.2)	0.9	(0.03)	Tr	(5)	(0.59)	(2)	0
1056	**Milky Way**	(28)	Tr	0.4	1.91	0.05	0.15	0.2	1.1	0.03	Tr	(4)	0.59	2	0
1057	**Smartie-type sweets**	5	28	Tr	0.80	0.08	0.79	0.3	1.7	0.03	Tr	4	0.67	2	0
1058	**Snickers**	(15)	N	Tr	0.97	0.08	0.18	3.4	N	0.19	0	(24)	(0.69)	(19)	0
1059	**Twix**	(15)	7	0.3	3.72	0.06	0.22	0.3	0.9	0.05	Tr	N	0.61	3	0

Composition of food per 100g edible portion

No.	Food	Description and main data sources	Edible conversion factor	Water	Total nitrogen	Protein	Fat	Carbo-hydrate	Energy value	
				g	g	g	g	g	kcal	kJ
Non-chocolate confectionery										
1060	**Boiled sweets**	6 samples	1.00	(16.6)	Tr	Tr	Tr	87.1	327	1394
1061	**Cereal chewy bar**	17 bars of different brands; assorted types	1.00	1.1	1.17	7.3	16.4	64.7[a]	419	1766
1062	**Cereal crunchy bar**	12 bars of different brands; assorted types	1.00	2.6	1.66	10.4	22.2	60.5[b]	468	1966
1063	**Chew sweets**	15 samples, 6 brands including Opal Fruits, Chewitts, Fruit-tella	1.00	7.6	0.16	1.0	5.6	87.0[c]	381	1616
1064	**Fruit gums/jellies**	11 samples, 10 brands; assorted flavours	1.00	14.0	1.04	6.5	0	79.5[d]	324	1383
1065	**Fruit pastilles**	6 samples of different brands; assorted flavours	1.00	9.1	0.45	2.8	0	84.2[e]	327	1395
1066	**Fudge**	Recipe	1.00	4.6	0.52	3.3	13.7	80.4	438	1849
1067	**Liquorice allsorts**	7 samples, 4 brands	1.00	8.4	0.59	3.7	5.2	76.7[f]	349	1483
1068	**Marshmallows**	7 samples of different brands	1.00	17.4	0.62	3.9	0	83.1[g]	327	1396
1069	**Peppermints**	Several samples of 6 different brands	1.00	0.2	0.08	0.5	0.7	102.7	393	1678
1070	**Sherbert sweets**	10 samples of different brands	1.00	0.2	0.10	0.6	0	93.9	355	1513
1071	**Toffees**, mixed	13 samples, 4 brands including cream and plain varieties	1.00	2.4	0.35	2.2	18.6	66.7[h]	426	1793
1072	**Turkish delight**, without nuts	7 assorted samples	1.00	16.1	0.10	0.6	0	77.9	295	1257
Savoury snacks										
1073	**Bombay Mix**	20 samples; savoury mix of gram flour, assorted peas, lentils, nuts and seeds	1.00	3.5	3.01	18.8	32.9	35.1	503	2099
1074	**Breadsticks**	10 samples, 3 brands	1.00	3.5	1.92	11.2	8.4	72.5	392	1661
1075	**Corn snacks**	20 samples, 7 brands including Wotsits, Monster Munch and Nik-Naks	1.00	3.3	1.12	7.0	31.9	54.3	519	2168

[a] Includes 6.4g maltodextrins
[b] Includes 4.4g maltodextrins
[c] Includes 32.0g maltodextrins
[d] Includes 18.9g maltodextrins
[e] Includes 21.5g maltodextrins
[f] Includes 4.9g maltodextrins
[g] Includes 14.1g maltodextrins
[h] Includes 21.9g maltodextrins

Composition of food per 100g edible portion

No.	Food	Starch	Total sugars	Individual sugars					Dietary fibre	Fatty acids				Cholest-erol
				Gluc	Fruct	Sucr	Malt	Lact	NSP	Satd	Mono-unsatd	Poly-unsatd	Trans	
		g	g	g	g	g	g	g	g	g	g	g	g	mg
	Non-chocolate confectionery													
1060	**Boiled sweets**	0.4	86.7	8.5	1.4	67.5	9.3	0	0	0	0	0	0	0
1061	**Cereal chewy bar**	25.6	32.7	7.4	5.0	10.6	8.2	1.5	3.2	5.0	8.7	1.8	3.2	N
1062	**Cereal crunchy bar**	28.2	27.9	2.3	1.5	21.8	1.5	0.8	4.8	4.5	11.3	5.4	1.3	Tr
1063	**Chew sweets**	Tr	55.0	7.7	0.8	39.4	7.1	Tr	1.0	3.0	2.2	0.2	1.3	0
1064	**Fruit gums/jellies**	1.9	58.7	6.3	Tr	46.4	6.0	Tr	N	0	0	0	0	0
1065	**Fruit pastilles**	3.4	59.3	6.5	2.1	45.4	5.3	Tr	N	0	0	0	0	0
1066	**Fudge**	0	80.4	0	0	76.8	0	3.6	0	8.7	3.6	0.5	0.4	39
1067	**Liquorice allsorts**	9.4	62.4	5.9	2.5	51.1	2.9	Tr	2.0	3.6	1.2	0.2	0.6	0
1068	**Marshmallows**	4.5	64.5	12.1	0.7	41.5	10.2	Tr	0	0	0	0	0	0
1069	**Peppermints**	0	102.7	1.0	0	101.7	0	0	0	N	N	N	N	0
1070	**Sherbert sweets**	Tr	93.9	0.2	Tr	93.7	Tr	Tr	Tr	0	0	0	0	0
1071	**Toffees**, mixed	Tr	44.8	5.2	0.6	32.4	4.5	2.0	0	9.5	7.5	0.7	3.3	17
1072	**Turkish delight**, without nuts	9.3	68.6	N	N	N	N	0	0	0	0	0	0	0
	Savoury snacks													
1073	**Bombay Mix**	32.8	2.3	0.1	0.1	2.2	0	0	6.2	4.0	16.2	11.3	0.2	0
1074	**Breadsticks**	67.5	5.0	0.8	0.5	0	3.7	0	2.8	5.9	1.3	0.9	0	0
1075	**Corn snacks**	49.7	4.6	0.3	0.1	0.7	3.5	0	1.0	11.8	12.9	5.8	0.2	0

Inorganic constituents per 100g edible portion

No.	Food	mg										µg	
		Na	K	Ca	Mg	P	Fe	Cu	Zn	Cl	Mn	Se	I
	Non-chocolate confectionery												
1060	**Boiled sweets**	25	8	5	2	12	0.4	0.09	N	68	N	Tr	N
1061	**Cereal chewy bar**	110	320	70	55	190	1.9	0.16	1.1	210	1.36	N	N
1062	**Cereal crunchy bar**	74	360	77	86	290	2.6	0.29	1.7	140	2.09	N	N
1063	**Chew sweets**	48	15	6	4	4	0.2	Tr	Tr	67	Tr	N	N
1064	**Fruit gums/jellies**	30	8	5	1	4	0.1	0.02	Tr	N	Tr	Tr	N
1065	**Fruit pastilles**	33	28	28	6	4	0.4	0.04	Tr	29	0.02	Tr	N
1066	**Fudge**	139	148	122	13	101	0.2	0.09	0.4	214	Tr	1	15
1067	**Liquorice allsorts**	57	600	170	76	44	7.3	0.34	0.5	N	1.14	N	N
1068	**Marshmallows**	29	2	4	2	4	0.3	Tr	Tr	36	Tr	N	N
1069	**Peppermints**	9	Tr	7	3	Tr	0.2	0.04	N	22	N	Tr	N
1070	**Sherbert sweets**	1050	15	42	69	Tr	0.2	0.04	Tr	6	0.02	N	N
1071	**Toffees**, mixed	340	110	73	8	62	0.2	0.02	0.3	500	Tr	N	N
1072	**Turkish delight**, without nuts	31	4	10	2	7	0.2	0.12	0.7	110	Tr	Tr	Tr
	Savoury snacks												
1073	**Bombay Mix**	770	770	58	100	290	3.8	0.62	2.5	1410	1.40	N	N
1074	**Breadsticks**	860	160	26	25	110	1.2	0.12	0.7	630	0.48	N	N
1075	**Corn snacks**	1130	200	68	18	140	0.8	0.04	0.5	1840	0.13	(3)	N

No.	Food	Retinol µg	Carotene µg	Vitamin D µg	Vitamin E mg	Thiamin mg	Riboflavin mg	Niacin mg	Trypt/60 mg	Vitamin B_6 mg	Vitamin B_{12} µg	Folate µg	Pantothenate mg	Biotin µg	Vitamin C mg
Non-chocolate confectionery															
1060	**Boiled sweets**	0	0	0	0	0	0	0	0	0	0	0	0	0	0
1061	**Cereal chewy bar**	0	N	0	N	0.24	0.17	1.2	1.9	0.13	Tr	11	N	N	Tr
1062	**Cereal crunchy bar**	0	Tr	0	3.84	0.24	0.12	2.3	4.4	0.14	0	15	N	N	Tr
1063	**Chew sweets**	0	315	0	0.91	Tr	Tr	N	N	Tr	0	Tr	Tr	Tr	0
1064	**Fruit gums/jellies**	0	N	0	0	0	0	0	0	0	0	0	0	0	0
1065	**Fruit pastilles**	0	0	0	0	0	0	0	0	0	0	0	0	0	0
1066	**Fudge**	155	107	1.2	0.24	0.02	0.17	0.1	0.7	0.03	0.3	4	0.33	2	0
1067	**Liquorice allsorts**	0	0	0	0	0	0	0	0.2	0	0	0	0	0	0
1068	**Marshmallows**	0	0	0	0	0	0	Tr	Tr	0	0	0	0	0	0
1069	**Peppermints**	0	0	0	0	0	0	0	0	0	0	0	0	0	0
1070	**Sherbert sweets**	0	0	0	0	0	0	Tr	Tr	0	0	0	0	0	0
1071	**Toffees**, mixed	0	0	0	N	0	0	0	0.4	0	0	0	0	0	0
1072	**Turkish delight**, without nuts	0	0	0	0	0.13	N	N	N	N	N	N	N	N	0
Savoury snacks															
1073	**Bombay Mix**	0	Tr	0	4.71	0.38	0.10	4.3	3.5	0.54	0	N	1.19	24	Tr
1074	**Breadsticks**	0	Tr	0	0.44	0.12	0.08	1.6	3.9	0.10	Tr	18	0.60	2	0
1075	**Corn snacks**	0	460	0	5.80	0.19	0.16	0.9	0.7	0.13	0	49	N	N	Tr

No.	Food	Description and main data sources	Edible conversion factor	Water	Total nitrogen	Protein	Fat	Carbo-hydrate	Energy value	
				g	g	g	g	g	kcal	kJ
	Savoury snacks *continued*									
1076	**Popcorn**, candied	Recipe	1.00	2.6	0.33	2.1	20.0	77.6	480	2018
1077	plain	Recipe	1.00	0.9	0.99	6.2	42.8	48.7	593	2468
1078	**Pork scratchings**	19 samples, 4 brands	1.00	2.1	7.66	47.9	46.0	0.2	606	2520
1079	**Potato crisps**	20 samples, 8 brands; mixed plain and flavoured	1.00	2.8	0.91	5.7	34.2	53.3	530	2215
1080	lower fat	20 samples of different brands; mixed plain and flavoured	1.00	1.1	1.06	6.6	21.5	63.5	458	1924
1081	**Potato rings**	18 samples, 3 brands; assorted flavours; Hula Hoop type	1.00	2.8	0.62	3.9	32.0	58.5	523	2186
1082	**Pot savouries**	6 samples including assorted flavours of noodles, rice and chilli	1.00	8.9	1.86	11.6	10.9	58.8[a]	365	1541
1083	*made up*	85g product made up with 215ml water	1.00	74.2	0.53	3.3	3.1	16.7[b]	103	437
1084	**Tortilla chips**	20 samples, 6 brands, maize chips	1.00	0.9	1.22	7.6	22.6	60.1	459	1927
1085	**Twiglets**	20 samples, savoury wholewheat sticks	1.00	3.2	1.98	11.3	11.7	62.0	383	1617

[a] Includes 3.7g maltodextrins
[b] Includes 1.1g maltodextrins

Composition of food per 100g edible portion

No.	Food	Starch	Total sugars	Individual sugars					Dietary fibre	Fatty acids				Cholest-erol
				Gluc	Fruct	Sucr	Malt	Lact	NSP	Satd	Mono-unsatd	Poly-unsatd	Trans	
		g	g	g	g	g	g	g	g	g	g	g	g	mg
	Savoury snacks *continued*													
1076	**Popcorn**, candied	15.5	62.1	Tr	Tr	62.1	0	0	N	2.0	6.8	9.2	N	18
1077	plain	47.6	1.1	0.1	0.1	0.9	0	0	N	4.3	14.5	19.7	N	0
1078	**Pork scratchings**	Tr	0.2	0.2	Tr	Tr	0	0	0.3	N	N	N	N	N
1079	**Potato crisps**	52.6	0.7	0.1	0	0.5	0	Tr	5.3	14.0	13.7	5.0	N	0
1080	lower fat	62.0	1.5	0.2	Tr	0.8	0	0.5	5.9	9.3	8.7	2.5	0.1	0
1081	**Potato rings**	58.0	0.5	Tr	Tr	0.4	0	Tr	2.6	13.9	12.7	4.0	0.2	0
1082	**Pot savouries**	46.9	8.2	1.3	1.9	3.9	0.8	0.3	N	N	N	N	N	0
1083	*made up*	13.3	2.3	0.4	0.5	1.1	0.2	0.1	N	N	N	N	N	0
1084	**Tortilla chips**	58.9	1.2	0.1	0.1	1.0	0	0	(6.0)	4.0	10.6	6.7	4.4	0
1085	**Twiglets**	60.9	1.1	Tr	Tr	1.1	0	Tr	10.3	4.9	4.4	1.8	N	0

Inorganic constituents per 100g edible portion

No.	Food	mg										µg	
		Na	K	Ca	Mg	P	Fe	Cu	Zn	Cl	Mn	Se	I
	***Savoury snacks** continued*												
1076	**Popcorn**, candied	56	75	6	26	58	0.4	N	0.7	100	0.10	N	3
1077	plain	4	220	10	81	170	1.1	N	1.7	8	0.32	N	2
1078	**Pork scratchings**	1320	300	32	18	180	2.4	0.20	1.6	2090	0.09	N	N
1079	**Potato crisps**	800[a]	1060	29	57	110	1.4	0.15	0.6	1310	0.37	(1)	N
1080	lower fat	730	1020	36	48	130	1.8	0.38	0.9	1200	0.37	(1)	N
1081	**Potato rings**	1070	540	22	28	100	1.0	0.16	0.7	(1650)	0.21	(1)	N
1082	**Pot savouries**	1310	640	180	76	210	4.1	0.36	1.4	210	1.03	N	N
1083	*made up*	370	180	51	22	59	1.2	0.10	0.4	61	0.29	N	N
1084	**Tortilla chips**	860	220	150	89	240	1.6	0.09	1.2	1400	0.43	(3)	N
1085	**Twiglets**	1340	460	45	81	370	2.9	0.32	2.0	2520	1.61	N	N

[a] Na content ranged from 600mg to 1500mg per 100g. Lightly salted crisps contain about 400mg Na per 100g and unsalted crisps a trace

Vitamins per 100g edible portion

No.	Food	Retinol µg	Carotene µg	Vitamin D µg	Vitamin E mg	Thiamin mg	Ribo-flavin mg	Niacin mg	Trypt/60 mg	Vitamin B_6 mg	Vitamin B_{12} µg	Folate µg	Panto-thenate mg	Biotin µg	Vitamin C mg
	Savoury snacks *continued*														
1076	**Popcorn**, candied	52	98	0.1	3.75	0.06	0.04	0.3	0.2	0.07	0	3	0.10	1	0
1077	plain	0	230	0	11.03	0.18	0.11	1.0	0.7	0.20	0	9	0.30	4	0
1078	**Pork scratchings**	0	0	Tr	N	0.56	0.20	4.2	2.5	0.05	N	N	N	N	0
1079	**Potato crisps**	0	2	0	6.00	0.21	0.08	3.2	1.3	0.81	0	30	0.93	N	35
1080	lower fat	0	(2)	0	3.47	0.19	0.14	5.0	1.6	0.46	0	48	N	N	14
1081	**Potato rings**	0	0	0	N	N	N	N	0.1	N	0	N	N	N	3
1082	**Pot savouries**	0	N	0	N	N	N	N	N	N	0	N	N	N	0
1083	made up	0	N	0	N	N	N	N	N	N	0	N	N	N	0
1084	**Tortilla chips**	0	455	0	1.94	0.17	0.09	1.8	0.8	0.31	0	19	N	N	Tr
1085	**Twiglets**	0	Tr	0	2.47	0.37	0.48	7.8	2.3	0.38	0	78	1.54	15	Tr

Section 2.12

Beverages

The data in this section of the Tables have been taken from the *Miscellaneous Foods* supplement. New data have been incorporated for a few powdered drinks. Foods for which new data are incorporated have been allocated a new food code and can thus easily be identified in the food index.

This section includes beverages that are made up and drunk with milk as well as carbonated drinks, squash, cordials and fruit juices. Values for drinking chocolate have been given made-up with whole milk and semi-skimmed milk. Examples of the amounts of powder/essence and liquid that have been used in previous supplements (*Milk Products and Eggs; Miscellaneous Foods*) to calculate the made-up or diluted form are given below. As it is difficult to cover the range of strengths in which instant coffee, squash and cordials are consumed, only one entry, for the undiluted form, is given in the main Tables.

Drink	*Conversion information for calculation of made-up/diluted form*
Bournvita powder	8g powder with 200ml milk
Build-up powder, shake	38g powder with 284ml milk
Cocoa powder	4g cocoa powder, 4g sugar, 200ml milk
Coffee and chicory essence	10g essence with 225ml water
Complan powder, savoury	57g powder with 200ml water
Complan powder, sweet	58g powder with 200ml water
Horlicks powder	25g powder with 200ml milk
Milkshake powder	15g powder with 200ml milk
Ovaltine powder	25g powder with 200ml milk

Losses of labile vitamins assigned to made-up powdered drinks have been estimated from figures in Section 4.3.

The vitamin composition of beverages may be different from that quoted in these Tables if manufacturers have added to or changed the fortification of products. Concentrations of vitamin C in many fruit-based drinks can vary widely depending on fortification practices. The user should check the label of any beverage of this type to establish its vitamin C composition.

As many beverages may be sold or measured by volume, typical specific gravities (densities) of some of these products are given below.

Carbonated beverages		*Squashes and Cordials*	
Cola	1.040	Blackcurrant fruit drinks, undiluted	1.280
Fruit juice drinks	1.040	Fruit drinks, undiluted	1.090–1.120
Lemonade	1.020	Fruit drinks, low calorie, undiluted	1.010–1.030
Lucozade	1.070	Fruit juice drinks, ready to drink	1.030–1.040
		Lime juice cordial, undiluted	1.102

Composition of food per 100g edible portion

No.	Food	Description and main data sources	Edible conversion factor	Water	Total nitrogen	Protein	Fat	Carbo-hydrate	Energy value	
				g	g	g	g	g	kcal	kJ
	Powdered drinks and essences									
1086	**Bournvita powder**	6 samples	1.00	1.5	1.23	7.7	1.5	79.0	341	1450
1087	**Build-up powder**, shake	Manufacturer's data (Nestlé). Average of chocolate, strawberry, lemon & lime, vanilla and neutral flavours	1.00	(3.0)	3.68	23.0	1.1	65.3	347	1477
1088	soup	Manufacturer's data (Nestlé). Average of chicken, mushroom and potato & leek flavours	1.00	(3.0)	3.29	20.3	8.2	59.2[a]	377	1596
1089	**Cocoa powder**	10 samples, 2 brands	1.00	3.4	3.70[b]	18.5[c]	21.7	11.5	312	1301
1090	**Coffee and chicory essence**	7 bottles of the same brand (CAMP)	1.00	36.9	0.33[d]	1.6[c]	0.2	56.0	218	931
1091	**Coffee**, infusion, average	Average of strong and weak infusions	1.00	98.3	0.03	0.2	Tr	0.3	2	8
1092	instant	10 jars, 2 brands	1.00	3.4	3.26[e]	14.6[c]	Tr	4.5	75	320
1093	**Coffeemate**	Analysis and manufacturer's data (Nestlé)	1.00	3.0	0.42	2.7	34.9	57.3[a]	540	2254
1094	**Complan powder**, original & sweet	6 flavours, manufacturer's data (Heinz)	1.00	3.5	2.46	15.7	14.8	65.2[a]	441	1858
1095	savoury	Chicken flavour, manufacturer's data (Heinz)	1.00	3.5	2.42	15.4	14.6	63.6[a]	432	1820
1096	**Drinking chocolate powder**	10 tins, 3 brands	1.00	2.1	1.02[f]	5.4[c]	5.8	79.7	373	1582
1097	*made up with whole milk*	Calculated from 18g powder to 200ml milk	1.00	80.5	0.56	3.5	4.0	10.7	90	388
1098	*made up with semi-skimmed milk*	Calculated from 18g powder to 200ml milk	1.00	82.3	0.58	3.6	2.0	10.9	73	310
1099	reduced fat	10 samples, 4 brands	1.00	2.0	1.02	6.4	2.3	82.1	354	1507
1100	**Horlicks LowFat Instant powder**	Manufacturer's data (SmithKlineBeecham)	1.00	N	2.19	13.7	4.1	73.0	365	1553

[a] Including oligosaccharides from the glucose syrup/maltodextrins in the product
[b] Includes 0.74g purine nitrogen
[c] (Total N – purine N) × 6.25
[d] Includes 0.08g purine nitrogen
[e] Includes 0.93g purine nitrogen
[f] Includes 0.16g purine nitrogen

No.	Food	Starch	Total sugars	Individual sugars					Dietary fibre	Fatty acids				Cholest-erol
				Gluc	Fruct	Sucr	Malt	Lact	NSP	Satd	Mono-unsatd	Poly-unsatd	Trans	
		g	g	g	g	g	g	g	g	g	g	g	g	mg
	Powdered drinks and essences													
1086	**Bournvita powder**	27.0	52.0	N	N	N	N	N	N	N	N	N	N	N
1087	**Build-up powder**, shake	Tr	65.3	N	N	N	N	34.7	Tr	0.6	0	Tr	Tr	(12)
1088	soup	24.0	32.8[a]	N	N	N	N	17.2	N	3.8	Tr	Tr	0	0
1089	**Cocoa powder**	11.5	Tr	0	0	0	0	0	12.1	12.8	7.2	0.6	N	0
1090	**Coffee and chicory essence**	2.2	53.8	2.9	3.4	47.5	0	0	0	Tr	Tr	Tr	Tr	0
1091	**Coffee**, infusion, average	0	0	0	0	0	0	0	0	Tr	Tr	Tr	Tr	0
1092	instant	4.5	0	0	0	0	0	0	0	Tr	Tr	Tr	Tr	0
1093	**Coffeemate**	Tr	9.8[a]	5.2	0	0	4.6	Tr	0	32.1	1.1	Tr	Tr	2
1094	**Complan powder**, original & sweet	Tr	46.4[ab]	0.5	0	5.3[b]	2.8	37.8	Tr	6.6	6.3	1.6	Tr	N
1095	savoury	5.6	7.0[a]	1.9	0.1	0.2	3.6	1.2	0.3	6.2	5.9	1.6	Tr	N
1096	**Drinking chocolate powder**	Tr	77.7	0	0	77.7	0	0	N	3.4	1.8	0.3	0	0
1097	*made up with whole milk*	Tr	10.6	0	0	6.4	0	4.4	Tr	2.6	1.1	0.2	0.1	13
1098	*made up with semi-skimmed milk*	Tr	10.7	0	0	6.4	0	4.3	Tr	1.3	0.5	0.1	0.1	5
1099	reduced fat	Tr	82.1	Tr	0.9	81.2	Tr	Tr	N	(1.3)	(0.7)	(0.1)	0	Tr
1100	**Horlicks LowFat Instant powder**	N	N	N	N	N	N	N	N	N	N	N	Tr	Tr

[a] Not Including oligosaccharides from the glucose syrup/maltodextrins in the product
[b] Dependent on variety

No.	Food	mg										µg	
		Na	K	Ca	Mg	P	Fe	Cu	Zn	Cl	Mn	Se	I
	Powdered drinks and essences												
1086	**Bournvita powder**	190	330	62	110	250	1.9	0.64	1.4	70	N	N	N
1087	**Build-up powder**, shake	400	1050	850	263	697	12.0	(4.90)	13.0	(700)	(2.60)	N	131
1088	soup	1567	767	652	243	652	11.3	Tr	12.0	N	Tr	Tr	124
1089	**Cocoa powder**	950	1500	130	520	660	10.5	3.90	6.9	460	N	N	N
1090	**Coffee and chicory essence**	65	75	30	39	90	0.7	0.60	N	85	N	N	N
1091	**Coffee**, infusion, average	Tr	92	3	8	7	0.1	Tr	Tr	3	0.05	Tr	Tr
1092	instant	81	3780	140	330	310	4.6	0.62	1.1	65	2.10	9	Tr
1093	**Coffeemate**	200	900	4	N	350	N	N	N	N	N	N	N
1094	**Complan powder**, original & sweet	230	730	570	79	470	6.7	0.53	4.2	520	0.61	20	61
1095	savoury	1300	470	360	79	380	6.7	0.53	4.2	1300	0.61	20	61
1096	**Drinking chocolate powder**	228	495	39	132	193	3.5	3.69	5.6	107	1.00	N	165
1097	*made up with whole milk*	58	183	112	21	101	0.3	0.30	0.9	90	0.09	1	42
1098	*made up with semi-skimmed milk*	58	184	114	21	102	0.3	0.30	0.9	89	0.08	1	42
1099	reduced fat	(228)	(495)	(39)	(132)	(193)	(3.5)	(3.69)	(5.6)	(107)	(1.00)	N	(165)
1100	**Horlicks LowFat Instant powder**	800	870	655	54	N	N	0.10	0.7	N	N	N	N

No.	Food	Retinol µg	Carotene µg	Vitamin D µg	Vitamin E mg	Thiamin mg	Ribo-flavin mg	Niacin mg	Trypt/60 mg	Vitamin B_6 mg	Vitamin B_{12} µg	Folate µg	Panto-thenate mg	Biotin µg	Vitamin C mg
	Powdered drinks and essences														
1086	**Bournvita powder**	Tr	Tr	Tr	Tr	N	N	N	1.7	N	Tr	N	N	N	0
1087	**Build-up powder**, shake	700	Tr	4.5	8.70	1.30	1.40	15.6	(5.8)	1.80	1.6	174	5.30	130	53
1088	soup	0	690	4.2	8.10	1.23	1.33	14.5	N	1.73	0.7	163	4.97	120	49
1089	**Cocoa powder**	0	(40)	0	0.68	0.16	0.06	1.7	3.9	0.07	0.4	38	N	N	0
1090	**Coffee and chicory essence**	0	N	0	N	0	0.03	2.8	N	N	0.4	N	N	N	0
1091	**Coffee**, infusion, average	0	0	0	Tr	Tr	0.01	0.7	0	Tr	0.4	Tr	Tr	3	0
1092	instant	0	N	0	Tr	0.04	0.21	24.8[a]	2.9	0.02	0.4	11	Tr	67	0
1093	**Coffeemate**	0	200	0	N	0	1.00	0	0.6	0	0.4	0	0	0	0
1094	**Complan powder**, original & sweet	310	Tr	4.4	3.50	0.78	0.58	7.0	3.5	0.61	0.7	170	2.30	50	44
1095	savoury	310	Tr	4.4	3.50	0.78	0.58	7.0	5.2	0.61	0.7	170	2.30	50	44
1096	**Drinking chocolate powder**	0	N	0	0.41	0.02	0.06	0.6	1.2	0.01	0.4	7	0.30	9	0
1097	*made up with whole milk*	32	18	4.2	0.11	0.03	0.20	0.2	0.7	0.05	0.8	6	0.51	3	1
1098	*made up with semi-skimmed milk*	17	8	Tr	0.06	0.03	0.20	0.1	0.8	0.05	0.4	5	0.29	3	1
1099	reduced fat	0	N	0	(0.16)	(0.02)	(0.06)	(0.6)	(1.2)	(0.01)	0.4	(7)	(0.30)	(9)	0
1100	**Horlicks LowFat Instant powder**	500	Tr	3.1	6.30	0.88	1.00	11.3	4.2	1.25	0.6	125	N	N	38

[a] Can be as high as 39mg per 100g. Decaffeinated instant coffee contains about the same

Composition of food per 100g edible portion

No.	Food	Description and main data sources	Edible conversion factor	Water	Total nitrogen	Protein	Fat	Carbo-hydrate	Energy value	
				g	g	g	g	g	kcal	kJ
Powdered drinks and essences *continued*										
1101	**Horlicks powder**	Manufacturer's data (SmithKline Beecham)	1.00	2.5	1.54	9.6	4.7	78.0	373	1585
1102	**Instant drinks powder**, chocolate, low calorie	11 samples, 6 brands, assorted flavours	1.00	4.5	2.56	16.0	11.1	52.0	359	1515
1103	malted	10 samples, 3 brands	1.00	3.4	1.87	11.7	9.5	75.7	416	1762
1104	**Milk shake powder**	6 samples (Nesquik), 3 flavours	1.00	0.5	0.21	1.3	1.6	98.3	388	1654
1105	**Ovaltine powder**	Manufacturer's data (Novartis)	1.00	2.0	1.14	7.3	1.9	81.4	352	1497
1106	**Tea**, black, infusion, average	15g leaves per litre water, strained after 5 minutes	1.00	99.5	Tr	0.1	Tr	Tr	Tr	
Carbonated drinks										
1107	**Cola**	10 samples, 6 brands	1.00	89.7	Tr	Tr	0	10.9	41	174
1108	diet	Calculated from Cola	1.00	99.8	Tr	Tr	0	Tr	1	2
1109	**Fruit juice drink**, carbonated, ready to drink	Mixed sample of different brands; bottles and cans; orange, lemon, apple and tropical fruit flavours e.g. Citrus Spring, Fanta, Orangina and Tango	1.00	89.7	Tr	Tr	Tr	10.3	39	165
1110	**Ginger ale**, dry	10 samples, 5 brands	1.00	95.9	0	0	0	3.9	15	62
1111	**Lemonade**	10 samples, 8 brands	1.00	93.8	Tr	Tr	0	5.8	22	93
1112	**Lucozade**	Analytical and manufacturer's data (SmithKline Beecham) including lemon, orange, tropical flavours	1.00	81.8	Tr	Tr	0	16.0[a]	60	256
1113	**Tonic water**	Ref. Cutrufelli and Matthews (1986)	1.00	91.1	0	0	0	8.8	33	141

[a] Includes 1.7g oligosaccharides

No.	Food	Starch	Total sugars	Individual sugars					Dietary fibre	Fatty acids				Cholest-erol
				Gluc	Fruct	Sucr	Malt	Lact	NSP	Satd	Mono-unsatd	Poly-unsatd	Trans	
		g	g	g	g	g	g	g	g	g	g	g	g	mg
	Powdered drinks and essences *continued*													
1101	**Horlicks powder**	25.0	53.0	N	N	N	N	N	4.0	N	N	N	Tr	N
1102	**Instant drinks powder**, chocolate, low calorie	Tr	33.0	0.6	Tr	Tr	Tr	32.9	N	8.1	1.7	0.9	0.1	3
1103	malted	Tr	33.8	2.3	0	6.6	8.8	16.1	N	8.7	0.2	0.1	0.1	5
1104	**Milk shake powder**	Tr	98.3	0.1	0.1	95.2	2.8	0.2	Tr	N	N	N	N	Tr
1105	**Ovaltine powder**	34.4	47.0	N	N	N	N	N	2.5	1.0	N	N	N	N
1106	**Tea**, black, infusion, average	0	Tr	0	0	0	0	0	0	Tr	Tr	Tr	Tr	0
	Carbonated drinks													
1107	**Cola**	Tr	10.9	3.5	3.4	4.0	0	0	0	0	0	0	0	0
1108	diet	0	Tr	Tr	Tr	Tr	0	0	0	0	0	0	0	0
1109	**Fruit juice drink**, carbonated, ready to drink	0	10.3	2.2	2.1	5.9	0.1	0	Tr	Tr	Tr	Tr	Tr	0
1110	**Ginger ale**, dry	0	3.9	1.7	1.6	0.5	0	0	0	0	0	0	0	0
1111	**Lemonade**	0	5.8	1.5	1.4	2.8	0.1	0	0	0	0	0	0	0
1112	**Lucozade**	Tr	14.3	7.5	4.6	0.2	2.0	0	0	0	0	0	0	0
1113	**Tonic water**	0	N	N	N	N	N	0	0	0	0	0	0	0

No.	Food	mg										µg	
		Na	K	Ca	Mg	P	Fe	Cu	Zn	Cl	Mn	Se	I
	Powdered drinks and essences *continued*												
1101	**Horlicks powder**	490	686	640	39	300	11.2	0.20	0.6	N	N	N	N
1102	**Instant drinks powder**, chocolate, low calorie	1513	1802	411	186	618	7.5	2.20	3.2	2399	1.10	N	178
1103	malted	(488)	(1191)	(349)	(67)	(470)	(0.6)	(7.57)	(5.1)	(753)	(0.10)	N	N
1104	**Milk shake powder**	20	150	8	27	53	2.0	0.10	0.4	27	0.20	N	32
1105	**Ovaltine powder**	130	156	800	300	N	28.0	N	Tr	136	N	N	N
1106	**Tea**, black, infusion, average	Tr	27	Tr	2	2	Tr	0.01	Tr	1	0.15	Tr	Tr
	Carbonated drinks												
1107	**Cola**	5	1	6	1	30	Tr	Tr	Tr	Tr	Tr	Tr	Tr
1108	diet	(5)	(1)	(6)	(1)	(30)	Tr	Tr	Tr	Tr	Tr	Tr	Tr
1109	**Fruit juice drink**, carbonated, ready to drink	8	27	7	7	2	Tr	Tr	Tr	3	Tr	Tr	Tr
1110	**Ginger ale**, dry	N	N	N	N	N	N	N	N	N	Tr	Tr	Tr
1111	**Lemonade**	7	15	5	1	Tr	Tr	Tr	Tr	2	Tr	Tr	Tr
1112	**Lucozade**	26	7	3	1	1	Tr	Tr	Tr	14	Tr	Tr	Tr
1113	**Tonic water**	4	0	1	0	0	Tr	Tr	Tr	Tr	Tr	Tr	Tr

Vitamins per 100g edible portion

No.	Food	Retinol µg	Carotene µg	Vitamin D µg	Vitamin E mg	Thiamin mg	Riboflavin mg	Niacin mg	Trypt/60 mg	Vitamin B_6 mg	Vitamin B_{12} µg	Folate µg	Pantothenate mg	Biotin µg	Vitamin C mg
	Powdered drinks and essences *continued*														
1101	**Horlicks powder**	640	0	4.0	N	1.12	1.28	14.4	3.0	N	0.8	160	N	N	48
1102	**Instant drinks powder**, chocolate, low calorie	Tr	16	0.8	0.74	0.10	0.80	1.1	200.0	0.04	0.2	13	1.70	21	0
1103	malted	0	(4)	Tr	(3.83)	(0.36)	(0.91)	(4.9)	(2.6)	(0.17)	Tr	(18)	(1.50)	(16)	Tr
1104	**Milk shake powder**	Tr	Tr	0	0.15	Tr	0.02	0.2	0.3	0.01	0.2	3	N	N	0
1105	**Ovaltine powder**	N	N	5.0	20.00	1.40	1.60	N	N	2.00	Tr	400	N	N	120
1106	**Tea**, black, infusion, average	0	0	0	N	Tr	0.02	0.2	0.2	Tr	0.2	3	0.04	1	0
	Carbonated drinks														
1107	**Cola**	0	0	0	0	0	0	0	0	0	0	0	0	0	0
1108	diet	0	0	0	0	0	0	0	0	0	0	0	0	0	0
1109	**Fruit juice drink**, carbonated, ready to drink	0	94	0	Tr	Tr	Tr	Tr	Tr	Tr	0	1	Tr	Tr	1[a]
1110	**Ginger ale**, dry	0	0	0	0	0	0	0	0	0	0	0	0	0	0
1111	**Lemonade**	0	Tr	0	Tr	Tr	Tr	Tr	Tr	Tr	0	Tr	Tr	Tr	Tr[b]
1112	**Lucozade**	0	835	0	0	Tr	Tr	Tr	Tr	Tr	0	1	Tr	Tr	8
1113	**Tonic water**	0	0	0	0	0	0	0	0	0	0	0	0	0	0

[a] Fortified product contains 14mg vitamin C per 100g

[b] 5-15mg vitamin C per 100g may be added to some brands

Composition of food per 100g edible portion

No.	Food	Description and main data sources	Edible conversion factor	Water	Total nitrogen	Protein	Fat	Carbo-hydrate	Energy value	
				g	g	g	g	g	kcal	kJ
	Squash and cordials									
1114	**Blackcurrant juice drink**, *undiluted*	Mixed sample and manufacturer's data (Ribena and own brands)	1.00	40.9	0.02	0.1	0	60.8[a]	228	975
1115	**Fruit drink/squash**, *undiluted*	Mixed sample; lemon, orange, apple and mixed fruit flavours	1.00	74.4	0.02	0.1	Tr	24.8[b]	93	399
1116	**Fruit drink**, low calorie, *undiluted*	10 samples, 2 brands; lemon, orange and mixed fruit flavours	1.00	97.3	0.02	0.1	Tr	0.8	3	15
1117	**Fruit juice drink**, ready to drink	Mixed sample; lemon, orange, apple and mixed fruit flavours	1.00	89.5	0.02	0.1	Tr	9.8	37	159
1118	low calorie, ready to drink	10 samples, 2 brands; mixed fruit flavours	1.00	96.6	0.03	0.2	Tr	2.5	10	43
1119	**Lime juice cordial**, *undiluted*	6 bottles of the same brand (Roses)	1.00	70.5	0.01	0.1	0	29.8	112	479
1120	**Sunny Delight**	Manufacturer's data. Average of Florida and California styles	1.00	84.5	Tr	Tr	0.2	10.0[a,d]	39[d]	166[d]
	Juices									
1121	**Apple juice**, unsweetened	10 samples; bottles and cartons	1.00	88.0	0.01	0.1	0.1	9.9	38	164
1122	**Cranberry juice**	Manufacturer's data (Ocean Spray) and literature	1.00	85.5	Tr	Tr	0	*14.4*	*61*	*259*
1123	**Grape juice**, unsweetened	10 samples, 6 brands; red and white juice	1.00	85.4	0.05	0.3	0.1	11.7	46	196
1124	**Grapefruit juice**, unsweetened	50 samples; cartons, canned, bottled and frozen[c]	1.00	89.4	0.07	0.4	0.1	8.3	33	140
1125	**Lemon juice**, *fresh*	Analysis and literature sources	1.00	91.4	0.05	0.3	Tr	1.6	7	31
1126	**Orange juice**, unsweetened	60 samples; fresh, canned, bottled and frozen	1.00	89.2	0.08	0.5	0.1	8.8	36	153
1127	**Orange juice concentrate**, unsweetened	17 samples, 58.4 Brix; imported commercial concentrate	1.00	41.6	0.46	2.9	0.5	44.9	185	786
1128	**Pineapple juice**, unsweetened	18 samples, cartons only	1.00	87.8	0.05	0.3	0.1	10.5	41	177
1129	**Tomato juice**	10 samples, 9 brands	1.00	93.8	0.13	0.8	Tr	3.0	14	62

[a] Includes oligosaccharides
[b] Includes 0.2g oligosaccharides
[c] Frozen samples were diluted as per manufacturers' instructions prior to analysis
[d] Orange outburst, blackcurrant blast, tropical tornado and apple and kiwi kick flavours contain 1.5g carbohydrate, 7kcal and 31kJ per 100g

Composition of food per 100g edible portion

No.	Food	Starch	Total sugars	Individual sugars					Dietary fibre	Fatty acids				Cholest-erol
				Gluc	Fruct	Sucr	Malt	Lact	NSP	Satd	Mono-unsatd	Poly-unsatd	Trans	
		g	g	g	g	g	g	g	g	g	g	g	g	mg
	Squash and cordials													
1114	**Blackcurrant juice drink**, *undiluted*	Tr	59.1	9.9	8.8	38.9	1.4	0	0	0	0	0	0	0
1115	**Fruit drink/squash**, *undiluted*	0	24.6	10.3	10.3	3.1	0.9	0	Tr	Tr	Tr	Tr	Tr	0
1116	**Fruit drink**, low calorie, *undiluted*	0	0.8	0.3	0.4	0.1	Tr	0	Tr	Tr	Tr	Tr	Tr	0
1117	**Fruit juice drink**, ready to drink	0	9.8	2.7	3.7	3.4	Tr	0	Tr	Tr	Tr	Tr	Tr	0
1118	low calorie, ready to drink	0	2.5	0.8	0.9	0.8	Tr	0	Tr	Tr	Tr	Tr	Tr	0
1119	**Lime juice cordial**, *undiluted*	Tr	29.8	11.5	11.0	5.9	1.4	0	0	0	0	0	0	0
1120	**Sunny Delight**	Tr	9.4	N	N	N	N	0	Tr	Tr	Tr	Tr	0	0
	Juices													
1121	**Apple juice**, unsweetened	0	9.9	2.6	6.3	1.1	0	0	Tr	Tr	Tr	0.1	Tr	0
1122	**Cranberry juice**	N	N	N	N	N	N	0	N	0	0	0	0	0
1123	**Grape juice**, unsweetened	0	11.7	5.5	6.2	Tr	0	0	0	Tr	Tr	Tr	Tr	0
1124	**Grapefruit juice**, unsweetened	0	8.3	3.0	3.3	2.0	0	0	Tr	Tr	Tr	Tr	Tr	0
1125	**Lemon juice**, *fresh*	0	1.6	0.5	0.9	0.2	0	0	0.1	Tr	Tr	Tr	Tr	0
1126	**Orange juice**, unsweetened	0	8.8	2.8	2.9	3.1	0	0	0.1	Tr	Tr	Tr	Tr	0
1127	**Orange juice concentrate**, unsweetened	0	44.9	11.7	12.3	20.9	0	0	Tr	0.1	0.1	0.2	Tr	0
1128	**Pineapple juice**, unsweetened	0	10.5	2.9	2.9	4.7	0	0	Tr	Tr	Tr	Tr	Tr	0
1129	**Tomato juice**	Tr	3.0	1.4	1.6	Tr	0	0	0.6	Tr	Tr	Tr	Tr	0

Inorganic constituents per 100g edible portion

No.	Food	mg										µg	
		Na	K	Ca	Mg	P	Fe	Cu	Zn	Cl	Mn	Se	I
	Squash and cordials												
1114	**Blackcurrant juice drink**, *undiluted*	16	92	8	2	3	0.2	0.01	0.1	2	Tr	Tr	Tr
1115	**Fruit drink/squash**, *undiluted*	40	27	6	1	2	Tr	Tr	Tr	4	Tr	Tr	Tr
1116	**Fruit drink**, low calorie, *undiluted*	40	31	5	1	2	Tr	Tr	Tr	3	Tr	Tr	Tr
1117	**Fruit juice drink**, ready to drink	5	44	6	3	2	Tr	Tr	Tr	3	Tr	Tr	Tr
1118	low calorie, ready to drink	5	48	5	3	3	Tr	Tr	Tr	2	0.03	Tr	Tr
1119	**Lime juice cordial**, *undiluted*	8	49	9	4	5	0.3	0.07	N	4	Tr	Tr	Tr
1120	**Sunny Delight**	38	25	1	N	33	Tr	Tr	Tr	N	Tr	Tr	Tr
	Juices												
1121	**Apple juice**, unsweetened	2	110	7	5	6	0.1	Tr	Tr	3	Tr	Tr	Tr
1122	**Cranberry juice**	N	N	N	N	N	N	N	N	N	N	0	N
1123	**Grape juice**, unsweetened	7	55	19	7	14	0.9	Tr	0.1	6	0.10	(1)	N
1124	**Grapefruit juice**, unsweetened	7	100	14	8	11	0.2	0.01	Tr	4	0.20	(1)	N
1125	**Lemon juice**, *fresh*	1	130	7	7	8	0.1	0.03	Tr	3	Tr	(1)	N
1126	**Orange juice**, unsweetened	10	150	10	8	13	0.2	Tr	Tr	9	0.10	(1)	(2)
1127	**Orange juice concentrate**, unsweetened	10	880	36	46	83	0.4	0.11	0.2	17	0.10	(5)	(11)
1128	**Pineapple juice**, unsweetened	8	53	8	6	1	0.2	0.02	0.1	15	0.70	Tr	Tr
1129	**Tomato juice**	230	230	10	10	19	0.4	0.06	0.1	400	0.10	Tr	(2)

No.	Food	Retinol µg	Carotene µg	Vitamin D µg	Vitamin E mg	Thiamin mg	Riboflavin mg	Niacin mg	Trypt/60 mg	Vitamin B_6 mg	Vitamin B_{12} µg	Folate µg	Pantothenate mg	Biotin µg	Vitamin C mg
	Squash and cordials														
1114	**Blackcurrant juice drink**, *undiluted*	0	N	0	N	Tr	Tr	7.8[a]	Tr	1.01[b]	3.0[b]	Tr	Tr	Tr	78[a]
1115	**Fruit drink/squash**, *undiluted*	0	690	0	N	Tr	Tr	0.1	Tr	0.01	0	2	Tr	Tr	25[c]
1116	**Fruit drink**, low calorie, *undiluted*	0	N	0	Tr	Tr	Tr	0.1	Tr	0.01	0	2	0.05	Tr	Tr
1117	**Fruit juice drink**, ready to drink	0	N	0	N	Tr	Tr	0.1	Tr	0.01	0	2	Tr	Tr	23[d]
1118	low calorie, ready to drink	0	Tr	0	Tr	0.02	Tr	0.1	Tr	0.01	0	2	0.06	Tr	5
1119	**Lime juice cordial**, *undiluted*	0	Tr	0	Tr	Tr	Tr	Tr	Tr	Tr	0	Tr	Tr	Tr	Tr
1120	**Sunny Delight**	0	720	0	Tr	0.21	Tr	Tr	Tr	0.30	0	Tr	Tr	Tr	30
	Juices														
1121	**Apple juice**, unsweetened	0	Tr	0	Tr	0.01	0.01	0.1	Tr	0.02	0	4	0.04	1	14
1122	**Cranberry juice**	0	0	0	0	Tr	Tr	Tr	Tr	Tr	0	Tr	Tr	Tr	30
1123	**Grape juice**, unsweetened	0	Tr	0	Tr	Tr	0.01	0.1	Tr	0.04	0	1	0.03	1	Tr
1124	**Grapefruit juice**, unsweetened	0	1	0	0.19	0.04	0.01	0.2	Tr	0.02	0	6	0.08	1	31
1125	**Lemon juice**, *fresh*	0	12	0	N	0.03	0.01	0.1	Tr	0.05	0	13	0.10	0	36
1126	**Orange juice**, unsweetened	0	17	0	0.17	0.08	0.02	0.2	0.1	0.07	0	18	0.13	1	39
1127	**Orange juice concentrate**, unsweetened	0	170	0	0.68	0.31	0.13	1.3	0.3	0.25	0	90	0.73	5	210
1128	**Pineapple juice**, unsweetened	0	8	0	0.03	0.06	0.01	0.1	0.1	0.05	0	8	0.07	Tr	11
1129	**Tomato juice**	0	200	0	1.01	0.02	0.02	0.7	0.1	0.06	0	10	(0.20)	(2)	8

[a] When fresh, provides 9.5mg niacin and 107mg vitamin C per 100g. Value will vary with brand and shelf life

[b] These are declared amounts and represent levels present at the end of shelf life

[c] Unfortified product contains 2mg vitamin C per 100g

[d] Unfortified product contains 7mg vitamin C per 100g

Section 2.13

Alcoholic beverages

The data in this section of the Tables have been taken from the *Miscellaneous Foods* (1994) supplement.

The values for wines were obtained on typical examples but, due to the variety of alcoholic strengths available, these should only be used as a guide, not as the definitive source for the composition of wines.

In contrast to foods in other parts of the Tables the data here represent composition per 100ml. The alcohol contents of a range of strengths 'by volume' is given below.

Alcohol contents of various strengths 'by volume'

% Alcohol by volume	*Alcohol (g/100 ml)*
5	4.0
10	7.9
15	11.9
20	15.8
25	19.8
30	23.7
35	27.7
40	31.6

For information regarding the specific gravity of drinks, see below.

Specific gravities of alcoholic beverages

Beers		***Ciders***		***Fortified wines***	
Beer, bitter, canned	1.008	**Cider**, dry	1.007	**Port**	1.026
Draught	1.004	sweet	1.012	**Sherry**, dry	0.988
Keg	1.001	vintage	1.017	medium	0.998
Mild, draught	1.009	***Wines***		sweet	1.009
Brown ale, bottled	1.008	**Red wine**	0.998	***Vermouths***	
Larger, bottled	1.005	**Rose wine**, medium	1.003	Vermouths, dry	1.005
Pale ale, bottled	1.003	**White wine**, dry	0.995	sweet	1.046
Stout, bottled	1.014	medium	1.005	***Liqueurs***	
Extra	1.002	sparkling	0.995	Advocaat	1.093
Strong ale	1.018	sweet	1.016	Cherry Brandy	1.093
				Curacao	1.052
				Spirits	
				40% volume	0.950

Composition of food per 100ml

No.	Food	Description and main data sources	Water	Alcohol	Total nitrogen	Protein	Fat	Carbo-hydrate	Energy value	
				g	g	g	g	g	kcal	kJ
Beers										
1130	**Beer**, bitter, average	5 samples from different brewers; canned, draught and bottled	(93.9)	2.9	0.05	0.3	Tr	2.2	30	124
1131	**Bitter**, best/premium	Mixed sample from different brewers	(93.0)	3.4	0.05	0.3	Tr	2.2	33	139
1132	**Brown ale**, bottled	Mixed sample from different brewers	(93.3)	2.5	0.04	0.3	Tr	3.0	30	126
1133	**Lager**	Mixed sample; Skol, Hofmeister, Tennents, Carling Black Label, Stella Artois and Fosters; canned and draught	(93.0)	4.0	0.05	0.3	Tr	Tr	29	121
1134	alcohol-free	10 samples; Kaliber and Barbican	96.3	Tr	0.06	0.4	Tr	1.5[a]	7	31
1135	low alcohol	10 samples; Carlton LA, Swan Light, Tennents LA	97.0	0.5	0.04	0.2	0	1.5[b]	10	41
1136	premium	10 samples; Carlsberg Special Brew and Heldenbrau Extra Special	(88.7)	6.9	0.05	0.3	Tr	2.4	59	244
1137	**Pale ale**, bottled	Mixed sample from different brewers	(93.9)	2.8	0.05	0.3	Tr	2.0	28	118
1138	**Shandy**	10 cans, 4 brands	(94.0)	0.7	Tr	Tr	0	5.0	24	100
1139	**Stout**, Guinness	10 samples; canned, bottled and draught	(90.2)	3.3	0.06	0.4	Tr	1.5	30	126
1140	**Strong ale/barley wine**	Mixed sample from different brewers	(86.3)	5.7	0.11	0.7	Tr	6.1	66	275
Ciders										
1141	**Cider**, dry	3 samples of different brands	(92.5)	3.8	Tr	Tr	0	2.6	36	152
1142	low alcohol	10 samples, 3 brands including Strongbow LA	94.9	0.6	Tr	Tr	0	3.6	17	74
1143	sweet	3 samples of different brands	(91.2)	3.7	Tr	Tr	0	4.3	42	176
1144	vintage	3 samples of the same brand	(80.6)	10.5	Tr	Tr	0	7.3	101	421

[a] Includes 0.3g oligosaccharides
[b] Includes 0.5g oligosaccharides

No.	Food	Starch	Total sugars	Individual sugars					Dietary fibre	Fatty acids			Cholest-erol
				Gluc	Fruct	Sucr	Malt	Lact	NSP	Satd	Mono-unsatd	Poly-unsatd	
		g	g	g	g	g	g	g	g	g	g	g	mg
	Beers												
1130	**Beer**, bitter, average	0	2.2	0	0	0	2.2	0	Tr	Tr	Tr	Tr	0
1131	**Bitter**, best/premium	0	2.2	0.3	0	0	1.9	0	Tr	Tr	Tr	Tr	0
1132	**Brown ale**, bottled	0	3.0	0.4	0.4	0.1	2.1	0	Tr	Tr	Tr	Tr	0
1133	**Lager**	0	Tr	Tr	0	0	0	0	Tr	Tr	Tr	Tr	0
1134	alcohol-free	0	1.2	0.6	0.4	Tr	0.2	0	Tr	Tr	Tr	Tr	0
1135	low alcohol	0	1.0	0.5	0.2	0	0.3	0	Tr	Tr	Tr	Tr	0
1136	premium	0	2.4	1.0	0	0	(1.4)	0	Tr	Tr	Tr	Tr	0
1137	**Pale ale**, bottled	0	2.0	0.7	Tr	0	1.3	0	Tr	Tr	Tr	Tr	0
1138	**Shandy**	0	5.0	1.6	1.7	1.7	0	0	Tr	0	0	0	0
1139	**Stout**, Guinness	0	1.5	Tr	Tr	0	(1.5)	Tr	N	Tr	Tr	Tr	0
1140	**Strong ale/barley wine**	0	6.1	Tr	Tr	0	6.1	0	Tr	Tr	Tr	Tr	0
	Ciders												
1141	**Cider**, dry	0	2.6	0.6	0.5	0.7	0.8	0	0	0	0	0	0
1142	low alcohol	0	3.6	0.7	1.4	1.4	0.1	0	0	0	0	0	0
1143	sweet	0	4.3	1.0	0.7	1.2	1.3	0	0	0	0	0	0
1144	vintage	0	7.3	1.8	1.3	2.0	2.3	0	0	0	0	0	0

Inorganic constituents per 100ml

No.	Food	mg										µg	
		Na	K	Ca	Mg	P	Fe	Cu	Zn	Cl	Mn	Se	I
Beers													
1130	**Beer**, bitter, average	6	32	8	7	14	0.1	0.01	0.1	24	0.03	Tr	N
1131	**Bitter**, best/premium	8	46	9	8	16	Tr	0.03	0.1	36	0.01	Tr	N
1132	**Brown ale**, bottled	16	33	7	6	11	0	0.07	0.3	37	Tr	Tr	N
1133	**Lager**	7	39	5	7	19	Tr	Tr	Tr	20	0.01	Tr	N
1134	alcohol-free	2	44	3	7	19	Tr	Tr	Tr	Tr	0.01	Tr	N
1135	low alcohol	12	56	8	12	10	Tr	Tr	Tr	1	0.01	Tr	N
1136	premium	7	39	5	7	19	Tr	Tr	Tr	20	0.01	Tr	N
1137	**Pale ale**, bottled	10	49	9	10	15	0	0.04	Tr	31	Tr	Tr	N
1138	**Shandy**	7	6	8	1	5	Tr	Tr	Tr	8	Tr	Tr	Tr
1139	**Stout**, Guinness	6	48	4	8	26	0.2	Tr	Tr	17	0.01	Tr	N
1140	**Strong ale/barley wine**	15	110	14	20	40	0	0.08	Tr	57	Tr	Tr	N
Ciders													
1141	**Cider**, dry	7	72	8	3	3	0.5	0.04	Tr	6	Tr	Tr	N
1142	low alcohol	3	81	7	2	4	0.1	0.03	Tr	2	0.01	Tr	Tr
1143	sweet	7	72	8	3	3	0.5	0.04	Tr	6	Tr	Tr	N
1144	vintage	2	97	5	4	9	0.3	0.02	Tr	5	Tr	Tr	N

No.	Food	Retinol µg	Carotene µg	Vitamin D µg	Vitamin E mg	Thiamin mg	Riboflavin mg	Niacin mg	Trypt/60 mg	Vitamin B_6 mg	Vitamin B_{12} µg	Folate µg	Pantothenate mg	Biotin µg	Vitamin C mg
	Beers														
1130	**Beer**, bitter, average	0	Tr	0	N	Tr	0.03	0.2	0.2	0.07	Tr	5	0.05	1	0
1131	**Bitter**, best/premium	0	Tr	0	N	Tr	0.04	0.8	0.2	0.09	Tr	8	0.07	1	0
1132	**Brown ale**, bottled	0	Tr	0	N	Tr	0.02	0.3	0.1	0.01	Tr	4	0.10	1	0
1133	**Lager**	0	Tr	0	N	Tr	0.04	0.7	0.3	0.06	Tr	12	0.03	1	0
1134	alcohol-free	0	Tr	0	N	Tr	0.02	0.6	0.4	0.03	Tr	5	0.09	Tr	0
1135	low alcohol	0	Tr	0	N	Tr	0.02	0.5	0.3	0.03	Tr	6	0.07	Tr	0
1136	premium	0	Tr	0	N	Tr	0.04	0.7	0.3	0.06	Tr	12	0.03	1	0
1137	**Pale ale**, bottled	0	Tr	0	N	Tr	0.02	0.4	0.2	0.01	Tr	4	(0.10)	(1)	0
1138	**Shandy**	0	Tr	0	N	Tr	Tr	0.1	Tr	0.01	Tr	1	0.02	Tr	0
1139	**Stout**, Guinness	0	Tr	0	N	Tr	0.03	0.8	0.2	0.08	Tr	6	0.04	1	0
1140	**Strong ale/barley wine**	0	Tr	0	N	Tr	0.06	0.8	0.4	0.04	Tr	9	N	N	0
	Ciders														
1141	**Cider**, dry	0	Tr	0	N	Tr	Tr	0	Tr	0.01	Tr	N	0.04	1	0
1142	low alcohol	0	Tr	0	N	Tr	Tr	0.1	Tr	Tr	Tr	2	0.07	Tr	0
1143	sweet	0	Tr	0	N	Tr	Tr	0	Tr	0.01	Tr	N	0.03	1	0
1144	vintage	0	Tr	0	N	Tr	Tr	0	Tr	(0.01)	Tr	N	(0.03)	(1)	0

Composition of food per 100ml

No.	Food	Description and main data sources	Water	Alcohol	Total nitrogen	Protein	Fat	Carbo-hydrate	Energy value	
				g	g	g	g	g	kcal	kJ
Wines										
1145	**Red wine**	Mixed sample from different countries	(88.4)	9.6[a]	0.03	0.1	0	0.2	68	283
1146	**Rose wine**, medium	5 samples from different countries	(87.3)	8.7	0.01	0.1	0	2.5	71	294
1147	**White wine**, dry	5 samples from different countries	(89.1)	9.1[b]	0.02	0.1	0	0.6	66	275
1148	medium	Mixed sample from different countries	(86.3)	8.9[c]	0.02	0.1	0	3.0	74	308
1149	sparkling	5 samples from different countries	(85.8)	7.6	0.04	0.3	0	5.1	74	307
1150	sweet	Mixed sample from different countries	(80.6)	10.2	0.03	0.2	0	5.9	94	394
Fortified wines										
1151	**Port**	2 samples	(71.1)	15.9	0.02	0.1	0	12.0	157	655
1152	**Sherry**, dry	1 sample	(81.0)	15.7	0.03	0.2	0	1.4	116	481
1153	medium	8 samples; including Spanish, British, Cyprus, own label	(78.8)	13.3	0.02	0.1	0	5.9	116	482
1154	sweet	1 sample	(74.8)	15.6	0.05	0.3	0	6.9	136	568
Vermouths										
1155	**Vermouth**, dry	5 samples of different brands	(82.1)	13.9	0.01	0.1	0	3.0	109	453
1156	sweet	5 samples of different brands	(70.6)	13.0	Tr	Tr	0	15.9	151	631

[a] Typical range 8.7g to 10.7g (11.0ml to 13.5ml) alcohol per 100ml

[b] Typical range 7.1g to 10.3g (9.0ml to 13.0ml) alcohol per 100ml

[c] Typical range 7.9g to 9.0g (10.0ml to 11.4ml) alcohol per 100ml

No.	Food	Starch	Total sugars	Individual sugars					Dietary fibre	Fatty acids			Cholest-erol
				Gluc	Fruct	Sucr	Malt	Lact	NSP	Satd	Mono-unsatd	Poly-unsatd	
		g	g	g	g	g	g	g	g	g	g	g	mg
Wines													
1145	**Red wine**	0	0.2	Tr	Tr	Tr	0	0	0	0	0	0	0
1146	**Rose wine**, medium	0	2.5	0.8	1.7	0	0	0	0	0	0	0	0
1147	**White wine**, dry	0	0.6	0.3	0.3	0	0	0	0	0	0	0	0
1148	medium	0	3.0	1.2	1.4	N	0	0	0	0	0	0	0
1149	sparkling	0	5.1	2.2	2.8	0.1	0	0	0	0	0	0	0
1150	sweet	0	5.9	2.6	3.3	0.1	0	0	0	0	0	0	0
Fortified wines													
1151	**Port**	0	12.0	4.6	4.6	2.8	0	0	0	0	0	0	0
1152	**Sherry**, dry	0	1.4	0.7	0.7	0	0	0	0	0	0	0	0
1153	medium	0	5.9	3.0	2.9	0	0	0	0	0	0	0	0
1154	sweet	0	6.9	3.6	3.5	0	0	0	0	0	0	0	0
Vermouths													
1155	**Vermouth**, dry	0	3.0	1.1	1.2	0.7	0	0	0	0	0	0	0
1156	sweet	0	15.9	6.1	6.1	3.7	0	0	0	0	0	0	0

Inorganic constituents per 100ml

No.	Food	mg										µg	
		Na	K	Ca	Mg	P	Fe	Cu	Zn	Cl	Mn	Se	I
	Wines												
1145	**Red wine**	7	110	7	11	13	0.9	0.06	0.1	11	0.10	Tr	N
1146	**Rose wine**, medium	4	75	12	7	6	1.0	0.02	Tr	7	0.10	Tr	N
1147	**White wine**, dry	4	61	9	8	6	0.5	0.01	Tr	10	0.10	Tr	N
1148	medium	11	81	12	8	8	0.8	Tr	Tr	3	0.10	Tr	N
1149	sparkling	5	58	9	7	9	0.5	0.01	Tr	7	0.04	Tr	N
1150	sweet	13	110	14	11	13	0.6	0.05	Tr	7	0.10	Tr	N
	Fortified wines												
1151	**Port**	4	97	4	11	12	0.4	0.10	N	8	Tr	Tr	N
1152	**Sherry**, dry	10	57	7	13	11	0.4	0.03	N	12	Tr	Tr	N
1153	medium	27	55	8	5	24	0.4	0.04	Tr	10	0.01	Tr	N
1154	sweet	13	110	7	11	10	0.4	0.11	N	14	Tr	Tr	N
	Vermouths												
1155	**Vermouth**, dry	11	34	7	6	6	0.3	0.03	Tr	7	Tr	Tr	N
1156	sweet	28	30	6	4	6	0.4	0.04	Tr	16	Tr	Tr	N

No.	Food	Retinol µg	Carotene µg	Vitamin D µg	Vitamin E mg	Thiamin mg	Riboflavin mg	Niacin mg	Trypt/60 mg	Vitamin B_6 mg	Vitamin B_{12} µg	Folate µg	Pantothenate mg	Biotin µg	Vitamin C mg
	Wines														
1145	**Red wine**	0	Tr	0	N	Tr	0.02	0.1	Tr	0.03	Tr	1	0.04	2	0
1146	**Rose wine**, medium	0	Tr	0	N	Tr	0.01	0.1	Tr	0.02	Tr	Tr	0.04	N	0
1147	**White wine**, dry	0	Tr	0	N	Tr	0.01	0.1	Tr	0.02	Tr	Tr	0.03	N	0
1148	medium	0	Tr	0	N	Tr	Tr	0.1	Tr	0.01	Tr	Tr	0.06	1	0
1149	sparkling	0	Tr	0	N	Tr	0.01	0.1	Tr	0.02	Tr	Tr	0.04	1	0
1150	sweet	0	Tr	0	N	Tr	0.01	0.1	Tr	0.01	Tr	Tr	0.03	N	0
	Fortified wines														
1151	**Port**	0	Tr	0	0	Tr	0.01	0.1	Tr	0.01	Tr	Tr	N	N	0
1152	**Sherry**, dry	0	Tr	0	0	Tr	0.01	0.1	Tr	0.01	Tr	Tr	N	N	0
1153	medium	0	Tr	0	0	Tr	0.01	0.1	Tr	0.02	Tr	Tr	0.02	1	0
1154	sweet	0	Tr	0	0	Tr	0.01	0.1	Tr	0.01	Tr	Tr	N	N	0
	Vermouths														
1155	**Vermouth**, dry	0	Tr	0	0	Tr	Tr	0	Tr	0.01	Tr	Tr	N	N	0
1156	sweet	0	Tr	0	0	Tr	Tr	0	Tr	Tr	Tr	Tr	N	N	0

Composition of food per 100ml

No.	Food	Description and main data sources	Water	Alcohol	Total nitrogen	Protein	Fat	Carbo-hydrate	Energy value	
				g	g	g	g	g	kcal	kJ
Liqueurs										
1157	**Cream liqueurs**	2 samples of Baileys Original Irish Cream	(44.4)	13.5	Tr	Tr	16.1	22.8	325	1352
1158	**Liqueurs**, high strength	5 samples including Pernod, Drambuie, Cointreau, Grand Marnier, Southern Comfort	(28.0)	31.8	Tr	Tr	0	24.4	314	1313
1159	low-medium strength	10 samples including Cherry brandy, Tia Maria and Creme de Menthe	(47.4)	19.8[a]	Tr	Tr	0	32.8	262	1099
Spirits										
1160	**Spirits**, 40% volume	Mean of brandy, gin, rum, whisky and vodka	(68.3)	31.7	Tr	Tr	0	Tr	222	919

[a] The alcohol content of Tia Maria is 20.9g, Pimms and Campari 19.8g, Malibu 19.0g, Monterez 13.8g per 100ml

No.	Food	Starch	Total sugars	Individual sugars					Dietary fibre	Fatty acids			Cholest-erol
				Gluc	Fruct	Sucr	Malt	Lact	NSP	Satd	Mono-unsatd	Poly-unsatd	
		g	g	g	g	g	g	g	g	g	g	g	mg
	Liqueurs												
1157	**Cream liqueurs**	Tr	22.8	0	0	22.0	0	0.8	0	N	N	N	N
1158	**Liqueurs**, high strength	0	24.4	2.6	2.3	17.1	2.4	0	0	0	0	0	0
1159	low-medium strength	0	32.8	6.3	6.1	20.4	0	0	0	0	0	0	0
	Spirits												
1160	**Spirits**, 40% volume	0	Tr	0	0	Tr	0	0	0	0	0	0	0

Inorganic constituents per 100ml

No.	Food	mg										µg	
		Na	K	Ca	Mg	P	Fe	Cu	Zn	Cl	Mn	Se	I
	Liqueurs												
1157	**Cream liqueurs**	89	19	18	2	38	0.1	Tr	0.2	25	Tr	Tr	N
1158	**Liqueurs**, high strength	6	3	Tr	Tr	Tr	Tr	Tr	Tr	4	Tr	Tr	Tr
1159	low-medium strength	12	34	5	2	7	0.1	0.02	Tr	20	0.02	N	N
	Spirits												
1160	**Spirits**, 40% volume	Tr	Tr	Tr	Tr	Tr	Tr	Tr	Tr	Tr	Tr	Tr	Tr

No.	Food	Retinol µg	Carotene µg	Vitamin D µg	Vitamin E mg	Thiamin mg	Ribo-flavin mg	Niacin mg	Trypt/60 mg	Vitamin B_6 mg	Vitamin B_{12} µg	Folate µg	Panto-thenate mg	Biotin µg	Vitamin C mg
Liqueurs															
1157	**Cream liqueurs**	190	91	Tr	0.57	N	N	N	N	N	0	Tr	N	N	0
1158	**Liqueurs**, high strength	0	Tr	0	0	Tr	Tr	Tr	Tr	Tr	Tr	Tr	Tr	Tr	0
1159	low-medium strength	0	Tr	0	0	Tr	Tr	Tr	Tr	Tr	Tr	Tr	Tr	Tr	0
Spirits															
1160	**Spirits**, 40% volume	0	0	0	0	0	0	0	0	0	0	0	0	0	0

Section 2.14

Soups, sauces and miscellaneous foods

The data in this section of the Tables have been taken from the *Miscellaneous Foods* (1994) supplement.

The foods in this group cover homemade, canned and packet soups; dairy sauces, salad sauces, dressings and pickles; non-salad sauces; and a selection of miscellaneous food items.

An entry for water has been included in the miscellaneous foods section, mainly for use in recipe calculations. There is considerable variation in the composition of tap water both by area of the country and source of supply. The local Water Company will be able to provide information on the composition of tap water from a specific area.

Dried soups as made up were corrected for evaporative loss.

Losses of labile vitamins assigned to recipes were estimated from figures in Section 4.3.

Composition of food per 100g edible portion

No.	Food	Description and main data sources	Edible conversion factor	Water	Total nitrogen	Protein	Fat	Carbo-hydrate	Energy value	
				g	g	g	g	g	kcal	kJ
	Canned soups									
1161	**Chicken soup**, cream of, canned	10 cans, 3 brands	1.00	87.9	0.27	1.7	3.8	4.5	58	242
1162	condensed	7 cans of the same brand	1.00	82.2	0.41	2.6	5.8	6.0	85	355
1163	condensed, *as served*	Diluted with an equal volume of water	1.00	91.1	0.20	1.3	2.9	3.0	43	177
1164	**Low calorie soup**, canned	7 cans, 3 brands; tomato, vegetable and minestrone varieties	1.00	93.3	0.14	0.9	0.2	4.0	20	87
1165	**Minestrone soup**, canned	Manufacturer's data (Heinz)	1.00	(92.1)	0.21	1.3	0.5	5.7	31	132
1166	**Mushroom soup**, cream of, canned	10 cans, 3 brands	1.00	90.4	0.20	1.1	3.0	3.9	46	192
1167	**Oxtail soup**, canned	10 cans, 3 brands	1.00	88.5	0.38	2.4	1.7	5.1	44	185
1168	**Tomato soup**, cream of, canned	10 cans, 3 brands	1.00	84.2	0.13	0.8	3.0	5.9	52	219
1169	condensed	7 cans, 2 brands	1.00	70.6	0.27	1.7	6.8	14.6	123	514
1170	condensed, *as served*	Diluted with an equal volume of water	1.00	85.3	0.14	0.9	3.4	7.3	62	258
1171	**Vegetable soup**, canned	10 cans, 3 brands	1.00	87.8	0.22	1.4	0.6	9.9	48	204
	Packet soups									
1172	**Chicken noodle soup**, *dried, as served*	Calculated from 35g soup powder to 570ml water	1.00	94.5	0.16	1.0	0.3	3.2[a]	19	79
1173	**Instant soup powder**, *dried*	10 packets, 3 brands; assorted flavours	1.00	4.1	1.04	6.5	14.3	64.4[b]	396	1670
1174	*made up with water*	Calculated from 37g powder to 190ml water	1.00	84.4	0.17	1.1	2.3	10.5[c]	64	270
1175	**Minestrone soup**, *dried, as served*	Calculated from 45g soup powder to 570ml water	1.00	92.6	0.17	0.7	0.4	4.2[d]	22	94
1176	**Tomato soup**, *dried, as served*	Calculated from 58g soup powder to 570ml water	1.00	91.2	0.07	0.4	1.3	6.0[e]	36	151
1177	**Vegetable soup**, *dried, as served*	Calculated from 45g soup powder to 570ml water	1.00	93.1	0.14	0.9	0.3	4.2[d]	22	92

[a] Includes 0.3g maltodextrins
[b] Includes 18.7g maltodextrins
[c] Includes 3.0g maltodextrins
[d] Includes 0.1g maltodextrins
[e] Includes 0.5g maltodextrins

No.	Food	Starch	Total sugars	Individual sugars Gluc	Fruct	Sucr	Malt	Lact	Dietary fibre NSP	Fatty acids Satd	Mono-unsatd	Poly-unsatd	Trans	Cholest-erol
		g	g	g	g	g	g	g	g	g	g	g	g	mg
	Canned soups													
1161	**Chicken soup**, cream of, canned	3.4	1.1	Tr	0.1	0.6	0	0.4	Tr	(0.6)	(2.0)	(1.0)	0.1	97
1162	condensed	4.6	1.4	Tr	0.2	0.4	0	0.8	Tr	0.8	3.0	1.4	N	(4)
1163	condensed, *as served*	2.3	0.7	Tr	0.1	0.2	0	0.4	0	0.4	1.5	0.7	N	(2)
1164	**Low calorie soup**, canned	2.0	2.0	1.0	1.0	Tr	0	0	N	Tr	Tr	Tr	Tr	0
1165	**Minestrone soup**, canned	4.5	1.2	N	N	N	N	0	0.7	0.3	0.1	0.1	Tr	1
1166	**Mushroom soup**, cream of, canned	3.1	0.8	Tr	0.1	0.3	0	0.4	0.1	0.5	1.6	0.9	0.1	1
1167	**Oxtail soup**, canned	4.2	0.9	0.2	0.2	0.5	Tr	0	0.1	0.6	0.6	0.2	N	(7)
1168	**Tomato soup**, cream of, canned	3.3	2.6	0.8	0.6	1.2	0	Tr	0.7	0.5	1.6	0.8	0.2	1
1169	condensed	3.4	11.2	2.4	1.8	6.2	0	0.8	1.0	1.0	2.6	3.0	N	(1)
1170	condensed, *as served*	1.7	5.6	1.2	0.9	3.1	0	0.4	0.5	0.5	1.3	1.5	N	Tr
1171	**Vegetable soup**, canned	4.8	5.1	1.4	1.6	2.1	0	0	1.5	N	N	N	N	N
	Packet soups													
1172	**Chicken noodle soup**, *dried, as served*	2.7	0.3	Tr	Tr	0.2	Tr	0	0.2	N	N	N	N	N
1173	**Instant soup powder**, *dried*	34.1	11.3	2.1	2.2	7.0	Tr	Tr	N	6.9	6.2	0.5	2.8	6
1174	*made up with water*	5.6	1.8	0.3	0.4	1.1	Tr	0	N	1.0	0.6	0.1	0.4	1
1175	**Minestrone soup**, *dried, as served*	2.7	1.4	0.2	0.3	0.9	Tr	0	N	N	N	N	N	0
1176	**Tomato soup**, *dried, as served*	2.1	3.4	0.4	0.5	2.5	Tr	Tr	N	0.6	0.3	Tr	Tr	Tr
1177	**Vegetable soup**, *dried, as served*	3.2	0.9	0.2	0.2	0.5	Tr	0	N	N	N	N	N	0

Inorganic constituents per 100g edible portion

No.	Food	mg										µg	
		Na	K	Ca	Mg	P	Fe	Cu	Zn	Cl	Mn	Se	I
	Canned soups												
1161	**Chicken soup**, cream of, canned	400	41	27	5	27	0.4	0.02	0.3	610	Tr	Tr	(2)
1162	condensed	710	(62)	(41)	(7)	(41)	(0.5)	(0.03)	(0.5)	1070	Tr	Tr	(4)
1163	condensed, *as served*	350	(31)	(20)	(4)	(20)	(0.3)	(0.02)	(0.3)	530	Tr	Tr	(2)
1164	**Low calorie soup**, canned	370	130	13	7	17	0.3	0.01	0.1	580	0.05	N	N
1165	**Minestrone soup**, canned	300	100	18	9	24	0.3	0.04	0.2	470	0.10	0	1
1166	**Mushroom soup**, cream of, canned	470	55	30	4	30	0.3	0.04	0.3	750	Tr	1	(3)
1167	**Oxtail soup**, canned	440	93	40	6	37	1.0	0.04	0.4	660	Tr	Tr	(1)
1168	**Tomato soup**, cream of, canned	400	190	17	8	20	0.4	0.06	0.2	640	0.10	Tr	(3)
1169	condensed	830	(360)	(32)	(15)	(38)	(0.7)	(0.11)	0.3	1320	0.10	Tr	(5)
1170	condensed, *as served*	410	(180)	(16)	(8)	(19)	(0.3)	(0.06)	0.2	660	0.10	Tr	(3)
1171	**Vegetable soup**, canned	430	110	12	8	29	0.4	0.04	0.2	660	0.07	N	(16)
	Packet soups												
1172	**Chicken noodle soup**, *dried, as served*	300	14	4	3	15	0.2	0.01	0.1	440	0.04	N	N
1173	**Instant soup powder**, *dried*	3440	610	48	27	200	1.7	0.17	0.7	4770	0.25	N	N
1174	*made up with water*	560	100	8	4	33	0.3	0.03	0.1	780	0.04	N	N
1175	**Minestrone soup**, *dried, as served*	470	49	11	4	21	0.2	0.02	0.1	630	0.04	Tr	(5)
1176	**Tomato soup**, *dried, as served*	290	78	9	4	26	0.1	0.03	0.1	460	0.03	N	N
1177	**Vegetable soup**, *dried, as served*	370	50	11	5	18	0.2	0.02	0.1	520	0.04	N	N

Vitamins per 100g edible portion

No.	Food	Retinol µg	Carotene µg	Vitamin D µg	Vitamin E mg	Thiamin mg	Riboflavin mg	Niacin mg	Trypt/60 mg	Vitamin B_6 mg	Vitamin B_{12} µg	Folate µg	Pantothenate mg	Biotin µg	Vitamin C mg
	Canned soups														
1161	**Chicken soup**, cream of, canned	(39)	(16)	Tr	(0.55)	0.01	0.03	0.2	0.3	0.01	Tr	(1)	(0.04)	0	0
1162	condensed	(96)	(39)	0	(0.93)	(0.02)	0.04	0.6	0.5	(0.01)	Tr	(1)	(0.06)	0	0
1163	condensed, *as served*	(48)	(20)	0	(0.46)	(0.01)	0.02	0.3	0.2	(0.01)	Tr	Tr	(0.03)	0	0
1164	**Low calorie soup**, canned	0	N	0	N	0.35	0.14	2.0	0.1	0.20	0	(10)	N	N	Tr
1165	**Minestrone soup**, canned	4	340	0	0.27	0.02	0.01	0.2	0.2	0.04	0	5	0.05	Tr	0
1166	**Mushroom soup**, cream of, canned	(40)	(16)	0	(0.54)	Tr	0.05	0.3	0.2	0.01	Tr	(2)	(0.10)	(1)	0
1167	**Oxtail soup**, canned	0	0	0	(0.20)	0.02	0.03	0.7	0.5	0.03	0	1	(0.05)	0	0
1168	**Tomato soup**, cream of, canned	(40)	210	0	(1.40)	0.03	0.02	0.5	0.1	0.06	Tr	12	(0.12)	(1)	Tr
1169	condensed	0	(400)	0	(3.49)	(0.06)	0.05	1.0	0.2	(0.10)	0	(10)	(0.24)	(1)	Tr
1170	condensed, *as served*	0	(200)	0	(1.75)	(0.03)	0.03	0.5	0.1	(0.05)	Tr	(5)	(0.12)	(1)	Tr
1171	**Vegetable soup**, canned	0	18	0	N	0.09	0.02	2.5	0.2	0.01	0	10	N	N	Tr
	Packet soups														
1172	**Chicken noodle soup**, *dried, as served*	Tr	0	0	N	0.01	0.01	0.2	0.2	N	Tr	N	N	N	0
1173	**Instant soup powder**, *dried*	0	N	0	N	(0.05)	(0.02)	(0.4)	1.0	N	0	N	N	N	0
1174	*made up with water*	0	N	0	N	(0.01)	Tr	(0.1)	0.2	N	0	N	N	N	0
1175	**Minestrone soup**, *dried, as served*	0	N	0	N	0.02	0.01	0.2	0.1	N	0	N	N	N	0
1176	**Tomato soup**, *dried, as served*	0	N	0	N	Tr	Tr	Tr	Tr	N	0	N	N	N	0
1177	**Vegetable soup**, *dried, as served*	0	N	0	N	Tr	Tr	Tr	0.2	N	0	N	N	N	Tr

Composition of food per 100g edible portion

No.	Food	Description and main data sources	Edible conversion factor	Water	Total nitrogen	Protein	Fat	Carbo-hydrate	Energy value	
				g	g	g	g	g	kcal	kJ
	Dairy sauces									
1178	**Bread sauce**, *made with whole milk*	Recipe	1.00	75.2	0.67	4.1	4.0	15.2	110	463
1179	*made with semi-skimmed milk*	Recipe	1.00	76.5	0.68	4.2	2.5	15.3	97	409
1180	**Cheese sauce**, *made with whole milk*	Recipe	1.00	66.8	1.27	8.1	14.8	8.7	198	824
1181	*made with semi-skimmed milk*	Recipe	1.00	68.6	1.29	8.2	12.8	8.8	181	754
1182	**Cheese sauce packet mix**, *made with whole milk*	Recipe	1.00	77.0	0.86	5.4	6.0	9.0	111	461
1183	*made with semi-skimmed milk*	Recipe	1.00	79.0	0.88	5.5	3.8	9.2	91	383
1184	**Onion sauce**, *made with whole milk*	Recipe	1.00	80.5	0.46	2.9	6.6	8.1	101	422
1185	*made with semi-skimmed milk*	Recipe	1.00	81.8	0.47	3.0	5.1	8.2	88	369
1186	**White sauce**, savoury, *made with whole milk*	Recipe	1.00	73.5	0.67	4.2	10.3	10.6	151	626
1187	*made with semi-skimmed milk*	Recipe	1.00	75.6	0.70	4.4	8.0	10.7	130	541
1188	sweet, *made with whole milk*	Recipe	1.00	68.1	0.62	3.9	9.5	18.3	171	714
1189	*made with semi-skimmed milk*	Recipe	1.00	70.0	0.64	4.0	7.4	18.5	152	636

Composition of food per 100g edible portion

No.	Food	Starch	Total sugars	Individual sugars					Dietary fibre	Fatty acids				Cholest-erol
				Gluc	Fruct	Sucr	Malt	Lact	NSP	Satd	Mono-unsatd	Poly-unsatd	Trans	
		g	g	g	g	g	g	g	g	g	g	g	g	mg
	Dairy sauces													
1178	**Bread sauce**, *made with whole milk*	9.9	4.8	0.5	0.4	0.5	Tr	3.1	0.6	1.9	1.1	0.6	0.1	10
1179	*made with semi-skimmed milk*	9.9	4.9	0.5	0.4	0.5	Tr	3.2	0.6	1.0	0.6	0.5	0.1	4
1180	**Cheese sauce**, *made with whole milk*	4.6	4.1	Tr	Tr	0	0	4.0	0.2	7.2	4.2	2.5	0.4	30
1181	*made with semi-skimmed milk*	4.6	4.2	Tr	Tr	0	0	4.1	0.2	5.9	3.7	2.4	0.3	23
1182	**Cheese sauce packet mix**, *made with whole milk*	3.9	5.1	Tr	Tr	0	Tr	5.0	N	N	N	N	N	N
1183	*made with semi-skimmed milk*	3.9	5.3	Tr	Tr	0	Tr	5.2	N	N	N	N	N	N
1184	**Onion sauce**, *made with whole milk*	3.6	4.1	Tr	Tr	0	0	3.0	0.4	2.5	1.9	1.8	0.1	10
1185	*made with semi-skimmed milk*	3.5	4.2	Tr	Tr	0	0	3.2	0.4	1.5	1.5	1.8	0.1	4
1186	**White sauce**, savoury, *made with whole milk*	5.6	4.9	Tr	Tr	0	0	4.8	0.2	4.0	3.1	2.9	0.2	15
1187	*made with semi-skimmed milk*	5.6	5.1	Tr	Tr	0	0	5.0	0.2	2.4	2.4	2.7	0.1	6
1188	sweet, *made with whole milk*	5.2	13.1	Tr	Tr	8.6	0	4.4	0.2	3.6	2.8	2.6	0.1	14
1189	*made with semi-skimmed milk*	5.2	13.3	Tr	Tr	8.6	0	4.6	0.2	2.2	2.2	2.5	0.1	6

Inorganic constituents per 100g edible portion

No.	Food	mg										µg	
		Na	K	Ca	Mg	P	Fe	Cu	Zn	Cl	Mn	Se	I
	Dairy sauces												
1178	**Bread sauce**, *made with whole milk*	399	164	104	13	87	0.5	0.04	0.4	627	0.11	1	22
1179	*made with semi-skimmed milk*	399	165	105	13	88	0.5	0.04	0.4	626	0.10	1	21
1180	**Cheese sauce**, *made with whole milk*	447	158	246	16	180	0.2	0.02	1.1	693	0.05	2	33
1181	*made with semi-skimmed milk*	446	160	247	16	181	0.2	0.02	1.1	691	0.04	2	32
1182	**Cheese sauce packet mix**, *made with whole milk*	452	190	165	14	173	0.1	Tr	0.7	642	0.04	N	N
1183	*made with semi-skimmed milk*	451	191	167	14	174	0.1	Tr	0.7	639	0.03	N	N
1184	**Onion sauce**, *made with whole milk*	245	145	93	10	74	0.2	0.03	0.4	392	0.03	1	22
1185	*made with semi-skimmed milk*	244	146	94	10	75	0.2	0.03	0.4	391	0.03	1	21
1186	**White sauce**, savoury, *made with whole milk*	386	177	137	13	108	0.2	0.01	0.5	618	0.05	1	34
1187	*made with semi-skimmed milk*	385	179	139	13	109	0.2	0.01	0.5	615	0.04	1	33
1188	sweet, *made with whole milk*	89	163	127	12	100	0.2	0.02	0.4	162	0.05	1	31
1189	*made with semi-skimmed milk*	89	165	129	12	101	0.2	0.02	0.4	160	0.04	1	30

No.	Food	Retinol µg	Carotene µg	Vitamin D µg	Vitamin E mg	Thiamin mg	Riboflavin mg	Niacin mg	Trypt/60 mg	Vitamin B_6 mg	Vitamin B_{12} µg	Folate µg	Pantothenate mg	Biotin µg	Vitamin C mg
	Dairy sauces														
1178	**Bread sauce**, *made with whole milk*	31	20	0.1	0.56	0.06	0.17	0.5	0.8	0.08	1	6	0.37	2	1
1179	*made with semi-skimmed milk*	22	13	0.1	0.53	0.08	0.17	0.5	0.9	0.09	Tr	9	0.24	2	2
1180	**Cheese sauce**, *made with whole milk*	135	64	0.5	2.15	0.04	0.28	0.2	1.9	0.07	1	7	0.50	3	1
1181	*made with semi-skimmed milk*	123	54	0.5	2.11	0.05	0.28	0.2	2.0	0.08	1	10	0.33	3	Tr
1182	**Cheese sauce packet mix**, *made with whole milk*	48	36	0.1	0.28	0.04	0.30	0.3	1.1	0.08	1	9	N	N	1
1183	*made with semi-skimmed milk*	34	26	0.1	0.24	0.04	0.30	0.2	1.2	0.08	1	8	N	N	Tr
1184	**Onion sauce**, *made with whole milk*	54	31	0.4	1.64	0.06	0.16	0.3	0.5	0.09	1	7	0.35	2	2
1185	*made with semi-skimmed milk*	44	24	0.4	1.61	0.07	0.16	0.3	0.7	0.08	Tr	6	0.21	2	1
1186	**White sauce**, savoury, *made with whole milk*	85	47	0.6	2.51	0.05	0.25	0.3	0.8	0.06	1	6	0.52	3	1
1187	*made with semi-skimmed milk*	69	35	0.6	2.47	0.06	0.26	0.2	1.0	0.06	Tr	5	0.30	2	1
1188	sweet, *made with whole milk*	78	43	0.5	2.32	0.04	0.23	0.3	0.7	0.06	1	5	0.48	3	1
1189	*made with semi-skimmed milk*	64	33	0.5	2.27	0.05	0.24	0.2	0.9	0.06	Tr	4	0.27	2	0

Composition of food per 100g edible portion

No.	Food	Description and main data sources	Edible conversion factor	Water	Total nitrogen	Protein	Fat	Carbo-hydrate	Energy value	
				g	g	g	g	g	kcal	kJ
	Salad sauces, dressings and pickles									
1190	**Apple chutney**	Recipe	1.44	47.7	0.14	0.9	0.2	49.2	190	810
1191	**Chutney**, mango, oily	10 assorted samples	1.00	34.8	0.06	0.4	10.9	49.5	285	1202
1192	tomato	9 samples, 5 brands	1.00	63.8	0.19	1.2	0.2	31.0	128[a]	542[a]
1193	**Dips**, sour-cream based	7 samples, 4 brands; assorted flavours	1.00	54.1	0.46	2.9	37.0	4.0[b]	360	1482
1194	**Dressing**, French, 'fat free'	Manufacturer's data (Kraft)	1.00	N	0.02	0.1	Tr	9.9	38	160
1195	blue cheese	7 samples, 2 brands	1.00	38.4	0.31	2.0	46.3	8.7	457	1886
1196	French	8 samples, 6 brands	1.00	33.3	0.02	0.1	49.4	4.5	462	1902
1197	thousand island	7 samples, 4 brands	1.00	47.6	0.18	1.1	30.2	12.5	323	1336
1198	**Mayonnaise**, *retail*		1.00	18.0	0.18	1.1	75.6	1.7	691	2843
1199	reduced calorie	12 samples, 8 brands	1.00	59.5	0.16	1.0	28.1	8.2	288	1188
1200	**Pickle**, sweet	9 samples, 4 brands	1.00	60.7	0.10	0.6	0.1	36.0	141[c]	604[c]
1201	**Salad cream**	3 samples, different brands	1.00	47.2	0.23	1.5	31.0	16.7	348	1440
1202	reduced calorie	Analysis and manufacturers' data	1.00	N	0.16	1.0	17.2	9.4	194	804
	Non-salad sauces									
1203	**Barbecue sauce**	Ref. Marsh (1980)	1.00	(75.5)	0.16	1.0	0.1	*23.4*	*93*	*395*
1204	**Brown sauce**, sweet	10 bottles, 4 brands	1.00	68.2	0.19	1.2	0.1	22.2	98[d]	418[d]
1205	**Cook-in-sauces**, canned	9 samples, 3 brands; assorted flavours	1.00	87.4	0.18	1.1	0.8	8.3	43	181
1206	**Curry sauce**, canned	10 samples, 4 brands; assorted flavours	1.00	81.4	0.24	1.5	5.0	7.1	78	324
1207	**Horseradish sauce**	8 samples, 5 brands; creamed and plain samples	1.00	64.0	0.40	2.5	8.4[e]	17.9[e]	153	640

[a] Includes 4 kcal, 18 kJ from acetic acid
[b] Includes 2.0g maltodextrins
[c] Includes 3 kcal, 14 kJ from acetic acid
[d] Includes 9 kcal, 39 kJ from acetic acid
[e] Creamed varieties have an average of 13g fat and 21g carbohydrate
Plain varieties have an average of 5g fat and 11g carbohydrate

No.	Food	Starch	Total sugars	Individual sugars Gluc	Fruct	Sucr	Malt	Lact	Dietary fibre NSP	Fatty acids Satd	Mono-unsatd	Poly-unsatd	Trans	Cholesterol
		g	g	g	g	g	g	g	g	g	g	g	g	mg
	Salad sauces, dressings and pickles													
1190	**Apple chutney**	0.1	48.4	4.3	5.7	38.4	0	0	1.3	Tr	Tr	0.1	Tr	0
1191	**Chutney**, mango, oily	0.4	49.1	N	N	N	0	0	0.9	N	N	N	N	0
1192	tomato	2.9	28.1	13.6	14.2	0.3	0	0	1.3	Tr	Tr	0.1	N	0
1193	**Dips**, sour-cream based	Tr	2.0	0.9	0.7	0.4	Tr	Tr	N	N	N	N	N	(60)
1194	**Dressing**, French, 'fat free'	0.6	9.3	N	N	N	0	0	N	Tr	Tr	Tr	Tr	0
1195	blue cheese	1.0	7.7	4.0	3.5	0.2	0	Tr	0	N	N	N	N	41
1196	French	0	4.5	2.0	2.2	0.3	0	0	0	8.0	10.6	28.4	0.3	0
1197	thousand island	1.7	10.8	5.0	3.3	2.5	0	0	0.4	N	N	N	N	29
1198	**Mayonnaise**, *retail*	0.4	1.3	0.1	0.1	1.1	0	0	0	11.4	18.2	42.4	1.2	75
1199	reduced calorie	3.6	4.6	1.1	1.0	2.5	0	0	0	4.2	6.9	15.7	0.5	22
1200	**Pickle**, sweet	2.1	33.9	11.2	11.8	10.9	0	0	1.2	Tr	Tr	Tr	Tr	0
1201	**Salad cream**	Tr	16.7	1.9	1.9	12.9	0	0	N	3.3	11.4	14.5	0.1	43
1202	reduced calorie	0.2	9.2	2.5	2.3	4.4	0	0	N	2.5	4.7	9.1	0	7
	Non-salad sauces													
1203	**Barbecue sauce**	0.1	(23.1)	N	N	N	0	0	0.5	0	0	0.1	0	0
1204	**Brown sauce**, sweet	2.6	19.6	7.3	8.3	4.0	0	0	0.7	Tr	Tr	Tr	0	0
1205	**Cook-in-sauces**, canned	3.3	5.0	1.1	1.3	2.6	0	Tr	N	0.1	0.4	0.2	0	Tr
1206	**Curry sauce**, canned	3.4	3.7	1.1	1.9	0.7	0	Tr	N	N	N	N	N	Tr
1207	**Horseradish sauce**	3.0	15.0	4.0	3.6	7.4	0	0[a]	2.5	1.1	3.8	3.2	0.2	14

[a] Creamed varieties contain lactose

Inorganic constituents per 100g edible portion

No.	Food	mg										µg	
		Na	K	Ca	Mg	P	Fe	Cu	Zn	Cl	Mn	Se	I
	Salad sauces, dressings and pickles												
1190	**Apple chutney**	166	186	20	9	23	0.8	0.10	0.2	263	0.11	1	1
1191	**Chutney**, mango, oily	1090	57	23	27	10	2.3	0.10	0.1	1720	0.10	N	N
1192	tomato	410	300	14	12	27	0.6	0.09	0.2	790	0.12	N	N
1193	**Dips**, sour-cream based	330	130	72	10	79	0.4	0.98	0.9	N	0.10	Tr	N
1194	**Dressing**, French, 'fat free'	1500	N	N	N	N	N	N	N	N	N	N	N
1195	blue cheese	1110	52	58	7	61	0.6	0.02	0.4	1400	0.10	1	6
1196	French	460	N	N	N	N	N	N	N	N	N	N	N
1197	thousand island	900	130	24	9	34	0.3	0.05	0.2	1390	0.07	1	5
1198	**Mayonnaise**, *retail*	450	16	8	1	27	0.3	0.02	0.1	750	Tr	N	35
1199	reduced calorie	(940)	N	N	N	N	N	N	N	(1450)	N	N	N
1200	**Pickle**, sweet	1610	94	15	6	12	0.6	Tr	0.1	1750	0.15	N	N
1201	**Salad cream**	1040	40	18	9	48	0.5	0.02	0.3	1620	0.10	N	11
1202	reduced calorie	N	N	N	N	N	N	N	N	N	N	N	N
	Non-salad sauces												
1203	**Barbecue sauce**	1190	240	17	23	27	0.6	0.11	0.2	1830	0.10	Tr	1
1204	**Brown sauce**, sweet	1420	(330)	(35)	(53)	(21)	(1.2)	(0.10)	(0.2)	1620	(0.34)	N	N
1205	**Cook-in-sauces**, canned	940	130	7	5	20	0.4	0.03	0.1	620	0.07	N	N
1206	**Curry sauce**, canned	980	180	30	18	31	1.1	0.05	0.2	760	0.20	N	N
1207	**Horseradish sauce**	910	220	43	18	42	0.6	0.05	0.4	1710	0.18	N	N

Vitamins per 100g edible portion

No.	Food	Retinol µg	Carotene µg	Vitamin D µg	Vitamin E mg	Thiamin mg	Riboflavin mg	Niacin mg	Trypt/60 mg	Vitamin B_6 mg	Vitamin B_{12} µg	Folate µg	Pantothenate mg	Biotin µg	Vitamin C mg
	Salad sauces, dressings and pickles														
1190	**Apple chutney**	0	11	0	0.20	0.05	0.01	0.3	0.2	0.09	0	4	0.04	1	5
1191	**Chutney**, mango, oily	0	130	0	N	0.02	0.03	0.1	Tr	N	0	N	N	N	1
1192	tomato	0	N	0	N	0.05	0.15	0.1	0.2	0.02	0	N	N	N	Tr
1193	**Dips**, sour-cream based	N	N	N	N	N	N	N	N	N	Tr	N	N	N	N
1194	**Dressing**, French, 'fat free'	0	0	0	N	Tr	Tr	Tr	Tr	Tr	0	Tr	Tr	Tr	Tr
1195	blue cheese	46	27	0.2	5.91	0.01	0.04	0	0.7	0.01	0	5	0.12	1	0
1196	French	0	0	0	20.49	0	0	0	0	0	0	0	0	0	0
1197	thousand island	14	170	0.1	8.10	0.01	0.02	0.1	0.2	0.02	0	4	0.10	1	Tr
1198	**Mayonnaise**, *retail*	86	100	0.3	16.87	0.02	0.07	Tr	0.3	0.01	1	4	N	N	Tr
1199	reduced calorie	Tr	57	Tr	8.33	N	N	N	N	N	0	N	N	N	0
1200	**Pickle**, sweet	0	250	0	N	0.03	0.01	0.1	0.1	0.01	0	Tr	N	Tr	Tr
1201	**Salad cream**	9	17	0.2	13.58	N	N	N	0.3	0.03	1	3	N	N	0
1202	reduced calorie	N	N	N	N	N	N	N	0.2	N	N	N	N	N	0
	Non-salad sauces														
1203	**Barbecue sauce**	0	505	0	0.91	0.03	0.02	0.4	0.1	0.04	0	5	0.10	1	3
1204	**Brown sauce**, sweet	0	(40)	0	N	(0.13)	(0.09)	(0.1)	(0.2)	(0.10)	(0)	(8)	N	N	Tr
1205	**Cook-in-sauces**, canned	Tr	N	0	N	Tr	0.01	0.1	0.1	0.03	0	1	N	N	Tr
1206	**Curry sauce**, canned	0	N	0	N	Tr	0.03	0.1	0.2	0.02	0	N	N	N	Tr
1207	**Horseradish sauce**	Tr	Tr	Tr	N	N	N	N	N	N	Tr	N	N	N	Tr

Composition of food per 100g edible portion

No.	Food	Description and main data sources	Edible conversion factor	Water	Total nitrogen	Protein	Fat	Carbo-hydrate	Energy value	
				g	g	g	g	g	kcal	kJ
	***Non-salad sauces** continued*									
1208	**Mint sauce**	8 samples, 4 brands	1.00	68.7	0.26	1.6	Tr	21.5	101[a]	432[a]
1209	**Pasta sauce**, tomato based	9 samples, 4 brands; assorted types	1.00	83.9	0.32	2.0	1.5	6.9	47	200
1210	**Piccalilli**	9 samples, 4 brands; mild, saucy and sweet varieties	1.00	79.1	0.16	1.0	0.5	17.6[b]	84[c]	360[c]
1211	**Relish**, burger/chilli/tomato	9 samples, 4 brands	1.00	68.4	0.19	1.2	0.1	27.6	114[d]	485[d]
1212	corn/cucumber/onion	9 samples, 5 brands	1.00	67.0	0.16	1.0	0.3	29.2	119[e]	510[e]
1213	**Soy sauce**	8 samples, 4 brands; light and dark varieties	1.00	68.6	0.48	3.0	Tr	8.2	43	182
1214	**Sweet and sour sauce**, canned	10 samples, 4 brands	1.00	81.7	0.06	0.4	0.1	10.6	44[f]	188[f]
1215	*takeaway*	7 samples purchased from Chinese restaurants	1.00	65.3	0.03	0.2	3.4	32.8	157[g]	666[g]
1216	**Tartare sauce**	10 samples, 4 brands	1.00	53.5	0.21	1.3	24.6	17.9	299[h]	1241[h]
1217	**Tomato ketchup**	10 samples, 5 brands	1.00	68.0	0.26	1.6	0.1	28.6	115	489
1218	**Worcestershire sauce**	7 samples, 3 brands	1.00	75.3	0.22	1.4	0.1	15.5	65	276
	Miscellaneous foods									
1219	**Baking powder**	6 samples of the same brand	1.00	6.3	0.91	5.2	0	37.8	163	693
1220	**Gelatine**	Literature sources and Ref. Lewis and English (1990)	1.00	13.0	15.2	84.4	0	0	338	1435
1221	**Gravy instant granules**	7 samples, 3 brands	1.00	4.0	0.70	4.4	32.5	40.6	462	1927
1222	*made up*	Calculated from 23.5g granules to 300ml water	1.00	93.0	0.05	0.3	2.4	3.0	34	142
1223	**Meat extract**	Mixed sample including Bovril and own brands	1.00	39.0	6.64	40.4	0.6	3.2	179	760
1224	**Mustard**, smooth	10 samples, 7 types including English and French	1.00	63.7	1.14	7.1	8.2	9.7	139	579
1225	wholegrain	9 samples, 5 brands	1.00	65.0	1.31	8.2	10.2	4.2	140	584

[a] Includes 14 kcal, 61 kJ from acetic acid
[b] Carbohydrate values range from 6g to 21g
[c] Includes 10 kcal, 43 kJ from acetic acid
[d] Includes 3 kcal, 14 kJ from acetic acid
[e] Includes 4 kcal, 18 kJ from acetic acid
[f] Includes 2 kcal, 8 kJ from acetic acid
[g] Includes 3 kcal, 12 kJ from acetic acid
[h] Includes 5 kcal, 22 kJ from acetic acid

No.	Food	Starch	Total sugars	Individual sugars					Dietary fibre	Fatty acids				Cholest-erol
				Gluc	Fruct	Sucr	Malt	Lact	NSP	Satd	Mono-unsatd	Poly-unsatd	Trans	
		g	g	g	g	g	g	g	g	g	g	g	g	mg
	Non-salad sauces *continued*													
1208	**Mint sauce**	0	21.5	4.9	4.8	11.8	0	0	N	Tr	Tr	Tr	Tr	0
1209	**Pasta sauce**, tomato based	1.2	5.7	2.9	2.8	Tr	0	0	N	0.2	0.3	0.8	N	0
1210	**Piccalilli**	2.8	14.8	6.5	6.8	1.5	0	0	1.0	0.1	0.1	0.3	N	0
1211	**Relish**, burger/chilli/tomato	2.5	25.1	6.5	6.8	11.8	0	0	1.3	Tr	Tr	Tr	Tr	0
1212	corn/cucumber/onion	3.6	25.6	8.1	8.5	9.0	0	0	1.2	Tr	0.1	0.2	N	0
1213	**Soy sauce**	0.9	7.3	2.2	0.9	4.2	0	0	0	0	0	0	0	0
1214	**Sweet and sour sauce**, canned	3.3	7.3	2.5	2.9	1.9	0	0	N	0.1	0	0	0	0
1215	*takeaway*	5.3	27.5	9.5	9.7	8.3	0	0	N	N	N	N	N	0
1216	**Tartare sauce**	1.7	16.2	6.5	6.3	3.4	0	0	Tr	N	N	N	N	49
1217	**Tomato ketchup**	1.1	27.5	5.9	6.4	15.2	0	0	0.9	Tr	Tr	Tr	Tr	0
1218	**Worcestershire sauce**	0.8	14.7	4.0	4.7	6.0	0	0	0	Tr	Tr	Tr	Tr	0
	Miscellaneous foods													
1219	**Baking powder**	37.8	0	0	0	0	0	0	0	0	0	0	0	0
1220	**Gelatine**	0	0	0	0	0	0	0	0	0	0	0	0	0
1221	**Gravy instant granules**	39.3	1.3	0.6	0.5	0.2	0	0	Tr	N	N	N	N	N
1222	*made up*	2.9	0.1	Tr	Tr	Tr	0	0	Tr	N	N	N	N	N
1223	**Meat extract**	2.8	0.4	0.2	0.2	Tr	0	0	0	N	N	N	N	N
1224	**Mustard**, smooth	1.9	7.8	3.4	2.9	1.5	0	0	N	0.5	5.8	1.6	N	0
1225	wholegrain	0.3	3.9	2.0	1.9	Tr	0	0	4.9	0.6	7.2	1.9	N	0

No.	Food	mg										µg	
		Na	K	Ca	Mg	P	Fe	Cu	Zn	Cl	Mn	Se	I
	***Non-salad sauces** continued*												
1208	**Mint sauce**	690	210	120	46	27	7.4	0.30	0.2	1120	0.86	Tr	Tr
1209	**Pasta sauce**, tomato based	410	490	23	21	42	0.7	0.16	0.2	830	0.10	N	N
1210	**Piccalilli**	1340	40	16	6	17	0.6	0.03	0.1	1330	0.10	N	N
1211	**Relish**, burger/chilli/tomato	480	290	13	12	26	0.3	0.07	0.1	980	0.10	N	N
1212	corn/cucumber/onion	340	110	13	9	24	0.3	0.07	0.2	660	0.07	N	N
1213	**Soy sauce**	7120	180	17	37	47	2.4	0.01	0.2	10640	0.18	N	N
1214	**Sweet and sour sauce**, canned	390	93	10	6	10	0.5	0.02	0.1	460	0.18	N	N
1215	*takeaway*	150	16	6	2	4	6.0	Tr	Tr	240	0.04	N	N
1216	**Tartare sauce**	800	42	15	17	36	0.5	0.03	0.3	1540	0.00	1	8
1217	**Tomato ketchup**	1630	350	13	19	31	0.3	0.05	0.1	1800	0.10	N	N
1218	**Worcestershire sauce**	1200	600	190	73	31	10.1	0.21	0.4	2090	0.98	(1)	(1)
	Miscellaneous foods												
1219	**Baking powder**	11800[a]	49	1130[a]	9	8430[a]	Tr	Tr	2.8	29	Tr	Tr	Tr
1220	**Gelatine**	330	7	250	15	32	2.1	0.05	0.2	N	0.13	19	6
1221	**Gravy instant granules**	6330	150	22	15	71	0.5	0.24	0.3	10000	0.40	N	N
1222	*made up*	460	10	1	1	5	Tr	0.02	Tr	730	Tr	N	N
1223	**Meat extract**	4370	970	37	65	400	8.1	0.26	1.5	6550	0.08	N	N
1224	**Mustard**, smooth	2950	200	70	82	190	2.9	0.19	1.0	3550	0.70	N	N
1225	wholegrain	1620	220	120	93	200	2.8	0.21	1.2	2210	0.70	N	N

[a] The sodium, calcium and phosphorus content will depend on the brand

Vitamins per 100g edible portion

No.	Food	Retinol µg	Carotene µg	Vitamin D µg	Vitamin E mg	Thiamin mg	Riboflavin mg	Niacin mg	Trypt/60 mg	Vitamin B_6 mg	Vitamin B_{12} µg	Folate µg	Pantothenate mg	Biotin µg	Vitamin C mg
	***Non-salad sauces** continued*														
1208	**Mint sauce**	0	Tr	0	Tr	Tr	Tr	Tr	0.3	Tr	0	Tr	Tr	Tr	Tr
1209	**Pasta sauce**, tomato based	0	(100)	0	N	0.06	0.50	0.1	0.3	0.06	0	10	N	N	Tr
1210	**Piccalilli**	0	N	0	N	Tr	0.02	0.1	0.2	0.01	0	N	N	Tr	Tr
1211	**Relish**, burger/chilli/tomato	0	N	0	N	0.06	0.05	0.2	0.2	N	0	N	N	N	N
1212	corn/cucumber/onion	0	N	0	N	N	N	N	0.2	N	0	N	N	N	N
1213	**Soy sauce**	0	0	0	N	0.05	0.13	3.4	1.4	N	0	11	N	N	0
1214	**Sweet and sour sauce**, canned	0	N	0	N	0.11	Tr	0	0.1	N	0	N	N	N	N
1215	*takeaway*	0	N	0	N	0.11	Tr	0	Tr	N	0	N	N	N	N
1216	**Tartare sauce**	24	150	0.2	10.10	0.02	0.02	Tr	0.3	0.02	0	4	0.12	1	2
1217	**Tomato ketchup**	0	473	0	N	1.00	0.09	2.1	0.2	0.03	0	1	N	N	2
1218	**Worcestershire sauce**	0	8	0	N	Tr	(0.01)	0.4	0.2	N	0	(1)	N	N	0
	Miscellaneous foods														
1219	**Baking powder**	0	0	0	Tr	Tr	Tr	Tr	1.0	Tr	0	Tr	Tr	Tr	0
1220	**Gelatine**	0	0	0	0	Tr	Tr	Tr	Tr	Tr	0	Tr	Tr	Tr	0
1221	**Gravy instant granules**	N	Tr	Tr	N	N	N	N	0.8	N	Tr	Tr	N	N	0
1222	*made up*	N	0	0	N	N	N	N	0.1	N	0	0	N	N	0
1223	**Meat extract**	N	0	0	N	9.70	8.50	87.0	3.0	0.57	8	1300	N	N	0
1224	**Mustard**, smooth	0	N	0	N	N	N	N	2.1	N	0	0	N	N	0
1225	wholegrain	0	N	0	N	N	N	N	2.4	N	0	0	N	N	0

Composition of food per 100g edible portion

No.	Food	Description and main data sources	Edible conversion factor	Water	Total nitrogen	Protein	Fat	Carbo-hydrate	Energy value	
				g	g	g	g	g	kcal	kJ
Miscellaneous foods *continued*										
1226	**Salt**	2 samples	1.00	Tr	0	0	0	0	0	0
1227	**Stock cubes**, beef	10 samples, 6 brands including Bovril, Oxo & own brands	1.00	6.1	2.85[a]	16.8[b]	9.2	N	N	N
1228	chicken	7 samples, 4 brands including Oxo	1.00	5.8	2.50[c]	15.4[b]	15.4	9.9	237	990
1229	vegetable	8 samples, 4 brands including Oxo	1.00	5.7	2.16	13.5	17.3	11.6	253	1055
1230	**Tomato puree**	8 samples, 4 brands	1.00	0.80	0.8	5.0	0.3	14.2	76	323
1231	**Vinegar**	4 samples including malt, cider and wine vinegar	1.00	N	0.07	0.4	0	0.6	22[d]	89[d]
1232	**Water**, *distilled*	Included for recipe calculation	1.00	100.0	0	0	0	0	0	0
1233	**Yeast extract**	Mixed sample including Marmite and own brands	1.00	26.7	6.78[e]	40.7[b]	0.4	3.5	180	763
1234	**Yeast**, bakers, *compressed*	Literature sources	1.00	70.0	2.02[c]	11.4[b]	0.4	1.1	53	226
1235	*dried*	Literature sources	1.00	5.0	6.32[c]	35.6[b]	1.5	3.5	169	717

[a] Includes 0.17g purine nitrogen
[b] (Total N – purine N) × 6.25
[c] Purine nitrogen forms about 10% of total nitrogen
[d] Includes 18 kcal, 73 kJ from acetic acid
[e] Includes 0.27g purine nitrogen

Composition of food per 100g edible portion

No.	Food	Starch	Total sugars	Individual sugars					Dietary fibre	Fatty acids				Cholest-erol
				Gluc	Fruct	Sucr	Malt	Lact	NSP	Satd	Mono-unsatd	Poly-unsatd	Trans	
		g	g	g	g	g	g	g	g	g	g	g	g	mg
	***Miscellaneous foods** continued*													
1226	**Salt**	0	0	0	0	0	0	0	0	0	0	0	0	0
1227	**Stock cubes**, beef	N	N	N	N	N	0	0	0	3.5	3.3	1.4	0.3	Tr
1228	chicken	7.9	2.0	Tr	0.3	1.6	0	0	0	N	N	N	N	Tr
1229	vegetable	9.4	2.2	0.1	0.7	1.4	0	0	Tr	N	N	N	N	0
1230	**Tomato puree**	0.1	14.1	6.5	7.6	Tr	0	0	2.8	Tr	0.1	0.1	N	0
1231	**Vinegar**	0	0.6	0.3	0.3	0	0	0	0	0	0	0	0	0
1232	**Water**, *distilled*	0	0	0	0	0	0	0	0	0	0	0	0	0
1233	**Yeast extract**	1.9	1.6	Tr	1.5	0.2	0	0	0	N	N	N	N	0
1234	**Yeast**, bakers, *compressed*	1.1	Tr	Tr	Tr	Tr	0	0	N	N	N	N	N	0
1235	*dried*	(3.5)	Tr	Tr	Tr	Tr	0	0	N	N	N	N	N	0

Inorganic constituents per 100g edible portion

No.	Food	mg										µg	
		Na	K	Ca	Mg	P	Fe	Cu	Zn	Cl	Mn	Se	I
	Miscellaneous foods *continued*												
1226	**Salt**	39300	89	(10)	76	(1)	0.3	0.08	(0.1)	59900	Tr	N	44[a]
1227	**Stock cubes**, beef	14560	490	40	32	240	1.2	0.70	0.8	21010	0.20	N	N
1228	chicken	16300	400	120	47	200	4.9	0.10	1.2	8850	0.27	N	N
1229	vegetable	16800	390	47	44	120	2.8	0.05	0.4	9550	0.26	N	N
1230	**Tomato puree**	240[b]	1200	35	26	94	1.4	0.53	0.5	550	0.24	N	N
1231	**Vinegar**	5	34	3	4	10	(0.1)	(0.01)	(0.1)	47	(0.01)	(1)	N
1232	**Water**, *distilled*	0	0	0	0	0	0	0	0	0	0	0	0
1233	**Yeast extract**	4300	2100	70	160	950	2.9	0.20	2.7	6630	0.19	N	49
1234	**Yeast**, bakers, *compressed*	16	610	25	59	390	5.0	1.60	3.2	20	N	N	N
1235	*dried*	(50)	(2000)	80	230	(1290)	20.0	5.00	8.0	N	N	N	N

[a] Iodised salt contains 3100µg iodine per 100g. Sea salt contains 50µg iodine per 100g

[b] The sodium content of unsalted tomato puree is approximately 20mg per 100g

Vitamins per 100g edible portion

No.	Food	Retinol µg	Carotene µg	Vitamin D µg	Vitamin E mg	Thiamin mg	Ribo-flavin mg	Niacin mg	Trypt/60 mg	Vitamin B_6 mg	Vitamin B_{12} µg	Folate µg	Panto-thenate mg	Biotin µg	Vitamin C mg
	Miscellaneous foods *continued*														
1226	**Salt**	0	0	0	0	0	1	0	0	0	0	0	0	0	0
1227	**Stock cubes**, beef	N	N	0	N	N	N	N	N	N	N	N	N	N	0
1228	chicken	N	N	0	N	N	N	N	N	N	N	N	N	N	0
1229	vegetable	0	N	0	N	N	N	N	N	N	Tr	N	N	N	0
1230	**Tomato puree**	0	1784	0	5.37	0.40	0.19	4.0	0.7	0.11	0	48	1.00	6	10
1231	**Vinegar**	0	0	0	0	0	0	0	0	0	0	0	0	0	0
1232	**Water**, *distilled*	0	0	0	0	0	0	0	0	0	0	0	0	0	0
1233	**Yeast extract**	0	0	0	N	4.10	11.9	64.0	9.0	1.60	1	2620	N	N	0
1234	**Yeast**, bakers, *compressed*	0	Tr	0	Tr	0.71	1.70	11.0	2.0	0.60	Tr	1250	3.50	60	Tr
1235	*dried*	0	Tr	0	Tr	2.33[a]	4.00	8.5	7.0	2.00	Tr	4000	11.00	200	Tr

[a] Value for bakers yeast. Brewers yeast contains 15.6mg thiamin per 100g

Additional Tables

3.1 PHYTOSTEROLS

Plants contain a number of phytosterols (plant sterols) which are distinct from cholesterol. In plant oils, the three most common sterols are β-sitosterol, campesterol and stigmasterol. There may also be measurable amounts of at least nine other phytosterols.

The amounts of the five main phytosterols are shown below for selected foods.

Phytosterols, mg per 100g edible portion

No.	Food	Brassica-sterol	Campe-sterol	Stigma-sterol	β-Sito-sterol	5-Avena-sterol	Other	Total phytosterols
Cereal products								
18	**Egg fried rice**, *takeaway*	1.8	6.8	0.3	19.3	0	0.5	28.7
40	**Brown bread**, average	0.3	5.8	0.9	17.1	1.0	3.3	28.4
45	**Garlic bread**, pre-packed, *frozen*	1.2	8.9	1.2	21.1	1.0	2.7	36.1
46	**Granary bread**	0.3	5.5	1.5	15.9	0.8	2.9	26.9
47	**Malt bread,** fruited	0.5	4.5	0.6	13.8	1.1	1.7	22.2
48	**Naan bread**	2.6	14.6	0.9	28.4	1.6	2.2	50.3
49	**Pappadums**, *takeaway*	18.7	75.3	11.4	116.6	4.7	6.4	233.1
52	**Wheatgerm bread**	0.4	12.2	1.7	32.2	2.1	4.5	53.1
53	**White bread**, sliced	0.2	3.2	0.4	11.5	0.5	1.6	17.4
56	farmhouse or split tin	0.2	4.6	0.4	13.7	0.5	2.0	21.4
57	French stick	0.2	5.4	0.8	15.8	0.7	2.6	25.5
62	**Wholemeal bread**, average	0.4	7.9	1.8	20.9	1.3	4.0	36.3
66	**Croissants**	0.7	10.1	2.1	21.8	1.0	2.8	38.5
69	**White rolls**, crusty	0.2	6.5	0.5	16.4	0.7	2.2	26.5
105	**Chocolate chip cookies**	0	6.3	3.3	20.4	1.0	0	31.0
136	**Gateau**, chocolate based, *frozen*	0	2.7	1.0	7.1	0	0	10.8
137	fruit, *frozen*	0	2.4	0	5.9	0	0	8.3
167	**Muffins**, English style, white	0.3	6.8	0.7	14.2	0.7	2.0	24.7
169	**Scones**, fruit, *retail*	2.6	14.5	1.1	27.3	1.4	2.2	49.1
172	**Scotch pancakes**, *retail*	4.0	17.3	0.6	28.0	2.4	1.3	53.6
191	**Prawn crackers**, *takeaway*	3.3	11.7	0.2	16.8	0.5	1.6	34.1
197/8	**Pizza**, cheese and tomato, deep pan/thin base	0.5	8.3	0	18.1	0	0	26.9
202	fish topped, *takeaway*	0	6.3	0	14.2	0	0	20.5
204	meat topped	1.0	7.3	0	14.9	0	0	23.2
Milk								
207	**Skimmed milk**, pasteurised, average	0	0	0	0.2	0	0.2	0.4

Phytosterols, mg per 100g edible portion *continued*

No.	Food	Brassica-sterol	Campe-sterol	Stigma-sterol	β-Sito-sterol	5-Avena-sterol	Other	Total phytosterols
***Milk** continued*								
212	**Semi-skimmed milk**, pasteurised, average	0.1	0	0	0	0	0.1	0.2
218	**Whole milk**, pasteurised, average	0.1	0.1	0	0.1	0	0.1	0.5
Cheeses								
257	**Brie**	0	0.9	0	0.6	0	0	1.5
259	**Cheddar**	0	0.7	0.1	0.3	0	0	1.1
262	**Cheese spread**, plain	0	0.4	0	0.1	0	0	0.5
269	**Edam**	0	0.4	0	0.1	0	0	0.5
271	**Goats milk soft cheese**, full fat, white rind	0	0.7	0	0.2	0	0	0.9
272	**Gouda**	0	0.5	0	0.1	0	0	0.6
275	**Parmesan**, *fresh*	0	0.8	0	0.5	0	0	1.3
280	**Spreadable cheese**, *soft white, full fat*	0	0.5	0	0.2	0	0	0.7
281	**Stilton**, blue	0	0.9	0	0.4	0	0	1.3
Milk products								
296	**Soya**, alternative to yogurt, fruit	0	0.1	0	0.2	0	0	0.3
316	**Cheesecake**, fruit, *individual*	0	0.4	0	1.1	0	0	1.5
317	**Chocolate dairy desserts**	0.1	1.1	2.0	4.8	0.2	0.1	8.3
328	**Mousse**, chocolate, reduced fat	0	0.4	0.3	1.2	0	0	1.9
335	**Torte**, fruit	0	2.5	0	6.1	0	0	8.6
337	**Trifle**, fruit	0	0.2	0	0.5	0	0	0.7
Fats and oils								
363	**Fat spread**, (60% fat), with olive oil	Tr	64.0	12.0	147.0	0	0	223.0
372	**Ghee**, vegetable	Tr	11.0	7.0	33.0	0	0	51.0
375	**Suet**, vegetable	0	30.2	8.0	72.2	0	6.4	116.8
Meat and meat products								
396	**Bacon rashers**, back, fat trimmed, *raw*	0	0	0	0.2	0	0	0.2
397	back, fat trimmed, *grilled*	0	0	0	0.1	0	0	0.1
400	streaky, *raw*	0	0.5	0	0.2	0	0	0.7
401	streaky, *grilled*	0	0.7	0	0.1	0	0	0.8
407	**Ham**, gammon joint, *boiled*	0	0.7	0	0.1	0	0	0.8
408	gammon rashers, *grilled*	0	0.7	0	0.1	0	0	0.8
416	**Beef**, mince, *raw*	0	0.4	0	0	0	0	0.4
418	mince, *stewed*	0	0.2	0	0.1	0	0	0.3
422	rump steak, fried, lean	0	1.9	0.4	5.6	0.2	0.2	8.3
445	**Lamb**, loin chops, grilled, lean	0.1	0.5	0	0.3	0	0	0.9
450	mince, stewed	0.2	0.1	0	0.2	0	0	1.0

Phytosterols, mg per 100g edible portion *continued*

No.	Food	Brassica-sterol	Campe-sterol	Stigma-sterol	β-Sito-sterol	5-Avena-sterol	Other	Total phytosterols
***Meat and meat products** continued*								
463	**Pork**, diced, *casseroled*, lean only	0	2.6	0.6	5.8	0.3	0.4	9.7
464	fillet strips, *stir-fried*, lean	0	6.7	2.1	19.1	1.1	1.4	30.4
471	loin chops, *grilled*, lean and fat	0.1	0.6	0	0.2	0	0	0.9
477	**Veal escalope**, *fried*	0.1	4.1	1.2	12.7	0.6	0.9	19.6
482	**Chicken**, breast, *grilled*, meat only	0.3	0.9	0	0.3	0.1	0	1.6
495	**Turkey**, breast fillet, *grilled*, meat only	0.2	0.4	0	0.1	0	0	0.7
497	roasted, dark meat	0.6	0.9	0	0.2	0.1	0	1.8
498	roasted, light meat	0.2	0.5	0	0.1	0	0	0.8
501	**Duck**, *raw*, meat only	0	2.0	0	1.0	0	0	3.0
503	*roasted*, meat only	0	1.0	0	0	0	0	1.0
508	**Rabbit**, *stewed*, meat only	0	0	0	1.0	0	0	1.0
530	**Economy burgers**, *frozen, grilled*	0.1	1.5	0	3.1	0.1	0.2	5.0
537	**Chicken pie**, *individual, chilled/frozen, baked*	1.1	8.2	0.9	17.1	0.7	0.4	28.4
540	**Cornish pastie**	0	7.0	2.0	15.7	0	0	24.8
559	**Saveloy**, unbattered, *takeaway*	0	2.0	1.0	7.0	0	0	10.0
560	**Scotch egg**, *retail*	1.3	6.6	0.7	12.1	0.2	0	20.8
592	**Cottage/Shepherd's pie**, *chilled/frozen, reheated*	0.1	1.0	0.2	2.3	0.2	0.2	4.0
593	**Doner kebabs**, meat only	0	0.6	0	0.6	0	0	1.2
595	**Faggots in gravy**, *chilled/frozen, reheated*	0.3	2.7	0.2	5.4	0.3	0.3	9.2
599	**Irish stew**, canned	0	0	0	1.5	0	0	1.5
605	**Lasagne**, *chilled/frozen, reheated*	0.2	2.0	1.1	5.2	0.3	0.2	8.9
607	**Moussaka**, *chilled/frozen/longlife, reheated*	2.0	8.6	0.5	12.8	0.8	0.3	25.0
611	**Shish kebab**, meat only	0.6	3.2	0	3.3	0	0.3	7.4
615	**Spring rolls**, meat, *takeaway*	3.6	16.0	2.4	25.4	0.6	2.2	50.2
Fish products								
686	**Curry**, prawn, *takeaway*	7.5	27.0	3.8	43.7	5.1	2.8	89.9
698	**Szechuan prawns with vegetables**, *takeaway*	2.1	8.4	1.1	14.2	0.8	1.4	28.0
699	**Taramasalata**	34.5	149.3	1.8	190.4	14.3	4.2	394.5
Vegetable products								
732	**Potato waffles**, *frozen, cooked*	3.0	4.0	2.0	10.0	0	0	19.0
750	**Hummus**	1.3	10.3	3.0	31.3	2.5	0	48.4
842	**Quorn**, *pieces, as purchased*	0.3	1.5	0	3.8	3.5	63.9	73.0
898	**Vegetable kiev**, *baked*	0	6.3	2.6	18.9	0	0	27.8

Phytosterols, mg per 100g edible portion *continued*

No.	Food	Brassica-sterol	Campe-sterol	Stigma-sterol	β-Sito-sterol	5-Avena-sterol	Other	Total phytosterols
	Vegetable products *continued*							
900	**Vegetables**, *stir-fried, takeaway*	2.9	12.8	2.5	13.7	0.3	1.4	33.6
901	**Vegetarian sausages**, *baked/ grilled*	0	4.8	3.1	12.1	0	0.2	20.2
	Herbs and spices							
909	**Mustard powder**	23.5	71.3	0	137.1	14.9	0	246.7
	Nuts and seeds							
1014	**Marzipan**, *retail*	0	0	0	29.0	0	0	29.0
1025	**Sunflower seeds**	0	20.9	15.8	140.4	23.8	21.6	222.5
	Sugars, preserves and snacks							
1029	**Chocolate nut spread**	0	4.5	2.9	21.8	1.0	0.3	30.4
1040	**Lemon curd**	1.0	4.0	1.0	6.0	0	0	12.0
1053	**Creme eggs**	0	3.0	6.0	14.0	0	0	23.0
1062	**Cereal crunchy bar**	0	6.9	6.9	27.2	4.0	0.9	45.8
1078	**Pork scratchings**	0	2.2	0	2.7	0	0	4.9
1081	**Potato rings**	0	4.0	0	11.0	0	0	15.0
1084	**Tortilla chips**	0	16.4	7.6	68.0	5.3	1.2	98.4
	Beverages							
1096	**Drinking chocolate powder**	0	1.3	2.9	8.6	0	0	12.8
1102	**Instant drinks powder**, chocolate, low calorie	0	0	0	2.8	0	0	2.8
1103	malted	0	0	0	3.7	0	0	3.7
	Soups, sauces and miscellaneous foods							
1161	**Chicken soup**, cream of, canned	1.0	6.0	0	8.0	0	0	15.0
1166	**Mushroom soup**, cream of, canned	1.0	6.0	0	8.0	0	0	15.0
1168	**Tomato soup**, cream of, canned	1.0	6.0	0	8.0	0	0	16.0
1173	**Instant soup powder**, *dried*	0	5.0	4.0	9.0	0	0	18.0
1195	**Dressing**, blue cheese	0.5	25.4	17.3	58.4	2.0	1.5	105.2
1197	thousand island	11.4	64.2	9.0	99.0	7.2	1.8	192.6
1199	**Mayonnaise**, reduced calorie	0	9.3	7.6	32.5	0	3.3	52.7
1201	**Salad cream**	0	12.3	8.9	71.8	0	2.9	95.8
1202	reduced calorie	0	5.8	2.4	25.8	0	0	34.1
1205	**Cook-in sauces**, canned	1.5	7.1	0	9.7	0.7	0	19.0
1207	**Horseradish sauce**	1.3	6.6	0.6	14.0	0.7	0.1	23.3

3.2 ALTERNATIVE WAYS OF MEASURING DIETARY FIBRE

The main Tables give fibre values as measured by the non-starch polysaccharides (NSP) method of Englyst and Cummings (1994,1992,1988). Previous editions and supplements have also included fibre measured by the Southgate method (Southgate 1969).

Total dietary fibre is often measured using the AOAC enzymatic-gravimetric method (AOAC method 985.29). For nutritional labelling purposes, it is recommended that fibre values obtained by AOAC methodology are used.

AOAC values are generally higher than NSP values because they include substances measuring as lignin and also include resistant starch.

Data included in this Table show a direct comparison between the two methods. These values are taken from recent MAFF analyses where the samples were analysed using both methods. This Table was prepared from available data and does not include all food groups or indicate importance as a source of fibre.

		Fibre, g per 100g edible portion	
No.	Food	Non-starch polysaccharides	Total Dietary fibre (AOAC)
	Cereals and cereal products		
18	**Egg fried rice**, *takeaway*	0.8	1.1
40	**Brown bread**, average	3.5	5.0
46	**Granary bread**	3.4	5.4
47	**Malt bread,** fruited	2.6	3.5
48	**Naan bread**	2.0	2.9
49	**Pappadums**, *takeaway*	5.8	6.3
50	**Pitta bread**, white	2.4	2.3
52	**Wheatgerm bread**	4.0	5.7
53	**White bread** sliced	1.9	2.5
56	farmhouse or split tin, freshly baked	2.1	2.9
57	French stick	2.4	3.3
58	Premium	1.9	2.8
59	Danish style	2.4	3.1
62	**Wholemeal bread**, average	5.0	7.0
65	**Brown rolls**, soft	3.8	4.3
66	**Croissants**	1.6	3.1
67	**Granary rolls**	3.6	4.4
69	**White rolls**, crusty	2.4	2.9
70	Soft	2.0	2.6
71	**Wholemeal rolls**	4.4	5.5
128	**Chocolate fudge cake**	0.9	1.9

Fibre, g per 100g edible portion *continued*

No.	Food	Non-starch polysaccharides	Total Dietary fibre (AOAC)
	***Cereals and cereal products** continued*		
145	**Sponge cake**, *with dairy cream with jam*	Tr	2.7
167	**Muffins**, English style, white	1.9	2.6
169	**Scones**, fruit, *retail*	2.0	2.9
172	**Scotch pancakes**, *retail*	1.5	1.9
176	**Crumble**, fruit	1.3	2.3
191	**Prawn crackers**, *takeaway*	1.2	0.9
	Milk and milk products		
240	**Soya**, non-dairy alternative to milk, *unsweetened*	0.2	0.5
288	**Low fat yogurt**, fruit	0.2	0.3
296	**Soya**, alternative to yogurt, fruit	0.3	0.7
301	**Fromage frais**, virtually fat free, fruit	0.4	0.7
304	**Cornetto-type ice-cream cone**	0.3	0.4
310	**Ice-cream**, non-dairy, vanilla	Tr	0.2
312	**Lollies**, with real fruit juice	Tr	Tr
313	**Sorbet**, fruit	Tr	1.0
314	**Banoffee pie**	2.5	1.8
315	**Cheesecake**, *frozen*	0.8	1.0
316	**Cheesecake**, fruit, *individual*	1.0	1.6
330	**Pavlova,** with fruit and cream	0.3	0.4
337	**Trifle**, fruit	2.1	2.4
	Meat dishes		
578	**Chicken chow mein**, *takeaway*	1.1	1.5
579	**Chicken curry**, average, *takeaway*	2.0	2.2
583	**Chicken satay**, *takeaway*	2.2	2.0
615	**Spring rolls**, meat, *takeaway*	1.9	2.0
	Fish dishes		
686	**Curry**, prawn, *takeaway*	2.0	2.5
698	**Szechuan prawns with vegetables**, *takeaway*	1.4	1.2
	Vegetable dishes		
900	**Vegetables**, *stir-fried, takeaway*	1.8	2.1

The following Table provides average values for the ten major food groups (Englyst *et al.*, 1996). Analyses were undertaken on composite samples from the 1994 Total Diet Study (Peattie *et al.*, 1983).

	Fibre, g per 100g edible portion	
Food Group	Non-starch polysaccharides	Total Dietary fibre (AOAC)
Bread	2.9	3.8
Other cereals	3.2	4.0
Meat products	0.5	0.5
Green vegetables	2.7	3.3
Potatoes	1.9	2.4
Other vegetables	1.8	3.2
Canned vegetables	2.0	3.0
Fresh fruit	1.4	1.9
Fruit products	0.5	0.7
Nuts	6.6	8.8

3.3 CAROTENOID FRACTIONS

β-Carotene is the main or only source of vitamin A activity in most fruit and vegetables. Carrots are the major exception and contain approximately 30% as α-carotene. When there are known to be significant amounts of other carotenoids these are shown below. The values for cryptoxanthins were often unspecified; the β form is likely to predominate with smaller amounts of the α form present. The β-carotene equivalent is the sum of the β-carotene and half of any α-carotene or cryptoxanthins present, and the retinol equivalent is one sixth of the β-carotene equivalent. Lycopene has no vitamin A activity. Absorption and utilisation of carotenes vary, for example with the amount of fat in the diet and β-carotene concentration (Brubacher and Weiser, 1985), and there is currently much debate about the use of retinol equivalents (Scott and Rodriguez-Amaya, 2000).

Carotene fractions, µg per 100g edible portion

No.	Food	Carotene fractions				
		α-carotene	β-carotene	β-cryptoxanthin	Carotene equiv	Retinol equiv
	Beans and lentils					
736	**Beansprouts**, mung, *raw*	20	20	20	40	7
742	**Broad beans**, *frozen, boiled in unsalted water*	12	220	0	225	37
749	**Green beans/French beans**, *frozen, boiled in unsalted water*	52	494	0	520	87
	Peas					
769	**Mushy peas**, canned, *re-heated*	Tr	Tr	0	Tr	Tr
770	**Peas**, *raw*	19	290	0	300	50
771	*boiled in unsalted water*	7	245	0	250	41
772	frozen, *boiled in unsalted water*	26	558	0	571	95
774	canned, *re-heated, drained*	15	526	0	534	89
775	**Petit pois**, *frozen, boiled in unsalted water*	(27)	(390)	0	(405)	(67)
	Vegetables, general					
777	**Asparagus**, *raw*	10	310	0	315	53
778	*boiled in salted water*	0	389	0	389	65
779	**Aubergine**, *raw*	60	40	0	70	12
780	*fried in corn oil*	110	70	0	125	21
781	**Beetroot**, *raw*	20	10	0	20	3
792	**Carrots,** old, *raw*	4070	10400	0	12500	2080
793	old, *boiled in unsalted water*	4170	11300	0	13400	2230
794	young, *raw*	3380	6120	0	7810	1300
795	young, *boiled in unsalted water*	3420	5990	0	7700	1280
796	canned, *re-heated, drained*	729	1710	0	2070	345

No.	Food	Carotene fractions				
		α-carotene	β-carotene	β-cryptoxanthin	Carotene equiv	Retinol equiv
	Vegetables, general *continued*					
802	**Courgette**, *raw*	0	550	0[a]	610	100
804	*fried in corn oil*	0	450	0[a]	500	83
806	**Curly kale**, *raw*	0	3130	32	3150	525
807	*boiled in salted water*	0	3350	33	3380	560
812	**Gourd**, karela, *raw*	95	295	0	345	57
815	**Lettuce**, average, *raw*	0	1020	0	1023	171
819	**Mixed vegetables**, *frozen, boiled in salted water*	705	2160	26	2520	420
824	**Okra**, *raw*	30	500	0	515	85
825	*boiled in unsalted water*	29	450	0	465	77
826	*stir-fried in corn oil*	35	545	0	560	94
834	**Peppers**, capsicum, green, *raw*	9	260	0	265	44
835	*boiled in salted water*	8	235	0	240	40
836	capsicum, red, *raw*	135	3170	1220	3840	640
837	*boiled in salted water*	133	3120	1200	3780	630
840	**Pumpkin**, *raw*	14	445	0	450	75
841	*boiled in salted water*	29	940	0	955	160
845	**Spinach**, *raw*	0	3520	35	3540	589
846	*boiled in unsalted water*	0	3820	39	3840	640
847	*frozen, boiled in unsalted water*	0	(3820)	(39)	(3840)	(640)
848	**Spring greens**, *raw*	0	2620	216	2630	438
849	*boiled in unsalted water*	0	2260	23	2270	378
856	**Sweetcorn**, kernels, canned, *re-heated, drained*	Tr	22	180	110	19
857	on-the-cob, whole, *boiled in unsalted water*	Tr	14	115	71	12
858	**Tomatoes**, *raw*	0	564	0[b]	564	94
859	*fried in corn oil*	0	740	42	765	125
860	*grilled*	0	1790	97	1840	306
861	canned, *whole contents*	0	362	0	362	60
900	**Vegetables**, *stir-fried, takeaway*	73	534	8	575	96
	Fruit					
923	**Apricots**, *raw*	2	405	0	405	67
926	canned in syrup	0	810	0	810	135
927	**Avocado**, average	8	540	0	545	91
929	**Blackberries**, *raw*	4	78	0	80	13
930	*stewed with sugar*	3	61	0	62	10
933	**Cherries**, *raw*	4	23	0	25	4
934	canned in syrup	4	15	0	17	3
936	**Cherry pie filling**	2	17	0	18	3
937	**Clementines**	5	73	0	75	13
949	**Fruit salad**, *homemade*	1	19	0	20	3

[a] Courgettes raw and fried in corn oil contain 120 and 100μg α-cryptoxanthin per 100g, respectively
[b] Raw tomatoes also contain 879μg lycopene per 100g

No.	Food	Carotene fractions			Carotene equiv	Retinol equiv
		α-carotene	β-carotene	β-cryptoxanthin		
***Fruit** continued*						
951	**Gooseberries**, *stewed with sugar*	3	40	0	41	7
952	**Grapefruit**, *raw*	9	12	0	17	3
956	**Guava**, *raw*	0	380	110	435	73
958	**Kiwi fruit**	0	40	0	40	7
963	**Mandarin oranges**, canned in juice	7	92	0	95	16
964	canned in syrup	7	105	0	105	18
965	**Mangoes**, ripe, *raw*	Tr	682	27	696	116
966	**Melon**, Canteloupe-type	19	1760	Tr	1770	294
969	watermelon	0	116	0	116	19
971	**Nectarines**	0	114	0	114	19
973	**Oranges**	26	14	39	47	8
974	**Passion fruit**	410	360	370	750	125
975	**Paw-paw**, *raw*	0	130	1365	810	135
977	**Peaches**, *raw*	0	119	0	119	19
980	**Pears**, average, *raw*	0	17	3	18	3
981	*raw, peeled*	0	18	3	19	3
987	**Plums**, average, *raw*	23	355	19	376	63
990	**Prunes**, canned in juice	15	135	0	140	23
992	ready-to-eat	27	125	0	140	3
999	**Satsumas**	5	73	0	75	13
1003	**Tangerines**	6	94	0	97	16
Nuts and seeds						
1027	**Trail mix**	3	45	0	47	8
Fruit juices						
1126	**Orange juice**, unsweetened	2	5	21	17	3

3.4 VITAMIN E FRACTIONS

The vitamin E activity of foods can be derived from a number of different tocopherols and tocotrienols. Where vitamin E is present, and the amount of each tocopherol was known, the values are shown below for selected foods. The total vitamin E activity is also shown as α-tocopherol equivalents, which has been taken as the sum of the α-tocopherol, 40% of the β-tocopherol, 10% of the γ-tocopherol, 1% of the δ-tocopherol, 30% of α-tocotrienol, 5% of the β-tocotrienol, 1% of the γ-tocotrienol and 1% of δ-tocotrienol (McClaughlin and Weihrauch 1979).

Vitamin E fractions, mg per 100g edible portion

No.	Food	α-Tocopherol	β-Tocopherol	γ-Tocopherol	δ-Tocopherol	Vitamin E equiv
	Cereals					
18	**Egg fried rice**, *takeaway*	0.76	Tr	1.23	0.08	0.88
33	**Pasta**, fresh, cheese and vegetable stuffed, *cooked*	0.68	0.21	0.53	0.13	0.82
40	**Brown bread**, average	0.01	0.01	0.01	Tr	0.01
45	**Garlic bread**, pre-packed, *frozen*	0.13	0.03	0.01	0.22	0.16
46	**Granary bread**	0.16	0.16	0.08	0.03	0.23
48	**Naan bread**	0.49	0.08	1.13	0.05	0.64
49	**Pappadums**, *takeaway*	3.27	Tr	4.29	0.22	3.70
56	**White bread**, farmhouse or split tin	0.13	0.11	0.43	0.02	0.22
58	premium	0.10	0.09	0.30	0.05	0.17
62	**Wholemeal bread**, average	0.19	0.12	0.52	0.02	0.29
66	**Croissants**	0.88	0.08	0.82	0.31	0.99
103	**Chocolate biscuits,** full coated	2.15	0.16	3.82	0.18	2.60
104	cream filled, full coated	0.90	0.07	2.39	0.19	1.16
108	**Crunch biscuits**, cream filled	1.45	0.10	1.32	0.22	1.63
116	**Sandwich biscuits**, jam filled	1.44	0.14	1.93	0.42	1.69
119	**Shortbread**	0.72	0.17	0.87	0.00	0.88
121	**Wafers,** filled, chocolate, full coated	0.72	0.03	2.99	0.11	1.03
135	**Gateau**, chocolate based, *frozen*	0.85	0.21	0.73	0.01	1.01
136	fruit, *frozen*	0.91	0.24	0.62	0.03	1.06
169	**Scones**, fruit, *retail*	0.10	0.01	0.10	0.02	0.11
172	**Scotch pancakes**, *retail*	1.12	0.12	1.82	0.26	1.35
191	**Prawn crackers**, *takeaway*	5.46	Tr	3.10	0.06	5.77
200	**Pizza**, cheese and tomato, *frozen*	1.83	0.45	0.95	Tr	2.10
204	meat topped	0.72	0.31	0.91	Tr	0.93
205	vegetarian	1.34	0.29	0.62	Tr	1.52

Vitamin E fractions, mg per 100g edible portion *continued*

No.	Food	α-Tocopherol	β-Tocopherol	γ-Tocopherol	δ-Tocopherol	Vitamin E equiv
Milk and milk products						
240	**Soya milk**, non-dairy alternative to milk, *unsweetened*	0.15	0.03	1.56	0.68	0.32
241	**Cream,** fresh, single	0.46	0.01	Tr	Tr	0.47
243	fresh, whipping	1.29	0.07	Tr	Tr	1.32
244	fresh, double	1.62	0.04	Tr	Tr	1.64
246	**Creme fraiche**	0.72	Tr	Tr	0.03	0.72
247	half fat	0.42	Tr	Tr	0.01	0.42
255	**Elmlea**, double	1.13	0.12	1.54	0.21	1.33
256	**Tip Top dessert topping**	0.13	Tr	0.16	0.04	0.14
257	**Brie**	0.81	0.01	0.01	Tr	0.81
259	**Cheddar cheese**	0.43	0.12	0.40	0.06	0.52
260	**Cheddar-type**, half fat	0.42	0.01	0.12	Tr	0.47
262	**Cheese spread**, plain	0.30	0.01	0.02	Tr	0.30
264	**Cottage Cheese**, plain	0.10	Tr	Tr	Tr	0.10
280	**Spreadable cheese**, *soft white, full fat*	0.24	Tr	0.02	0.09	0.24
281	**Stilton**, blue	0.60	Tr	0.02	0.01	0.60
284	**Whole milk yogurt**, fruit	0.18	Tr	Tr	Tr	0.18
296	**Soya**, alternative to yoghurt, fruit	1.71	Tr	1.84	1.79	1.91
298	**Fromage frais**, plain	0.15	Tr	Tr	Tr	0.15
316	**Cheesecake**, fruit, *individual*	1.28	Tr	0.14	0.16	1.29
317	**Chocolate dairy desserts**	0.43	Tr	0.86	0.04	0.52
321	**Custard**, ready-to-eat	0.28	Tr	0.16	Tr	0.29
328	**Mousse**, chocolate, reduced fat	0.75	Tr	0.35	Tr	0.79
333	**Rice pudding**, canned	0.16	Tr	Tr	Tr	0.16
335	**Tortes**, fruit	1.26	0.26	0.70	0.02	1.43
337	**Trifle**, fruit	0.64	Tr	0.16	0.16	0.66
Fats						
351	**Butter**	1.82	0.07	0.08	0.02	1.85
352	spreadable	2.92	Tr	Tr	Tr	2.90
353	**Blended spread,** (70–80% fat)	10.08	0.07	11.64	1.22	11.28
354	(40% fat)	2.24	0.26	14.86	4.76	3.88
356	**Margarine**, hard, animal and vegetable fats	4.28	0	1.56	0.27	4.44
358	soft, not polyunsaturated	11.59	Tr	7.40	0.54	12.34
359	soft, polyunsaturated	31.15	1.29	8.95	3.46	32.60
360	**Fat spread,** (70–80% fat), not polyunsaturated	2.22	0.07	2.17	6.04	2.53
362	(60% fat), polyunsaturated	29.81	1.20	4.43	1.27	30.75
364	(40% fat), not polyunsaturated	7.27	0	7.29	0.42	8.01
366	(20–25% fat), not polyunsaturated	4.71	0	3.98	0.45	5.11
375	**Suet**, vegetable	17.97	0	0	0	17.97

Vitamin E fractions, mg per 100g edible portion *continued*

No.	Food	α-Tocopherol	β-Tocopherol	γ-Tocopherol	δ-Tocopherol	Vitamin E equiv
Oils						
376	**Coconut oil**	0.50	0	0	0.60	0.70[a]
378	**Corn oil**	11.20	0	60.20	1.80	17.24
380	**Olive oil**	5.10	0	Tr	0	5.10
381	**Palm oil**	25.60	0	31.60	7.00	33.10[b]
382	**Peanut (Groundnut) oil**	13.00	0	21.40	2.10	15.20
383	**Rapeseed oil**	18.40	0	38.00	1.20	22.20
384	**Safflower oil**	38.70	0	17.40	24.00	40.70
386	**Soya oil**	10.10	0	59.30	26.40	16.06
387	**Sunflower oil**	48.70	0	5.10	0.80	49.22
390	**Wheatgerm oil**	133.70	0	26.00	27.10	136.70[c]
Meat dishes						
578	**Chicken chow mein**, *takeaway*	0.83	Tr	1.29	Tr	0.96
579	**Chicken curry**, average, *takeaway*	1.88	Tr	1.81	0.75	2.12
583	**Chicken satay**	1.30	Tr	1.09	Tr	1.41
606	**Meat samosa**, *takeaway*	0.44	Tr	1.09	0.14	0.55
615	**Spring rolls**, meat, *takeaway*	1.23	Tr	2.35	0.15	1.47
616	**Sweet and sour chicken**, *takeaway*	1.88	Tr	2.54	0.15	2.14
Fish dishes						
686	**Curry**, prawn, *takeaway*	2.88	Tr	2.75	0.17	3.16
698	**Szechuan Prawns with vegetables**, *takeaway*	1.86	Tr	1.27	0.08	1.99
Vegetable products and dishes						
731	**Potato fritters**, *battered, frozen*	0.87	0.03	0.08	0.03	0.88
896	**Vegetable and cheese grill/burger**, in crumbs, *baked/grilled*	0.94	0.12	1.30	0.39	1.12
898	**Vegetable kiev**, *baked*	0.78	0.09	2.16	0.42	1.03
900	**Vegetables**, *stir-fried, takeaway*	1.11	Tr	1.45	0.12	1.26
Fruit						
929	**Blackberries**, *raw*	2.05	0	2.90	2.75	2.37
932	*stewed with sugar*	1.60	0	2.30	2.22	1.85
987	**Plums**, average, *raw*	0.60	0	0.07	0	0.61
994	**Raspberries**, *raw*	0.30	0	1.50	2.70	0.48
995	canned in syrup	0.10	0	0.50	0.90	0.15
Nuts						
1004	**Almonds**	23.77	0.26	0.81	0	23.96
1005	**Brazil nuts**	5.72	0.15	13.87	0.17	7.18
1006	**Cashew nuts**, *roasted and salted*	0.77	0.04	5.09	0.38	1.30

[a] Includes contribution from 0.50mg α-tocotrienol
[b] Includes contribution from 14.30mg α-tocotrienol
[c] Includes contribution from 2.60mg α-tocotrienol

Vitamin E fractions, mg per 100g edible portion *continued*

No.	Food	α-Tocopherol	β-Tocopherol	γ-Tocopherol	δ-Tocopherol	Vitamin E equiv
1007	**Chestnuts**	0.50	0	7.00	0	1.20
1008	**Coconut**, *creamed block*	1.34	0	0.57	0	1.40
1009	*desiccated*	1.21	0	0.52	0	1.26
1011	**Hazelnuts**	24.20	0.80	4.33	0.22	24.98
1014	**Marzipan**, *retail*	6.13	0.07	0.21	0	6.18
1016	**Peanut butter**, smooth	4.70	Tr	2.90	Tr	4.99
1018	**Peanuts**, *plain*	9.21	0.23	7.91	0.37	10.09
1019	*dry roasted*	0.70	0.18	3.30	0.53	1.11
1020	*roasted and salted*	0.41	0.14	1.90	0.37	0.66
1028	**Walnuts**	1.35	0.09	24.46	2.29	3.85
Confectionery						
1048	**Chocolate covered caramels**	2.03	0.08	3.14	0.19	2.37
1049	**Chocolate,** fancy and filled	1.35	0.05	2.73	0.31	1.65
1050	milk	0.38	0.02	0.66	0.06	0.45
1051	plain	0.86	0.18	5.11	0.28	1.44
1052	white	0.61	Tr	5.26	0.21	1.14
1054	**Kit Kat**	0.58	0.04	4.06	0.22	1.03[a]
1055	**Milky Way**	1.36	0.03	2.10	0.13	1.91[b]
1057	**Smartie-type sweets**	0.46	Tr	3.20	0.18	0.80[c]
1059	**Twix**	2.74	0.09	2.74	0.18	3.72[d]
1061	**Cereal crunchy bar**	3.37	0.16	4.07	0.35	3.84
1063	**Chew sweets**	0.86	0.04	0.38	0.11	0.91
Savoury snacks						
1075	**Corn snacks**	5.38	0.11	4.09	0.54	5.80
1079	**Potato crisps**	5.42	0.22	2.88	0.73	5.83
1080	lower fat	3.36	0.09	0.62	0.06	3.47
1084	**Tortilla chips**	1.72	0.06	1.97	0.15	1.94
Beverages						
1096	**Drinking chocolate powder**	0.28	Tr	1.36	0.09	0.41
1102	**Instant drinks powder**, chocolate, low calorie	0.54	0.18	1.27	0.07	0.74
1103	malted	3.63	0.15	1.30	0.03	3.83
1126	**Orange juice**, unsweetened	0.17	0	0.01	0	0.17
Sauces						
1198	**Mayonnaise,** *retail*	14.80	Tr	40.80	11.30	18.99
1199	reduced calorie	7.74	0.06	5.66	0.44	8.33

[a] Includes contribution from 0.08mg α-tocotrienol, 0.13mg γ-tocotrienol
[b] Includes contribution from 1.04mg α-tocotrienol, 1.2mg γ-tocotrienol, 0.23mg δ-tocotrienol
[c] Includes contribution from 0.07mg α-tocotrienol, 0.06mg γ-tocotrienol
[d] Includes contribution from 2.13mg α-tocotrienol, 2.35mg γ-tocotrienol, 0.52mg δ-tocotrienol

3.5 VITAMIN K_1

Information on the phylloquinone content of foods has been accumulating over a number of years by direct analysis using HPLC in redox mode with electro-chemical or UV detection (McCarthy *et al.*, 1997; Shearer *et al.*, 1996). These analyses have enabled recipe calculations and led to a report of 'provisional' vitamin K_1 food composition data (Bolton-Smith *et al.*, 2000). Results from this work, with further new direct analytical data, are reported in the Table. Whilst the phylloquinone content of many of the foods included have been determined on pooled (n 4–7) UK-representative samples, others are the result of single sample, or 2–3 non-UK-representative sample analyses. The latter have been included when evidence supporting these values was available from work in other countries (e.g. Booth *et al.*, 1996a; Koivu *et al.*, 1997). The phylloquinone content of individual retail products and mixed dishes will vary considerably depending on the oil type used. Additionally, currently unquantified amounts of 2',3'-dihydro-phylloquinone and menaquinones may also be present in some foods (see Introduction paragraph 1.4.11).

Vitamin K_1, μg per 100g

No.	Food	Vitamin K_1
	Cereals and cereal products	
1	**Bran**, wheat	10.4
8	**Soya flour**, full fat	25.3
12	**Wheat flour**, white, plain	0.76
16	**Brown rice**, raw	0.82
23	**White rice**, easy cook, raw	0.4
35	**Spaghetti**, white, raw	0.21
36	**Spaghetti,** white, boiled	0.05
48	**Naan bread**	3.8
53	**White bread**, sliced	0.42
62	**Wholemeal bread**, average	2
72	**Sandwich,** Bacon, lettuce and tomato, white bread	7.8
73	**Sandwich,** Cheddar cheese and pickle, white bread	4
75	**Sandwich,** Egg mayonnaise, white bread	4.6
76	**Sandwich,** Ham salad, white bread	12.9
77	**Sandwich,** Tuna mayonnaise, white bread	4.2
83	**Corn Flakes**	0.06
102	**Weetabix**	1.7
103	**Chocolate biscuits**, full coated	3.46
110	**Digestive biscuits**, plain	1.51
112	**Gingernut biscuits**	1.6
115	**Sandwich biscuits**, cream filled	3.8
117	**Semi-sweet biscuits**	1.71

No.	Food	Vitamin K_1
Cereals and cereal products *continued*		
118	**Short sweet biscuits**	4
131	**Fruit cake**, plain, retail	7.26
142	**Sponge cake**	18.1
198	**Pizza**, cheese and tomato, frozen, thin base	2.56
Milk and milk products		
206	**Skimmed milk**, average	0.02
217	**Whole milk**, average	0.6
223	**Channel Island milk** whole, pasteurised	0.87
228	**Condensed milk**, whole sweetened	0.36
231	**Evaporated milk**, whole	0.5
236	**Goats milk**, pasteurised	0.53
239	**Soya**, non-dairy alternative to milk, sweetened, calcium enriched	0.69
240	**Soya**, non-dairy alternative to milk, unsweetened	1.71
244	**Cream**, double	6.4
257	**Brie**	2.4
259	**Cheddar cheese**	4.7
268	**Danish blue**	4.1
276	**Processed cheese**, plain	1.6
280	**Spreadable cheese**, soft white, full fat	4.7
287	**Low fat yogurt**, plain	0.03
288	**Low fat yogurt**, fruit	0.03
299	**Fromage frais**, fruit	0.34
310	**Ice cream**, non-dairy, vanilla	0.8
Eggs		
338	**Eggs**, chicken, raw	0.28
339	**Eggs**, chicken, white, raw	Tr
340	**Eggs**, chicken, yolk, raw	0.87
Fats and oils		
351	**Butter**	7.4
356	**Margarine**, hard, animal and vegetable fats	9
358	**Margarine**, soft, not polyunsaturated	25
360	**Fat spread** (70-80% fat), not polyunsaturated	12
363	**Fat spread** (60% fat), polyunsaturated, with olive oil	56
364	**Fat spread** (40% fat), not polyunsaturated	36
369	**Compound cooking fat**	13.8
370	**Dripping**, beef	24.5
376	**Coconut oil**	1
377	**Cod liver oil**	0.3
378	**Corn oil**	3
380	**Olive oil**[a]	57.5
381	**Palm oil**	7.9
383	**Rapeseed oil**	112.5
384	**Safflower oil**	3.4
385	**Soya oil**	131
387	**Sunflower oil**	6.3

[a] Mean of extra virgin and standard olive oils

No.	Food	Vitamin K_1
Meat and meat products		
396	**Bacon rashers, back,** fat trimmed, raw	Tr
404	**Bacon**, fat only, cooked, average	Tr
418	**Beef, mince**, stewed	7.18
420	**Beef, rump steak**, raw, lean and fat	0.8
431	**Beef, topside**, roasted well-done, lean[a]	0.19
445	**Lamb loin chops**, grilled, lean[b]	0.28
460	**Pork**, average, trimmed fat, raw	0.18
461	**Pork**, average, fat, cooked	0.35
468	**Pork, loin chops**, raw, lean and fat	0.03
469	**Pork, loin chops**, grilled, lean[c]	0.16
478	**Chicken, dark meat**, raw	0.05
487	**Chicken**, roasted, light meat	0.04
539	**Corned beef**, canned	2
547	**Paté**, liver	0.95
557	**Salami**	1.11
558	**Sausage rolls**, puff pastry	0.82
562	**Steak and kidney/Beef pie**, individual, chilled/frozen, baked	3.82
574	**Beef stew (recipe)**[d]	0.43
585	**Chicken tikka masala**, retail	0.28
592	**Cottage/Shepherd's pie**, chilled/frozen, re-heated	1.09
607	**Lasagne**, chilled/frozen, re-heated	2.66
Fish and fish products		
618	**Cod**, raw	0.01
624	**Cod**, in crumbs, frozen, fried in blended oil[e]	10.12
660	**Salmon**, steaks, steamed, flesh only	0.16
668	**Tuna**, canned in brine, drained	0.25
674	**Prawns**, boiled	0.08
Potatoes		
702	**New potatoes**, boiled in unsalted water	1.16
705	**Old potatoes**, average, raw	0.94
706	**Old potatoes**, baked, flesh and skin	0.94
707	**Old potatoes**, baked, flesh only	0.94
711	**Old potatoes**, roasted in corn oil	1.28
714	**Chips**, homemade, fried in corn oil	1.1
718	**Chips**, retail, fried in vegetable oil	14.9
724	**Chips**, fine cut, frozen, fried in corn oil	3.51
727	**Oven chips**, frozen, baked	5.5
728	**Instant potato powder**, made up with water	0.9

[a] Beef, topside, roasted, fat only contains 3.56 µg/100 g vitamin K_1.
[b] Lamb loin chops, grilled, fat only contains 1.42 µg/100 g vitamin K_1.
[c] Pork chops, grilled, fat only contains 0.35 µg/100 g vitamin K_1.
[d] Vitamin K_1 value for Beef Stew with dumplings, retail.
[e] Vitamin K_1 value for crumbed and baked fish, type unspecified.

No.	Food	Vitamin K_1
Potatoes *continued*		
730	**Potato croquettes**, fried in blended oil	15.7
732	**Potato waffles**, frozen, cooked	2.5
Beans and lentils		
734	**Baked beans**, canned in tomato sauce, re-heated	2.72
736	**Beansprouts**, mung, raw	Tr
737	**Beansprouts**, stir-fried in blended oil	6.4
742	**Broad beans**, frozen, boiled in unsalted water	11.4
745	**Chick peas**, whole, dried, raw	8.92
746	**Chick peas**, dried, boiled in unsalted water	2.24
748	**Green beans/French beans**, raw	39
749	**Green beans/French beans**, frozen, boiled in unsalted water	7.8
760	**Runner beans**, raw	26
761	**Runner beans**, boiled in unsalted water	26
Peas		
767	**Mange-tout peas**, boiled in salted water	15
769	**Mushy peas**, canned, re-heated	22.5
770	**Peas**, raw	39
771	**Peas**, boiled in unsalted water	39
774	**Peas**, canned, re-heated, drained	30.4
775	**Petit pois**, frozen, boiled in unsalted water	28.3
Vegetables, general		
778	**Asparagus**, boiled in salted water	51.82
779	**Aubergine**, raw	6.1
780	**Aubergine**, fried in corn oil	10.3
784	**Broccoli**, green, raw	185
785	**Broccoli**, green, boiled in unsalted water	135.0
786	**Brussels sprouts**, raw	153
787	**Brussels sprouts**, boiled in unsalted water	127.0
788	**Brussels sprouts**, frozen, boiled in unsalted water	119.5
789	**Cabbage**, raw, average	242
790	**Cabbage**, boiled in unsalted water, average	201
791	**Cabbage**, white, raw[a]	60
792	**Carrots**, old, raw	5.5
793	**Carrots**, old, boiled in unsalted water	5.5
794	**Carrots**, young, raw	9.2
795	**Carrots**, young, boiled in unsalted water	9.2
797	**Cauliflower**, raw	31
798	**Cauliflower**, boiled in unsalted water	28.5
799	**Celery**, raw	4.9
800	**Celery**, boiled in unsalted water	4.9
802	**Courgette**, raw	3.3
804	**Courgette**, fried in corn oil	3.69

[a] Cabbage, white, outer leaves contain 137 μg/100 g vitamin K_1

No.	Food	Vitamin K_1
Vegetables, general *continued*		
805	**Cucumber**, raw	20.9
806	**Curly kale**, raw	623
809	**Fennel**, Florence, boiled in salted water	4.9
813	**Leeks**, raw	10.1
814	**Leeks**, boiled in unsalted water	9.5
815	**Lettuce**, average, raw	129
818	**Marrow**, boiled in unsalted water	2.4
820	**Mushrooms**, common, raw	0.28
822	**Mushrooms**, fried in corn oil	0.82
823	**Mustard and cress**, raw	88
828	**Onion**, fried in corn oil	3
831	**Parsnip**, raw[a]	0.05
834	**Peppers**, capsicum, green, raw	6.4
835	**Peppers**, capsicum, green, boiled in salted water	6.4
836	**Peppers**, capsicum, red, raw	1.6
837	**Peppers**, capsicum, red, boiled in salted water	1.6
838	**Plantain**, boiled in unsalted water	0.06
839	**Plantain**, ripe, fried in vegetable oil	10.4
841	**Pumpkin**, boiled in salted water	2
845	**Spinach**, raw	394
846	**Spinach**, boiled in unsalted water	575
847	**Spinach**, frozen, boiled in unsalted water	840
849	**Spring greens**, boiled in unsalted water	393
851	**Swede**, raw	2
855	**Sweetcorn**, baby, canned and drained	0.2
857	**Sweetcorn**, on-the-cob, whole, boiled in unsalted water	0.37
858	**Tomatoes**, raw	6
859	**Tomatoes**, fried in corn oil	6.8
860	**Tomatoes**, grilled	12.5
861	**Tomatoes**, canned, whole contents	6
862	**Turnip**, raw	0.15
863	**Turnip**, boiled in unsalted water	0.15
864	**Watercress**, raw	315
873	**Coleslaw**, with mayonnaise, retail	45.1
877	**Flan,** vegetable	30.9
882	**Pakora/bhaji**, vegetable, retail	72.1
895	**Vegeburger,** retail, grilled[b]	2.18
Herbs and spices		
912	**Parsley**, fresh	548
Fruit		
921	**Apples**, eating, average, raw[c]	5.6
928	**Bananas**	0.06

[a] Parsnip, roasted in corn oil, estimated to contain 0.2 µg/100 g vitamin K_1
[b] Vitamin K_1 value for Vegeburger mix, made up with water and grilled.
[c] Value refers to Cox's apples

Vitamin K_1, μg per 100g *continued*

No.	Food	Vitamin K_1
Fruit *continued*		
941	**Dates**, raw	5.6
952	**Grapefruit**, raw	0.01
955	**Grapes**, average	8.6
965	**Mangoes**, ripe, raw	0.52
968	**Melon**, Honeydew	0.13
969	**Melon**, watermelon	0.3
971	**Nectarines**	2.5
973	**Oranges**	0.05
977	**Peaches**, raw	5.8
980	**Pears**, average, raw	3.6
984	**Pineapple**, raw	0.21
987	**Plums**, average raw[a]	7.5
995	**Raisins**	3.7
996	**Rhubarb**, raw	4.3
1000	**Strawberries**, raw	3
Nuts and seeds		
1019	**Peanuts**, dry roasted	0.31
Sugars, preserves and snacks		
1037	**Jam**, fruit with edible seeds	0.9
1051	**Chocolate**, plain	2.3
1054	**Kit Kat**	4.5
1055	**Mars bar**	4.8
1075	**Corn snacks**	15.5
1079	**Potato crisps**	9.6
1084	**Tortilla chips**	0.18
Beverages		
1089	**Cocoa powder**	1.5
1091	**Coffee**, infusion, average	0.06
1092	**Coffee**, instant	4.3
1093	**Coffeemate**	6
1106	**Tea**, black, infusion, average	0.27
1126	**Orange juice**, unsweetened	0.06
Alcoholic beverages		
1130	**Beer**, bitter, average	Tr
1133	**Lager**	Tr
1139	**Stout**, Guinness	Tr
1145	**Red wine**	Tr
1160	**Spirits**, 40% volume	Tr
Soups, sauces and miscellaneous foods		
1173	**Instant soup powder**, dried[b]	7.43

[a] Value refers to Victoria plum

[b] Average of vegetable and meat varieties.

The Appendices

4.1 ANALYTICAL TECHNIQUES USED FOR THE TABLES

The methods which have been used for the analysis of foods in the Tables are shown; usually the first reference given is the most recent.

The nutrient values quoted in the Tables have been determined by a variety of methods. Although most give results of the same order of accuracy, with new methods merely improving the efficiency of analysis, some methods give different results and these have been documented in the Tables only where they appear to be substantial.

The following abbreviations are used in the text:-

GLC	Gas liquid chromatography
HPLC	High performance liquid chromatography
ICPOES	Inductively coupled plasma optical emission spectrophotometry

Nutrient	Method	
Water	Freeze drying Vacuum drying at 70°C Air drying at 100°C	
Nitrogen	Kjeldahl procedure Dumas combustion method	
Fat	Werner Schmidt Weibuhl Stoldt Rose-Gottlieb Weibull-Berntrop	Egan *et al.* (1981) Egan *et al.* (1981) Egan *et al.* (1981) IDF 125A (1988)
Fatty acids Trans fatty acids	GLC of methyl esters (IUPAC, 1976) (IUPAC, 1979)	
Cholesterol	GLC	
Phytosterols	GLC (IUPAC, 1987; AOAC, 1977)	
Alcohol	Standard Inland Revenue distillation method	

Nutrient	Method
Carbohydrates	
Total sugars (as monosaccharides)	Boehringer enzyme kit (Egan *et al.*, 1981) HPLC (Southgate *et al.*, 1978; Dean, 1978) Colorimetry (Southgate, 1976)
Starch	Enzymatic hydrolysis and measurement of glucose (Dean, 1978) Polarimetry (Egan *et al.*, 1981)
Fibre	
Non-starch polysaccharides	Englyst *et al.* (1994) Englyst *et al.* (1992) Englyst and Cummings (1988) Englyst and Cummings (1984) Englyst *et al.* (1982)
AOAC	Official method 985.29
Inorganics	
Sodium	ICPOES Emission spectrometry (Moxon, 1983) Atomic absorption spectrophotometry Flame photometry
Potassium	ICPOES Emission spectrometry Atomic absorption spectrophotometry Flame photometry
Calcium	ICPOES Atomic absorption spectrophotometry Titrimetry
Magnesium Copper Iron Zinc	ICPOES Atomic absorption spectrophotometry Colorimetry
Phosphorus	ICPOES Colorimetry
Chloride	Colorimetry Titrimetry
Manganese	ICPOES Atomic absorption spectrophotometry
Selenium	Hydride generation atomic absorption spectroscopy (Tinggi *et al.*, 1995) Fluorimetry (Michie *et al.*, 1978)
Iodine	Spectrophotometry (Moxon, 1980) GLC

Nutrient	Method
Vitamins	
Fat soluble vitamins	
Retinol β-Carotene Other carotenoids	HPLC Chromatographic separation and absorption spectrophotometry
Vitamin D	Reverse phase HPLC Biological assay and spectrophotometry GLC
Vitamin E	Normal phase HPLC Reverse phase HPLC Colorimetry combined with GLC (Christie *et al.*, 1973)
Vitamin K	HPLC in redox mode with electrochemical or UV detection (Bolton-Smith *et al.*, 2000; Shearer, 1996)
Water soluble vitamins	
Thiamin	HPLC with fluorimetric detection (Finglas and Faulks, 1984) Fluorimetry (Society of Public Analysts and Other Analytical Chemists: Analytical Methods Committee, 1951) Microbiological assay (Bell, 1974)
Riboflavin	HPLC with fluorimetric detection (Finglas and Faulks, 1984) Microbiological assay (Bell, 1974)
Niacin	HPLC (Kwiatowska *et al.*, 1989) Microbiological assay (Bell, 1974)
Vitamin B_6	HPLC with fluorimetric detection (Kwiatowska *et al.*, 1989; Brubacher *et al.*, 1985) Microbiological assay (Bell, 1974)
Vitamin B_{12} Pantothenate Biotin	Microbiological assay (Bell, 1974)
Folate	Microbiological assay (Phillips and Wright, 1983; Bell, 1974)
Vitamin C	
Ascorbic acid	HPLC with fluorimetric detection (Finglas and Faulks, 1984)
Ascorbic acid and dehydroascorbic acid	Titrimetry (AOAC, 1975) Fluorimetry (AOAC, 1975)

4.2 CALCULATION OF NUTRIENT CONTENTS FOR FOODS 'AS PURCHASED' OR 'AS SERVED'

Many foods are purchased or served with material that is clearly inedible or material that might be discarded as inedible by some individuals. For the purposes of this publication 'waste' encompasses both types of material, which might include for example

- outer leaves or stalks of vegetables
- stones, pips or peel of fruit
- nut shells
- fish skin and bones
- meat fat and bones
- liquid contents of canned foods.

Previous editions of the food tables have included data for foods both with and without waste. All the nutrient values in this edition apply to the edible part of the food and are expressed per 100g of edible portion. The proportion of edible matter in the food is given in the Tables and allows calculation of the nutrient content of foods when weighed with waste. For raw foods the edible proportion factor refers to the edible material remaining after the inedible waste has been trimmed away, e.g. the outer leaves of a cabbage. If the quantity of food consumed (including waste) is known, this can be multiplied by the edible proportion enabling the nutrient values given in the Tables (as per 100g edible portion) to be used in calculations of nutrient content.

For canned foods, such as vegetables and fish, the proportion of the edible contents after the liquid has been drained off is given in the edible proportion column and the values in the Tables are for the drained contents only unless otherwise stated in the description, e.g. tomatoes, canned, whole contents.

For canned fruits whose media may be consumed, the proportion of the edible contents after the liquid has been drained off is given in the description. The values in the Tables for canned fruit are for the fruit together with the syrup or juice in which it was canned.

For the calculation of the composition of cooked foods from raw materials, users should refer to Section 4.3 Cooked foods and dishes.

The nutrient content of a food 'as purchased' is calculated by multiplying the nutrient content 'as consumed' by the edible conversion factor. Worked examples showing the calculation of nutrients in foods when weighed with waste (such as foods 'as purchased' or 'as served') are given below.

Example 1: Carbohydrate content of Bananas weighed with skin

Carbohydrate in Bananas (flesh only)	=	23.2g/ 100g
Edible conversion factor for Bananas weighed with skin	=	0.66
Carbohydrate in Bananas weighed with skin	=	23.2×0.66
	=	15.3g/ 100g bananas 'as purchased'

Example 2: Protein content of Lamb, loin chop, grilled, lean, weighed with fat and bone

Protein in Lamb, loin chop, grilled, lean (as consumed)	=	29.2g/ 100g
Edible conversion factor for Lamb, loin chop, grilled, lean, weighed with fat and bone	=	0.61
Protein in Lamb, loin chop, grilled, lean weighed with fat and bone	=	29.2×0.61
	=	17.8 g/ 100g lamb, loin chop, grilled 'as served'

Example 3: Carbohydrate content of Apples, eating, average, raw, weighed with core

Carbohydrate in Apples, eating, average, raw (flesh and skin only)	=	11.8g/ 100g
Edible conversion factor for Apples, eating, average, raw, weighed with core	=	0.89
Carbohydrate in Apples, eating, average, raw, weighed with core	=	11.8×0.89
	=	10.5g/ 100g apples, eating, raw, 'as purchased/served'

Example data

No.	Food	Edible conversion factor	Water (g)	Protein (g)	Fat (g)	Carbo-hydrate (g)
445	**Lamb, loin chops,** grilled, lean	0.61	59.6	29.2	10.7	0
	weighed with fat and bone		36.4	17.8	6.5	0
921	**Apples**, eating, average, raw	0.89	84.5	0.4	0.1	11.8
	weighed with core		75.2	0.4	0.1	10.5
928	**Bananas**, flesh only	0.66	75.1	1.2	0.3	23.2
	weighed with skin		49.6	0.8	0.2	15.3

4.3 COOKED FOODS AND DISHES

i Weight changes on preparation of foods

The figures below show the percentage changes in weight recorded during the cooking of foods included in this edition. The values were obtained by Holland *et al.* (1991), Wiles *et al.* (1980), Paul and Southgate (1977), McCance and Shipp (1933), and from previously unpublished determinations where a measure of weight change was available. The values should be treated as guidelines only; for more accurate figures users should make their own determinations. The weight changes during cooking of recipe dishes have been included with each recipe.

The majority of changes result from the loss or gain of water, but for many meats and fried foods there will also have been a loss or gain of fat. The values have been calculated as:

$$\frac{\text{Weight of cooked food or dish} - \text{Weight of raw food(s)}}{\text{Weight of raw food(s)}} \times 100$$

A value of +200 thus means not that the food doubled its weight, but that it gained twice its original weight on cooking (i.e. tripled in weight), because:

$$\frac{300 - 100}{100} \times 100 = +200$$

A plus sign (+) indicates that the food or dish gained weight on cooking while a minus sign (–) shows that it lost weight. The value given is a mean weight change, but where data is available for the range of losses, it is given in brackets.

For root and leafy vegetables boiled in water there is little difference in weight between the raw and cooked food.

CEREALS AND CEREAL PRODUCTS

	% weight change mean		% weight change mean
Rice and pasta		***Bread***	
Brown rice, boiled	+153	White bread, fried	–29
Savoury rice, cooked	+197	toasted	–18
White rice, easy cook, boiled	+177	'with added fibre', toasted	–16
Macaroni, dried, boiled	+146	Wholemeal bread, toasted	–15
Spaghetti, white, dried, boiled	+113		
egg, white, fresh, boiled	+82	***Buns and pastries***	
wholewheat, dried, boiled	+130	Crumpets, toasted	–11

CEREALS AND CEREAL PRODUCTS *continued*

	% weight change mean		% weight change mean
***Rice and pasta** continued*		***Buns and pastries** continued*	
Fusilli, dried, boiled	+123	Muffins, English style, toasted	–13
fresh, boiled	+82	Teacakes, toasted	–10
Tagliatelle, dried, boiled	+127		
egg, fresh, boiled	+83		

MEAT AND MEAT PRODUCTS

	% weight change mean (range)		% weight change mean (range)
Bacon and ham		***Pork***	
Bacon rashers, back, dry-fried	–33 (23–40)	Belly joint/slices, grilled	–36 (29–61)
back, grilled	–32 (22–44)	Diced, casseroled	–37 (31–410
back, grilled crispy	–53 (28–68)	Fillet strips, stir-fried	–33 (26–40)
back, microwaved	–39 (26–48)	Leg joint, roasted medium	–35 (23–41)
back, fat trimmed, grilled	–33 (25–44)	Loin chops, barbecued	–28 (15–48)
back, reduced salt, grilled	–32 (24–38)	grilled	–32 (21–40)
middle, grilled	–38 (27–49)	microwaved	–32 (20–40)
streaky, grilled	–35 (21–51)	roasted	–38 (24–57)
Ham, gammon joint, boiled	–29 (15–40)	Steaks, grilled	–38 (28–47)
Gammon rashers, grilled	–34 (32–50)		
		Veal	
Beef		Escalope, fried	–38 (18–46)
Braising steak, braised	–40 (36–46)		
Fore-rib/rib-roast, roasted	–34 (28–41)	***Chicken***	
Mince, microwaved	–28 (19–43)	Breast, casseroled, meat only	–25
Mince, stewed	–18 (4–25)	grilled, without skin, meat only	–25
extra lean, stewed	–18 (9–29)	Breast, strips, stir-fried	–21 (17–29)
Rump steak, barbecued	–31 (18–47)	Breast in crumbs, fried	–5
fried	–27 (20–36)	Drumsticks, roasted, meat and skin	–26 (14–50)
grilled	–28 (18–35)		
strips, stir-fried	–29 (17–40)	***Turkey***	
Silverside, salted, boiled	–39 (29–50)	Breast fillet, grilled	–32 (23–46)
Stewing steak, stewed	–36 (26–47)	strips, stir-fried	–23 (18–27)
Topside, roasted, well-done	–42 (34–52)		

MEAT AND MEAT PRODUCTS *continued*

	% weight change mean (range)		% weight change mean (range)
Lamb		***Burgers and grillsteaks***	
Best-end neck cutlets, grilled	–32 (15–54)	Beefburgers, chilled/ frozen, fried	–38
Breast, roasted	–28 (20–41)	grilled	–34
Leg, whole, roasted, medium	–31 (20–35)	Economy burgers, frozen, grilled	–17 (11–23)
Loin chops, grilled	–31 (15–52)	Grillsteaks, beef, chilled/frozen, grilled	–25
microwaved	–33 (24–41)		
roasted	–37 (17–57)		
Mince, stewed	–28 (22–33)	***Meat products***	
Shoulder, whole, roasted medium	–32 (21–40)	Black pudding, dry-fried	–12
Stewing lamb, pressure-cooked	–28 (22–40)	Chicken pie, chilled/frozen, baked	–5 (3–9)
stewed	–27 (21–41)	Sausages, beef, chilled, grilled	–25
		pork, frozen, fried	–20
		pork, chilled, grilled	–24
		pork, reduced fat, chilled, grilled	–24
		premium, chilled, grilled	–24

MEAT DISHES

	% weight change mean (range)
Beef curry, chilled/frozen, baked	–5
microwaved	–16
Chicken curry, chilled/frozen, baked	–1
microwaved	–24
Chicken tandoori, chilled, baked	–18
microwaved	–13
Chicken tikka masala, chilled/frozen, baked	–3 (Tr–4)
microwaved	–6 (2–9)
Chicken, stir-fried with rice and vegetables, frozen, reheated	–19 (15–23)
Chilli con carne, chilled/frozen, reheated, meat and sauce only	–6 (0–18)
Cottage/Shepherds pie, chilled/frozen, baked	–13 (1–48)
microwaved	–21 (1–40)
Faggots in gravy, chilled/frozen, baked	–6 (4–10)
microwaved	–2 (2–3)

MEAT DISHES *continued*

	% weight change mean (range)
Lamb/Beef hot pot with potatoes, chilled/frozen, baked	–13 (6–21)
microwaved	–8 (1–14)
Lasagne, chilled/frozen, baked	–17 (7–47)
Moussaka, chilled/frozen/longlife, baked	–9 (4–13)
microwaved	–7 (5–11)
Spaghetti bolognaise, chilled/frozen, baked	–7 (6–8)
microwaved	–15 (11–19)

FISH AND FISH PRODUCTS

	% weight change mean (range)		% weight change mean (range)
White fish		***Crustacea***	
Cod, baked fillets	–19	Crab, boiled	–20
poached fillets	–14	Lobster, boiled	–19
frozen, grilled steaks	–15	Scampi in breadcrumbs, fried	–23
dried, salted, boiled	+19		
Haddock, smoked, steamed	–15		
Lemon sole, steamed	–13	***Molluscs***	
Plaice, in crumbs, fried in blended oil	+21	Mussels, boiled	–33
Coley, steamed	–24	***Fish products and dishes***	
Whiting, steamed	+16	Fish cakes, fried	+2
in crumbs, fried in blended oil	+13	Fish fingers, fried	–10
		grilled	–7
Fatty fish		Roe, cod, hard, fried	–7
Herring, grilled	–9		
Salmon, steamed	–10		
Whitebait, fried	–23		

VEGETABLES AND VEGETABLE PRODUCTS

	% weight change mean (range)
Beans and lentils	
Aduki beans, soaked and boiled	+155
Black gram, urad gram, soaked and boiled	+208
Blackeye beans, soaked and boiled	+164
Chick peas, whole, soaked and boiled	+163
Lentils, green and brown, boiled	+139
red split, boiled	+227
Mung beans, whole, soaked and boiled	+199
Red kidney beans, soaked and boiled	+161
Soya beans, soaked and boiled	+156

ii Calculation of cooked edible matter from raw foods

The Tables show the edible conversion factors for the edible portion and Appendix 4.2 gives examples of calculation of the nutrient content of foods weighed with waste. It is sometimes necessary to estimate the amount of cooked edible material that would be obtained from a known weight of the raw food 'as purchased'. This is done by combining the percentage weight loss on cooking with the edible matter as a proportion of the cooked food, as follows:

Cooked edible matter as a proportion of raw 'as purchased' food

$$= \frac{\text{Edible proportion of cooked food} \times (100 - \%\text{ weight loss on cooking})}{100}$$

For example, the weight loss on grilling lamb chops is 31% and the edible proportion of grilled lamb chops, lean and fat, weighed with bone is 0.81. 200g of raw lamb chops with lean, fat and bone will therefore yield

$$200 \times \frac{0.81 \times (100-31)}{100} = 112\text{g cooked lamb (lean and fat) to eat}$$

iii Calculation of the composition of dishes prepared from recipes

The composition of cooked dishes in this book has been calculated, as in previous editions, from the recipes listed in Appendix 4.4, based on the composition of the ingredients, the changes in weight on cooking and the vitamin losses on cooking.

The change in weight on cooking is usually due only to the evaporation of water or to its gain by absorption. The composition of dishes where the method of preparation also involves a change in fat content cannot be calculated directly in this way. In these cases the cooked dishes were either analysed for fat and water before the calculations were made or the weight change was corrected for fat uptake measured after preparation.

The method of calculation was as follows. The weights of the raw ingredients were used to calculate the total amounts of nutrients in the raw dish. A correction to allow for any wastage due to ingredients left on utensils and in the vessels used in preparation was made at this stage. The weight of the raw dish was then measured, using a scale accurate to about 1g. The dish was then cooked as specified and re-weighed. (A minor correction to allow for the difference between weighing the dish hot and at room temperature is not usually necessary). Where the difference in weight was accounted for by water alone, the nutrient composition and the water content of the cooked dish were calculated as follows:

$$\text{Nutrient content of cooked dish per 100 grams} = \frac{\text{Total nutrient content of raw ingredients}}{\text{Weight of cooked dish}} \times 100$$

$$\text{Water content of cooked dish per 100 grams} = \frac{\text{Water in raw ingredients} - \text{weight loss on cooking}}{\text{Weight of cooked dish}} \times 100$$

An example of this calculation is shown in Table 13 below.

Table 13 *Custard, made up with whole milk*

Ingredient		Amount in recipe g	Amounts contributed: Water g	Protein g	Fat g	Carbo-hydrate g
Milk, whole (500ml)		515	451	17.0	20.1	23.2
Custard powder		25	3.1	0.1	0.2	23.0
Sugar		25	0	0	0	26.3
Total in recipe (a)		565	454.1	17.1	20.3	72.5
Cooked weight (b)		447				
Weight loss on cooking (c)	= a – b	118				
% weight loss on cooking (d)	= c/a × 100	*20.9*				
Nutrient content of cooked dish (per 100g) (e)	= a/b × 100			3.8	4.5	16.2
Water content of cooked dish (per 100g) (f)	= $\frac{a-c}{b}$ × 100		75.2			

If a recipe is to be calculated from the ingredients, but the weight of the cooked dish is not known, this may be estimated by using the % weight change from a similar recipe as follows (provided that all the weight change can be attributed to water):

$$\text{Weight of cooked dish} = \frac{\text{Weight of raw ingredients} \times (100 - \%\text{ weight loss of similar dish})}{100}$$

For recipes which gain weight on cooking, for example dumplings:

$$\text{Weight of cooked dish} = \frac{\text{Weight of raw ingredients} \times (100 + \%\text{ weight gain of similar dish})}{100}$$

iv Vitamin loss estimation in foods and recipe calculations

The losses of heat- and water-labile vitamins in cooked recipe dishes were estimated by assigning a set of factors for percentage vitamin losses to each ingredient in the recipe, according to its food group and the method of cooking. Vitamin losses were not assigned to minor ingredients such as herbs, spices and salt. The percentage vitamin losses used for each food group and cooking method are shown in the Tables below. This is a change from the approach used for previous editions, in which vitamin loss factors were applied to the whole recipe dish according to the major ingredient. The values in the Tables should be treated as guidelines only. Vitamin losses will vary according to the length, temperature and method of cooking, and the nature of the ingredients. For more accurate information the foods or composite dish should be analysed.

An example of the method of calculating the vitamin content in a cooked product taking into account the percentage vitamin loss is given below.

$$\text{Vitamin content of cooked dish per 100 grams} = \text{Vitamin content of uncooked dish per 100 grams} - \frac{\text{Vitamin content of uncooked dish} \times \text{\% vitamin loss on cooking}}{100}$$

For example, the thiamin content of brown bread, average, is 0.22 mg per 100 grams and the percentage thiamin loss on toasting bread is 15%. Therefore the thiamin content of toasted brown bread is calculated as

$$0.22 - \frac{(0.22 \times 15)}{100} = 0.19 \text{ mg}$$

Table 14 *Cereals: typical percentage losses of vitamins on cooking*

	% vitamin losses	
	Boiling	Baking
Thiamin	40	25[a]
Riboflavin	40	15
Niacin	40	5
Vitamin B_6	40	25
Folate	50	50
Pantothenate	40	25
Biotin	40	0

[a] 15% in bread-making and toasting

Table 15 *Milk: typical percentage losses of vitamins on cooking*

	% vitamin losses		
	Boiling[a]	Sauces[b]	Baked dishes
Vitamin E	20	(20)	ND
Thiamin	10	20	25
Riboflavin	10	(10)	15
Niacin	(0)	(0)	5
Vitamin B_6	10	20	25
Vitamin B_{12}	5	(5)	ND
Folate	20	50	50
Pantothenate	10	20	25
Vitamin C	50	50	ND

[a] In milk-based drinks, custards, etc.
ND = Not determined
[b] For example, for cheese sauce
Values in brackets are estimates

Table 16 *Eggs: typical percentage losses of vitamins on cooking*

	% vitamin losses		
	Scrambled	Omelette	Baked dishes
Thiamin	5	5	15
Riboflavin	20	20	15
Niacin	5	5	5
Vitamin B_6	15	15	25
Folate	30	30	50
Pantothenate	15	15	25

Table 17 *Meats: typical percentage losses of vitamins on cooking*

	% vitamin losses	
	Meat, grilled or fried	Meat dishes[a]
Vitamin A	0	0
Vitamin E	20	20
Thiamin	20	20
Riboflavin	20	20
Niacin	20	20
Vitamin B_6	20	20
Vitamin B_{12}	20	20
Folate	ND[b]	50
Pantothenate	20	20
Biotin	10	10
Vitamin C	ND[b]	50

[a] Some vitamins are lost on heating but the vitamins (and minerals and fat) that leach into the liquor during cooking will not be lost if the sauce or the gravy is eaten as part of the dish. On average therefore, the losses in meat dishes are no higher than from grilled or fried meat even though the cooking times are longer.

[b] The amounts of folate and vitamin C in meat are too low to make meaningful calculations of losses.

ND = Not determined

Table 18 *Fish: typical percentage losses of vitamins on cooking*[a]

	% vitamin losses			
	Poaching	Baking	Grilling	Frying
Vitamin A	0	0	0	0
Vitamin D	0	0	0	0
Vitamin E	0	0	0	0
Thiamin	10	30	10	20
Riboflavin	0	20	10	20
Niacin	10	20	10	20
Vitamin B_6	0	10	10	20
Vitamin B_{12}	0	10	0	0
Folate	0	20	0	0
Pantothenate	20	20	5	20
Biotin	10	10	0	10

[a] Apart from grilling, the losses are mainly based on those found on cooking cod.

Table 19 *Vegetables: typical percentage losses of vitamins on cooking*

	% vitamin losses		
	Boiling	Frying	Cooked dishes
Carotene	ND	ND	0
Vitamin E	0	0	0
Thiamin	35	20	20
Riboflavin	20	0	20
Niacin	30	0	20
Vitamin B_6	40	25	20
Folate	40	55	50
Pantothenate	ND	ND	20
Biotin	ND	ND	20
Vitamin C	45	30	50

ND = Not determined

Table 20 *Fruit: typical percentage losses of vitamins on stewing*

	% vitamin loss
Carotene	(0)
Thiamin	25
Riboflavin	25
Niacin	25
Vitamin B_6	20
Folate	80
Pantothenate	25
Biotin	25
Vitamin C	25

Values in brackets are estimates

4.4 RECIPES

All recipes in these tables that were previously published in the 5th edition or in a supplement have been recalculated using updated ingredient composition and standard portion sizes.

Where a recipe source indicated a portion but not the quantity of an ingredient the portion size was taken from Food Portion Sizes (MAFF, 1993) or weighed during recipe testing.

Portion sizes

Baking powder	1 tsp = 3.5g
Banana, without skin, medium	100g
Egg	50g
Egg white	32g
Egg yolk	18g
Flour	1 level tbsp = 20g, heaped = 30g
Garlic	1 clove = 4g
Herbs, dried	1 tsp = 1g
Marmite/Yeast extract	1 tsp = 9g
Mustard powder	1 tsp = 3g
Pepper	1 tsp = 1g
Salt	1 tsp = 5g
Soy sauce	1 tsp = 5g
Spices, dried	1 tsp = 3g
Sugar	1 tsp = 4g
Vegetable oil	1 tbsp = 11g

The amounts of beer, lemon juice, milk, stock, vinegar, water and wine are given in millilitres but for beer, milk and wine, the millilitre measures were converted to gram weights for the purposes of recipe calculation. Stock was made up using 6g stock cube to 190ml water.

Quantities have not been included for recipes obtained in confidence from manufacturers. The ingredients have however, been listed in quantity order.

For a number of recipes obtained from dietary survey records only the major ingredients were recorded. These recipes do not contain a measure for salt, spices or other 'lesser' ingredients and these were not therefore, included in the recipe calculation

Unless specified, all the recipe items used were raw. Whole pasteurised milk, Cheddar cheese, non-dairy vanilla ice cream, plain white flour and distilled water were used. The bacon was without rind, the carrots, onions, potatoes and root ginger were peeled, the chilli peppers were deseeded, and, except where otherwise specified, the turkey and chicken were skinless and boneless, and the beef, lamb and pork included both lean and fat.

Where canned fruit were used as ingredients, the nutrient composition was an average of the fruit canned in syrup and juice. Where canned tomatoes were used, the nutrient composition included the juice as well.

The type of fat used in the recipes has been specified. The vegetable oil was a retail blended vegetable oil. Margarine was soft, polyunsaturated. Butter was salted. For fried dishes, the fat used during frying has been included at the end of the ingredients list with the quantity absorbed shown in brackets.

The baking powder used was a proprietary preparation whose composition is listed in these Tables (No 1223). Use of another brand could result in a different composition in the cooked dish with respect to sodium, calcium and phosphorus.

19 **Pilau, plain**

200g raw white rice
45ml water absorbed on soaking
25g butter ghee
1 tsp cumin seeds
90g sliced onion
500ml boiling water
1 tsp salt
½ tsp turmeric

Soak the rice. Heat the ghee and add cumin seeds. Add the onion and fry for 5 minutes. Reduce heat, add remaining ingredients, cover and cook until all the water has been absorbed and the rice is soft.

Weight loss: 17%

25 **White rice, fried**

550g boiled rice
168g chopped onion
2 tbsp vegetable oil
21g garlic
2g salt
¼ tsp pepper
1g spices

Fry onion and garlic until soft. Add boiled rice and seasoning. Fry until oil has been absorbed and rice is fully coated.

Weight loss: 5.6%

28 **Macaroni cheese**

280g cooked macaroni
350ml milk
25g margarine
25g flour
100g grated cheese
½ tsp salt

Boil the macaroni and drain well. Make a white sauce from the margarine, flour and milk. Add 75g of the cheese and season. Add the macaroni and put in a pie dish. Sprinkle with remaining cheese and brown under grill or in a hot oven at 220°C/mark 7.

Weight loss: 9.4%

72 **Bacon, lettuce and tomato sandwich**

86g white bread
29g grilled bacon
20g tomato
16g lettuce
11g mayonnaise
7g fat spread

Average weights from University of North London survey of commercial sandwiches

73 **Cheese and pickle sandwich**

75g white bread
43g cheddar cheese
16g sweet pickle
7g fat spread

Average weights from University of North London survey of commercial sandwiches

74 **Chicken salad sandwich**

89g white bread
46g cooked chicken
20g tomato
12g lettuce
12g cucumber
7g fat spread

Average weights from University of North London survey of commercial sandwiches

75 **Egg mayonnaise sandwich**

94g white bread
42g boiled egg
10g mayonnaise
7g fat spread

Average weights from University of North London survey of commercial sandwiches

76 **Ham salad sandwich**

88g white bread
35g ham
17g tomato
11g lettuce
9g cucumber
7g fat spread

Average weights from University of North London survey of commercial sandwiches

77 **Tuna mayonnaise sandwich**

93g white bread
56g tuna, canned in brine, drained
15g mayonnaise
7g fat spread

Average weights from University of North London survey of commercial sandwiches

91 **Porridge, made with water**

60g oatmeal
7g salt
500ml water

Weight loss: 14%

92 **Porridge, made with whole milk**

60g oatmeal
7g salt
500ml milk

Weight loss: 14%

111 **Flapjacks**

120g rolled oats
90g margarine
60g golden syrup
60g brown sugar
2g ginger

Melt fat, add sugar and syrup. Work in the oats. Press into a greased sandwich tin and bake at 170°C/mark 3 for 30 minutes.

Weight loss: 5%

123 **Wholemeal crackers**

105g fat
11.9g salt
15.4g bakers yeast
630g plain flour
210g wholemeal flour
2.1g bicarbonate of soda

Recipe from Flour Milling and Baking Research Association.

Weight loss: 11%

124 **Banana bread**

125g margarine
250g self-raising flour
175g caster sugar
2 eggs
500g bananas (weighed with skin)
125g glacé cherries
125g sultanas
60g walnuts

Recipe from *Mary Berry's Complete Cookbook*, p. 475. Reproduced by permission of Dorling Kindersley Ltd. © 1995 Dorling Kindersley Ltd, text © 1995 Mary Berry

Pour the flour into a bowl, add the margarine and rub in until the mixture resembles fine breadcrumbs. Add the caster sugar, sultanas, chopped walnuts, and glacé cherries and mix well. Add the eggs and mashed bananas and beat the mixture until well blended. Spoon the mixture into a greased 2lb loaf tin bottom lined with greaseproof paper. Bake in a pre-heated oven at 160°C/mark 3 for about 1¼ hours until well risen.

Weight loss: 8.6%

125 **Battenburg cake**

100g flour
75g margarine
120g sugar
90g eggs
58g marzipan
67.5g water
8g skimmed milk powder
5g baking powder
2g salt
8g jam

Recipe from Flour Milling and Baking Research Association.

Weight loss: 10% cake

127 **Carrot cake**

225g self-raising flour
150g light muscavado sugar
2 eggs
100g carrots
2 ripe bananas
50g walnuts
150g sunflower oil
2 tsp baking powder

Topping:
175g soft white full fat spreadable cheese
100g icing sugar
50g margarine
few drops of vanilla essence

Recipe based on *Mary Berry's Ultimate Cakes*. Reproduced with the permission of BBC Worldwide Limited. © Mary Berry 1994

Mix all the cake ingredients in a large bowl until thoroughly blended and smooth. Add the mixture to an 8 inch deep round tin lined with greased greaseproof paper. Level the mixture and bake in a pre-heated oven for about 50–60 minutes at 180°C/mark 4 until the cake is risen. Allow to cool and turn out onto a wire rack. Measure the topping ingredients into a bowl and mix until smooth. Spread over the top of the cake and chill before serving to allow the topping to harden.

Weight loss: 9.5%

129 **Crispie cakes**

112g plain chocolate
33g crisp rice cereal
33g corn flake type cereal

Melt chocolate in a bowl over hot water. Stir in cereals. Place in cases and allow to cool and set.

Weight loss: 0%

132 **Fruit cake, rich**

200g margarine
200g brown sugar
4 eggs
20g black treacle
20ml brandy
250g flour
1/4 tsp salt
750g mixed fruit
150g mixed glacé fruit, chopped
1 tsp mixed spice

Cream the fat and sugar. Beat the eggs, treacle and brandy. Fold in the sifted flour and spices and mix in the fruit. Turn into a 20cm cake tin. Bake for 4 hours at 150°C/mark 2.

Weight loss: 5%

133 **Fruit cake, rich, iced**

1680g fruit cake, rich
70g apricot jam
410g marzipan
Royal icing:
300g icing sugar
1 egg white
1 tsp lemon juice

Make the cake as in Rich Fruit Cake (No. 132) recipe. When cold spread with a thin layer of apricot jam and cover with marzipan. Make the icing by beating the egg whites and icing sugar; finally add lemon juice.

135 **Fruit cake, wholemeal**

200g margarine
200g brown sugar
3 eggs
200g plain flour
2 tsp baking powder
4g mixed spice
200g wholemeal flour
200g mixed fruit
100ml milk

Cream the fat and sugar, beat in eggs. Sift white flour, baking powder and spice, add creamed mixture together with wholemeal flour and fruit . Add milk until soft. Bake for 1½–2 hours at 180°C/mark 4.

Weight loss: 5%

137 **Jaffa cakes**

33.1% baked sponge base
39.6% orange jelly
27.3% plain chocolate

Recipe from Flour Milling and Baking Research Association.

139 **Muffins, American style, chocolate chip**

150g self raising flour
110ml milk
1 egg
120g plain chocolate
50g butter
50g sugar
½ tsp baking powder

Melt the butter in a saucepan and allow to cool. Sieve the flour into a large mixing bowl, add the sugar and baking powder and mix. Break the egg into a small bowl and whisk. Add the egg and the milk to the saucepan of melted butter and mix well. Pour the mixture into the flour and mix together quickly. Mix in the chocolate chips and divide the mixture into bun cases. Bake for 20–25 minutes at 200°C/mark 6.

Weight loss: 11.2%

142 **Sponge cake**

150g flour
1 tsp baking powder
150g margarine
150g caster sugar
3 eggs

Cream the fat and sugar until light and fluffy. Add the beaten egg a little at a time and beat well. Fold in the sifted flour and baking powder. Bake for about 20 minutes at 190°C/mark 5.

Weight loss: 12.9%

143 **Sponge cake, made without fat**

4 eggs
100g caster sugar
100g flour

Whisk the eggs and sugar in a basin over hot water until stiff. Fold in flour. Bake for 25 minutes at 190°C/mark 5.

Weight loss: 13.8%

147/148 **Flaky pastry**

200g flour
75g margarine
75g lard
½ tsp salt
85ml water
10ml lemon juice

Divide fat into 4 portions. Sift flour and salt, rub in one portion of fat. Mix with water and lemon juice, then knead until smooth and leave for 15 minutes. Roll out, dot two thirds with another fat portion and fold into 3. Roll out and repeat process with remaining 2 fat portions. Bake at 220°C/mark 7.

Weight loss: 24.3%

149/150 **Shortcrust pastry**

200g flour
50g margarine
50g lard
½ tsp salt
30ml water

Rub the fat into the flour, mix to a stiff dough with the water, roll out and bake at 200°C/mark 6.

Weight loss:13.8%

151/152 Wholemeal pastry

200g wholemeal flour
50g margarine
50g lard
2g salt
30ml water

Rub the fat into the flour, mix to a stiff dough with the water, roll out and bake at 220°C/mark 7.

Weight loss: 13.6%

154 Chelsea buns

200g strong flour
85g skimmed milk
65g margarine
45g sugar
35g eggs
15g yeast
55g currants
2g salt

Recipe from Flour Milling and Baking Research Association.

Weight loss: 15%

161 Eccles cakes

212g frozen puff pastry
1 egg white
25g mixed peel, chopped
½ tsp ground mixed spice
25g butter
50g demerara sugar
100g currants
Caster sugar for sprinkling

Recipe from *The Dairy Book of British Food*, by the Milk Marketing Board, published by Ebury Press. Reprinted by permission of The Random House Group Ltd.

Melt the butter in a saucepan, then stir in the currants, peel, sugar and spice and mix thoroughly. On a lightly floured surface, roll out the pastry very thinly and cut out eight 12.5cm (5 inch) circles using a saucer as a guide. Divide the fruit mixture between the circles, damp the edges of the pastry and draw them into the centre, sealing well together. Turn the cakes over and roll gently into circles with a rolling pin. Brush with egg white and sprinkle with caster sugar. Make 3 diagonal cuts across the top of each. Place on dampened baking sheets and bake at 220°C/mark 7 for about 15 minutes, until light golden brown.

Weight loss: 9.1%

164 Hot cross buns

450g strong flour
28g fresh yeast
1 egg
pinch salt
56g margarine
112g currants
1g cinnamon
1g nutmeg
2g mixed spice
45g peel
150ml milk
60ml water
56g caster sugar

Glaze:
45g sugar
30ml milk
30ml water

Cream yeast with milk and add salt. Add to flour and eggs, and mix. Knead for 10 mins. Sprinkle with sugar, dot with fat and leave for 30 mins. Mix fat, sugar and fruit into mixture and mould. Cut a cross on each bun, glaze and bake for 15 minutes at 250°C/mark 9.

Weight loss: 15%

166 **Mince pies, individual**

300g raw short crust pastry (149)
200g mincemeat

Roll out the pastry and cut into rounds. Place half of the rounds in tart tins. Fill with mincemeat and cover with remaining pastry. Bake for about 20 minutes at 190°C/mark 5.

Weight loss 12.6%

170 **Scones, plain**

200g flour
4 tsp baking powder
¼ tsp salt
50g margarine
10g sugar
125ml milk

Sift the flour, sugar and baking powder and rub in fat. Mix in the milk. Roll out and cut into rounds. Bake in a hot oven at 220°C/mark 7 for about 10 minutes.

Weight loss: 18.5%

171 **Scones, wholemeal**

200g wholemeal flour
14g baking powder
1g salt
50g margarine
10g sugar
125ml milk

Method as recipe for plain scones (No.170).

Weight loss: 18.5%

174 **Bread pudding**

225g white bread
275ml milk
50g melted butter
75g demarara sugar
4g mixed spice
1 beaten egg
175g dried fruit

Break bread into pieces, cover with milk and leave for 30 mins. Add remaining ingredients, mix well and bake for 1¼ hours at 180°C/mark 4.

Weight loss: 24%

176/177 **Crumble, fruit, plain or wholemeal**

400g prepared fruit
50g margarine
100g plain or wholemeal flour
100g sugar

Prepare fruit. Arrange in a dish and sprinkle with sugar. Rub together the other ingredients and pile on top. Bake for 40 minutes at 190°C/mark 5.

Weight loss: 7.4%

178/182 **Fruit pie, one crust, plain or wholemeal**

200g raw plain or wholemeal shortcrust pastry (149/151)
450g prepared fruit
80g sugar

Place fruit in pie dish and cover with pastry. Bake for 10–15 minutes at 200°C/mark 6 to set pastry, then about 20 minutes at 180°C/mark 4 to cook fruit.

Weight loss: 4.2%

179/183 **Fruit pie, pastry top and bottom, plain or wholemeal**

450g raw plain or wholemeal shortcrust pastry (149/151)
450g fruit (eg. apple, gooseberry, rhubarb, plum)
80g sugar

Line a pie dish with half the pastry. Fill with prepared fruit and sugar and cover with remaining pastry. Bake for 10–15 minutes at 220°C/mark 7 to set pastry, then for about 20–30 minutes at 180°C/mark 4 to cook the fruit.

Weight loss: 4.2%

181 **Fruit pie, blackcurrant, pastry top and bottom**

450g raw shortcrust pastry (149)
450g blackcurrants
80g sugar

Method as for fruit pie, pastry top and bottom (No. 179).

Weight loss: 4.2%

185 **Pancakes, sweet**

100g flour
250ml whole milk
1 egg
50g lard (for pan)
50g sugar

Sieve the flour into a basin, add the egg and about 100ml of the milk, stirring until smooth. Add the rest of the milk and beat to a smooth batter. Heat a little of the lard in a frying pan and pour in enough batter to cover the bottom. Cook both sides and turn onto sugared paper. Dredge lightly with sugar. Repeat until all the batter is used, to give about 10 pancakes.

Weight loss: 20%

187 **Treacle tart**

300g raw shortcrust pastry (149)
250g golden syrup
50g fresh breadcrumbs

Line shallow tins with pastry, pour in the syrup and sprinkle with the breadcrumbs.

Bake for 20–30 minutes at 200°C/mark 6.

Weight loss: 0%

189 **Dumplings**

100g flour
45g suet
75g water
1 tsp baking powder
½ tsp salt

Mix the dry ingredients together with the cold water to form a dough. Divide into balls, flour them and place in boiling water. Boil for 30 minutes.

Weight gain: + 52.7%

190 **Pancakes, savoury**

112g flour
300ml whole milk
1 egg
56g lard (for pan)
¼ tsp salt

Method as for sweet pancakes (No. 185).

Weight loss: 20%

192 **Risotto**

224g long grain rice
550g stock
84g chopped onion
56g margarine
1 tsp salt
1g pepper

Melt margarine, add onion and fry until soft. Add washed rice and stir over low heat for 10 minutes. Pour in stock, bring to boil and simmer until all is absorbed.

Weight loss: 37%

194 **Stuffing, sage and onion**

224g onion
112g white breadcrumbs
4g fresh sage, chopped
¼ tsp salt
¼ tsp pepper
1 egg
56g margarine

Parboil onions, drain and chop, mix with breadcrumbs, add sage. Melt margarine and add to stuffing. Mix thoroughly. Stir in egg and seasoning.

Weight loss: 19.0%

195 **Yorkshire pudding**

100g flour
1 tsp salt
1 egg
250ml milk
20g dripping

Sieve flour and salt into a basin. Break in the egg and add about 100ml milk, stirring until smooth. Add the rest of the milk and beat to a smooth batter. Pour into a tin containing the hot dripping. Bake for about 40 minutes at 220°C/mark 7.

Weight loss: 16%

297 **Tzatziki**

250g Greek cows milk yogurt
5g fresh garlic
fresh chopped mint
213g cucumber
4g salt

Recipe from yogurt manufacturers.

303 **Chocolate nut sundae**

115g ice cream
45ml double cream
70g chocolate sauce
6g chopped nuts
wafer (1g)

Cover ice cream with whipped cream and chocolate sauce. Sprinkle with nuts, add wafer.

319/320 **Custard made up with milk**

500ml whole or semi-skimmed milk
25g custard powder
25g sugar

Blend custard powder with a little of the milk. Add sugar to remainder of milk and bring to the boil. Pour immediately over paste, stirring all the time. Return to pan, bring back to boiling point while stirring.

Weight loss: 20.9%

323 **Jelly made with water**

130g jelly cubes
440ml water

Dissolve jelly cubes in hot water. Add rest of the cold water. Pour into a mould and allow to set.

324 **Meringue**

4 egg whites
200g caster sugar

Whisk egg whites until stiff. Fold in the sugar. Pipe onto baking sheet and bake at 130°C/mark ½ for 3 hours.

Weight loss: 33.3%

325 **Meringue with cream**

40% meringue | 60% whipping cream

Proportions derived from a number of shop-bought samples.

326 **Milk pudding**

500ml whole milk
50g rice, sago, semolina or tapioca
25g sugar

Simmer until cooked or bake in a moderate oven at 180°C/mark 4.

Weight loss: 19.1%

336 **Trifle**

75g sponge
25g jam
50g fruit juice
75g tinned fruit
25ml sherry
250g custard
25ml double cream
10g mixed nuts
10g angelica and cherries

Slit sponge cake, spread with jam and sandwich together. Cut into 4cm cubes. Soak in fruit juice and sherry. Mix with fruit, cover with cold custard and decorate with whipped cream, nuts and angelica.

344 **Scrambled eggs, with milk**

2 eggs
15ml milk
20g butter
½ tsp salt

Melt butter in pan, stir in beaten egg, milk and seasoning. Cook over gentle heat until mixture thickens.

Weight loss: 10.9%

346 **Omelette**

2 eggs
10ml water
10g butter
½ tsp salt
pepper

Beat eggs with salt and water. Heat butter in an omelette pan. Pour in the mixture and stir until it begins to thicken evenly. While still creamy, fold the omelette and serve.

Weight loss: 5.7%

347 **Omelette, cheese**

115g omelette, cooked
60g Cheddar cheese

Porportions are derived from recipe review.

348/349 **Quiche, cheese and egg, plain or wholemeal**

200g raw plain or wholemeal shortcrust pastry (149/151)
150g cheese
150g milk
3 eggs

Line a 20cm flan ring with the shortcrust pastry. Fill with grated cheese. Beat eggs in the warmed milk and pour into pastry case. Bake for 10 minutes at 200°C/mark 6 and then 30 minutes at 180°C/mark 4.

Weight loss: 10%

350 **Quiche Lorraine**

200g raw shortcrust pastry (149)
2 eggs
200ml milk
100g streaky bacon
100g cheese

Line a 20cm flan ring with shortcrust pastry. Fill with the fried, chopped bacon and grated cheese. Beat the eggs in warmed milk and pour into the pastry case. Bake for 10 minutes at 200°C/mark 6, then for 30 minutes at 180°C/mark 4.

Weight loss: 25.5%

542 **Game pie**

shortcrust pastry
venison
middle bacon rashers
rabbit
chicken livers
pork spare-rib
pheasant
redcurrant jelly
onions
red wine

Proportions of main ingredients obtained from manufacturers.

561 **Steak and kidney pie, single crust, homemade**

400g stewing beef, diced
200g lamb's kidneys, diced
2 tsp salt
15g flour
100ml water
350g flaky pastry (147)

Place the meat and kidneys rolled in seasoned flour in a pie dish with the water. Cover with pastry. Bake for 20 minutes at 200°C/mark 6 then lower the heat to 150°C/mark 2 and cover with greaseproof paper. Cook for a further 2–2½ hours.

Weight loss: 21%

567 **Beef bourguignonne**

1 tbsp vegetable oil
100g button onions
1 clove garlic, crushed
500g stewing beef, diced
50g streaky bacon rashers, chopped
15g flour
150g button mushrooms
5g tomato purée
1 tsp dried mixed herbs
250ml red wine
250ml stock
½ tsp salt
¼ tsp pepper

Brown the onions, garlic, meat and bacon in oil. Stir in flour, tomato purée, mixed herbs, wine, stock and seasoning. Bring to the boil, cover and simmer for 1 hour, stirring occasionally. Add mushrooms and cook for a further 30 minutes.

Weight loss: 33%

568 Beef bourguignonne, made with lean beef

As for beef bourguignonne (No. 567), except made with lean stewing steak and back bacon rashers.

569 Beef casserole, made with canned cook-in-sauce

500g braising steak, diced
390g cook-in-sauce, canned

Cook the meat with sauce in a covered casserole dish for 1½ hours at 180°C/mark 4.

Weight loss: 20%

573 Beef curry, reduced fat

1 clove garlic crushed
60g onions, chopped
500g lean braising steak, diced
1 tbsp vegetable oil
1 tbsp ground coriander
1 tsp chilli powder
½ tsp ground cumin
½ tsp ground turmeric
8g root ginger, ground
300ml water
½ tsp salt
5ml lemon juice
1 tsp garam masala

Brown the garlic, onions and meat in oil. Add spices and ginger. Stir in water, salt and lemon juice, cover and bring to the boil. Cook for 1½ hours stirring occasionally. Add garam masala.

Weight loss: 34%

574 Beef stew

500g stewing beef, diced
150g onions, chopped
1 tbsp vegetable oil
30g flour
500ml stock
150g carrots, chopped
½ tsp salt
¼ tsp pepper

Brown the meat and onions in oil, add flour and cook for 1 minute. Blend in the stock, add carrots and seasoning, transfer to a dish, cover and cook in the oven for 2 hours at 180°C/mark 4.

Weight loss: 27%

575 **Beef, stir-fried with green peppers**

500g rump steak, thinly sliced
2 tbsp vegetable oil
400g green peppers, sliced
60g spring onions, sliced
20g root ginger, grated
20g cornflour
¼ tsp salt
¼ tsp pepper

Marinade:
4 tsp sugar
1 red chilli, finely chopped
2 tbsp soy sauce
2 tbsp sherry

Marinade the steak for 30 minutes. Stir-fry the peppers, spring onions and ginger in oil for a few minutes, then add meat and stir-fry for 6 minutes.

Weight loss: 16%

576 **Bolognese sauce (with meat)**

1 clove garlic, crushed
60g onions, chopped
500g minced beef
40g carrots, chopped
30g celery, chopped
10g tomato pureé
397g canned tomatoes
250ml stock
2 tsp vegetable oil
½ tsp salt
¼ tsp pepper
¼ tsp dried mixed herbs

Brown the garlic, onions and mince in oil, add carrots and celery. Stir in the other ingredients and simmer for 40 minutes with the lid on.

Weight loss: 32% (whole dish)

577 **Chicken chasseur**

800g chicken breast (weighed with bone)
150g shallots
1 tbsp vegetable oil
1 tbsp flour
300ml dry white wine
300ml stock
2 bay leaves
1 tsp dried mixed herbs
½ tsp salt
¼ tsp pepper
1 clove garlic, crushed
15g tomato purée
1 tsp brown sugar
100g button mushrooms

Brown the chicken and shallots in oil. Remove and transfer to a casserole dish. Add the flour to the pan and gradually blend in the wine and stock and bring to the boil. Add remaining ingredients and stir. Pour over the chicken, cover and cook for 1 hour at 180°C/mark 4.

Weight loss: 18%

581 **Chicken curry, made with canned curry sauce**

1 tbsp vegetable oil
500g chicken breast
385g curry sauce, canned

Brown the chicken in oil. Add sauce, cover and simmer for 45 minutes.

Weight loss: 30%

588 **Chilli con carne**

500g minced beef
150g onions, chopped
100g green peppers, chopped
1 tbsp vegetable oil
1 tsp salt
¼ tsp pepper
15ml vinegar
1 tsp sugar
30g tomato purée
397g canned tomatoes
150ml stock
115g red kidney beans, canned, drained

Brown the mince, onions and peppers in oil. Blend the other ingredients and stir into the meat. Cover and simmer gently for 40 minutes. Add the kidney beans and continue cooking for a further 10 minutes.

Weight loss: 15%

590 **Coq au vin**

100g back bacon rashers, chopped
1000g chicken leg quarters (weighed with bone)
50g butter
50g flour
½ tsp salt
600ml red wine
¼ tsp pepper
100g shallots
1 tsp dried mixed herbs
100g button mushrooms

Brown the bacon and chicken coated in seasoned flour, in butter. Add the shallots, mixed herbs and red wine, cover and simmer for 35–45 minutes. Add the mushrooms and cook for another 20 minutes.

Weight loss: 16% (with bone), 19% (without bone).

591 **Coronation chicken**

300g mayonnaise
1 tbsp curry paste
2 tbsp apricot jam
500g cooked light and dark chicken meat, diced

Mix the ingredients together.

596 **Goulash**

300g onions, chopped
500g stewing beef, diced
2 tbsp vegetable oil
2 cloves garlic, crushed
2 tsp paprika
2g caraway seeds, crushed
397g canned tomatoes
150g green peppers, chopped
1 tsp salt
1 litre stock
500g potatoes, diced

Brown the onions and meat in oil. Add remaining ingredients except for the potatoes. Cover and simmer for 1 hour. Add potatoes and simmer for a further 25 minutes.

Weight loss: 32%

597 **Irish stew**

500g lamb neck fillet, diced
150g onions, sliced
200g carrots, sliced
500g potatoes, sliced
1 tbsp fresh parsley, chopped
1 tsp dried mixed herbs
15g flour
½ tsp salt
¼ tsp pepper
300ml stock

Make layers of meat, vegetables, herbs, flour and seasoning in a casserole dish, ending with a top layer of potatoes. Pour in stock and cover. Bake for 1 hour at 170°C/mark 3, remove lid and cook for a further 30 minutes.

Weight loss: 13%

598 **Irish stew, made with lean lamb**

As for Irish stew (No. 597), except made with lean lamb neck fillet.

600 **Lamb curry, made with canned curry sauce**

500g stewing lamb, diced
1 tbsp vegetable oil
385g curry sauce, canned

Brown the lamb in oil. Add the sauce, cover and simmer for 45 minutes.

Weight loss: 30%

601 **Lamb kheema**

6 tbsp vegetable oil
75g onions, finely chopped
2 garlic cloves, crushed
500g minced lamb
8g root ginger, grated
2 green chillies, deseeded, finely chopped
1 tsp coriander seeds, crushed
1 tsp cayenne pepper
200ml water
200g peas, frozen
2 tbsp fresh coriander leaves, chopped
1 tsp salt
2 tsp garam masala
220g canned tomatoes
1 tsp ground cumin

Brown the onions, garlic and mince in oil. Add the ginger and spices. Stir in 150ml of the water, cover and simmer for 30 minutes. Add the remaining ingredients and bring back to the boil. Cover and cook for a further 10 minutes.

Weight loss: 21%

603 **Lancashire hotpot**

500g stewing lamb, diced
½ tsp salt
¼ tsp pepper
100g carrots, sliced
100g turnip, chopped
100g onions, sliced
500g potatoes, sliced
300ml stock
2 tsp vegetable oil

Season the meat and mix with carrots, turnip and onions. Layer this with the potatoes in a casserole, beginning and ending with potatoes. Add stock and brush the top with oil. Cover and bake for 2 hours at 150°C/mark 2. Remove lid to brown the potatoes for the last 30 minutes.

Weight loss: 11%

604 Lasagne

Meat sauce:
1 tbsp vegetable oil
50g streaky bacon rashers, chopped
50g onions, chopped
50g carrots, chopped
30g celery, chopped
300g minced beef
220g canned tomatoes
375ml stock
1 clove garlic, crushed
½ tsp salt
¼ tsp pepper
½ tsp marjoram
1 bay leaf
50g mushrooms, sliced

Cheese sauce:
30g margarine
30g flour
400ml milk
75g cheese, grated

200g lasagne, raw

To top:
25g cheese, grated

Brown the bacon, onions, carrots, celery and mince in the oil. Stir in the remaining ingredients for the meat sauce and simmer for 15 minutes. For the cheese sauce, melt the margarine, add flour and cook for a few minutes, stir in the milk and cheese and cook gently until mixture thickens. In a dish, add alternative layers of lasagne, meat and cheese sauce ending with a layer of lasagne and cheese sauce. Sprinkle with cheese and bake for 1 hour at 190°C/mark 5.

Weight loss: 26%

608 Pasta with meat and tomato sauce

340g minced beef
900g boiled pasta
475g pasta sauce, tomato-based, canned

Brown the mince in a pan. Add pasta sauce and simmer for 20 minutes. Stir in pasta.

Weight loss: 17%

609 Pork casserole, made with canned cook-in sauce

675g pork steaks
390g cook-in sauce

Pour the sauce over the pork steaks and cook in a covered casserole dish for 1½ hours at 180°C/mark 4.

Weight loss: 20%

610 **Sausage casserole**

400g diced pork
150g onions, chopped
200g streaky bacon rashers, chopped
1 tbsp vegetable oil
200g pork sausage, chopped
227g baked beans, in tomato sauce, canned
1 bay leaf
1 tsp dried mixed herbs
300ml stock
½ tsp salt
¼ tsp pepper

Brown the pork, onions and bacon in the oil, add the remaining ingredients and bake, uncovered, for 1½ hours at 170°C/mark 3.

Weight loss: 15%

617 **Sweet and sour pork**

400g diced pork

Marinade:
½ tsp salt
1 tbsp soy sauce
2 tbsp sherry
tsp sugar

Batter:
20g cornflour
1 tbsp water
½ egg
vegetable oil (16g)

Sauce:
1 tbsp vegetable oil
1 clove garlic, crushed
7g root ginger, grated
100g onions, chopped
75g green peppers, sliced
75g red peppers, sliced
30g sugar
30ml vinegar
5g cornflour
1 tbsp soy sauce
1 tbsp sherry
1 tbsp tomato purée
5 tbsp water

Marinade the pork for 1 hour. Coat the pork with batter ingredients and deep-fry for 4 minutes. For the sauce, stir-fry garlic, ginger and onions in oil, add the remaining ingredients and cook until thickened. Add pork, stir and heat through.

Weight loss: 28%

685 **Curry, fish, Bangladeshi**

450g boal, raw
450g rohu, raw
225g onions, chopped
75g vegetable oil
2 tsp salt
2 tsp chilli powder
2 tsp coriander powder
½ tsp cumin powder
1½ tsp turmeric
300g water

Cut the fish into 1 inch slices and sprinkle with some of the chilli, turmeric and coriander. Add 2 tbsps of water and mix. Heat half the oil and fry the fish for 6 to 8 minutes then remove from pan and set aside. Fry the onions in the remaining oil until brown, add remaining spices and the remaining water and cook for 6 minutes. Add the fish and salt and cook for 4–5 minutes. Add the water, cover and cook for 10 minutes.

Weight loss: 21%

693 **Kedgeree**

200g smoked haddock, cooked
100g boiled white rice
2 eggs
25g margarine
½ tsp salt

Hard boil one egg. Melt the margarine and stir in the haddock, rice, salt and one beaten egg. Stir in chopped hard boiled egg and heat thoroughly.

Weight loss: 10%

867 **Beanburger, soya, fried in vegetable oil**

120g chopped onion
10g vegetable oil
320g boiled soya beans
75g porridge oats
10g chopped fresh parsley
1 tsp mixed herbs
20g soya sauce
35g tomato purée
1 egg
vegetable oil absorbed on frying (20g)

Fry onion in oil until brown. Mix beans and onions together with remaining ingredients, Form into 6–8 shapes approximately 1cm thick. Fry for 3 minutes either side.

Weight loss: 9%

868 **Bubble and squeak, fried in vegetable oil**

46% boiled cabbage
46% boiled potato
vegetable oil absorbed on frying (8%)

Fry the cabbage and potato together.

Weight loss: 10%

869 **Cannelloni, vegetable**

30% milk
25% boiled pasta
10% Ricotta cheese
5% tomatoes
4% boiled onions
4% vegetable oil
3% flour
2% boiled carrots
2% boiled spinach
2% boiled courgettes
2% boiled cabbage
2% boiled leeks
2% breadcrumbs
2% butter
2% parmesan cheese
1.5% cornflour
0.5% salt

Proportions are derived from dietary survey records.

870 **Casserole, vegetable**

240g diced potato
120g sliced carrot
120g diced onion
120g diced swede
120g diced parsnip
90g canned sweetcorn
90g frozen peas
90g chopped tomatoes
450g canned tomatoes
1 tsp marmite

Place all ingredients in a casserole and stir. Cover and cook for approximately 1 hour at 190°C/mark 5.

Weight loss: 15%

871 **Cauliflower cheese, made with semi-skimmed milk**

100g grated cheese
1 small cauliflower (700g)
100ml cauliflower water
½ level tsp salt
25g margarine
25g flour
250ml semi-skimmed milk
pepper

Boil cauliflower until just tender, break into florets. Drain saving 100ml water, place in a dish and keep warm. Make a white sauce from the margarine, flour, milk and cauliflower water. Add 75g cheese and season. Pour over the cauliflower and sprinkle with the remaining cheese. Brown under a grill or in a hot oven, 220°C/mark 7.

Weight loss: 15%

872 **Chilli, vegetable**

120g onion
240g carrots
240g parsnips
120g pepper
180g courgette
400g canned tomatoes
440g boiled or canned red kidney beans
10g chilli powder
330g canned sweetcorn
14g oxo
568ml water
5g salt

Quantities are derived from dietary survey records.

Weight loss: 15%

873 **Coleslaw, with mayonnaise, retail**

cabbage
mayonnaise
carrot
onion

Proportions obtained from a manufacturer in confidence

874 **Coleslaw, with reduced calorie dressing, retail**

cabbage
reduced calorie dressing
carrot
onion

Proportions obtained from a manufacturer in confidence

875 **Curry, chick pea dahl**

225g dry chick pea dahl
200ml water absorbed on soaking
28g vegetable oil
60g chopped onion
2g crushed garlic
1 tsp chilli powder
½ tsp garam masala
7g chopped green chilli
100g chopped tomato
415ml water

Soak the chick pea dahl overnight. Fry the onion and garlic until brown. Add a little water together with spices and tomatoes. Stir and cook until dry. Add dahl and water, simmer until cooked.

Weight loss: 35%

877 **Flan, vegetable**

30.8% shortcrust pastry
15.4% boiled carrot
15.4% boiled broccoli
15.4% boiled onion
15.4% white sauce made with skimmed milk
7.7% cheese

Proportions are derived from dietary survey records.

878 **Garlic mushrooms**

250g mushrooms
2g garlic
40g butter

Clean mushrooms and remove stems. Crush the garlic and sauté in butter. Fill mushroom caps with the garlic butter mixture and grill for 5–7 minutes.

Weight loss: 19%

881 **Nut roast**

90g chopped onion
11g vegetable oil
20g flour
140ml water
225g chopped mixed nuts
115g wholemeal breadcrumbs
1 tsp marmite
1 tsp mixed herbs

Fry onion in the oil. Add flour and water and thicken. Mix in nuts, breadcrumbs, marmite and herbs. Pack into a loaf tin and cover with foil. Bake at 190°C/mark 5 for 35–45 minutes.

Weight loss: 13%

882 **Pakora/bhajia, vegetable, retail**

potato
onion
chick pea flour
rapeseed oil
water
cauliflower
spinach
self-raising flour
mixed spices

Proportions obtained from manufacturer in confidence

883 **Pancakes, stuffed with vegetables**

320g prepared pancakes
Filling:

50g chopped mushrooms	90g chopped onion
200g canned tomatoes	1 tsp mixed herbs

Prepare the filling by cooking all the ingredients for approximately 15 minutes. Fill and roll up pancakes. Place under grill to reheat if necessary.

Weight loss: 20% for filling

884 **Pasty, vegetable**

50% cooked shortcrust pastry	6.3% boiled parsnip
15% boiled potato	6.3% boiled onion
7.5% water	6.3% boiled cabbage
6.3% boiled carrot	2.5% flour

Proportions are derived from dietary survey records

885 **Pie, vegetable**

100g chopped onion	200g canned tomatoes
100g sliced carrot	100ml water
100g sliced courgettes	2 tsp cornflour
60g chopped celery	1 tsp mixed herbs
50g sliced mushrooms	1 tsp marmite
80g chopped red pepper	300g raw shortcrust pastry
100g potatoes	

Place vegetables in a pan, together with herbs and marmite. Bring to the boil and simmer for 20–25 minutes. Make cornflour into a paste, add to pan, boil and stir until mixture thickens. Pour into pie dish and leave to cool. Roll pastry to fit dish size. Cut an additional 1 inch strip from remaining pastry, wet and place around the edge of the dish. Cover with pastry top and seal edges. Bake at 200°C/mark 6 for 30–40 minutes.

Weight loss: 15%

887 **Risotto, vegetable**

90g chopped onion	60g thinly sliced celery
44g vegetable oil	160g diced red pepper
175g raw white rice	250g sliced mushrooms
15g crushed garlic	270g canned red kidney beans
600ml water	15g soya sauce
1 tsp salt	50g cashew nuts

Fry onion in half the oil, add rice and some of the garlic, cook with stirring for 3 minutes. Add water and salt, bring to the boil, cover and simmer for 30–40 minutes until all the water has been absorbed. Fry celery, pepper and mushrooms in the remaining oil until soft, add the rest of the garlic. Add rice mixture, kidney beans, soya sauce and nuts. Cook until the beans are heated through.

Weight loss: 31%

888 **Salad, green**

150g shredded lettuce
230g sliced cucumber
160g sliced green pepper
30g sliced celery

Toss all ingredients together.

889 **Salad, potato with mayonnaise, retail**

potato
mayonnaise
onion

Proportions obtained from a manufacturer in confidence

890 **Salad, rice**

720g boiled white rice
240g spring onion
90g sweetcorn
60g cashew nuts
60g raisins
40g soya sauce
60g vegetable oil
80g green pepper

Quantities are derived from dietary survey records.

893 **Shepherd's pie, vegetable, retail**

boiled potatoes
tomatoes
boiled onion
water
boiled lentils
boiled carrots
boiled courgettes
single cream
boiled pearl barley
butter
soya oil
tomato purée
corn starch
salt

Proportions obtained from a manufacturer in confidence

Weight loss: 11% on re-heating

894 **Tagliatelle, with vegetables, retail**

water
boiled tagliatelle
tomatoes
boiled onions
boiled courgettes
cream
milk
boiled aubergines
soya oil
modified starch
garlic purée
sugar
starch

Proportions obtained from a manufacturer in confidence

Weight loss: 9% on re-heating

897 **Vegetable bake**

210g carrots
120g courgettes
120g onions
210g potatoes
45g margarine
30g flour
426ml milk
60g Leicester cheese
½ tsp mustard powder
45g white breadcrumbs

Quantities are derived from dietary survey records.

Weight loss: 15%

908 **Mixed herbs**

25g marjoram
25g parsley
25g sage
25g thyme

949 **Fruit salad**

400g eating apples
113g grapes
320g oranges
40ml lemon juice
200g bananas
120g kiwi fruit
113g strawberries

syrup
57g caster sugar
114ml water

Dissolve the sugar in the water in a pan over a low heat. Bring to the boil and simmer for a minute, then remove from the heat and allow to cool. Prepare fruit and sprinkle with lemon juice. Mix fruit with the cool syrup and refrigerate.

1013 **Marzipan**

300g ground almonds
150g caster sugar
150g icing sugar
1 egg
20ml lemon juice

Mix almonds and sugar, add beaten egg and knead all ingredients until smooth.

1066 **Fudge**

450g granulated sugar
175ml evaporated milk
150ml milk
75g butter
few drops of vanilla essence

Dissolve sugar in milks and add butter. Bring to the boil and boil gently to 125°C. Remove from heat, add vanilla essence. Beat mixture until thick and grainy. Pour into tin and cut into squares when almost set.

Weight loss: 28%

1076 **Popcorn, candied**

45ml vegetable oil
75g popping corn

glaze
45ml water
200g caster sugar
25g butter

Prepare corn as for plain popcorn (No. 1077). Heat glaze ingredients until sugar has dissolved, boil to soft ball stage (125°C). Add the popped corn and stir until coated.

Weight loss: 7% (popcorn), 15.2% (glaze)

1077 **Popcorn, plain**

45ml vegetable oil
75g popping corn

Heat oil gently in a saucepan until test corn pops. Remove from heat, add corn, cover and return to heat until all corn has popped.

1178/1179 **Bread sauce**

250ml whole or semi-skimmed milk
50g fresh breadcrumbs
5g margarine
½ tsp salt
2 cloves
mace
1 small onion

Put milk and onion, stuck with cloves, in a saucepan and bring to the boil. Add breadcrumbs, and simmer for about 20 minutes over gentle heat. Remove onion, stir in margarine and season.

Weight loss: 6.8%

1180/1181 **Cheese sauce**

350ml whole or semi-skimmed milk
75g cheese
½ level tsp salt
cayenne pepper
25g flour
25g margarine

Melt the fat in a pan, add flour and cook gently for a few minutes stirring all the time. Add milk and cook until mixture thickens, stirring continually. Add grated cheese and seasoning. Reheat to soften the cheese, serve immediately.

Weight loss: 15.2%

1182/1183 **Cheese sauce, packet mix, made up**

1 pkt cheese sauce mix (33g)
284ml whole or semi-skimmed milk

Prepared as packet directions.

Weight loss: 9.1%

1184/1185 **Onion sauce**

350ml whole or semi-skimmed milk
200g cooked onion
1 level tsp salt
pepper
25g flour
25g margarine

Make the white sauce (as Nos. 1186/1187), add the chopped onion and seasoning.

Weight loss: 12.6%

1186/1187 **White sauce, savoury**

350ml whole or semi-skimmed milk
25g flour
½ level tsp salt
25g margarine

Melt fat in a pan. Add flour and cook for a few minutes stirring constantly. Add milk and salt, and cook gently until mixture thickens.

Weight loss: 18.1%

1188/1189 **White sauce, sweet**

350ml whole or semi-skimmed milk
25g flour
25g margarine
30g sugar

As savoury white sauce (Nos. 1186/1187) except adding sugar and omitting salt.

Weight loss: 16.7%

1190 **Apple chutney**

500g cooking apples
400g onions
100g raisins
400ml vinegar
450g sugar
1 level teaspoon salt
1 level teaspoon curry powder
½ level teaspoon mustard
½ level teaspoon pepper
½ level teaspoon ground ginger

Peel and core the apples and peel the onions and chop into small pieces. Mix all the ingredients except the sugar and boil gently until soft. Add the sugar and boil for a further 30 minutes.

Weight loss: 32.1%

4.5 ALTERNATIVE AND TAXONOMIC NAMES

- Foods are listed below in the same order as in the main Tables.
- The alternative names listed in the left-hand column below are those that were most frequently encountered during data collection and are included to help in identifying foods. It is important to recognise that in some cases such names may be used for more than one food and that all such usages may not appear in this list.
- To see if a name is listed, the food index should be consulted first. If the term is included as an alternative name, a cross reference entry indicates the food name to which it refers. This allows all alternatives to be listed together.
- Taxonomic names listed in the right-hand column refer as specifically as possible to the data used.
- Where two or more taxonomic names are listed, the data are representative of a mixture of these varieties.
- The abbreviation 'var' is used to indicate the specific variety or unspecified variety(ies); 'sp' and 'spp' are used to indicate that one or more than one species of the specified Genus is included.

	Alternative names	Food names	Taxonomic names
Cereals			
		Oats	*Avena sativa*
		Rye	*Secale cereale*
		Wheat	*Triticum aestivum*
		Rice	*Oryza sativa*
		Pasta wheat	*Triticum durum*
Meat			
		Beef	*Bos taurus*
		Lamb	*Ovis aries*
		Pork	*Sus scrofa*
		Veal	*Bos taurus*
Poultry			
		Chicken	*Gallus domesticus*

Alternative names	Food names	Taxonomic names
	Turkey	*Meleagris gallopavo*
	Duck	*Anas platyrhynchos*
	Goose	*Anser anser*
	Pheasant	*Phasianus colchicus*
Game		
	Rabbit	*Lepus cuniculus*
	Venison	*Cervus* spp
Fish		
White fish		
	Cod	*Gadus morhua*
Coalfish Saithe	**Coley**	*Pollachius virens*
	Haddock	*Melanogrammus aeglefinus*
	Halibut	*Hippoglossus hippoglossus*
	Lemon sole	*Microstomus kitt*
	Plaice	*Pleuronetes platessa*
Rock eel Dogfish	**Rock salmon**	Probably *Squalus acanthias*
	Skate	*Raja* spp
	Whiting	*Merlangius merlangus*
Fatty fish		
	Anchovies	*Engraulis encrasicholus*
	Herring	*Clupea harengus*
	Kipper	*Clupea harengus*
	Mackerel	*Scomber scombrus*
	Pilchards	*Sardinops sagex ocellata*
	Salmon, Atlantic red	*Salmo salar* *Oncorhynchus nerka*

Alternative names	Food names	Taxonomic names
	Sardines	*Sardina pilchardus*
	Tuna	*Euthynnus* sp *Katsuwonus pelamis*
	Whitebait	Young of *Clupea harengus* and *Sprattus sprattus*
Crustacea		
	Crab	*Cancer pagurus*
	Lobster	*Homarus vulgaris*
	Prawns	*Paleamon serratus*
	Scampi	*Nephrops norvegicus*
	Shrimps	*Crangon crangon* *Pandalus montagui* *Pandalus borealis*
Molluscs		
	Cockles	*Cardium edule*
	Mussels	*Mytilus edulis*
	Squid	*Loligo vulgaris*
	Whelks	*Buccinum undatum*
	Winkles	*Littorina littorea*
Vegetables		
Potatoes		
Aloo Batata	**Potatoes**	*Solanum tuberosum*
Beans and lentils		
Adzuki beans	**Aduki beans**	*Vigna angularis*
	Baked beans	*Phaseolus vulgaris* (navy beans)
	Beansprouts, mung	*Phaseolus aureus*
Alad Urad	**Black gram,** urad gram	*Vigna mungo*

Alternative names	Food names	Taxonomic names
Blackeye peas Cowpeas Chori Lobia	**Blackeye beans**	*Vigna unguiculata*
	Broad beans	*Vicia faba*
Lima beans	**Butter beans**	*Phaseolus lunatus*
Channa Common gram Garbanzo Yellow gram	**Chick peas**	*Cicer arietinum*
Fansi	**Green beans/ French beans**	*Phaseolus vulgaris*
Continental lentils Masur	**Lentils,** green and brown	*Lens esculenta*
Masur dahl Masoor dahl	**Lentils,** red	*Lens esculenta*
Green gram Golden gram Moong beans	**Mung beans**	*Phaseolus aureus*
	Red kidney beans	*Phaseolus vulgaris*
	Runner beans	*Phaseolus coccineus*
	Soya beans	*Glycine max*
Peas		
Snowpeas	**Mange-tout peas**	*Pisum sativum* var *macrocarpum*
Badla Mattar Vatana	**Peas**	*Pisum sativum*
Other vegetables		
	Asparagus	*Asparagus officinalis* var *altilis*

Alternative names	Food names	Taxonomic names
Baingan Brinjal Eggplant Jew's apple Ringana	**Aubergine**	*Solanum melongerna* var *ovigerum*
	Beetroot	*Beta vulgaris*
Calabrese	**Broccoli**, green	*Brassica oleracea* var *botrytis*
Chote bund gobhi Nhanu kobi	**Brussel sprouts**	*Brassica oleracea* var *gemmifera*
Bund gobhi Kobi	**Cabbage**	*Brassica oleracea*
	Cabbage, January King	*Brassica oleracea* var *capitata*
	Cabbage, white	*Brassica oleracea* var
Gajjar	**Carrots**	*Daucas carota*
Pangoli Phool gobhi	**Cauliflower**	*Brassica oleracea* var *botrytis*
	Celery	*Apium graveolens* var *dulce*
Belgian chicory Witloof	**Chicory**	*Cichorium intybus*
Zucchini	**Courgette**	*Cucurbita pepo*
Kakdi Khira	**Cucumber**	*Cucumis sativus*
Borecole Kale	**Curly kale**	*Brassica oleracea* var *acephala*
	Fennel, Florence	*Foeniculum vulgare* var *dulce*
Lassan Lehsan	**Garlic**	*Allium sativum*
	Gherkins	*Cucumis sativus*
Bitter gourd Balsam apple	**Gourd**, karela	*Momordica charantia*

Alternative names	Food names	Taxonomic names
	Leeks	*Allium ampeloprasum* var *porrum*
	Lettuce	*Lactuca sativa*
	Marrow	*Cucurbita pepo*
	Mushrooms, common	*Agaricus campestris*
	Mustard and cress	*Brassica and Lepidium* spp
Bhendi Bhinda Bhindi Gumbo Lady's fingers	**Okra**	*Hibiscus esculentus*
Dungli Kanda Piyaz	**Onions**	*Allium cepa*
	Parsnip	*Pastinaca sativa*
Pimento	**Peppers**,capsicum, chilli, green	*Capiscum annuum* var *grossum*
Bell peppers Motamircha Simila mirch Sweet peppers	**Peppers,** capsicum (green/red)	*Capsium annuum* var *grossum*
	Plantain	*Musa paradisiaca*
Kumra Lal kaddu Lal phupala	**Pumpkin**	*Cucurbita* sp
	Quorn, myco-protein	*Fusarium graminearum*
	Radish, red	*Raphanus sativus*
Palak Saag	**Spinach**	*Spinacia oleracea*
	Spring greens	*Brassica oleracea* var
	Spring onions	*Allium cepa*
Neeps (England) Rutabaga Yellow turnip	**Swede**	*Brassica napus* var *napobrassica*

Alternative names	Food names	Taxonomic names
Shakaria Yam (USA)	**Sweet potato**	*Ipomoea batatas*
	Sweetcorn	*Zea mays*
	Tomatoes	*Lycopersicon esculentum*
Neeps (Scotland) Shalgam	**Turnip**	*Brassica rapa* var *rapifera*
	Watercress	*Nasturtium officinale*
	Yam	*Dioscorea* sp

Herbs and spices

Alternative names	Food names	Taxonomic names
	Cinnamon	*Cinnamomum verum* *Cinnamomum aromaticum*
	Mint	*Mentha spicata*
	Mustard	*Sinapis alba* *Brassica hirta*
	Nutmeg	*Myristica fragrans*
	Paprika	*Capsicum annuum*
	Parsley	*Petroselinum crispum*
	Pepper, black	*Piper nigrum*
	Pepper, white	*Piper nigrum*
	Rosemary	*Rosmarinus officinalis*
	Sage	*Salvia officinalis*
	Thyme	*Thymus vulgaris*

Fruit

Alternative names	Food names	Taxonomic names
Tarel	**Apples**	*Malus pumila*
	Apricots	*Prunus armeniaca*
	Avocado	*Persea americana*
Kula	**Bananas**	*Musa* spp
	Blackberries	*Rubus ulmifolius*

Alternative names	Food names	Taxonomic names
	Blackcurrants	*Ribes nigrum*
	Cherries	*Prunus avium*
	Clementines	*Citrus reticulata* var *Clementine*
	Currants	*Vitis vinifera*
	Damsons	*Prunus domestica* subsp *institia*
	Dates	*Phoenix dactylifera*
Gullar	**Figs**	*Ficus carica*
	Gooseberries	*Ribes grossularia*
	Grapefruit	*Citrus paradisi*
	Grapes	*Vitis vinifera*
	Guava	*Psidium guajava*
Chinese gooseberry	**Kiwi fruit**	*Actinidia chinensis*
	Lemons	*Citrus limon*
Chinese cherry Lichee Lichi Litchee Litchi	**Lychees**	*Litchi chinensis*
	Mandarin oranges	*Citrus reticulata*
	Mangoes	*Mangifera indica*
	Melon, Canteloupe-type	*Cucumis melo* var *cantaloupensis*
	Melon, Galia	*Cucumis melo* var *reticulata*
	Melon, Honeydew	*Cucumis melo* var *indorus*
	Nectarines	*Prunus persica* var *nectarina*
	Olives	*Olea europaea*
	Oranges	*Citrus sinensis*
Purple grenadillo	**Passion fruit**	*Passiflora edulis* f *edulis*

Alternative names	Food names	Taxonomic names
Papai Papaya	**Paw-paw**	*Carica papaya*
	Peaches	*Prunus persica*
	Pears	*Pyrus communis*
	Pineapple	*Ananas comosus*
	Plums	*Prunus domestica* subsp *domestica*
	Prunes	*Prunus domestica*
	Raisins	*Vitis vinifera*
	Raspberries	*Rubus idaeus*
	Rhubarb	*Rheum rhaponticum*
	Satsumas	*Citrus reticulata*
	Strawberries	*Fragaria* sp
	Sultanas	*Vitis vinifera*
	Tangerines	*Citrus reticulata*

Nuts and seeds

Alternative names	Food names	Taxonomic names
Badam	**Almonds**	*Prunus amygdalus*
	Brazil nuts	*Bertholletia excelsa*
Kaju	**Cashew nuts**	*Anacardium occidentale*
	Chestnuts	*Castanea vulgaris*
	Coconut	*Cocos nucifera*
	Hazelnuts	*Corylus avellana* *Corylus maxima*
Queensland nuts	**Macadamia nuts**	*Macadamia integrifolia* *Macadamia tetraphylla*
Groundnuts Monkey nuts	**Peanuts**	*Arachis hypogaea*
Hickory nuts	**Pecan nuts**	*Carya illinoensis*

Alternative names	Food names	Taxonomic names
Indian nuts Pignolias Pine kernels	**Pine nuts**	*Pinus pinea* *Pinus edulis*
Pista	**Pistachio nuts**	*Pistacia vera*
Benniseed Gingelly Til	**Sesame seeds**	*Sesamum indicum*
	Sunflower seeds	*Helianthus annuus*
Akhrot Madeira nuts	**Walnuts**	*Juglans regia*

4.6 REFERENCES

Publications in '*The Composition of Foods*' series

Paul, A.A., Southgate, D.A.T. and Russell, J. (1980) *Amino acid composition (mg per 100g food) and fatty acid composition (g per 100g food).* First supplement to 4th edition of *McCance and Widdowson's The Composition of Foods.* HMSO, London

Tan, S.P., Wenlock, R.W. and Buss, D.H. (1985) *Immigrant Foods.* Second supplement to 4th edition of *McCance and Widdowson's The Composition of Foods.* HMSO, London

Holland, B., Unwin, I.D. and Buss, D.H. (1988) *Cereals and Cereal Products.* Third supplement to 4th edition of *McCance and Widdowson's The Composition of Foods.* The Royal Society of Chemistry, Nottingham

Holland, B., Unwin, I.D. and Buss, D.H. (1989) *Milk and Milk Products.* Fourth supplement to 4th edition of *McCance and Widdowson's The Composition of Foods.* The Royal Society of Chemistry, Cambridge

Holland, B., Unwin, I.D. and Buss, D.H. (1991) *Vegetables, Herbs and Spices.* Fifth supplement to 4th edition of *McCance and Widdowson's The Composition of Foods.* The Royal Society of Chemistry, Cambridge

Holland, B., Welch, A.A., Unwin, I.D., Buss, D.H., Paul, A.A. and Southgate, D.A.T. (1991) *McCance and Widdowson's The Composition of Foods*, 5th edition. The Royal Society of Chemistry, Cambridge

Holland, B., Unwin, I.D. and Buss, D.H. (1992) *Fruit and Nuts.* First supplement to 5th edition of *McCance and Widdowson's The Composition of Foods.* The Royal Society of Chemistry, Cambridge

Holland, B., Welch, A.A. and Buss, D.H. (1992) *Vegetable Dishes.* Second supplement to 5th edition of *McCance and Widdowson's The Composition of Foods.* The Royal Society of Chemistry, Cambridge

Holland, B., Brown, J. and Buss, D.H. (1993) *Fish and Fish Products.* Third supplement to 5th edition of *McCance and Widdowson's The Composition of Foods.* The Royal Society of Chemistry, Cambridge

Chan, W., Brown, J. and Buss, D.H. (1994) *Miscellaneous Foods.* Fourth supplement to 5th edition of *McCance and Widdowson's The Composition of Foods.* The Royal Society of Chemistry, Cambridge

Chan, W., Brown, J., Lee, S.M. and Buss, D.H. (1995) *Meat, Poultry and Game.* Fifth supplement to 5th edition of *McCance and Widdowson's The Composition of Foods.* The Royal Society of Chemistry, Cambridge

Chan, W., Brown, J., Church, S.M. and Buss, D.H. (1996) *Meat Products and Dishes*. Sixth supplement to 5th edition of *McCance and Widdowson's The Composition of Foods*. The Royal Society of Chemistry, Cambridge

Ministry of Agriculture, Fisheries and Food. (1998) *Fatty Acids*. Seventh supplement to 5th edition of *McCance and Widdowson's The Composition of Foods*. The Royal Society of Chemistry, Cambridge.

References to the Introduction, Tables and Appendices

Allen, L.H. (1982) Calcium bioavailability and absorption: a review. *Am. J. Clin. Nutr.* **35,** 783–808

Anderson, B.A., Kinsella, J.A. and Watt, B.K. (1975) Comprehensive evaluation of fatty acids in foods. II. Beef products. *J. Am. Diet. Assoc.* **67**, 35–41

Anderson, B.A. (1976) Comprehensive evaluation of fatty acids in foods. VII. Pork products. *J. Am. Diet. Assoc.* **69**, 44–49

AOAC (1975) *Official methods of analysis, 12th edition.* Association of Official Analytical Chemists, Washington DC

AOAC (2000) *Official methods of analysis, 17th edition.* Association of Official Analytical Chemists, Gaithersburg, MD

Barclay, M.N.I., MacPherson, A. and Dixon, J. (1995) Selenium content of a range of UK foods. *J. Food Comp. Anal.* **8,** 307–318

Bell, J.G. (1974) Microbiological assay of vitamins of the B group in foodstuffs. *Lab. Pract.* **23,** 235–242, 252

Berry, M. (1994) *Mary Berry's ultimate cake book.* BBC Books, London

Berry, M. (1995) *Mary Berry's complete cookbook.* Dorling Kindersley, London

Biesalski, H.K. (1997). Bioavailability of vitamin A. *Eur. J. Clin. Nutr.* **51**, S71–S75

Bingham, S.A. (1987) The dietary assessment of individuals; methods, accuracy, new techniques and recommendations. *Nutr. Abs. Rev.* (Series A) **57,** 705–742

Bingham, S.A. and Day, K.C. (1987) Average portion weights of foods consumed by a randomly selected British population sample. *Hum. Nutr.: Appl. Nutr.* **41A,** 258–264

Bolton-Smith, C., Price, R.J.G., Fenton, S.T., Harrington, D.J. and Shearer, M.J. (2000) Compilation of a provisional UK database for the phylloquinone (vitamin K_1) content of foods. *Br. J. Nutr.* **83**, 389–399

Booth, S.L., Pennington, J.A.T. and Sadowski, J.A. (1996) Dihydro-vitamin K_1: primary food sources and estimated dietary intakes in the American diet. *Lipids* **31**, 715–720

The Bread and Flour Regulations (1998) *Statutory Instrument No. 141.* The Stationery Office, London

Brubacher, G., Müller-Mulot, W. and Southgate, D.A.T. (1985) *Methods for the determination of vitamins in food.* Elsevier Applied Science Publishers Ltd, London

Brubacher, G.B. and Weiser H. (1985) The vitamin A activity of β-carotene. *Internat. J. Vit. Res.* **55,** 5–15

Cameron, M.E., and van Staveren, W.A. (1988) *Manual on methodology for food consumption studies.* Oxford University Press

Cashel, K., English, R. and Lewis J. (1989) *Composition of Foods, Australia.* Volume 1. Department of Community Services and Health, Canberra

Christie, A.A., Dean, A.C., and Millburn, B.A. (1973) The determination of vitamin E in food by colorimetry and gas-liquid chromatography. *Analyst* **98**, 161–167

Chughtai, M.I.D. and Waheed Khan, A. (1960) *Nutritive value of food-stuffs and planning of satisfactory diets in Pakistan, Part 1. Composition of raw food-stuffs,* Punjab University Press, Lahore

Cohn, W. (1997) Bioavailability of vitamin E. *Eur. J. Clin. Nutr.* **51**, S80–S85

Cutrufelli, R. and Matthews, R.H. (1986) *Composition of foods: beverages, raw, processed and prepared.* Agriculture Handbook No. 8–14, US Department of Agriculture, Washington DC

Cutrufelli, R. and Pehrsson, P.R. (1991) *Composition of foods: snacks and sweets, raw, processed and prepared.* Agriculture Handbook No. 8–19, US Department of Agriculture, Washington DC

Davidson, K.W., Booth, S.L., Dolnikowski, G.G. and Sadowski, S.A. (1995) Characterisation of dihydro-phylloquinone; a hydrogenated form of phylloquinone in foods containing partially hydrogenated vegetable oils. *FASEB J.* **9**, A474

Davies, J. and Dickerson J. (1991) *Nutrient content of food portions.* Royal Society of Chemistry, Cambridge

Dean, A.C. (1978) Method for the estimation of available carbohydrate in foods. *Food Chem.* **3,** 241–250

Department of Health (1991) *Dietary reference values for food energy and nutrients for the United Kingdom.* Report on Health and Social Subjects No. 41, HMSO, London

Egan, H., Kirk, R.S. and Sawyer, R. (1981) *Pearson's Chemical Analysis of Foods,* 8th edition. Churchill Livingstone, Edinburgh

Englyst, H.N., Wiggins, H.S. and Cummings, J.H. (1982) Determination of the non-starch polysaccharides in plant foods by gas-liquid chromatography of constituent sugars as alditol acetates. *Analyst* **107,** 307–318

Englyst, H.N. and Cummings, J.H. (1984) Simplified method for the measurement of total non-starch polysaccharides by gas-liquid chromatography of constituent sugars as alditol acetates. *Analyst* **109,** 937–942

Englyst, H.N. Cummings, J.H. (1988) An improved method for the measurement of dietary fibre as the non-starch polysaccharides in plant foods. *J. Assoc. Off. Anal. Chem.* **71,** 808–814

Englyst, H.N., Quigley, M.E., Hudson, J.G. and Cummings, J.H. (1992) Determination of dietary fiber as nonstarch polysaccharides by gas-liquid-chromatography. *Analyst* **117**, 1707–1714

Englyst, H.N., Quigley, M.E. and Hudson, J.G. (1994) Determination of dietary fiber as nonstarch polysaccharides with gas-liquid-chromatographic, high-performance liquid-chromatographic or spectrophotometric measurement of constituent sugars. *Analyst* **119**, 1497–1509

Englyst, H.N, Quigley, M.E., Englyst, K.N., Bravo, L. and Hudson, G.J. (1996) Dietary Fibre. Report of a study commissioned by the Ministry of Agriculture, Fisheries and Food. *J. Assoc. Publ. Analysts* **32,** 1–38

Exler, J., Kinsella, J.E. and Watt, B.K. (1975) Lipids and fatty acids of important finfish. New data for nutrient tables. *J. Am. Oil Chem.* Soc. **52**, 154–159

Fairweather-Tait, S.J. (1998) Trace element bioavailability. In: *Role of trace elements for health promotion and disease prevention.* Edited by Sandstrom, B. and Walter, P. *Bibliotheca Nutritio et Dieta* **54**, 29–39

Fairweather-Tait**,** S.J. and Hurrell, R.F. (1996) Bioavailability of minerals and trace elements. *Nutr. Res. Rev.* **9**, 295–324

Fairweather-Tait, S.J. (1999) The importance of trace element speciation in nutritonal sciences. *Fresenius' J. Anal. Chem.* **363**, 536–540

FAO/WHO (1973) *Energy and protein requirements.* Report of a Joint FAO/WHO *Ad Hoc* Expert Committee. FAO Nutrition Meetings Report Series, No. 52; WHO Technical Report Series, No. 522

FAO (1998) Carbohydrates in human nutrition. Report of a Joint FAO/WHO Expert Consultation. Food and Nutrition Paper No. 66. Food and Agriculture Organization of the United Nations, Rome

Faulks, R.M. and Southon, S. (1997) Dietary carotenoids. *Nutr. Food Sci.* 246–250

Ferland, G. and Sadowski, J.A. (1992) The vitamin k_1 (phylloquinone) content of green vegetables: effects of plant maturation and geographical growth location. *J. Agric. Food Chem.* **40**, 1874–1877

Finglas, P.M. and Faulks, R.M. (1984) The HPLC analysis of thiamin and riboflavin in potatoes. *Food Chem.* **15**, 37–44

Finglas, P.M. and Faulks, R.M. (1984) Nutritional composition of UK retail potatoes, both raw and cooked. *J. Sci. Fd. Agric.* **35**, 1347–1356

Gopalan, C., Rama Sastri, B.V. and Balasubramanian, S.C. (1980) *Nutritive value of Indian foods,* National Institute of Nutrition, Indian Council of Medical Research, Hyderabad

Gregory, J.F. (1997a). Bioavailability of folate. *Eur. J. Clin. Nutr.* **51**, S54–S59.

Gregory, J.F. (1997b). Bioavailability of vitamin B_6. *Eur. J. Clin. Nutr.* **51**, S43–48

Greenfield, H. and Southgate, D.A.T. (1994) *Food composition data: Production, management and use.* Chapman and Hall, London

Haytowitz, D.B. and Matthews, R.H. (1986) *Composition of foods: legumes and legume products, raw, processed and prepared.* Agriculture Handbook No. 8–11, US Department of Agriculture, Washington DC

Holland, B., Welch, A.A., Unwin, I.D., Buss, D.H., Paul, A.A. and Southgate, D.A.T. (1991) *McCance and Widdowson's The Composition of Foods*, 5th edition. The Royal Society of Chemistry, Cambridge

International Dairy Federation (1988) *Edible ices and ice mixes: Determination of fat content: Weibull-Berntrop gravimetric method (Reference method).* Joint IDF/ISO/AOAC Publication, Brussels

IUPAC (1976) *Standard methods for the analysis of oils, fats and soaps. 4th supplement to the 5th edition.* Method II D.19 Preparation of fatty acid methyl esters. Method II D.25 Gas liquid chromatography of fatty acid methyl esters

IUPAC (1979) *Standard methods for the analysis of oils, fats and derivatives. 6th Edition.* Pergamon Press, Oxford

IUPAC (1987) *Standard methods for the analysis of oils, fats and derivatives. 7th Edition.* Blackwell Scientific Publications

Judd, P. A., Kassam-Khamis, T. and Thomas, J.E. (2000) *The composition and nutrient content of foods commonly consumed by South Asians in the UK.* The Aga Khan Health Board for the United Kingdom, London

Koivu, T.J., Piironen, V.I., and Mattila, P.H. (1998) Phylloquinone (vitamin K_1) in cereal products. *Cereal Chem.* **75**, 113–116

Kwiatkowska, C.A., Finglas, P.M. and Faulks, R.M. (1989) The vitamin content of retail vegetables in the UK. *J. Hum. Nut. Diet.* **2,** 159–172.

Laboratory of the Government Chemist (1993a) Analytical survey of nutrient composition of retail cuts of beef. (Food Surveillance Information Sheet No. 39)

Laboratory of the Government Chemist (1993b) Analytical survey of nutrient composition of retail cuts of lamb. (Food Surveillance Information Sheet No. 49)

Laboratory of the Government Chemist (1994) Analytical survey of nutrient composition of retail cuts of pork. (Food Surveillance Information Sheet No. 40)

Laboratory of the Government Chemist (1996) Individual folates in foodstuffs. (Food Surveillance Information Sheet No. 92)

Laboratory of the Government Chemist (2001) The determination of sodium content of bread. (Food Survey Information Sheet No. 19/01)

Lewis, J. and English, R. (1990) *Composition of foods, Australia. Volume 5, nuts and legumes, beverages, miscellaneous foods.* Department of Community Services and Health, Canberra

Livesey, G., Buss, D., Coussement, P., Edwards, D.G., Howlett, J., Jonas, D.A., Kleiner, J.E., Muller, D. and Sentko, A. (2000) Suitability of traditional energy values for novel foods and food ingredients. *Food control* **11**, (4), 249–289

Marsh, A.C., Moss, M.K. and Murphy, E.W. (1977) *Composition of foods: spices and herbs, raw, processed and prepared.* Agriculture Handbook No. 8–2, US Department of Agriculture, Washington DC

Marsh, A.C., (1980) *Composition of foods: soups, sauces, and gravies, raw, processed and prepared.* Agriculture Handbook No. 8–6, US Department of Agriculture, Washington DC

McCance, R.A. and Shipp, H.L. (1933) *The chemistry of flesh foods and their losses on cooking.* Medical Research Council Special Report Series, No. 187. HMSO, London

McCance, R.A. and Widdowson, E.M. (1940) *The chemical composition of foods.* Medical Research Council Spec. Rep. Ser. No. 235. HMSO, London

McCarthy, M.A. and Matthews, R.H. (1984) *Composition of foods: nut and seed products, raw, processed and prepared.* Agriculture Handbook No. 8–12, US Department of Agriculture, Washington DC

McCarthy, P.T., Harrington, D. and Shearer, M.J. (1997) Assay of phylloquinone in plasma. *Methods in Enzymology* **282**, 421–433

McClaughlin, P.J. and Weihrauch, J.L. (1979) Vitamin E content of foods. *J. Am. Diet. Assoc.* **75,** 647–665

Michie, N.D., Dixon E.J. and Bunton, N.G. (1978) Critical review of AOAC fluorimetric method for determining selenium in foods. *J. Assoc. Off. Analyt. Chem.* **61**, 48–51

Milk Marketing Board (1988) *The dairy book of British food.* Ebury Press, London.

Ministry of Agriculture, Fisheries and Food. (1993) *Food portion sizes,* second edition HMSO, London

Ministry of Agriculture, Fisheries and Food. (1998) *Fatty acids.* Seventh supplement to 5th edition of *McCance and Widdowson's The Composition of Foods.* The Royal Society of Chemistry, Cambridge

Moxon, R.E.D. and Dixon, E.J. (1980) Semi-automated method for the determination of total iodine in food. *Analyst* **105,** 344–352

Moxon, R.E.D. (1983) A rapid method for the determination of sodium in butter. *J. Assoc. Publ. Analysts* **21,** 83–87

Nelson, M., Atkinson, M. and Meyer, J. (1997). *Food portion sizes. A photographic atlas.* Ministry of Agriculture, Fisheries and Food Publications, London

Nelson, M., Black, A.E., Morris, J.A. and Cole, T.J. (1989) Between and within subject variation in nutrient intake from infancy to old age: estimating the number of days required to rank dietary intakes with desired precision. *Am. J. Clin. Nutr.* **50,** 155–167.

Nelson, M. and Bingham, S.A. (1997) Assessment of food composition and nutrient intake. In: *Design concepts in nutritional epidemiology, second edition.* Edited by B.M. Margetts and M. Nelson. Oxford University Press, Oxford

Ollilainen, V., Van den Berg, H., Finglas, P.M. and de Froidmont-Goertz, I. (2001). Certification of B-group vitamins (B_1, B_2, B_6 and B_{12}) in four food reference materials. *J. Agric. Food Chem.* **48** (12), 6325–6331

Olson, J.A. (1989) Pro vitamin A function of carotenoids: The conversion of β-carotene into vitamin *A. J. Nutr.* **119,** 105–108

Paul, A.A. and Southgate, D.A.T. (1977) A study on the composition of retail meat: dissection into lean, separable fat and inedible portion. *J. Hum. Nutr.* **31**, 259–272

Paul, A.A. and Southgate, D.A.T. (1978) *McCance and Widdowson's The Composition of Foods Fourth edition.* HMSO, London

Peattie, M.E., Buss, D.H., Lindsay, D.G. and Smart, G.Q. (1983) Reorganisation of the British Total Diet Study for monitoring food constituents from 1981. *Food Chem. Toxicol.* **21,** 503–507

Phillips, D.R. and Wright A.J.A. (1983) Studies on the response of *Labtobacillus casei to folate vitamin in foods. Br. J. Nutr.* **49,** 181–186

Phillips, D.I.W., Nelson, M., Baker, D.J.P., Morris, J.A. and Wood, T.J. (1988) Iodine in milk and the incidence of thyrotoxicosis in England. *Clin. Endocrinol.* **28,** 61–66

Posati, L.P., Kinsella, J.E. and Watt, B.K. (1975) Comprehensive evaluation of fatty acids in foods. III. Eggs and egg products. *J. Am. Diet. Assoc.* **67**, 111–115

Posati, L.P. and Orr, M.L. (1976) *Composition of foods: dairy and egg products, raw processed and prepared.* Agriculture Handbook No. 8–1, US Department of Agriculture, Washington DC

Rich, G., Fillery-Travis, A. and Parker, M. (1998) Low pH enhances the transfer of carotene from carrot juice to olive oil. *Lipids* **33,** 985–992

Royal Society (1972) *Metric units, conversion factors and nomenclature in nutritional and food sciences.* Report of the subcommittee on metrication of the British National Committee for Nutritional Sciences

Scott, K.J. and Rodriguez-Amaya, D. (2000). Pro-vitamin A carotenoid conversion factors: retinol equivalents – fact or fiction? *Food Chem.* **69**, 125–127

Shearer, M.J., Bach, A. and Kohlmeier, M. (1996) Chemistry, nutritional sources, tissue distribution and metabolism of vitamin K with special reference to bone health. *J. Nutr.* **126**, S1181–S1186

Sivell, L.M., Bull, N.L., Buss, D.H., Wiggins, R.A., Scuffam, D., and Jackson, P.A. (1984) Vitamin A activity in foods of animal origin. *J. Sci. Food Agric.* **35,** 931–939

Smith, D. (1992) *Delia Smith's Complete Cookery Course.* BBC Books, London

Society of Public Analysts and Other Analytical Chemists: Analytical Methods Committee (1951) The chemical assay of aneurine in foodstuffs. *Analyst* **76,** 127–133

Southgate, D.A.T. (1969) Determination of carbohydrates in foods II Unavailable carbohydrates. *J. Sci. Food Agric.* **20**, 331–335

Southgate, D.A.T. and Durnin, J.V.G.A. (1970) Calorie conversion factors: an experimental reassessment of the factors used in the calculations of the energy value of human diets. *Br. J. Nutr.* **24,** 517–535

Southgate, D.A.T. (1974) *Guidelines for the preparation of tables of food composition.* S. Karger, Basel

Southgate, D.A.T. (1976) *Determination of food carbohydrates.* Applied Science Publishers, London

Southgate, D.A.T., Paul, A.A., Dean, A.C. and Christie, A.A. (1978) Free sugars in foods. *J. Hum. Nutr.* **32,** 335–347

Tinggi, U., Reilly, C. and Patterson, C.M. (1992) Determination of selenium in foodstuffs using spectrofluorimetry and hydride generation atomic absorption spectroscopy. *J. Food Comp. Anal.* **5,** 269–280

U.S. Department of Agriculture, Agricultural Research Service. 1998. *USDA Nutrient Database for Standard Reference, Release 12.* Nutrient Data Laboratory Home Page, http://www.nal.usda.gov/fnic/foodcomp

Van den Berg, H. (1997) Bioavailability of niacin. *Eur. J. Clin. Nutr.* **51**, S64–S65

Weihrauch, J.L., Kinsella, J.E. and Watt, B.K. (1976) Comprehensive evaluation of fatty acids in foods. VI. Cereal products. *J. Am. Diet. Assoc.* **68**, 335–340

Wharton, P.A., Eaton, P.M. and Day, K.C. (1983) Sorrento Asian food tables: food tables, recipes and customs of mothers attending Sorrento Maternity Hospital, Birmingham, England. *Hum. Nutr.: Appl. Nutr.,* **37A,** 378–402

Widdowson, E.M. and McCance, R.A. (1943) Food tables. Their scope and limitations. *Lancet* **i**, 230–2

Wiles, S.J., Nettleton, P.A., Black A.E. and Paul, A.A. (1980) The nutrition composition of some cooked dishes eaten in Britain: A supplementary food composition table. *J. Hum. Nutr.* **34,** 189–223

Wu Leung, W.T., Butrum, R.R., Chang, F.H., Narayana Rao, M. and Polacchi, W. (1972) *Food composition table for use in East Asia.* Food and Agriculture Oranization and US Department of Health, Education and Welfare, Bethesda

---

Ministry of Agriculture, Fisheries and Food Analytical Reports (now Department for Environment, Food and Rural Affairs)

Laboratory of the Government Chemist (1992) Nutritional analysis of foods for pre-school children. (Food Surveillance Information Sheet No. 18)

Milk Marketing Board (1992) Fat, fatty acids and Vitamin E content of a selection of biscuits, cakes and desserts. (Food Surveillance Information Sheet No. 14)

RHM Research and Engineering Ltd (1993) Fatty acids in foods. (Food Surveillance Information Sheet No. 7)

Laboratory of the Government Chemist (1993a) Analytical survey of nutrient composition of retail cuts of beef. (Food Surveillance Information Sheet No. 39)

Laboratory of the Government Chemist (1993b) Analytical survey of nutrient composition of retail cuts of lamb. (Food Surveillance Information Sheet No. 49)

Laboratory of the Government Chemist (1994) Analytical survey of nutrient composition of retail cuts of pork. (Food Surveillance Information Sheet No. 40)

Laboratory of the Government Chemist (1994) Nutrient analysis of foods important in elderly people. (Food Surveillance Information Sheet No.42)

Laboratory of the Government Chemist (1995) Added folic acid in supplements and fortified foods.(Food Surveillance Information Sheet No. 67)

Laboratory of the Government Chemist (1995) Nutrient analysis of foods commonly consumed by schoolchildren. (Food Surveillance Information Sheet No. 84)

RHM Technology (1995) Nutrient analysis of pizzas. (Food Surveillance Information Sheet No. 68)

RHM Technology (1995) Nutrient analysis of selected foods. (Food Surveillance Information Sheet No. 70)

ADAS Laboratory Services (1996) Nutrient analysis of pasteurised liquid milk. (Food Surveillance Information Sheet No. 128)

Bolton-Smith, C., Price, R.J.G., Fenton, S.T., Harrington, D.J. and Shearer, M.J. (1996) Compilation of a provisional UK database for the phylloquinone (vitamin K_1) content of foods.

Laboratory of the Government Chemist (1996) Individual folates in foodstuffs. (Food Surveillance Information Sheet No. 92)

Aspland and James Ltd (1997) Nutrient analysis of ethnic and takeaway foods. (Food Surveillance Information Sheet No. 147)

Laboratory of the Government Chemist (1997) Determination of 25-OH vitamin D in selected foodstuffs. (Food Surveillance Information Sheet No. 101)

Laboratory of the Government Chemist (1997) Determination of cis carotenoids in foodstuffs. (Food Surveillance Information Sheet No. 104)

Laboratory of the Government Chemist (1997) The determination of different forms of Iron in foodstuffs. (Food Surveillance Information Sheet No. 117)

Leatherhead Food Research Association (1997) Trans fatty acids in frying oils and fried foods. (Food Surveillance Information Sheet No. 112)

RHM Technology (1997) Nutrient analysis of manufactured foods for vegetarians. (Food Surveillance Information Sheet No. 118)

University of London School of Life Sciences (1997) Analysis of composition of commercial sandwiches. (Food Surveillance Information Sheet No. 111)

Campden and Chorleywood Food Research Association (1998) Nutrient analysis of yoghurts, fromage frais and chilled desserts. (Food Surveillance Information Sheet No. 148)

Laboratory of the Government Chemist (1998) Nutrient analysis of 'other' milk and cream. (Food Surveillance Information Sheet No. 178)

ADAS Laboratories (1999) Nutrient analysis of ice creams and desserts. (Food Surveillance Information Sheet No. 195)

Laboratory of the Government Chemist (1999) Nutrient analysis of bread and morning goods. (Food Surveillance Information Sheet No. 194)

Laboratory of the Government Chemist (1999) Nutrient analysis of cheese. (Food Surveillance Information Sheet No. 196)

Bolton-Smith, C. and Shearer, M.J. (2000) The extension and verification of the provisional UK phylloquinone (vitamin K_1) food composition database. Report to the Food Standards Agency.

Laboratory of the Government Chemist (2001) Determination of vitamin D_3 and 25OH-vitamin D_3 in imported fish and other samples. Report to the Food Standards Agency.

Laboratory of the Government Chemist (2001) The determination of sodium content of bread. (Food Survey Information Sheet No. 19/01)

Laboratory of the Government Chemist (2002) The Determination of vitamin D_3 and 25-Hydroxy Vitamin D_3 in fish and other foods and the effect of dietary intake on an 'at-risk' population. Report to the Food Standards Agency.

Note: Copies of the analytical reports can be consulted at the Department for Environment, Food and Rural Affairs library. Food Surveillance Information Sheets (which can be accessed on the Food Standards Agency website www.food.gov.uk) do not, in most cases, contain the full set of analytical results. Copies of the reports to the Food Standards Agency can be consulted at the Agency's Dr Elsie Widdowson Library.

4.7 FOOD INDEX

Foods are indexed by their publication number and, for ease of reference, each food has been assigned a consecutive publication number for the purposes of this edition only. In addition, each food has a unique food code number which will allow read-across to the supplements or the fifth edition, where appropriate.

For foods that have already been included in supplements or in the fifth edition and for which there are no new data, their food code number (including the unique 2 digit prefix) has been repeated. These prefixes are 11- *Cereals and Cereal Products*, 12 – *Milk Products and Eggs*, 13 – *Vegetables, Herbs and Spices*, 14 – *Fruit and Nuts*, 15 – *Vegetable Dishes*, 16 – *Fish and Fish Products*, 17 – *Miscellaneous Foods*, 18 – *Meat, Poultry and Game*, 19 – *Meat Products and Dishes*, and 50 – *Fifth Edition*. Foods that have not previously been included have been given a new food code number in the supplement using that prefix (e.g. plain bagel (11-534)). Where new data have been incorporated for an existing food, a new food code has been allocated but with the same supplement prefix (e.g. beef bourguignonne was 19-161, now 19-330). For ease of use the original food code number is given alongside the new food code. These are the numbers that will be used in nutrient databank applications.

The index includes two kinds of cross-reference. The first is the normal coverage of alternative names (e.g. Back bacon see **Bacon rashers, back**). The second is to common examples of components of generically described foods, including brand names, which although not part of the food name have in general been included in the product description (e.g. Anchor half fat butter see **Blended spread, 40% fat**).

	Publication number	New food code	Old food code
American style muffins, chocolate chip	139	11-608	
Anacardium occidentale	See **Cashew nuts**		
Ananas comosus	See **Pineapple**		
Anas platyrhynchos	See **Duck**		
Anchor half fat butter	see **Blended spread (40% fat)**		
Anchovies, canned in oil, drained	648	16-323	16-168
Anser anser	See **Goose**		
Apium graveolens var dulce	See **Celery**		
Apple chutney	1190	17-531	17-341
Apple juice, unsweetened	1121		14-271
Apples, cooking, raw, peeled	918		50-852
Apples, cooking, stewed with sugar	919		50-854
Apples, cooking, stewed without sugar	920		50-855
Apples, eating, average, raw	921		50-856
Apples, eating, average, raw, peeled	922		50-858
Apricots, canned in juice	925	14-302	50-864
Apricots, canned in syrup	926	14-290	50-863
Apricots, raw	923		50-860
Apricots, ready-to-eat	924		50-862
Arachis hypogaea	See **Peanuts**		
Asparagus, boiled in salted water	778	13-442	50-738
Asparagus, raw	777		50-737
Asparagus officinalis var altilis	See **Asparagus**		
Aubergine, fried in corn oil	780		50-740
Aubergine, raw	779		50-739
Avena sativa	See **Oats**		
Avocado, average	927		50-865
Bacon rashers, back, dry-fried	392		19-002
Bacon rashers, back, fat trimmed, grilled	397		19-008
Bacon rashers, back, fat trimmed, raw	396		19-007
Bacon rashers, back, grilled	393		19-003
Bacon rashers, back, grilled crispy	394		19-004
Bacon rashers, back, microwaved	395		19-005
Bacon rashers, back, raw	391		19-001
Bacon rashers, back, reduced salt, grilled	398		19-009
Bacon rashers, middle, grilled	399		19-015
Bacon rashers, streaky, fried	402		19-017
Bacon rashers, streaky, grilled	401		19-018
Bacon rashers, streaky, raw	400		19-016
Bacon, fat only, average, cooked	404		50-339
Bacon, fat only, average, raw	403		50-338
Bacon, lettuce and tomato sandwich, white bread	72	11-563	
Badam	See **Almonds**		
Badla	See **Peas**		
Bagels, plain	153	11-534	

	Publication number	New food code	Old food code
Baileys Original Irish cream	See **Cream Liqueurs**		
Baingan	See **Aubergine**		
Baked beans, canned in tomato sauce, re-heated	734		50-694
Baked beans, canned in tomato sauce, reduced sugar, reduced salt	735		50-695
Baking powder	1219		17-355
Balsam apple	See **Gourd, karela**		
Banana bread	124	11-573	
Bananas	928		50-867
Banoffee pie	314	12-394	
Barbecue sauce	1203		17-289
Barley wine/strong ale	1140		17-221
Batata	See **Old potatoes**		
Bath buns	See **Chelsea buns**		
Battenburg cake	125	11-574	50-109
Beanburger, soya, fried in vegetable oil	867	15-366	15-008
Beans, aduki, dried, boiled in unsalted water	See **Aduki beans**		
Beans, baked	See **Baked beans**		
Beans, blackeye	See **Blackeye beans**		
Beans, broad	See **Broad beans**		
Beans, butter	See **Butter beans**		
Beans, French	See **Green beans/French beans**		
Beans, green	See **Green beans/French beans**		
Beans, mung	See **Mung beans**		
Beans, red kidney	See **Red kidney beans**		
Beans, runner	See **Runner beans**		
Beans, soya	See **Soya beans**		
Beansprouts, mung, raw	736	13-426	50-696
Beansprouts, mung, stir-fried in blended oil	737	13-427	50-697
Beef bourguignonne	567	19-330	19-161
Beef bourguignonne, made with lean beef	568	19-331	19-162
Beef casserole, made with canned cook-in sauce	569	19-332	19-164
Beef chow mein, retail, reheated	570		19-165
Beef curry, chilled/frozen, reheated	571		19-169
Beef curry, chilled/frozen, reheated, with rice	572		19-170
Beef curry, reduced fat	573	19-333	19-167
Beef sausages, chilled, grilled	552		19-077
Beef stew	574	19-334	19-175
Beef, average, fat, cooked	411		18-005
Beef, average, trimmed fat, raw	410		18-003
Beef, average, trimmed lean, raw	409	18-468	18-001
Beef, braising steak, braised, lean	412		18-008
Beef, braising steak, braised, lean and fat	413		18-009
Beef, fore-rib/rib-roast, raw, lean and fat	414		18-029
Beef, fore-rib/rib-roast, roasted, lean and fat	415		18-034

	Publication number	New food code	Old food code
Beef, mince, extra lean, stewed	419		18-041
Beef, mince, microwaved	417		18-037
Beef, mince, raw	416	18-469	18-036
Beef, mince, stewed	418	18-470	18-038
Beef, rump steak, barbecued, lean	421		18-045
Beef, rump steak, fried, lean	422	18-473	18-047
Beef, rump steak, fried, lean and fat	423	18-472	18-048
Beef, rump steak, from steakhouse, lean	425		18-050
Beef, rump steak, grilled, lean	424	18-474	18-049
Beef, rump steak, raw, lean and fat	420	18-471	18-044
Beef, rump steak, strips, stir-fried, lean	426		18-052
Beef, silverside, salted, boiled, lean	427		18-060
Beef, stewing steak, raw, lean and fat	428		18-077
Beef, stewing steak, stewed, lean and fat	429		18-081
Beef, stir-fried with green peppers	575	19-335	19-180
Beef, topside, raw, lean and fat	430		18-085
Beef, topside, roasted well-done, lean	431		18-090
Beef, topside, roasted well-done, lean and fat	432		18-091
Beefburgers, chilled/frozen, fried	524		19-029
Beefburgers, chilled/frozen, grilled	525		19-030
Beefburgers, chilled/frozen, raw	523	19-309	19-028
Beer, bitter, average	1130	17-506	17-207
Beer, bitter, best/premium	1131		17-208
Beer, lager	1133		17-211
Beer, lager, alcohol-free	1134		17-212
Beer, lager, low alcohol	1135		17-213
Beer, lager, premium	1136		17-214
Beetroot, boiled in salted water	782		50-742
Beetroot, pickled, drained	783		50-743
Beetroot, raw	781		50-741
Belgian chicory	See **Chicory**		
Bell peppers	See **Peppers, capsicum, green**		
Belly joint/slices	See **Pork, belly joint/slices**		
Benniseed	See **Sesame seeds**		
Bertholletia excelsa	See **Brazil nuts**		
Best end neck cutlets	See **Lamb, best end neck cutlets**		
Beta vulgaris	See **Beetroot**		
Bhajia	See **Pakora/Bhajia**		
Bhendi	See **Okra**		
Bhinda	See **Okra**		
Bhindi	See **Okra**		
Big Mac	526	19-310	19-039
Biscuits, chocolate chip cookies	105	11-508	
Biscuits, chocolate, cream filled, full coated	104	11-507	
Biscuits, chocolate, full coated	103	11-506	50-093

	Publication number	New food code	Old food code
Biscuits, crunch, cream filled	108	11-520	
Biscuits, digestive, chocolate	109	11-512	50-096
Biscuits, digestive, plain	110	11-513	50-097
Biscuits, gingernut	112	11-514	50-099
Biscuits, oat based	113	11-517	
Biscuits, sandwich, cream filled	115	11-519	50-103
Biscuits, sandwich, jam filled	116	11-516	
Biscuits, semi-sweet	117	11-521	50-104
Biscuits, short sweet	118	11-522	50-105
Biscuits, wafer, filled	120	11-524	50-107
Biscuits, wafers, filled, chocolate, full coated	121	11-509	
Biscuits, water	122		11-187
Bitter gourd	see **Gourd, karela**		
Bitter, beer, average	1130	17-506	17-207
Bitter, best/premium	1131		17-208
Black gram, urad gram, dried, boiled in unsalted water	739		50-699
Black gram, urad gram, dried, raw	738		50-698
Black pudding, dry-fried	535		19-114
Blackberries, raw	929		50-869
Blackberries, stewed with sugar	930		50-870
Blackcurrant juice drink, undiluted	1114		17-187
Blackcurrants, raw	931		50-872
Blackcurrants, stewed with sugar	932		50-873
Blackeye beans, dried, boiled in unsalted water	741		50-701
Blackeye beans, dried, raw	740		50-700
Blackeye peas	See **Blackeye beans**		
Blended spread (70–80% fat)	353		17-015
Blended spread, (40% fat)	354		17-016
Blue Band	See **Margarine, soft, polyunsaturated**		
Blue cheese dressing	1195		17-300
Boiled sweets	1060		17-101
Bolognese sauce (with meat)	576	19-352	19-183
Bombay Mix	1073		50-1034
Borecole	See **Curly kale**		
Bos taurus	See **Beef, veal**		
Bounty bar	1047	17-546	17-082
Bournvita powder	1086		50-1043
Bovril	See **Meat extract**		
Braising steak	See **Beef, braising steak**		
Bran Flakes	79	11-486	50-066
Bran, wheat	1		50-001
Brandy	See **Spirits**		
Brassica and Lepidium	See **Mustard and cress**		
Brassica hirta	See **Mustard**		

	Publication number	New food code	Old food code
Brassica nappus var napobrassica	See **Swede**		
Brassica oleracea	See **Cabbage**		
Brassica oleracea var	See **Spring greens**		
Brassica oleracea var acephala	See **Curly kale**		
Brassica oleracea var botrytis	See **Broccoli, Cauliflower**		
Brassica oleracea var gemmifera	See **Brussels sprouts**		
Brassica rapa var rapifera	See **Turnip**		
Brazil nuts	1005	14-871	50-974
Bread pudding	174	11-594	50-151
Bread sauce, made with semi-skimmed milk	1179	17-520	50-1145
Bread sauce, made with whole milk	1178	17-519	50-1144
Bread, banana	124	11-573	
Bread, brown, average	40	11-456	50-033
Bread, ciabatta	43	11-609	
Bread, currant	44		50-037
Bread, garlic, pre-packed, frozen	45	11-460	
Bread, granary	46	11-461	50-039
Bread, malt, fruited	47	11-462	50-042
Bread, naan	48	11-463	50-043
Bread, pitta, white	50	11-465	50-045
Bread, rye	51		50-046
Bread, wheatgerm	52	11-467	
Bread, white, Danish style	59	11-466	11-112
Bread, white, farmhouse or split tin, freshly baked	56	11-470	11-101
Bread, white, French stick	57	11-471	11-107
Bread, white, fried in lard	54	11-469	50-051
Bread, white, premium	58	11-474	
Bread, white, sliced	53	11-468	50-049
Bread, white, toasted	55	11-475	50-052
Bread, white, 'with added fibre'	60	11-472	50-054
Bread, white, 'with added fibre', toasted	61	11-473	50-055
Bread, wholemeal, average	62	11-476	50-056
Bread, wholemeal, toasted	63	11-611	50-057
Breadsticks	1074		17-123
Breakfast milk, pasteurised, average	224	12-321	
Breakfast milk, summer	225	12-322	
Breakfast milk, winter	226	12-323	
Brie	257	12-344	50-226
Brinjal	See **Aubergine**		
Broad beans, frozen, boiled in unsalted water	742	13-428	50-702
Broccoli, green, boiled in unsalted water	785		50-745
Broccoli, green, raw	784		50-744
Brown ale, bottled	1132		17-210
Brown bread, average	40	11-456	50-033
Brown lentils	See **Lentils, green and brown**		

	Publication number	New food code	Old food code
Brown rice, boiled	17	11-443	50-019
Brown rice, raw	16	11-442	50-018
Brown rolls, crusty	64	11-477	50-058
Brown rolls, soft	65	11-478	50-059
Brown sauce, sweet	1204		17-293
Brussels sprouts, boiled in unsalted water	787		50-747
Brussels sprouts, frozen, boiled in unsalted water	788	13-443	50-748
Brussels sprouts, raw	786		50-746
Bubble and squeak, fried in vegetable oil	868	15-383	15-054
Buccinum undatum	See **Whelks**		
Build-up powder, shake	1087	17-534	
Build-up powder, soup	1088	17-535	
Bund gobhi	See **Cabbage**		
Buns, Chelsea	154	11-588	50-130
Buns, currant	156	11-536	50-133
Buns, hot cross	164	11-590	50-141
Burger, bean, soya, fried in vegetable oil	867	15-366	15-008
Burger, chicken, takeaway	528	19-315	19-041
Burgers, economy, frozen, grilled	530		19-043
Burgers, economy, frozen, raw	529		19-042
Burgers, hamburger	See also **Beefburgers, Hamburger, Big Mac, Cheeseburger, Quarterpounder** and **whopper burger**		
Butter	351	17-485	17-013
Butter, spreadable	352	17-486	17-014
Butter beans, canned, re-heated, drained	743	13-429	50-703
Cabbage, boiled in unsalted water, average	790	13-444	50-750
Cabbage, raw, average	789	13-468	50-749
Cabbage, white, raw	791	13-445	50-753
Cake mix, made up	126	11-525	11-192
Cake, banana bread	124	11-573	
Cake, Battenburg	125	11-574	50-109
Cake, carrot	127	11-616	
Cake, chocolate fudge cake	128	11-527	
Cake, Eccles	161	11-589	50-138
Cake, fruit, plain, retail	131	11-529	50-113
Cake, fruit, rich	132	11-577	50-114
Cake, fruit, rich, iced	133	11-578	50-115
Cake, fruit, wholemeal	134	11-579	50-116
Cake, Madeira	138	11-531	50-118
Cake, sponge	142	11-580	50-119
Cake, sponge, jam filled	144		50-121
Cake, sponge, made without fat	143	11-581	50-120
Cake, sponge, with dairy cream and jam	145	11-532	
Cake, reduced fat	140		11-617

	Publication number	New food code	Old food code
Calabrese	See **Broccoli, green**		
Calf liver, fried	514		18-410
Camembert	258	12-345	50-227
Cancer pagurus	See **Crab**		
Canned anchovies, in oil, drained	648	16-323	16-168
Canned apricots, in juice	925	14-302	50-864
Canned apricots, in syrup	926	14-290	50-863
Canned baked beans in tomato sauce, re-heated	734		50-694
Canned baked beans in tomato sauce, reduced sugar, reduced salt	735		50-695
Canned butter beans, re-heated, drained	743	13-429	50-703
Canned carrots, re-heated, drained	796	13-450	50-758
Canned cherries, in syrup	934		50-878
Canned chick peas, re-heated, drained	747		50-706
Canned chicken in white sauce	582		19-194
Canned cook-in-sauces	1205		17-295
Canned corned beef	539		19-128
Canned crab, in brine, drained	672		16-234
Canned cream, sterilised	249		50-217
Canned curry sauce	1206		17-298
Canned fruit cocktail, in juice	946		50-891
Canned fruit cocktail, in syrup	947		50-892
Canned grapefruit, in juice	953		50-901
Canned grapefruit, in syrup	954		50-902
Canned guava, in syrup	957		50-907
Canned luncheon meat	545		19-135
Canned lychees, in syrup	962		50-913
Canned mandarin oranges, in juice	963		50-914
Canned mandarin oranges, in syrup	964		50-915
Canned mushy peas, re-heated	769	13-437	50-728
Canned new potatoes, re-heated, drained	704		50-663
Canned paw-paw, in juice	976		50-937
Canned peaches, in juice	978		50-940
Canned peaches, in syrup	979		50-941
Canned pears, in juice	982		50-945
Canned pears, in syrup	983		50-946
Canned peas, re-heated, drained	774	13-441	50-733
Canned pilchards, in tomato sauce	657		16-201
Canned pineapple, in juice	985		50-948
Canned pineapple, in syrup	986		50-949
Canned plums, in syrup	989		50-954
Canned processed peas, re-heated, drained	776		50-736
Canned prunes, in juice	990		50-955
Canned prunes, in syrup	991		50-956
Canned raspberries, in syrup	995		50-960

	Publication number	New food code	Old food code
Canned ravioli, in tomato sauce	34		11-621
Canned red kidney beans, re-heated, drained	759	13-435	50-718
Canned rhubarb, in syrup	998		50-964
Canned rice pudding	333	12-406	50-287
Canned rice pudding, low fat	334	12-407	
Canned salmon, pink, in brine, flesh only, drained	662		16-208
Canned sardines, in brine, drained	663	16-328	16-215
Canned sardines, in oil, drained	664	16-329	16-216
Canned sardines, in tomato sauce	665		16-217
Canned shrimps, in brine, drained	676		16-247
Canned spaghetti, in tomato sauce	39		50-176
Canned sponge pudding	186	11-549	11-328
Canned strawberries, in syrup	1001		50-968
Canned sweetcorn, baby, drained	855		50-823
Canned sweetcorn, kernels, re-heated, drained	856	13-459	50-824
Canned tomatoes, whole contents	861	13-461	50-832
Canned tuna, in brine, drained	668		16-229
Canned tuna, in oil, drained	669		16-230
Cannelloni, vegetable	869	15-367	15-059
Capsicum annum	See **Paprika**		
Capsicum annum var grossum	See **Peppers**		
Cardium edule	See **Cockles**		
Carica papaya	See **Paw-paw**		
Carrots, canned, re-heated, drained	796	13-450	50-758
Carrots, old, boiled in unsalted water	793	13-447	50-755
Carrots, old, raw	792	13-446	50-754
Carrots, young, boiled in unsalted water	795	13-449	50-757
Carrots, young, raw	794	13-448	50-756
Carya illinoensis	See **Pecan nuts**		
Cashew nuts, roasted and salted	1006		50-976
Casserole, beef, made with canned cook-in sauce	569	19-332	19-164
Casserole, pork, made with canned cook-in sauce	609	19-348	19-256
Casserole, sausage	610	19-351	19-269
Casserole, vegetable	870	15-368	15-063
Castanea vulgaris	See **Chestnuts**		
Cauliflower cheese, made with semi-skimmed milk	871	15-369	15-065
Cauliflower, boiled in unsalted water	798		50-760
Cauliflower, raw	797		50-759
Celery, boiled in salted water	800		50-762
Celery, raw	799	13-451	50-761
Cereal chewy bar	1061	17-494	17-102
Cereal crunchy bar	1062		17-103
Channa	See **Chick peas, whole**		
Channel Island milk, whole, pasteurised	223		50-194

	Publication number	New food code	Old food code
Chapati flour, brown	2	11-433	50-002
Chapati flour, white	3	11-434	50-003
Chapatis, made with fat	41	11-458	50-035
Chapatis, made without fat	42	11-459	50-036
Cheddar cheese	259	12-346	50-228
Cheddar cheese and pickle sandwich, white bread	73	11-564	
Cheddar type, half fat	260	12-348	50-230
Cheddar, vegetarian	261	12-347	50-229
Cheerios	80	11-623	
Cheese and onion rolls, pastry	188	11-550	
Cheese sauce, made with semi-skimmed milk	1182	17-522	50-1147
Cheese sauce, made with whole milk	1181	17-521	50-1146
Cheese sauce, packet mix, made up with semi-skimmed milk	1183	17-524	50-1149
Cheese sauce, packet mix, made up with whole milk	1182	17-523	50-1148
Cheese spread, plain	262	12-349	50-231
Cheese spread, reduced fat	263	12-350	
Cheese, Brie	257	12-344	50-226
Cheese, Camembert	258	12-345	50-227
Cheese, Cheddar	259	12-346	50-228
Cheese, Cheddar type, half fat	260	12-348	50-230
Cheese, Cheddar, vegetarian	261	12-347	50-229
Cheese, cottage, plain	264	12-351	50-232
Cheese, cottage, plain, reduced fat	265	12-352	50-234
Cheese, cottage, plain, with additions	266		50-233
Cheese, cream	267	12-353	50-235
Cheese, Danish blue	268	12-354	50-236
Cheese, Edam	269	12-355	50-237
Cheese, Feta	270	12-356	50-238
Cheese, goats milk soft, full fat, white rind	271	12-357	12-162
Cheese, Gouda	272	12-358	50-243
Cheese, hard, average	273	12-359	50-244
Cheese, Mozzarella, fresh	274	12-360	12-170
Cheese, Parmesan, fresh	275	12-361	50-247
Cheese, processed, plain	276	12-362	50-248
Cheese, processed, slices, reduced fat	277	12-363	
Cheese, spreadable, soft white, full fat	280	12-364	50-242
Cheese, spreadable, soft white, low fat	278	12-366	
Cheese, spreadable, soft white, medium fat	279	12-365	50-246
Cheese, Stilton, blue	281	12-367	50-249
Cheese, white, average	282	12-368	50-250
Cheeseburger, takeaway	527	19-314	19-040
Cheesecake, frozen	315	12-395	50-274
Cheesecake, fruit, individual	316	12-396	

	Publication number	New food code	Old food code
Chelsea buns	154	11-588	50-130
Cherries, canned in syrup	934		50-878
Cherries, glace	935		50-879
Cherries, raw	933		50-876
Cherry pie filling	936		50-880
Chestnuts	1007		50-977
Chew sweets	1063		17-104
Chick pea flour/besan flour	744		13-073
Chick peas, canned, re-heated, drained	747		50-706
Chick peas, whole, dried, boiled in unsalted water	746	13-430	50-705
Chick peas, whole, dried, raw	745		50-704
Chicken breast in crumbs, chilled, fried	484		19-118
Chicken burger, takeaway	528	19-315	19-041
Chicken chasseur	577	19-350	19-186
Chicken chow mein, takeaway	578	19-321	
Chicken curry, average, takeaway	579	19-322	
Chicken curry, chilled/frozen, reheated, with rice	580		19-189
Chicken curry, made with canned curry sauce	581	19-336	19-190
Chicken in white sauce, canned	582		19-194
Chicken liver, fried	515		18-412
Chicken noodle soup, dried, as served	1172		17-254
Chicken nuggets, takeaway	536		19-124
Chicken pie, individual, chilled/frozen, baked	537		19-055
Chicken roll	538		19-125
Chicken salad sandwich, white bread	74	11-565	
Chicken satay	583	19-323	
Chicken soup, cream of, canned	1161		17-250
Chicken soup, cream of, canned, condensed	1162		17-251
Chicken soup, cream of, canned, condensed, as served	1163		17-252
Chicken tandoori, chilled, reheated	584		19-127
Chicken tikka masala, retail	585	19-325	
Chicken wings, marinated, chilled/frozen, barbecued	586		19-204
Chicken, breast, casseroled, meat only	481		18-307
Chicken, breast, grilled without skin, meat only	482		18-323
Chicken, breast, strips, stir-fried	483		18-326
Chicken, dark meat, raw	478		18-289
Chicken, drumsticks, roasted, meat and skin	485		18-335
Chicken, light meat, raw	479		18-290
Chicken, meat, average, raw	480	18-488	18-291
Chicken, roasted, dark meat	487		18-329
Chicken, roasted, leg quarter, meat and skin	489		18-337
Chicken, roasted, light meat	488		18-330
Chicken, roasted, meat, average	486		18-331

	Publication number	New food code	Old food code
Chicken, roasted, wing quarter, meat and skin	490		18-339
Chicken, skin, dry, roasted/grilled	491		18-332
Chicken, stir-fried with rice and vegetables, frozen, reheated	587		19-201
Chicken, sweet and sour, takeaway	616	19-324	
Chicory, raw	801		50-763
Chilli con carne	588	19-337	19-206
Chilli con carne, chilled/frozen, reheated, with rice	589		19-209
Chilli powder	902		50-838
Chilli, vegetable	872	15-370	15-073
Chinese 5 spice	903		13-813
Chinese cherry	See **Lychees**		
Chinese gooseberry	See **Kiwi fruit**		
Chips, fine cut, frozen, fried in blended oil	723		50-684
Chips, fine cut, frozen, fried in corn oil	724		50-685
Chips, fine cut, frozen, fried in dripping	725		50-686
Chips, French fries, retail	See **French fries, retail**		
Chips, homemade, fried in blended oil	713		50-674
Chips, homemade, fried in corn oil	714		50-675
Chips, homemade, fried in dripping	715		50-676
Chips, microwave, cooked	726		13-028
Chips, oven, frozen, baked	727		50-687
Chips, retail, fried in blended oil	716		50-677
Chips, retail, fried in dripping	717		50-678
Chips, retail, fried in vegetable oil	718	13-422	50-679
Chips, straight cut, frozen, fried in blended oil	720		50-681
Chips, straight cut, frozen, fried in corn oil	721		50-682
Chips, straight cut, frozen, fried in dripping	722		50-683
Choc ice	302	12-384	50-263
Chocolate biscuits, cream filled, full coated	104	11-507	
Chocolate biscuits, full coated	103	11-506	50-093
Chocolate chip cookies	105	11-508	
Chocolate covered caramels	1048	17-492	17-083
Chocolate dairy desserts	317	12-398	
Chocolate fudge cake	128	11-527	
Chocolate mousse	327	12-400	50-285
Chocolate mousse, reduced fat	328	12-401	
Chocolate nut spread	1030		17-070
Chocolate nut sundae	303	12-411	50-264
Chocolate spread	1029		17-069
Chocolate, fancy and filled	1049		17-088
Chocolate, milk	1050		17-089
Chocolate, plain	1051	17-491	17-090
Chocolate, white	1052		17-091
Chori	See **Blackeye beans**		

	Publication number	New food code	Old food code
Chote bund gobhi	See **Brussels sprouts**		
Chow mein, beef, retail, reheated	570		19-165
Chow mein, chicken, takeaway	578	19-321	
Christmas pudding, retail	175		50-153
Chutney, apple	1190	17-531	17-341
Chutney, mango, oily	1191		17-342
Chutney, tomato	1192		17-345
Ciabatta	43	11-609	
Cicer arietinum	See **Chick peas**		
Cichorium intybus	See **Chicory**		
Cider, dry	1141		17-222
Cider, low alcohol	1142		17-223
Cider, sweet	1143		17-224
Cider, vintage	1144		17-225
Cinnamomum aromaticum	See **Cinnamon**		
Cinnamomum verum	See **Cinnamon**		
Cinnamon, ground	904		50-839
Citrus limon	See **Lemon**		
Citrus paradisi	See **Grapefruit**		
Citrus reticulata	See **Mandarin oranges, Tangerines**		
Citrus reticulata var Clementine	See **Clementines**		
Citrus sinensis	See **Oranges**		
Clementines	937	14-291	50-881
Clotted cream, fresh	245		50-216
Clover	See **Blended spread (70–80% fat)**		
Clover Extra Light	See **Blended spread (40% fat)**		
Clupea harengus	See **Herring, Kipper, Whitebait**		
Clusters	81	11-487	
Coalfish	See **Coley**		
Cockles, boiled	678		16-252
Coco Pops	82	11-488	50-067
Cocoa powder	1089		50-1050
Coconut milk	1010		14-820
Coconut oil	376		17-031
Coconut, creamed block	1008	14-872	50-978
Coconut, desiccated	1009	14-873	50-979
Cocos nucifera	See **Coconut**		
Cod liver oil	377	17-488	17-032
Cod, baked	619		16-013
Cod, dried, salted, boiled	626		50-572
Cod, frozen, grilled	622		16-020
Cod, frozen, raw	621		16-019
Cod, in batter, fried in blended oil	623		16-021
Cod, in crumbs, frozen, fried in blended oil	624		16-027
Cod, in parsley sauce, frozen, boiled	625		16-030

	Publication number	New food code	Old food code
Cod, poached	620		16-015
Cod, raw	618		16-012
Coffee and chicory essence	1090	17-545	17-162
Coffee, infusion, average	1091		17-152
Coffee, instant	1092		17-158
Coffeemate	1093		50-1056
Cola	1107		17-175
Cola, diet	1108	17-505	
Coleslaw, with mayonnaise, retail	873		15-077
Coleslaw, with reduced calorie dressing, retail	874		15-078
Coley, raw	627		16-031
Coley, steamed	628	16-340	16-032
Common gram	See **Chick peas, whole**		
Complan powder, original and sweet	1094	17-540	50-1059
Complan powder, savoury	1095	17-541	50-1057
Compound cooking fat	369		17-004
Condensed milk, skimmed, sweetened	227		50-198
Condensed milk, whole, sweetened	228		50-199
Cook-in-sauces, canned	1205		17-295
Cooking apples, raw, peeled	918		50-852
Cooking apples, stewed with sugar	919		50-854
Cooking apples, stewed without sugar	920		50-855
Coq au vin	590	19-338	19-210
Corn chips	See **Tortilla chips**		
Corn Flakes	83	11-490	50-069
Corn Flakes, crunchy nut	84	11-491	50-070
Corn oil	378		17-033
Corn snacks	1075		17-125
Corn, sweet	See **Sweetcorn**		
Corned beef, canned	539		19-128
Cornetto-type ice-cream cone	304	12-386	50-265
Cornflour	4	11-435	50-004
Cornish pastie	540	19-316	19-056
Coronation chicken	591	19-339	19-213
Corylus avellana	See **Hazelnuts**		
Corylus maxima	See **Hazelnuts**		
Cottage cheese, plain	264	12-351	50-232
Cottage cheese, plain, reduced fat	265	12-352	50-234
Cottage cheese, plain, with additions	266		50-233
Cottage/Shepherd's pie, chilled/frozen, reheated	592		19-216
Courgette, boiled in unsalted water	803		50-765
Courgette, fried in corn oil	804		50-766
Courgette, raw	802		50-764
Cowpeas	See **Blackeye beans**		
Crab, boiled	671	16-331	16-232

	Publication number	New food code	Old food code
Crab, canned in brine, drained	672		16-234
Crabsticks	684		16-273
Crackers, cream	106	11-510	50-094
Crackers, wholemeal	123	11-572	50-108
Cranberry juice	1122	17-537	
Crangon crangon	See **Shrimps**		
Cream cheese	267	12-353	50-235
Cream crackers	106	11-510	50-094
Cream liqueurs	1157		17-242
Cream, dairy, extra thick	248	12-337	
Cream, dairy, UHT, canned spray	250	12-338	50-218
Cream, dairy, UHT, canned spray, half fat	251	12-339	
Cream, fresh, clotted	245		50-216
Cream, fresh, double	244	12-334	50-215
Cream, fresh, single	241	12-332	50-212
Cream, fresh, soured	242		50-213
Cream, fresh, whipping	243	12-333	50-214
Cream, sterilised, canned	249		50-217
Creme caramel	318	12-397	50-275
Creme egg	1053	17-544	17-092
Creme fraiche	246	12-335	
Creme fraiche, half fat	247	12-336	
Cress, mustard and	See **Mustard and cress**		
Cress, water	See **Watercress**		
Crispbread, rye	107	11-511	50-095
Crispie cakes	129	11-576	50-111
Crisps	See **Potato crisps**		
Croissants	66	11-480	50-060
Crumble, fruit	176	11-546	50-154
Crumble, fruit, wholemeal	177	11-595	50-155
Crumpets, toasted	155	11-535	50-132
Crunch biscuits, cream filled	108	11-520	
Crunchy Nut Corn Flakes	84	11-491	50-070
Cucumber, raw	805		50-767
Cucumis melo var indorus	See **Melon, Honeydew**		
Cucumis melo var cantatoupensis	See **Melon, Canteloupe-type**		
Cucumis melo var reticulata	See **Melon, Galia**		
Cucumis sativus	See **Cucumber, Gherkins**		
Cucurbita	See **Pumpkin**		
Cucurbita pepo	See **Courgette, Marrow**		
Curly kale, boiled in salted water	807		50-769
Curly kale, raw	806		50-768
Currant bread	44		50-037
Currant buns	156	11-536	50-133
Currants	938		50-883

	Publication number	New food code	Old food code
Curry powder	905		50-840
Curry sauce, canned	1206		17-298
Curry, beef, chilled/frozen, reheated	571		19-169
Curry, beef, chilled/frozen, reheated, with rice	572		19-170
Curry, beef, reduced fat	573	19-333	19-167
Curry, chick pea dahl	875	15-371	15-099
Curry, chicken, average, takeaway	579	19-322	
Curry, chicken, chilled/frozen, reheated, with rice	580		19-189
Curry, chicken, made with canned curry sauce	581	19-336	19-190
Curry, fish, Bangladeshi	685	16-336	16-274
Curry, lamb, made with canned curry sauce	600	19-344	19-227
Curry, prawn, takeaway	686	16-333	
Curry, vegetable, retail, with rice	876		15-155
Custard powder	5		50-005
Custard tarts, individual	157	11-537	50-134
Custard, ready-to-eat	321	12-399	50-278
Custard, made up with semi-skimmed milk	320	12-413	12-223
Custard, made up with whole milk	319	12-412	50-276
Dairy cream, extra thick	248	12-337	
Dairy cream, UHT, canned spray	250	12-338	50-218
Dairy cream, UHT, canned spray, half fat	251	12-339	
Dairy desserts, chocolate	317	12-398	
Dairy spread (40% fat)	355		17-017
Damsons, raw	939		14-077
Damsons, stewed with sugar	940		14-079
Danish blue	268	12-354	50-236
Danish pastries	158	11-538	50-135
Dates, dried	942		14-085
Dates, raw	941		14-083
Daucas carota	See **Carrots**		
Demerara sugar	See **Sugar, demerara**		
Digestive biscuits, chocolate	109	11-512	50-096
Digestive biscuits, plain	110	11-513	50-097
Dips, sour-cream based	1193		17-299
Discorea	See **Yam**		
Dogfish	See **Rock salmon/Dogfish**		
Doner kebab in pitta bread with salad	594		19-130
Doner kebabs, meat only	593		19-129
Double cream, fresh	244	12-334	50-215
Doughnuts, jam	159		50-136
Doughnuts, ring	160	11-539	50-137
Dream Topping, made up with semi-skimmed milk	252		50-221
Dressing, blue cheese	1195		17-300
Dressing, French	1196	17-509	17-302

	Publication number	New food code	Old food code
Dressing, French 'fat free'	1194	17-538	
Dressing, thousand island	1197		17-306
Dried mixed fruit	943		50-888
Dried skimmed milk	229		50-200
Dried skimmed milk, with vegetable fat	230		50-201
Drinking chocolate powder	1096	17-498	50-1064
Drinking chocolate powder, made up with semi-skimmed milk	1098	17-532	
Drinking chocolate powder, made up with whole milk	1097	17-533	
Drinking chocolate powder, reduced fat	1099	17-499	
Drinking yogurt	293		50-251
Dripping, beef	370	17-487	17-006
Drumsticks, chicken, roasted, meat and skin	485		18-335
Duck, crispy, Chinese style	502	18-490	
Duck, raw, meat only	501	18-489	18-369
Duck, roasted, meat only	503		18-372
Duck, roasted, meat, fat and skin	504		18-374
Dumplings	189	11-603	50-167
Dungli	See **Onions**		
Eating apples, average, raw	921		50-856
Eating apples, average, raw, peeled	922		50-858
Eccles cake	161	11-589	50-138
Echo	See **Margarine, hard, animal and vegetable fats**		
Eclairs, frozen	162		50-139
Economy burgers, frozen, grilled	530		19-043
Economy burgers, frozen, raw	529		19-042
Edam	269	12-355	50-237
Eel, jellied	649		16-174
Egg fried rice, takeaway	18	11-444	50-298
Egg mayonnaise sandwich, white bread	75	11-567	
Egg noodles, boiled	30		50-028
Egg noodles, raw	29		50-027
Eggplant	See **Aubergine**		
Eggs, chicken, boiled	341		50-293
Eggs, chicken, fried in vegetable oil	342	12-919	50-294
Eggs, chicken, poached	343		50-295
Eggs, chicken, raw	338	12-918	50-290
Eggs, chicken, scrambled, with milk	344	12-926	50-296
Eggs, chicken, white, raw	339		50-291
Eggs, chicken, yolk, raw	340		50-292
Eggs, duck, whole, raw	345	12-920	50-297
Elmlea, double	255	12-342	50-224
Elmlea, single	253	12-340	50-222

	Publication number	New food code	Old food code
Elmlea, whipping	254	12-341	50-223
Engraulis encrasicholus	See **Anchovies**		
Euthynnus	See **Tuna**		
Evaporated milk, light, 4% fat	232	12-324	
Evaporated milk, whole	231		50-202
Evening primrose oil	379		17-035
Faggots in gravy, chilled/frozen, reheated	595		19-131
Fancy iced cakes, individual	130	11-528	50-112
Fansi	See **Green beans/French beans**		
Fat spread (20–25% fat), not polyunsaturated	366	17-553	17-028
Fat spread (20–25% fat), polyunsaturated	367		17-029
Fat spread (35–40% fat), polyunsaturated	365		17-027
Fat spread (40% fat), not polyunsaturated	364	17-552	17-026
Fat spread (5% fat)	368	17-554	17-030
Fat spread (60% fat), polyunsaturated	362		17-024
Fat spread (60% fat), with olive oil	363		17-025
Fat spread (70% fat), polyunsaturated	361	17-551	17-023
Fat spread (70–80% fat), not polyunsaturated	360		17-022
Fennel, Florence, boiled in salted water	809		50-771
Fennel, Florence, raw	808		50-770
Feta	270	12-356	50-238
Ficus carica	See **Figs**		
Figs, dried	944		50-889
Figs, ready-to-eat	945		50-890
Fish balls, steamed	687		16-279
Fish cakes, fried in blended oil	688		16-282
Fish curry, Bangladeshi	685	16-336	16-274
Fish fingers, cod, fried in blended oil	689		16-289
Fish fingers, cod, grilled	690		16-288
Fish paste	691	16-334	16-293
Fisherman's pie, retail	692		16-295
Flaky pastry, cooked	148	11-583	50-125
Flaky pastry, raw	147	11-582	50-124
Flan, vegetable	877	15-372	15-175
Flapjacks	111	11-571	50-098
Flavoured milk, pasteurised	233	12-326	
Flavoured milk, pasteurised, chocolate	234	12-325	
Flora	See **Fat spread (70% fat), polyunsaturated**		
Flora Extra light	See **Fat spread (40% fat), polyunsaturated**		
Flour, chapati, brown	2	11-433	50-002
Flour, chapati, white	3	11-434	50-003
Flour, chick pea/besan	744		13-073
Flour, corn	4	11-435	50-004
Flour, rye, whole	7		50-007
Flour, soya, full fat	8		50-009

	Publication number	New food code	Old food code
Flour, soya, low fat	9		50-010
Flour, wheat, white, self-raising	13	11-440	50-015
Flour, wheat, brown	10	11-437	50-012
Flour, wheat, white, breadmaking	11	11-438	50-013
Flour, wheat, white, plain	12	11-439	50-014
Flour, wheat, wholemeal	14	11-441	50-016
Foeniculum vulgare var dulce	See **Fennel, Florence**		
Fore-rib	See **Beef, fore-rib/rib roast**		
Fragaria	See **Strawberries**		
Frankfurter	541		19-100
French beans	See **Green beans/French beans**		
French dressing	1196	17-509	17-302
French dressing, 'fat free'	1194	17-538	
French fries, retail	719	13-423	50-680
Fromage frais, fruit	299	12-370	50-239
Fromage frais, plain	298	12-369	50-240
Fromage frais, virtually fat free, fruit	301	12-372	
Fromage frais, virtually fat free, natural	300	12-371	
Frosties	85	11-492	50-071
Frozen ice-cream desserts	305	12-385	50-266
Fruit cake, plain, retail	131	11-529	50-113
Fruit cake, rich	132	11-577	50-114
Fruit cake, rich, iced	133	11-578	50-115
Fruit cake, wholemeal	134	11-579	50-116
Fruit cocktail, canned in juice	946		50-891
Fruit cocktail, canned in syrup	947		50-892
Fruit drink, low calorie, undiluted	1116		17-191
Fruit drink/squash, undiluted	1115		17-189
Fruit gums/jellies	1064		17-107
Fruit juice drink, carbonated, ready to drink	1109		17-177
Fruit juice drink, low calorie, ready to drink	1118		17-196
Fruit juice drink, ready to drink	1117		17-195
Fruit mousse	329	12-402	50-286
Fruit 'n Fibre	86	11-493	50-072
Fruit pastilles	1065		17-108
Fruit pie filling	948		50-893
Fruit pie, individual	180	11-547	50-158
Fruit pie, one crust	178	11-596	50-156
Fruit pie, pastry top and bottom	179	11-597	50-157
Fruit pie, pastry top and bottom, blackcurrant	181	11-598	50-150
Fruit pie, wholemeal, one crust	182	11-599	50-159
Fruit pie, wholemeal, pastry top and bottom	183	11-600	50-160
Fruit salad, homemade	949		50-894
Fruit spread	1031		17-071
Fudge	1066	17-518	17-109

	Publication number	New food code	Old food code
Fusarium graminearum	See **Quorn, myco-protein**		
Gadus morhua	See **Cod**		
Gajar	See **Carrots**		
Gallus domesticus	See **Chicken**		
Game pie	542		19-058
Garam masala	906		50-841
Garbanzo	See **Chick peas**		
Garlic bread, pre-packed, frozen	45	11-460	
Garlic mushrooms (not coated)	878	15-373	15-179
Garlic, raw	810		50-772
Gateau, chocolate based, frozen	135	11-526	
Gateau, fruit, frozen	136	11-530	
Gelatine	1220		17-360
Ghee, butter	371		17-007
Ghee, vegetable	372		17-009
Gherkins, pickled, drained	811		50-773
Gin	See **Spirits**		
Gingelly	See **Sesame seeds**		
Ginger ale, dry	1110		17-178
Gingernut biscuits	112	11-514	50-099
Glucose liquid, BP	1032		17-049
Glycine max	See **Soya beans**		
Goats milk soft cheese, full fat, white rind	271	12-357	12-162
Goats milk, pasteurised	236	12-328	50-204
Gobhi, bund	See **Cabbage**		
Gobhi, chote bund	See **Brussels sprouts**		
Gobhi, phool	See **Cauliflower**		
Gold	See **Fat spread (40% fat), not polyunsaturated**		
Gold Lowest	See **Fat spread (20–25% fat), not polyunsaturated**		
Gold Sunflower spread	See **Fat spread (40% fat), polyunsaturated**		
Golden gram	See **Mung beans**		
Goose, roasted, meat, fat and skin	505		18-376
Gooseberries, cooking, raw	950		50-895
Gooseberries, cooking, stewed with sugar	951		50-896
Gouda	272	12-358	50-243
Goulash	596	19-340	19-221
Gourd, bitter	See **Gourd, karela**		
Gourd, karela, raw	812		50-774
Gram, black	See **Black gram, urad gram**		
Gram, common	See **Chick peas**		
Gram, golden	See **Mung beans**		
Gram, green	See **Mung beans**		
Gram, yellow	See **Chick peas**		
Granary bread	46	11-461	50-039

	Publication number	New food code	Old food code
Granary rolls	67	11-479	
Grape juice, unsweetened	1123		14-273
Grapefruit juice, unsweetened	1124		14-275
Grapefruit, canned in juice	953		50-901
Grapefruit, canned in syrup	954		50-902
Grapefruit, raw	952	14-292	50-899
Grapes, average	955		50-903
Gravy instant granules	1221		17-310
Gravy instant granules, made up	1222		17-311
Greek pastries	163		50-140
Greek style yogurt, fruit	292	12-377	
Greek style yogurt, plain	291	12-376	
Greek yogurt, sheep	294	12-420	50-253
Green beans/French beans, frozen, boiled in unsalted water	749	13-432	50-708
Green beans/French beans, raw	748	13-431	50-707
Green gram	See **Mung beans**		
Greens, spring	See **Spring greens**		
Grillsteaks	See also **Beefburgers**		
Grillsteaks, beef, chilled/frozen, grilled	531		19-046
Groundnuts	See **Peanuts**		
Guava, canned in syrup	957		50-907
Guava, raw	956		50-905
Guinness	1139		17-219
Gullar	See **Figs**		
Haddock, in crumbs, frozen, fried in blended oil	632		16-063
Haddock, raw	629		16-044
Haddock, smoked, steamed	631		16-068
Haddock, steamed	630		16-049
Haggis, boiled	543		19-132
Halibut, grilled	633		16-074
Ham	405	19-308	19-023
Ham salad sandwich, white bread	76	11-566	
Ham, gammon joint, boiled	407		19-021
Ham, gammon joint, raw	406		19-020
Ham, gammon rashers, grilled	408		19-022
Hamburger	See also **Big Mac, Cheeseburger, Quarterpounder** and **whopper burger**		
Hamburger buns	68	11-481	50-061
Hamburger, takeaway	532	19-311	19-047
Hard cheese, average	273	12-359	50-244
Hazelnuts	1011	14-874	50-980
Heart, lamb, roasted	510	18-492	18-397
Helianthus annus	See **Sunflower seeds**		
Herbs, mixed, dried	908	13-871	

	Publication number	New food code	Old food code
Herring, grilled	651		16-176
Herring, raw	650		16-175
Hibiscus esculentes	See **Okra**		
Hickory nuts	See **Pecan nuts**		
Himarus vulgaris	See **Lobster**		
Hippoglossus hippoglossus	See **Halibut**		
Honey	1033		17-050
Honeycomb	1034		17-051
Horlicks LowFat Instant powder	1100	17-502	50-1067
Horlicks powder	1101	17-503	50-1069
Horseradish sauce	1207		17-314
Hot cross buns	164	11-590	50-141
Hot pot, lamb/beef with potatoes, chilled/ frozen, retail, reheated	602		19-231
Hotpot, Lancashire	603	19-345	19-236
Hula Hoops	See **Potato rings**		
Human milk, mature	237		50-207
Hummus	750	13-433	50-709
Ice-cream bar, chocolate coated	306	12-391	
Ice-cream desserts, frozen	305	12-385	50-266
Ice-cream sauce, topping	1035		17-053
Ice-cream wafers	307		50-272
Ice-cream, dairy, premium	309	12-392	
Ice-cream, dairy, vanilla	308	12-387	50-267
Ice-cream, non-dairy, vanilla	310	12-388	50-269
Indian nuts	See **Pine nuts**		
Instant dessert powder	322		50-279
Instant drinks powder, chocolate, low calorie	1102	17-500	
Instant drinks powder, malted	1103	17-501	
Instant potato powder, made up with water	728		50-688
Instant potato powder, made up with whole milk	729		50-689
Instant soup powder, dried	1173	17-507	17-259
Instant soup powder, dried, made up with water	1174	17-508	17-260
Ipomoea batatas	See **Sweet potato**		
Irish stew	597	19-341	19-222
Irish stew, canned	599		19-224
Irish stew, made with lean lamb	598	19-342	19-223
Jaffa cakes	137	11-515	50-101
Jaggery	1036		17-058
Jam tarts, retail	165	11-540	50-143
Jam, fruit with edible seeds	1037		17-073
Jam, reduced sugar	1038		17-075
Jam, stone fruit	1039		17-074
Jelly, made with water	323		50-282
Jew's apple	see **Aubergine**		

	Publication number	New food code	Old food code
Juglans regia	See **Walnuts**		
Juice, apple, unsweetened	1121		14-271
Juice, cranberry	1122	17-537	
Juice, grape, unsweetened	1123		14-273
Juice, grapefruit, unsweetened	1124		14-275
Juice, lemon, fresh	1125		14-277
Juice, orange, unsweetened	1126	14-301	14-283
Juice, pineapple, unsweetened	1128		14-286
Juice, tomato	1129		50-1093
Kaju	See **Cashew nuts**		
Kakdi	See **Cucumber**		
Kale	See **Curley kale**		
Kanda	See **Onions**		
Kebab, doner, in pitta bread with salad	594		19-130
Kebab, doner, meat only	593		19-129
Kebab, shish, in pitta bread with salad	612		19-151
Kebab, shish, meat only	611		19-150
Kedgeree	693	16-337	16-296
Ketchup, tomato	1217	17-513	17-338
Kheema, lamb	601	19-343	19-228
Khira	See **Cucumber**		
Kidney beans	See **Red kidney beans**		
Kidney, lamb, fried	511	18-493	18-403
Kidney, ox, stewed	512		18-405
Kidney, pig, stewed	513		18-408
Kiev, vegetable, baked	898		15-362
Kipper, grilled	653		16-188
Kipper, raw	652		16-187
Kit Kat	1054	17-493	17-093
Kiwi fruit	958	14-293	50-908
Kobi	See **Cabbage**		
Kobi, nhanu	See **Brussels sprouts**		
Krona Gold	See **Fat spread (70–80% fat), not polyunsaturated**		
Kula	See **Bananas**		
Kumra	See **Pumpkin**		
Lactuca sativa	See **Lettuce**		
Lady's fingers	See **Okra**		
Lager	1133		17-211
Lager, alcohol-free	1134		17-212
Lager, low alcohol	1135		17-213
Lager, premium	1136		17-214
Lal kaddu	See **Pumpkin**		
Lal phupala	See **Pumpkin**		
Lamb curry, made with canned curry sauce	600	19-344	19-227

	Publication number	New food code	Old food code
Lamb kheema	601	19-343	19-228
Lamb, average, trimmed fat, cooked	435		18-100
Lamb, average, trimmed fat, raw	434		18-098
Lamb, average, trimmed lean, raw	433	18-475	18-097
Lamb, best end neck cutlets, grilled, lean	437		18-107
Lamb, best end neck cutlets, grilled, lean and fat	438		18-109
Lamb, best end neck cutlets, raw, lean and fat	436		18-101
Lamb, breast, roasted, lean	439		18-113
Lamb, breast, roasted, lean and fat	440		18-114
Lamb, heart, roasted	510	18-492	18-397
Lamb, kidney, fried	511	18-493	18-403
Lamb, leg, average, raw, lean and fat	441	18-478	18-123
Lamb, leg, whole, roasted medium, lean	442	18-479	18-135
Lamb, leg, whole, roasted medium, lean and fat	443	18-480	18-136
Lamb, liver, fried	516	18-494	18-414
Lamb, loin chops, grilled, lean	445		18-141
Lamb, loin chops, grilled, lean and fat	446	18-477	18-143
Lamb, loin chops, microwaved, lean and fat	447		18-147
Lamb, loin chops, raw, lean and fat	444	18-476	18-139
Lamb, loin chops, roasted, lean and fat	448		18-151
Lamb, mince, raw	449	18-481	18-158
Lamb, mince, stewed	450		18-159
Lamb, neck fillet, strips, stir-fried, lean	451		18-164
Lamb, shoulder, diced, kebabs, grilled, lean and fat	453		18-172
Lamb, shoulder, raw, lean and fat	452		18-170
Lamb, shoulder, whole, roasted, lean	454		18-179
Lamb, shoulder, whole, roasted, lean and fat	455		18-180
Lamb, stewing, pressure cooked, lean	456		18-184
Lamb, stewing, stewed, lean	457		18-186
Lamb, stewing, stewed, lean and fat	458		18-187
Lamb/Beef hot pot with potatoes, chilled/ frozen, retail, reheated	602		19-231
Lancashire hotpot	603	19-345	19-236
Lard	373		17-010
Lasagne	604	19-346	19-237
Lasagne, chilled/frozen, reheated	605		19-238
Lasagne, vegetable, retail	879		15-189
Lassan	See **Garlic**		
Lassi, sweetened	295	12-373	
Leeks, boiled in unsalted water	814	13-452	50-776
Leeks, raw	813	13-466	50-775
Leg joint, pork	See **Pork, leg joint**		
Leg lamb	See **Lamb, leg**		
Lehsan	See **Garlic**		

	Publication number	New food code	Old food code
Lemon curd	1040	17-490	17-076
Lemon juice, fresh	1125		14-277
Lemon meringue pie	184	11-548	50-161
Lemon peel	959		14-127
Lemon sole, goujons, baked	636		16-087
Lemon sole, goujons, fried in blended oil	637		16-088
Lemon sole, raw	634		16-082
Lemon sole, steamed	635		16-085
Lemonade	1111		17-179
Lemons, whole, without pips	959		50-910
Lentils, green and brown, whole, dried, boiled in salted water	752		50-711
Lentils, green and brown, whole, dried, raw	751		50-710
Lentils, red, split, dried, boiled in unsalted water	754	13-434	50-713
Lentils, red, split, dried, raw	753		50-712
Lens esculenta	See **Lentils**		
Lettuce, average, raw	815	13-453	50-777
Lettuce, Iceberg, raw	816		50-779
Lichee	See **Lychees**		
Lichi	See **Lychees**		
Lima beans	See **Butter beans**		
Lime juice cordial, undiluted	1119		17-200
Liqueurs, cream	1157		17-242
Liqueurs, high strength	1158		17-244
Liqueurs, low-medium strength	1159		17-245
Liquorice allsorts	1067		17-112
Litchee	See **Lychees**		
Litchi	See **Lychees**		
Litchi chinensis	See **Lychees**		
Littorina littoria	See **Winkles**		
Liver pate	547	19-317	19-143
Liver sausage	544		19-106
Liver, calf, fried	514		18-410
Liver, chicken, fried	515		18-412
Liver, lamb, fried	516	18-494	18-414
Liver, ox, stewed	517		18-416
Liver, pig, stewed	518		18-418
Lobia	See **Blackeye beans**		
Lobster, boiled	673	16-332	16-236
Loin chops, lamb	See **Lamb, loin chops**		
Loin chops, pork	See **Pork, loin chops**		
Loligo vulgaris	See **Squid**		
Lollies, containing ice-cream	311	12-390	
Lollies, with real fruit juice	312	12-389	
Low calorie soup, canned	1164		17-265

	Publication number	New food code	Old food code
Low fat yogurt, fruit	288	12-380	50-257
Low fat yogurt, plain	287	12-379	50-255
Lucozade	1112	17-543	17-180
Luncheon meat, canned	545		19-135
Lychees, canned in syrup	962		50-913
Lychees, raw	961		50-911
Lycopersicon esculentum	See **Tomatoes**		
M & M's	See **Smartie-type sweets**		
Macadamia integrifolia	See **Macadamia nuts**		
Macadamia nuts, salted	1012		50-982
Macadamia tetraphylla	See **Macadamia nuts**		
Macaroni cheese	28	11-562	50-168
Macaroni, boiled	27	11-448	50-026
Macaroni, raw	26	11-447	50-025
Mackerel, grilled	655	16-325	16-194
Mackerel, raw	654	16-324	16-191
Mackerel, smoked	656		16-196
Madeira cake	138	11-531	50-118
Madeira nuts	See **Walnuts**		
Maize chips	See **Tortilla chips**		
Maize oil	See **Corn oil**		
Malt bread, fruited	47	11-462	50-042
Malus pumila	See **Apples**		
Mandarin oranges, canned in juice	963		50-914
Mandarin oranges, canned in syrup	964		50-915
Mangifera indica	See **Mangoes**		
Mange-tout peas, boiled in salted water	767	13-436	50-726
Mange-tout peas, raw	766		50-725
Mange-tout peas, stir-fried in blended oil	768		50-727
Mango chutney, oily	1191		17-342
Mangoes, ripe, raw	965	14-294	50-916
Margarine, hard, animal and vegetable fats	356		17-018
Margarine, hard, vegetable fats only	357	17-539	17-019
Margarine, soft, not polyunsaturated	358		17-020
Margarine, soft, polyunsaturated	359		17-021
Marmalade	1041		17-078
Marmite	See **Yeast extract**		
Marrow, boiled in unsalted water	818		50-781
Marrow, raw	817		50-780
Mars bar	1055	17-547	17-094
Marshmallows	1068		17-114
Marzipan, home-made	1013	14-881	50-983
Marzipan, retail	1014	14-875	50-984
Masoor dahl	See **Lentils, red**		
Masur	See **Lentils, green and brown**		

	Publication number	New food code	Old food code
Masur dahl	See **Lentils, red**		
Mattar	See **Peas**		
Mayonnaise, reduced calorie	1199	17-511	17-318
Mayonnaise, retail	1198	17-510	17-316
Meat extract	1223	17-514	17-361
Meat pate, reduced fat	548		19-145
Meat samosas, takeaway	606	19-326	
Meat spread	546		19-139
Melanogrammus aeglefinus	See **Haddock**		
Meleagris gallopavo	See **Turkey**		
Melon, Canteloupe-type	966	14-295	50-919
Melon, Galia	967		50-921
Melon, Honeydew	968		50-923
Melon, watermelon	969	14-296	50-925
Mentha spicata	See **Mint**		
Meringue	324	12-414	50-299
Meringue, with cream	325	12-415	50-300
Merlangius merlangus	See **Whiting**		
Microstomus kitt	See **Lemon sole**		
Microwave chips, cooked	726		13-028
Milk chocolate	1050		17-089
Milk pudding, made with whole milk	326	12-416	50-283
Milk shake powder	1104		50-1073
Milk, breakfast, pasteurised, average	224	12-321	
Milk, breakfast, summer	225	12-322	
Milk, breakfast, winter	226	12-323	
Milk, Channel Island, whole, pasteurised	223		50-194
Milk, coconut	1010		14-820
Milk, condensed, skimmed, sweetened	227		50-198
Milk, condensed, whole, sweetened	228		50-199
Milk, evaporated, light, 4% fat	232	12-324	
Milk, evaporated, whole	231		50-202
Milk, flavoured, pasteurised	233	12-326	
Milk, flavoured, pasteurised, chocolate	234	12-325	
Milk, goats, pasteurised	236	12-328	50-204
Milk, human, mature	237		50-207
Milk, semi-skimmed, average	211	12-312	50-185
Milk, semi-skimmed, pasteurised, average	212	12-313	50-186
Milk, semi-skimmed, pasteurised, fortified plus SMP	215		50-187
Milk, semi-skimmed, pasteurised, summer	213	12-418	
Milk, semi-skimmed, pasteurised, winter	214	12-419	
Milk, semi-skimmed, UHT	216	12-314	50-188
Milk, sheeps, raw	238	12-329	50-208
Milk, skimmed, average	206	12-306	50-181

	Publication number	New food code	Old food code
Milk, skimmed, dried	229		50-200
Milk, skimmed, dried, with vegetable fat	230		50-201
Milk, skimmed, pasteurised, average	207	12-307	50-182
Milk, skimmed, pasteurised, fortified plus SMP	208		50-183
Milk, skimmed, sterilised	209	12-311	12-007
Milk, skimmed, UHT	210	12-310	50-184
Milk, soya, non-dairy alternative, sweetened, calcium enriched	239	12-330	
Milk, soya, non-dairy alternative, unsweetened	240	12-331	50-209
Milk, whole, average	217	12-315	50-189
Milk, whole, pasteurised, average	218	12-316	50-190
Milk, whole, pasteurised, summer	219	12-317	50-191
Milk, whole, pasteurised, winter	220	12-318	50-192
Milk, whole, sterilised	221	12-319	50-193
Milk, whole, UHT	222	12-320	12-016
Milkshake, thick, takeaway	235	12-327	50-1072
Milky Way	1056	17-548	17-095
Mince pies, individual	166	11-591	50-144
Minced beef	See **Beef, mince**		
Minced lamb	See **Lamb, mince**		
Mincemeat	1042		17-080
Minestrone soup, canned	1165	17-542	17-266
Minestrone soup, dried, as served	1175		17-269
Mint sauce	1208		17-319
Mint, fresh	907		50-842
Mixed fruit, dried	943		50-888
Mixed herbs, dried	908	13-871	
Mixed nuts	1015		50-985
Mixed peel	970		50-926
Mixed vegetables, frozen, boiled in salted water	819		50-782
Momordica charantia	See **Gourd, karela**		
Monkey nuts	See **Peanuts**		
Moong beans	See **Mung beans**		
Motamircha	See **Peppers, capsicum, green**		
Moussaka, chilled/frozen/longlife, reheated	607		19-248
Moussaka, vegetable, retail	880		15-206
Mousse, chocolate	327	12-400	50-285
Mousse, chocolate, reduced fat	328	12-401	
Mousse, fruit	329	12-402	50-286
Mozzarella, fresh	274	12-360	12-170
Muesli, swiss style	87	11-494	50-073
Muesli, with no added sugar	88	11-495	50-074
Muffins, American style, chocolate chip	139	11-608	
Muffins, English style, white	167	11-541	
Muffins, English style, white, toasted	168	11-542	

	Publication number	New food code	Old food code
Mung beans, whole, dried, boiled in unsalted water	756		50-715
Mung beans, whole, dried, raw	755		50-714
Musa	See **Bananas**		
Musa paradisiaca	See **Plantain**		
Mushroom soup, cream of, canned	1166		17-270
Mushrooms, common, fried in butter	821		50-786
Mushrooms, common, fried in corn oil	822		50-787
Mushrooms, common, raw	820		50-783
Mushrooms, garlic (not coated)	878	15-373	15-179
Mushy peas, canned, re-heated	769	13-437	50-728
Mussels, boiled	679		16-256
Mustard and cress, raw	823		50-788
Mustard powder	909		50-843
Mustard, smooth	1224		17-364
Mustard, wholegrain	1225		17-365
Myristica fragrans	See **Nutmeg**		
Mytilus edulis	See **Mussels**		
Naan bread	48	11-463	50-043
Nasturtium officinale	See **Watercress**		
Nectarines	971	14-297	50-927
Neeps (England)	See **Swede**		
Neeps (Scotland)	See **Turnip**		
Nephrops norvegicus	See **Scampi**		
New potatoes, average, raw	701		50-660
New potatoes, boiled in unsalted water	702		50-661
New potatoes, canned, re-heated, drained	704		50-663
New potatoes, in skins, boiled in unsalted water	703	13-420	50-662
Nhanu kobi	See **Brussels sprouts**		
Noodles, egg, boiled	30		50-028
Noodles, egg, raw	29		50-027
Nut roast	881	15-374	15-213
Nutmeg, ground	910		50-844
Nutri-Grain	89	11-612	11-140
Nuts, almonds	1004	14-870	50-972
Nuts, brazil	1005	14-871	50-974
Nuts, cashew, roasted and salted	1006		50-976
Nuts, chestnuts	1007		50-977
Nuts, hazelnuts	1011	14-874	50-980
Nuts, macadamia, salted	1012		50-982
Nuts, mixed	1015		50-985
Nuts, peanuts, dry roasted	1019	14-878	50-989
Nuts, peanuts, plain	1018	14-877	50-987
Nuts, peanuts, roasted and salted	1020		50-990
Nuts, pecan	1021		50-991

	Publication number	New food code	Old food code
Nuts, pine	1022		50-992
Nuts, pistachio, roasted and salted	1023		14-840
Nuts, walnuts	1028	14-879	50-979
Oat based biscuits	113	11-517	
Oat Bran Flakes, with raisins	90	11-489	50-068
Oatcakes, retail	114	11-518	50-102
Oatmeal, quick cook, raw	6		50-006
Oil, coconut	376		17-031
Oil, cod liver	377	17-488	17-032
Oil, corn	378		17-033
Oil, evening primrose	379		17-035
Oil, olive	380		17-038
Oil, palm	381		17-039
Oil, peanut (Groundnut)	382		17-040
Oil, rapeseed	383		17-041
Oil, safflower	384		17-042
Oil, sesame	385		17-043
Oil, soya	386		17-044
Oil, sunflower	387		17-045
Oil, vegetable, blended, average	388	17-489	17-046
Oil, walnut	389		17-047
Oil, wheatgerm	390		17-048
Okra, raw	824		50-789
Okra, boiled in unsalted water	825		50-790
Okra, stir-fried in corn oil	826		50-791
Olea europaea	See **Olives**		
Old potatoes, average, raw	705		50-664
Old potatoes, baked, flesh and skin	706		50-665
Old potatoes, baked, flesh only	707		50-666
Old potatoes, boiled in unsalted water	708	13-421	50-668
Old potatoes, mashed with butter	709		50-669
Old potatoes, roast in blended oil	710		50-671
Old potatoes, roast in corn oil	711		50-672
Old potatoes, roast in lard	712		50-673
Olive oil	380		17-038
Olives, in brine	972		50-929
Omelette, cheese	347	12-922	50-302
Omelette, plain	346	12-921	50-301
Onion sauce, made with semi-skimmed milk	1185	17-526	50-1151
Onion sauce, made with whole milk	1184	17-525	50-1150
Onions, fried in corn oil	828		50-795
Onions, pickled, cocktail/silverskin, drained	830		50-798
Onions, pickled, drained	829		50-797
Onions, raw	827		50-792
Onions, spring	See **Spring onions**		

	Publication number	New food code	Old food code
Orange juice concentrate, unsweetened	1127		14-284
Orange juice, unsweetened	1126	14-301	14-283
Oranges	973	14-298	50-931
Oryza sativa	See **Rice**		
Ovaltine powder	1105	17-504	50-1076
Oven chips, frozen, baked	727		50-687
Ovis aries	See **Lamb**		
Ox kidney, stewed	512		18-405
Ox liver, stewed	517		18-416
Oxtail, stewed	519		18-420
Oxtail soup, canned	1167		17-272
Pakora/bhajia, vegetable, retail	882		15-232
Palak	See **Spinach**		
Paleamon serratus	See **Prawns**		
Pale ale, bottled	1137		17-216
Palm oil	381		17-039
Pancakes, savoury, made with whole milk	190	11-604	50-169
Pancakes, Scotch, retail	172	11-544	
Pancakes, stuffed with vegetables	883	15-376	15-233
Pancakes, sweet, made with whole milk	185	11-601	50-162
Pandulus borealis	See **Shrimps**		
Pandulus montagui	See **Shrimps**		
Pangoli	See **Cauliflower**		
Papai	See **Paw-paw**		
Papaya	See **Paw-paw**		
Pappadums, takeaway	49	11-464	50-044
Paprika	911		50-845
Parmesan, fresh	275	12-361	50-247
Parsley, fresh	912		50-846
Parsnip, boiled in unsalted water	832	13-454	50-800
Parsnip, raw	831		50-799
Passiflora edulis f edulis	See **Passion fruit**		
Passion fruit	974		50-933
Pasta sauce, tomato based	1209		17-323
Pasta with meat and tomato sauce	608	19-347	19-252
Pasta, fresh, cheese and vegetable stuffed, cooked	33	11-451	
Pasta, plain, fresh, cooked	32	11-450	
Pasta, plain, fresh, raw	31	11-449	
Paste, fish	691	16-334	16-293
Pastilles, fruit	1065		17-108
Pastinaca sativa	See **Parsnip**		
Pastries, Danish	158	11-538	50-135
Pastries, Greek	163		50-140
Pastry, flaky, cooked	148	11-583	50-125

	Publication number	New food code	Old food code
Pastry, flaky, raw	147	11-582	50-124
Pastry, shortcrust, cooked	150	11-585	50-127
Pastry, shortcrust, raw	149	11-584	50-126
Pastry, wholemeal, cooked	152	11-587	50-129
Pastry, wholemeal, raw	151	11-586	50-128
Pastie, Cornish	540	19-316	19-056
Pasty, vegetable	884	15-377	15-236
Pate, liver	547	19-317	19-143
Pate, meat, reduced fat	548		19-145
Pate, tuna	700		16-308
Pavlova, no fruit	331	12-403	
Pavlova, with fruit and cream	330	12-404	
Paw-paw, canned in juice	976		50-937
Paw-paw, raw	975		50-935
Peaches, canned in juice	978		50-940
Peaches, canned in syrup	979		50-941
Peaches, raw	977	14-299	50-938
Peanut butter, smooth	1016		50-986
Peanut (Groundnut) oil	382		17-040
Peanuts and raisins	1017	14-882	50-1036
Peanuts, dry roasted	1019	14-878	50-989
Peanuts, plain	1018	14-877	50-987
Peanuts, roasted and salted	1020		50-990
Pears, average, raw	980		50-942
Pears, average, raw, peeled	981		50-944
Pears, canned in juice	982		50-945
Pears, canned in syrup	983		50-946
Peas, blackeye	See **Blackeye beans**		
Peas, boiled in unsalted water	771	13-439	50-730
Peas, canned, re-heated, drained	774	13-441	50-733
Peas, chick pea	See **Chick peas**		
Peas, frozen, boiled in salted water	772	13-465	50-731
Peas, frozen, boiled in unsalted water	773	13-440	50-732
Peas, mange-tout	See **Mange-tout peas**		
Peas, mushy	See **Mushy peas**		
Peas, petit pois	See **Petit pois**		
Peas, processed	See **Processed peas**		
Peas, raw	770	13-438	50-729
Pecan nuts	1021		50-991
Peel, mixed	970		50-926
Pepper, black	913		50-847
Pepper, white	914		50-848
Peppermints	1069		17-117
Peppers, capsicum, chilli, green, raw	833		50-801
Peppers, capsicum, green, boiled in salted water	835		50-803

	Publication number	New food code	Old food code
Peppers, capsicum, green, raw	834		50-802
Peppers, capsicum, red, boiled in salted water	837		50-805
Peppers, capsicum, red, raw	836		50-804
Petit pois, frozen, boiled in unsalted water	775		50-735
Petroselinum crispum	See **Parsley**		
Phaseolus aureus	See **Beansprouts, mung**		
Phaseolus coccineus	See **Runner beans**		
Phaseolus lunatus	See **Butter beans**		
Phaseolus vulgaris	See **Green beans/French beans, Baked beans, Red kidney beans**		
Phasianus colchicus	See **Pheasant**		
Pheasant, roasted, meat only	506		18-383
Phoenix dactylifera	See **Dates**		
Phool gobhi	see **Cauliflower**		
Piccalilli	1210		17-347
Pickle, sweet	1200		17-352
Pickled onions, cocktail/silverskin, drained	830		50-798
Pickled onions, drained	829		50-797
Pie filling, cherry	936		50-880
Pie filling, fruit	948		50-893
Pie, chicken, individual, chilled/frozen, baked	537		19-055
Pie, cottage/shepherd's, chilled/frozen, reheated	592		19-216
Pie, fisherman's, retail	692		16-295
Pie, fruit, blackcurrant, pastry top and bottom	181	11-598	50-150
Pie, fruit, individual	180	11-547	50-158
Pie, fruit, one crust	178	11-596	50-156
Pie, fruit, pastry top and bottom	179	11-597	50-157
Pie, fruit, wholemeal, one crust	182	11-599	50-159
Pie, fruit, wholemeal, pastry top and bottom	183	11-600	50-160
Pie, game	542		19-058
Pie, lemon meringue	184	11-548	50-161
Pie, mince, individual	166	11-591	50-144
Pie, pork, individual	550		19-063
Pie, steak and kidney, single crust, homemade	561	19-329	19-070
Pie, steak and kidney/beef, individual, chilled/ frozen, baked	562		19-069
Pie, vegetable	885	15-379	15-243
Pig kidney, stewed	513		18-408
Pig liver, stewed	518		18-418
Pignolias	See **Pine nuts**		
Pilau, plain	19	11-561	15-250
Pilchards, canned in tomato sauce	657		16-201
Pimento	See **Peppers, capsicum, chilli, green**		
Pine kernels	See **Pine nuts**		
Pine nuts	1022		50-992

	Publication number	New food code	Old food code
Pineapple juice, unsweetened	1128		14-286
Pineapple, canned in juice	985		50-948
Pineapple, canned in syrup	986		50-949
Pineapple, raw	984		50-947
Pinus edulis	See **Pine nuts**		
Pinus pinea	See **Pine nuts**		
Piper nigrum	See **Pepper**		
Pista	See **Pistachio nuts**		
Pistacia vera	See **Pistachio nuts**		
Pistachio nuts, roasted and salted	1023		14-840
Pisum sativum	See **Peas**		
Pisum sativum var macrocarpum	See **Mange-tout peas**		
Pitta bread, white	50	11-465	50-045
Piyaz	See **Onions**		
Pizza base, raw	196	11-552	
Pizza, cheese and tomato, deep pan	197	11-613	
Pizza, cheese and tomato, french bread	199	11-554	
Pizza, cheese and tomato, frozen	200	11-553	50-171
Pizza, cheese and tomato, thin base	198	11-614	
Pizza, chicken topped, chilled	201	11-559	
Pizza, fish topped, takeaway	202	11-560	
Pizza, ham and pineapple, chilled	203	11-558	
Pizza, meat topped	204	11-556	
Pizza, vegetarian	205	11-557	
Plaice, frozen, steamed	639		16-108
Plaice, goujons, baked	642		16-119
Plaice, goujons, fried in blended oil	643		16-120
Plaice, in batter, fried in blended oil	640		16-110
Plaice, in crumbs, fried in blended oil	641		16-114
Plaice, raw	638		16-102
Plain chocolate	1051	17-491	17-090
Plantain, boiled in unsalted water	838		50-807
Plantain, ripe, fried in vegetable oil	839		50-808
Pleuronetes platessa	See **Plaice**		
Plums, average, raw	987	14-300	50-950
Plums, average, stewed with sugar	988		14-215
Plums, canned in syrup	989		50-954
Pollachius virens	See **Coley**		
Polony	549		19-109
Popcorn, candied	1076		17-130
Popcorn, plain	1077		17-131
Porage	See **Porridge**		
Pork casserole, made with canned cook-in sauce	609	19-348	19-256
Pork pie, individual	550		19-063
Pork sausages, chilled, fried	554		19-079

	Publication number	New food code	Old food code
Pork sausages, chilled, grilled	555		19-080
Pork sausages, raw, average	553	19-318	19-081
Pork sausages, reduced fat, chilled/frozen, grilled	556		19-086
Pork scratchings	1078		17-132
Pork, average, trimmed lean, raw	459		18-201
Pork, belly joint/slices, grilled, lean and fat	462		18-209
Pork, diced, casseroled, lean only	463	18-482	18-219
Pork, fat, cooked	461		18-205
Pork, fillet strips, stir-fried, lean	464		18-228
Pork, leg joint, raw, lean and fat	465	18-483	18-236
Pork, leg joint, roasted medium, lean	466	18-484	18-240
Pork, leg joint, roasted medium, lean and fat	467	18-485	18-241
Pork, loin chops, barbecued, lean and fat	469		18-249
Pork, loin chops, grilled, lean	470		18-251
Pork, loin chops, grilled, lean and fat	471		18-252
Pork, loin chops, microwaved, lean and fat	472		18-254
Pork, loin chops, raw, lean and fat	468		18-246
Pork, loin chops, roasted, lean and fat	473		18-256
Pork, steaks, grilled, lean and fat	475		18-286
Pork, steaks, raw, lean and fat	474		18-284
Pork, sweet and sour	617	19-349	19-276
Pork, trimmed fat, raw	460		18-203
Porridge, made with water	91	11-569	50-076
Porridge, made with whole milk	92	11-570	50-077
Port	1151		17-234
Pot noodles	See **Pot savouries**		
Pot savouries	1082		17-143
Pot savouries, made up	1083		17-144
Potato chips	See **Chips**		
Potato crisps	1079	17-495	17-133
Potato crisps, lower fat	1080	17-496	17-136
Potato croquettes, fried in blended oil	730		50-690
Potato fritters, battered, cooked	731	13-424	
Potato powder, instant, made up with water	728		50-688
Potato powder, instant, made up with whole milk	729		50-689
Potato rings	1081		17-142
Potato waffles, frozen, cooked	732		50-691
Potato, sweet	See **Sweet potato**		
Potatoes, new	See **New potatoes**		
Potatoes, old	See **Old potatoes**		
Prawn crackers, takeaway	191	11-551	
Prawn curry, takeaway	686	16-333	
Prawns, boiled	674		16-239
Prawns, Szechuan with vegetables, takeaway	698	16-335	
Processed cheese, plain	276	12-362	50-248

	Publication number	New food code	Old food code
Processed cheese, slices, reduced fat	277	12-363	
Processed peas, canned, re-heated, drained	776		50-736
Profiteroles with sauce	332	12-405	
Prunes, canned in juice	990		50-955
Prunes, canned in syrup	991		50-956
Prunes, ready-to-eat	992		50-957
Prunus amygdalus	See **Almonds**		
Prunus armeniaca	See **Apricots**		
Prunus avium	See **Cherries**		
Prunus domestica subsp domestica	See **Plums**		
Prunus domestica subsp institia	See **Damsons**		
Prunus persica	See **Peaches**		
Prunus persica var nectarina	See **Nectarines**		
Psidium guajava	See **Guava**		
Pudding, black, dry-fried	535		19-114
Pudding, bread	174	11-594	50-151
Pudding, Christmas, retail	175		50-153
Pudding, milk, made with whole milk	326	12-416	50-283
Pudding, rice, canned	333	12-406	50-287
Pudding, rice, canned, low fat	334	12-407	
Pudding, sponge, canned	186	11-549	11-328
Pudding, white	566		19-159
Pudding, Yorkshire	195	11-607	50-180
Puffed Wheat	93		50-078
Pumpkin, boiled in salted water	841		50-810
Pumpkin, raw	840		50-809
Puree, tomato	1230	17-516	17-374
Purple grenadillo	See **Passion fruit**		
Pyrus communis	See **Pears**		
Quarterpounder, takeaway	533	19-312	19-048
Queensland nuts	see **Macadamia nuts, salted**		
Quiche Lorraine	350	12-925	12-285
Quiche, cheese and egg	348	12-923	50-303
Quiche, cheese and egg, wholemeal	349	12-924	50-304
Quorn, pieces, as purchased	842	13-455	50-811
Rabbit, raw, meat only	507		18-387
Rabbit, stewed, meat only	508		18-388
Radish, red, raw	843		50-812
Raisins	993		50-958
Raja	See **Skate**		
Rapeseed oil	383		17-041
Raphanus sativa	See **Radish, red**		
Raspberries, canned in syrup	995		50-960
Raspberries, raw	994		50-959
Ratatouille, retail	886		15-264

	Publication number	New food code	Old food code
Ravioli, canned in tomato sauce	34	11-621	
Ready Brek	94	11-496	50-080
Red kidney beans, canned, re-heated, drained	759	13-435	50-718
Red kidney beans, dried, boiled in unsalted water	758		50-717
Red kidney beans, dried, raw	757		50-716
Red wine	1145		17-228
Reduced fat cake	140	11-617	
Relish, burger/chilli/tomato	1211		17-354
Relish, corn/cucumber/onion	1212		17-353
Rheum rhaponticum	See **Rhubarb**		
Rhubarb, canned in syrup	998		50-964
Rhubarb, raw	996		50-961
Rhubarb, stewed with sugar	997		50-962
Ribena	See **Blackcurrant juice drink**		
Ribes grossularia	See **Gooseberries**		
Rice cakes	141	11-618	
Rice Krispies	95	11-497	50-081
Rice pudding, canned	333	12-406	50-287
Rice pudding, canned, low fat	334	12-407	
Rice, brown, boiled	17	11-443	50-019
Rice, brown, raw	16	11-442	50-018
Rice, egg fried, takeaway	18	11-444	50-298
Rice, pilau, plain	19	11-561	15-250
Rice, savoury, cooked	21	11-620	50-021
Rice, savoury, raw	20		50-020
Rice, white, basmati, raw	22		11-041
Rice, white, easy cook, boiled	24	11-446	50-023
Rice, white, easy cook, raw	23	11-445	50-022
Rice, white, fried	25	11-610	11-045
Ricicles	96	11-498	50-082
Ringana	See **Aubergine**		
Risotto, plain	192	11-605	50-173
Risotto, vegetable	887	15-378	15-275
Roast chicken	See **Chicken, roasted**		
Roast duck	See **Duck, roasted**		
Roast turkey	See **Turkey, roasted**		
Rock eel	See **Rock Salmon/Dogfish**		
Rock Salmon/Dogfish, in batter, fried in blended oil	644		16-134
Roe, cod, hard, fried in blended oil	694		16-300
Rolls, brown, crusty	64	11-477	50-058
Rolls, brown, soft	65	11-478	50-059
Rolls, granary	67	11-479	
Rolls, white, crusty	69	11-482	50-062
Rolls, white, soft	70	11-483	50-063

	Publication number	New food code	Old food code
Rolls, wholemeal	71	11-484	50-064
Rose wine, medium	1146		17-229
Rosmarinus officinalis	See **Rosemary**		
Rosemary, dried	915		50-849
Rubus idaeus	See **Raspberries**		
Rubus nigrum	See **Blackcurrants**		
Rubus ulmifolius	See **Blackberries**		
Rump steak beef, barbecued, lean	421		18-045
Rump steak beef, fried, lean	422	18-473	18-047
Rump steak beef, fried, lean and fat	423	18-472	18-048
Rump steak beef, from steakhouse, lean	425		18-050
Rump steak beef, grilled, lean	424	18-474	18-049
Rump steak beef, raw, lean and fat	420	18-471	18-044
Rump steak beef, strips, stir-fried, lean	426		18-052
Runner beans, boiled in unsalted water	761		50-720
Runner beans, raw	760		50-719
Rutabaga	See **Swede**		
Rye bread	51		50-046
Rye flour, whole	7	11-436	50-007
Saag	See **Spinach**		
Safflower oil	384		17-042
Sage, dried, ground	916		50-850
Saithe	See **Coley**		
Salad cream	1201	17-512	17-326
Salad cream, reduced calorie	1202		17-327
Salad, green	888	15-380	15-292
Salad, potato, with mayonnaise, retail	889		15-297
Salad, rice	890	15-381	15-299
Salami	551		19-110
Salmon en croute, retail	695		16-304
Salmon, grilled	659	16-327	16-203
Salmon, pink, canned in brine, flesh only, drained	662	16-338	16-208
Salmon, raw	658	16-326	16-202
Salmon, smoked	661		16-207
Salmon, steamed	660		16-205
Salmo salar	See **Salmon**		
Salt	1226		17-367
Salvia officinalis	See **Sage**		
Samosas, meat, takeaway	606	19-326	
Samosas, vegetable, retail	891		15-305
Sandwich biscuits, cream filled	115	11-519	50-103
Sandwich biscuits, jam filled	116	11-516	
Sandwich, Bacon, lettuce and tomato, white bread	72	11-563	
Sandwich, Cheddar cheese and pickle, white bread	73	11-564	
Sandwich, Chicken salad, white bread	74	11-565	

	Publication number	New food code	Old food code
Sandwich, Egg mayonnaise, white bread	75	11-567	
Sandwich, Ham salad, white bread	76	11-566	
Sandwich, Tuna mayonnaise, white bread	77	11-568	
Sardina pilchardus	See **Sardines**		
Sardines, canned in brine, drained	663	16-328	16-215
Sardines, canned in oil, drained	664	16-329	16-216
Sardines, canned in tomato sauce	665		16-217
Sardinops sagex ocellata	See **Pilchards**		
Satay, chicken	583	19-323	
Satsumas	999		50-965
Sauce, barbecue	1203		17-289
Sauce, bread, made with semi-skimmed milk	1179	17-520	50-1145
Sauce, bread, made with whole milk	1178	17-519	50-1144
Sauce, brown, sweet	1204		17-293
Sauce, cheese, made with semi-skimmed milk	1181	17-522	50-1147
Sauce, cheese, made with whole milk	1182	17-521	50-1146
Sauce, cheese, packet mix, made up with semi-skimmed milk	1183	17-524	50-1149
Sauce, cheese, packet mix, made up with whole milk	1184	17-523	50-1148
Sauce, curry, canned	1206		17-298
Sauce, horseradish	1207		17-314
Sauce, ice-cream topping	1035		17-053
Sauce, mint	1208		17-319
Sauce, onion, made with semi-skimmed milk	1185	17-526	50-1151
Sauce, onion, made with whole milk	1184	17-525	50-1150
Sauce, pasta, tomato based	1209		17-323
Sauce, soy	1213		17-334
Sauce, sweet and sour, canned	1214		17-335
Sauce, sweet and sour, take-away	1215		17-336
Sauce, tartare	1216		17-337
Sauce, white, savoury, made with semi-skimmed milk	1187	17-528	50-1153
Sauce, white, savoury, made with whole milk	1186	17-527	50-1152
Sauce, white, sweet, made with semi-skimmed milk	1189	17-530	50-1155
Sauce, white, sweet, made with whole milk	1188	17-529	50-1154
Sauce, Worcestershire	1218		17-340
Sauerkraut	892		13-336
Sausage casserole	610	19-351	19-269
Sausage rolls, puff pastry	558		19-066
Sausages, beef, chilled, grilled	552		19-077
Sausages, pork, chilled, fried	554		19-079
Sausages, pork, chilled, grilled	555		19-080
Sausages, pork, raw, average	553	19-318	19-081

	Publication number	New food code	Old food code
Shortcrust pastry, cooked	150	11-585	50-127
Shortcrust pastry, raw	149	11-584	50-126
Shoulder lamb	See **Lamb, shoulder**		
Shredded Wheat	97	11-499	50-083
Shreddies	98	11-500	50-084
Shrimps, canned in brine, drained	676		16-247
Shrimps, frozen	677		16-248
Silverside	See **Beef, silverside**		
Sinapis alba	See **Mustard**		
Single cream, fresh	241	12-332	50-212
Skate, in batter, fried in blended oil	645		16-146
Skimmed milk, average	206	12-306	50-181
Skimmed milk, dried	229		50-200
Skimmed milk, dried, with vegetable fat	228		50-201
Skimmed milk, pasteurised, average	207	12-307	50-182
Skimmed milk, pasteurised, fortified plus SMP	208		50-183
Skimmed milk, sterilised	209	12-311	12-007
Skimmed milk, UHT	210	12-310	50-184
Smartie-type sweets	1057		17-096
Snickers	1058	17-549	17-097
Snowpeas	See **Mange-tout peas**		
Solanum melongerna var ovigerum	See **Aubergine**		
Solanum tuberosum	See **Potatoes**		
Sorbet, fruit	313	12-393	50-273
Soup powder, instant, dried	1173	17-507	17-259
Soup powder, instant, dried, made up with water	1174	17-508	17-260
Soup, chicken noodle, dried, as served	1172		17-254
Soup, chicken, cream of, canned	1161		17-250
Soup, chicken, cream of, canned, condensed	1162		17-251
Soup, chicken, cream of, canned, condensed, as served	1163		17-252
Soup, low calorie, canned	1164		17-265
Soup, Minestrone, canned	1165	17-542	17-266
Soup, Minestrone, dried, as served	1175		17-269
Soup, mushroom, cream of, canned	1166		17-270
Soup, oxtail, canned	1167		17-272
Soup, tomato, cream of, canned	1168		17-278
Soup, tomato, cream of, canned, condensed	1169		17-279
Soup, tomato, cream of, canned, condensed, as served	1170		17-282
Soup, tomato, dried, as served	1176		17-282
Soup, vegetable, canned	1171		17-284
Soup, vegetable, dried, as served	1177		17-286
Soured cream, fresh	242		50-213
Soy sauce	1213		17-334

	Publication number	New food code	Old food code
Soya beans, dried, boiled in unsalted water	763		50-722
Soya beans, dried, raw	762		50-721
Soya flour, full fat	8		50-009
Soya flour, low fat	9		50-010
Soya oil	386		17-044
Soya, alternative to yogurt, fruit	296	12-381	50-258
Soya, non-dairy alternative to milk, sweetened, calcium enriched	239	12-330	
Soya, non-dairy alternative to milk, unsweetened	240	12-331	50-209
Spaghetti bolognese, chilled/frozen, reheated	613	19-328	19-273
Spaghetti bolognese, chilled/frozen, reheated, with spaghetti	614	19-353	
Spaghetti, canned in tomato sauce	39		50-176
Spaghetti, white, boiled	36	11-453	50-030
Spaghetti, white, raw	35	11-452	50-029
Spaghetti, wholemeal, boiled	38	11-455	50-032
Spaghetti, wholemeal, raw	37	11-454	50-031
Special K	99	11-501	50-086
Spinach, boiled in unsalted water	846	13-457	50-814
Spinach, frozen, boiled in unsalted water	847	13-458	50-815
Spinach, raw	845	13-456	50-813
Spinacia oleracea	See **Spinach**		
Spirits, 40% volume	1160		17-247
Sponge cake	142	11-580	50-119
Sponge cake, jam filled	144		50-121
Sponge cake, made without fat	143	11-581	50-120
Sponge cake, with dairy cream and jam	145	11-532	
Sponge pudding, canned	186	11-328	
Spread, cheese, plain	262	12-349	50-231
Spread, cheese, reduced fat	263	12-350	
Spread, chocolate	1029		17-069
Spread, chocolate nut	1030		17-070
Spread, fruit	1031		17-071
Spread, meat	546		19-139
Spreadable cheese, soft white, full fat	280	12-364	50-242
Spreadable cheese, soft white, low fat	278	12-366	
Spreadable cheese, soft white, medium fat	279	12-365	50-246
Spring greens, boiled in unsalted water	849		50-817
Spring greens, raw	848		50-816
Spring onions, bulbs and tops, raw	850		50-818
Spring rolls, meat, takeaway	615	19-327	
Sprouts, Brussels	see **Brussels sprouts**		
Squalus acanthias	See **Rock salmon/Dogfish**		
Squid, frozen, raw	680		16-264
Squid, in batter, fried in blended oil	681		16-265

	Publication number	New food code	Old food code
Steak and kidney pie, single crust, homemade	561	19-329	19-070
Steak and kidney/Beef pie, individual, chilled/ frozen, baked	562		19-069
Steak, braising	See **Beef, braising steak**		
Steak, rump, barbecued, lean	421		18-045
Steak, rump, fried, lean	422	18-473	18-047
Steak, rump, fried, lean and fat	423	18-472	18-048
Steak, rump, from steakhouse, lean	425		18-050
Steak, rump, grilled, lean	424	18-474	18-049
Steak, rump, raw, lean and fat	420	18-471	18-044
Steak, rump, strips, stir-fried, lean	426		18-052
Steak, stewed with gravy, canned	563		19-152
Steaks, pork	See **Pork, steaks**		
Stew, beef	574	19-334	19-175
Stew, Irish	597	19-341	19-222
Stew, Irish, canned	599		19-224
Stew, Irish, made with lean lamb	598	19-342	19-223
Stewed apples, cooking, stewed without sugar	920		50-855
Stewed apples, cooking, with sugar	919		50-854
Stewed blackberries, with sugar	930		50-870
Stewed blackcurrants, with sugar	932		50-873
Stewed damsons, with sugar	940		14-079
Stewed gooseberries, cooking, with sugar	951		50-896
Stewed plums, average, with sugar	988		14-215
Stewed rhubarb, with sugar	997		50-962
Stewed steak with gravy, canned	563		19-152
Stewing lamb	See **Lamb, stewing**		
Stewing steak	See **Beef, stewing steak**		
Stilton, blue	281	12-367	50-249
Stir-fried beef with green peppers	575	19-335	19-180
Stir-fried chicken, with rice and vegetables, frozen, reheated	587		19-201
Stir-fried lamb, neck fillets lean	451		18-164
Stir-fried pork, fillet strips, lean	464		18-228
Stir-fried turkey, breast, strips,	496		18-357
Stir-fried vegetables, takeaway	900	15-364	
Stir fry mix, vegetables, fried in vegetable oil	899		15-346
Stock cubes, beef	1227	17-515	17-368
Stock cubes, chicken	1228		17-369
Stock cubes, vegetable	1229		17-370
Stork	See **Margarine, hard, animal and vegetable fats**		
Stork SB	See **Margarine, soft, not polyunsaturated**		
Stout, Guinness	1139		17-219
Strawberries, canned in syrup	1001		50-968

	Publication number	New food code	Old food code
Strawberries, raw	1000		50-967
Strong ale/barley wine	1140		17-221
Stuffing mix, dried	193		17-371
Stuffing, sage and onion	194	11-606	17-373
Suet, shredded	374		17-011
Suet, vegetable	375		17-012
Sugar Puffs	100	11-503	50-088
Sugar, Demerara	1043		17-061
Sugar, white	1044		17-063
Sultana Bran	101	11-504	50-089
Sultanas	1002		50-969
Sunflower oil	387		17-045
Sunflower seeds	1025		50-995
Sunny Delight	1120	17-536	
Sus scrofa	See **Pork**		
Swede, boiled in unsalted water	852		50-820
Swede, raw	851		50-819
Sweet and sour chicken, takeaway	616	19-324	
Sweet and sour pork	617	19-349	19-276
Sweet and sour sauce, canned	1214		17-335
Sweet and sour sauce, take-away	1215		17-336
Sweet peppers	See **Peppers, capsicum, green**		
Sweet potato, boiled in salted water	854	13-464	50-822
Sweet potato, raw	853	13-463	50-821
Sweetcorn, baby, canned, drained	855		50-823
Sweetcorn, kernels, canned, re-heated, drained	856	13-459	50-824
Sweetcorn, on-the-cob, whole, boiled in unsalted water	857		13-370
Sweets, boiled	1060		17-101
Sweets, chew	1063		17-104
Sweets, sherbert	1070		17-119
Sweets, Smartie-type	1057		17-096
Swiss roll, chocolate, individual	146	11-533	50-123
Swordfish, grilled	666		16-222
Syrup, golden	1045		17-065
Szechuan prawns with vegetables, takeaway	698	16-335	
Tagliatelle, with vegetables, retail	894		15-317
Tahini paste	1026		50-996
Tandoori, chicken, chilled, reheated	584		19-127
Tangerines	1003		50-970
Taramasalata	699		16-307
Tarel	See **Apples**		
Tart, treacle	187	11-602	50-165
Tarts, custard, individual	157	11-537	50-134
Tarts, jam, retail	165	11-540	50-143

	Publication number	New food code	Old food code
Tartare sauce	1216		17-337
Tea, black, infusion, average	1106		17-165
Teacakes, toasted	173	11-545	50-149
Thousand island dressing	1197		17-306
Thyme, dried, ground	917		50-851
Thymus vulgaris	See **Thyme**		
Tikka masala, chicken, retail	585	19-325	
Til	See **Sesame seeds**		
Tip Top dessert topping	256	12-343	50-225
Toffees, mixed	1071		17-120
Tofu, soya bean, steamed	764		50-723
Tofu, soya bean, steamed, fried	765		50-724
Tomato chutney	1192		17-345
Tomato juice	1129		50-1093
Tomato ketchup	1217	17-513	17-338
Tomato puree	1230	17-516	17-374
Tomato soup, cream of, canned	1168		17-278
Tomato soup, cream of, canned, condensed	1169		17-279
Tomato soup, cream of, canned, condensed, as served	1170		17-280
Tomato soup, dried, as served	1176		17-282
Tomatoes, canned, whole contents	861	13-461	50-832
Tomatoes, fried in corn oil	859		50-829
Tomatoes, grilled	860	13-467	50-831
Tomatoes, raw	858	13-460	50-827
Tongue slices	564		19-154
Tongue, sheep, stewed	520		18-427
Tonic water	1113		17-184
Topside	See **Beef, topside**		
Torte, fruit	335	12-408	
Tortilla chips	1084	17-497	17-149
Trail Mix	1027		50-1041
Treacle tart	187	11-602	50-165
Treacle, black	1046		17-068
Trifle	336	12-417	50-288
Trifle, fruit	337	12-409	
Tripe, dressed, raw	521		18-428
Triticum aestivum	See **Wheat**		
Trotters and tails, boiled	522		18-429
Trout, rainbow, grilled	667	16-330	16-226
Tuna mayonnaise sandwich, white bread	77	11-568	
Tuna pate	700		16-308
Tuna, canned in brine, drained	668	16-339	16-229
Tuna, canned in oil, drained	669		16-230
Turkey roll	565		19-156

	Publication number	New food code	Old food code
Turkey, breast, fillet, grilled, meat only	495		18-356
Turkey, breast, strips, stir-fried	496		18-357
Turkey, dark meat, raw	492		18-348
Turkey, light meat, raw	493		18-349
Turkey, meat, average, raw	494		18-350
Turkey, roasted, dark meat	497		18-358
Turkey, roasted, light meat	498		18-359
Turkey, roasted, meat, average	499		18-361
Turkey, skin, dry, roasted	500		18-362
Turkish delight, without nuts	1072		17-122
Turnip, boiled in unsalted water	863		50-834
Turnip, raw	862		50-833
Turnip, yellow	See **Swede**		
Twiglets	1085		17-150
Twix	1059	17-550	17-100
Tzatziki	297	12-410	50-259
Urad	See **Black gram, urad gram**		
Vatana	See **Peas**		
Veal, escalope, fried	477	18-487	18-093
Veal, escalope, raw	476	18-486	18-092
Vegeburger, retail, grilled	895		15-331
Vegetable and cheese grill/burger, in crumbs, baked/grilled	896	15-363	
Vegetable bake	897	15-382	15-341
Vegetable cannelloni	869	15-367	15-059
Vegetable casserole	870	15-368	15-063
Vegetable chilli	872	15-370	15-073
Vegetable curry, retail, with rice	876		15-155
Vegetable flan	877	15-372	15-175
Vegetable kiev, baked	898		15-362
Vegetable lasagne, retail	879		15-189
Vegetable moussaka, retail	880		15-206
Vegetable oil, blended, average	388	17-489	17-046
Vegetable pasty	884	15-377	15-236
Vegetable pie	885	15-379	15-243
Vegetable risotto	887	15-378	15-275
Vegetable shepherd's pie, retail	893		15-313
Vegetable soup, canned	1171		17-284
Vegetable soup, dried, as served	1177		17-286
Vegetable stir fry mix, fried in vegetable oil	899		15-346
Vegetable, casserole	867	15-368	15-063
Vegetables, mixed, frozen, boiled in salted water	819		50-782
Vegetables, stir-fried, takeaway	900	15-364	
Vegetarian sausages, baked/grilled	901		15-365
Venison, roast	509	18-491	18-391

	Publication number	New food code	Old food code
Vermouth, dry	1155		17-239
Vermouth, sweet	1156		17-240
Vicia faba	See **Broad beans**		
Vigna angularis	See **Aduki beans**		
Vigna mungo	See **Black gram, urad gram**		
Vigna unguiculata	See **Blackeye beans**		
Vinegar	1231		17-339
Virtually fat free/diet yogurt, fruit	290	12-382	
Virtually fat free/diet yogurt, plain	289	12-383	
Vitalite Light	See **Fat spread (60% fat), polyunsaturated**		
Vitis vinifera	See **Grapes, Currants, Raisins, Sultanas**		
Vodka	See **Spirits**		
Wafer biscuits, filled	120	11-524	50-107
Wafers, filled, chocolate, full coated	121	11-509	
Walnut oil	389		17-047
Walnuts	1028	14-879	50-979
Water biscuits	122		11-187
Water, distilled	1232		17-377
Watercress, raw	864	13-462	50-835
Weetabix	102	11-505	50-090
Wheat flour, brown	10	11-437	50-012
Wheat flour, white, breadmaking	11	11-438	50-013
Wheat flour, white, plain	12	11-439	50-014
Wheat flour, white, self-raising	13	11-440	50-015
Wheat flour, wholemeal	14	11-441	50-016
Wheatgerm	15	11-622	50-017
Wheatgerm bread	52	11-467	
Wheatgerm oil	390		17-048
Whelks, boiled	682		16-268
Whipping cream, fresh	243	12-333	50-214
Whiskey	See **Spirits**		
White bread, Danish style	59	11-466	11-112
White bread, farmhouse or split tin, freshly baked	56	11-470	11-101
White bread, French stick	57		11-471
White bread, fried in lard	54	11-469	50-051
White bread, premium	58	11-474	
White bread, sliced	53	11-468	50-049
White bread, toasted	55	11-475	50-052
White bread, 'with added fibre'	60	11-472	50-054
White bread, 'with added fibre', toasted	61	11-473	50-055
White cheese, average	282	12-368	50-250
White chocolate	1052		17-091
White pudding	566		19-159
White rice, basmati, raw	22		11-041
White rice, easy cook, boiled	24	11-446	50-023

	Publication number	New food code	Old food code
White rice, easy cook, raw	23	11-445	50-022
White rice, fried	25	11-610	11-045
White rolls, crusty	69	11-482	50-062
White rolls, soft	70	11-483	50-063
White sauce, savoury, made with semi-skimmed milk	1187	17-528	50-1153
White sauce, savoury, made with whole milk	1186	17-527	50-1152
White sauce, sweet, made with semi-skimmed milk	1189	17-530	50-1155
White sauce, sweet, made with whole milk	1188	17-529	50-1154
White wine, dry	1147		17-230
White wine, medium	1148		17-231
White wine, sparkling	1149		17-232
White wine, sweet	1150		17-233
Whitebait, in flour, fried	670		16-231
Whiting, in crumbs, fried in blended oil	647		16-162
Whiting, steamed	646		16-160
Whole milk yogurt, fruit	284	12-375	50-261
Whole milk yogurt, plain	283		50-260
Whole milk yogurt, infant, fruit flavour	285	12-378	
Whole milk yogurt, twinpot, thick and creamy with fruit	286	12-374	
Whole milk, average	217	12-315	50-189
Whole milk, pasteurised, average	218	12-316	50-190
Whole milk, pasteurised, summer	219	12-317	50-191
Whole milk, pasteurised, winter	220	12-318	50-192
Whole milk, sterilised	221	12-319	50-193
Whole milk, UHT	222	12-320	12-016
Wholemeal bread, average	62	11-476	50-056
Wholemeal bread, toasted	63	11-611	50-057
Wholemeal crackers	123	11-572	50-108
Wholemeal pastry, cooked	152	11-587	50-129
Wholemeal pastry, raw	151	11-586	50-128
Wholemeal rolls	71	11-484	50-064
Whopper burger	534	19-313	19-050
Wine, red	1145		17-228
Wine, rose, medium	1146		17-229
Wine, white, dry	1147		17-230
Wine, white, medium	1148		17-231
Wine, white, sparkling	1149		17-232
Wine, white, sweet	1150		17-233
Winkles, boiled	683		16-270
Witloof	See **Chicory**		
Worcestershire sauce	1218		17-340
Yam (USA)	See **Sweet potato**		

	Publication number	New food code	Old food code
Yam, boiled in unsalted water	866		50-837
Yam, raw	865		50-838
Yeast extract	1233	17-517	17-380
Yeast, bakers, compressed	1234		17-378
Yeast, dried	1235		17-379
Yellow gram	See **Chick peas**		
Yellow turnip	See **Swede**		
Yogurt, drinking	293		50-251
Yogurt, Greek style, fruit	292	12-377	
Yogurt, Greek style, plain	291	12-376	
Yogurt, Greek, sheep	294	12-420	50-253
Yogurt, low fat, fruit	288	12-380	50-257
Yogurt, low fat, plain	287	12-379	50-255
Yogurt, soya alternative, fruit	296	12-381	50-258
Yogurt, virtually fat free/diet, fruit	290	12-382	
Yogurt, virtually fat free/diet, plain	289	12-383	
Yogurt, whole milk, infant, fruit flavour	285	12-378	
Yogurt, whole milk, fruit	284	12-375	50-261
Yogurt, whole milk, plain	283		50-260
Yogurt, whole milk, twinpot, thick and creamy with fruit	286	12-374	
Yorkshire pudding	195	11-607	50-180
Zea mays	See **Sweetcorn**		
Zucchini	See **Courgette**